The Subnuclear Series • Volume 30

Proceedings of the International School of Subnuclear Physics

FROM SUPERSTRINGS TO THE REAL SUPERWORLD

THE SUBNUCLEAR SERIES

Series Editor: ANTONINO ZICHICHI, *European Physical Society, Geneva, Switzerla*

1. 1963 STRONG, ELECTROMAGNETIC, AND WEAK INTERACTIONS
2. 1964 SYMMETRIES IN ELEMENTARY PARTICLE PHYSICS
3. 1965 RECENT DEVELOPMENTS IN PARTICLE SYMMETRIES
4. 1966 STRONG AND WEAK INTERACTIONS
5. 1967 HADRONS AND THEIR INTERACTIONS
6. 1968 THEORY AND PHENOMENOLOGY IN PARTICLE PHYSICS
7. 1969 SUBNUCLEAR PHENOMENA
8. 1970 ELEMENTARY PROCESSES AT HIGH ENERGY
9. 1971 PROPERTIES OF THE FUNDAMENTAL INTERACTIONS
10. 1972 HIGHLIGHTS IN PARTICLE PHYSICS
11. 1973 LAWS OF HADRONIC MATTER
12. 1974 LEPTON AND HADRON STRUCTURE
13. 1975 NEW PHENOMENA IN SUBNUCLEAR PHYSICS
14. 1976 UNDERSTANDING THE FUNDAMENTAL CONSTITUENTS OF MATTER
15. 1977 THE WHYS OF SUBNUCLEAR PHYSICS
16. 1978 THE NEW APSECTS OF SUBNUCLEAR PHYSICS
17. 1979 POINTLIKE STRUCTURES INSIDE AND OUTSIDE HADRONS
18. 1980 THE HIGH-ENERGY LIMIT
19. 1981 THE UNITY OF THE FUNDAMENTAL INTERACTIONS
20. 1982 GAUGE INTERACTIONS: Theoy and Experiment
21. 1983 HOW FAR ARE WE FROM THE GAUGE FORCES
22. 1984 QUARKS, LEPTONS, AND THEIR CONSTITUENTS
23. 1985 OLD AND NEW FORCES OF NATURE
24. 1986 THE SUPERWORLD I
25. 1987 THE SUPERWORLD II
26. 1988 THE SUPERWORLD III
27. 1989 THE CHALLENGING QUESTIONS
28. 1990 PHYSICS UP TO 200 TeV
29. 1991 PHYSICS AT THE HIGHEST ENERGY AND LUMINOSITY: To Understand the Origin of Mass
30. 1992 FROM SUPERSTRINGS TO THE REAL SUPERWORLD

Volume 1 was published by W. A. Benjamin, Inc., Nerw York; 2–8 and 11–12 by Academic P New York and London; 9–10 by Editrice Compositori, Bologna; 13–29 by Plenum Press, New Y and London, 30 by World Scientific, Singapore.

The Subnuclear Series • Volume 30

Proceedings of the International School of Subnuclear Physics

FROM SUPERSTRINGS TO THE REAL SUPERWORLD

Edizioni Scientifiche Galileiane

Edited by

Antonino Zichichi

European Physical Society
Geneva, Switzerland

World Scientific
Singapore • New Jersey • London • Hong Kong

Published by

World Scientific Publishing Co. Pte. Ltd.

P O Box 128, Farrer Road, Singapore 9128

USA office: Suite 1B, 1060 Main Street, River Edge, NJ 07661

UK office: 73 Lynton Mead, Totteridge, London N20 8DH

Library of Congress Cataloging-in-Publication Data

International School of Subnuclear Physics (30th : 1992 : Erice, Italy)

From superstrings to the real superworld : proceedings of the International School of Subnuclear Physics / edited by Antonino Zichichi.

p. cm. -- (Subnuclear series ; v. 30)

The 30th International School of Subnuclear Physics, July 1992, Erice, Italy.

Includes bibliographical references.

ISBN 9810216319 :

1. Nuclear physics--Congresses. 2. Standard model (Nuclear physics)--Congresses. 3. Supersymmetry--Congresses. I. Zichichi, Antonino. II. Title. III. Series.

QC770.I6453 1992

539.7--dc20 93-41590

CIP

Printed in Singapore.

PREFACE

During July 1992, a group of 88 physicists from 55 laboratories in 18 countries met in Erice for the 30th Course of the International School of Subnuclear Physics. The countries represented were: Belgium, Czechoslovakia, Denmark, France, Germany, Greece, Israel, Italy, Norway, Poland, Portugal, Romania, Russia, Spain, Sweden, Switzerland, United Kingdom and the United States of America.

The School was sponsored by the European Physical Society(EPS), the Italian Ministry of Public Education(MPI), the Italian Ministry of Scientific and Technological Research (MRSI), the Sicilian Regional Government(ERS) and the Weizmann Institute of Science.

The School was focussed on Supersymmetry with special reference to the implications of the high precision LEP data on the possible existence of a SUSY threshold within existing facilities such as Fermilab, LEP and HERA. The Standard model has been once again scrutinized together with the difficult task of trying to understand how to go beyond it. The reader can find a set of interesting lectures on many details concerning QCD and the Electroweak components of the Standard Model, including new phenomena below the MegaTeV range. A set of comprehensive review lectures on specialised topics such as the neutrino mass, the TOP search and other results from large experimental facilities like Fermilab complete the Volume.

The most interesting part of the School are the discussion sessions which have been reported as close as possible to the real happening, thanks to the work of scientific secretaries. Let me hope that the reader will enjoy the discussions as much as the participants. As a method of reading this book, I would like to suggest reading the discussions first. In one case the lecture has been omitted but not the discussion.

Antonino Zichichi

CONTENTS

OPENING LECTURE

CLOSING LECTURE

CLOSING CEREMONY

WHERE WE STAND WITH THE REAL SUPERWORLD

by

A. Zichichi
CERN, Geneva, Switzerland

During the last year or so, a topic of interest to me more than ten years ago [1] came to my attention. The novelty being the high precision LEP data. The problem: can SUSY threshold be "predicted"?.

The basis for this new frontier in physics - i.e. the existence of a Superworld - is the Renormalization Group Equations (RGEs) which allow to span an energy range as large as 14-17 orders of magnitude. In fact the high precision LEP data are at 10^2 GeV and we would like to understand what happens all the way up to E_{GUT} (the energy where the three gauge forces $SU(3)_C$, $SU(2)_L$ and $U(1)_Y$ - characterized by the three couplings $\alpha_1,\alpha_2,\alpha_3$ - unify).

Recently a great deal of confusion has been raised by the authors of Ref. [2] who claimed that it was possible to predict the Supersymmetry threshold on the basis of a χ^2-test on the convergence of the couplings $\alpha_1,\alpha_2,\alpha_3$ at E_{GUT}. This paper has many weak points [3-9], the weakest one being a logical inconsistency. In fact if the source of our knowledge about the Supersymmetry threshold is the "convergence" at E_{GUT}, the top priority problem is to study what happens at E_{GUT}. This means the study of the threshold effects in the very high energy limit (10^{15}-10^{17} GeV) because what happens at E_{GUT} is supposed to have consequences in the energy range many orders of magnitude below. Moreover, it is contradictory to work out a χ^2-test for the geometrical convergence at E_{GUT} of the three gauge couplings ($\alpha_1,\alpha_2,\alpha_3$) and then let them diverge again [2], above E_{GUT}. Physics is not Euclidean geometry. The synthesis of this logical inconsistency is shown in Fig. 1 (which is the key figure of the paper quoted above [2]).

The reason for my interest in this paper is because it produced a lot of discouragement in the physics community, including my young collaborators engaged in searching for a supersymmetric signal with existing facilities. We have put order in this field [3-9] and the conclusion is that, in addition to all weak points and logical inconsistencies, the quantity M_{SUSY} is meaningless. We have worked out the spectra expected [6] and found that the lightest detectable supersymmetric particle could be as light as 50 GeV in mass, with the M_{SUSY} parameter more than one order of magnitude far away. Furthermore we have pointed out that the evolution of masses needs to be included. Of special relevance is the evolution of the gaugino masses, the so-called EGM effect [5].

A great development in our way of thinking [10] is that, not only the gauge couplings evolve with energy, but the masses as well. A careful analysis shows [5] how interesting are the consequences of this conceptual development: the M_{SUSY} threshold goes from 21 TeV down to the present LEP energy scale range. This is shown in Fig. 2. Let me emphasize again that M_{SUSY} is a parameter. It does not correspond to any physical mass. We keep it for the sake of comparison with other calculations without EGM effect [11].

The physically interesting results are those concerning the spectra of the lightest supersymmetric particles: charginos ($\chi^{\pm}{}_{1,2}$), neutralinos ($\chi^{0}{}_{1,2,3,4}$), gluinos, sleptons and squarks. An example of mass spectra prediction [12] is shown in Fig. 3. Moreover, to account for the light threshold (ΔT_L), the heavy threshold (ΔT_H) and the radiative effects due to the evolution of masses is perfectly possible and it allows the gauge couplings to converge at E_{GUT}, not to diverge above E_{GUT} and have the lightest supersymmetric observable signal in the energy range of present existing facilities (Fig.4). This figure brings me to a very interesting point: i.e. E_{GUT} and E_{SU}. The first, E_{GUT}, is obtained from "below"; the second, E_{SU}, the String Unification Energy level, is obtained from "above". Taken at their central value these two quantities are two orders of magnitude apart. Nevertheless it should be realised that many uncertainties come into the derivation of E_{GUT}. We have performed a detailed study of this problem [8] and the results are shown in Fig. 5 . Indeed it is possible to get E_{GUT} as high as 10^{18} GeV when the various parameters are chosen accordingly. For the sake of discussion suppose E_{GUT} within the minimal SU(5) model was definitely determined to be at 10^{16} GeV. Then the best candidate to fill the gap between E_{GUT} and E_{SU} would be SU(5)×U(1), the so called "flipped" SU(5) whose most interesting feature is that it is derivable from strings. On the other hand Nature, for the electroweak interactions has decided not to choose SU(2) but SU(2)×U(1). The most natural extension of the minimal SU(5) supergravity model is SU(5)×U(1). A detailed comparison of these two supergravity models, SU(5) and SU(5)×U(1), will be presented at the School [13].

Let me close this lecture with a graphical representation of the three coupled differential non linear equations describing the evolution, versus Q^2, of the three gauge couplings $\alpha_1,\alpha_2,\alpha_3$. The standard RGEs at two loops are used. The results are shown in Fig. 6 where the correlation among the three gauge couplings is shown in steps - each one being an order of magnitude above the previous one. The first point is at $\sqrt{Q^2}=0.1$ TeV; the last point at $10^{16.2}$ GeV, which is the value of E_{GUT} obtained with the input values for α_3 and $\sin^2\theta_W$ as indicated. If Nature would have followed the apparently simplest way (the straight line) we could not be here. Nature has followed the road illustrated by the sequence of the big dots in Fig. 6 and these predictions could not be there without the knowledge of the RGEs, the number of families and the values of $\alpha_1,\alpha_2,\alpha_3$ measured at LEP.

References

[1] A. Zichichi, Closing Lecture at the EPS Conference (York, UK, September 25-29, 1978); also Opening Lecture at the EPS Conference (Geneva, CH, 27 June - 4 July, 1979) - CERN Proceedings; and "New Developments in Elementary Particle Physics", Rivista Nuovo Cimento, Vol 2, No. 14 (1979) 1. The statement on page 2 of this paper, "*Unification of all forces needs first a Supersymmetry. This can be broken later, thus generating the sequence of the various forces of nature as we observe them* ", was based on a work by A. Petermann and A. Zichichi where the renormalization group running of the couplings using Supersymmetry was studied with the result that at one loop the convergence of the three couplings was theoretically granted. This work was not published, but known to a few.

[2] U. Amaldi, W. de Boer, H. Fürstenau, Phys. Lett. B 260 (1991) 447.

[3] F. Anselmo, L. Cifarelli, A. Peterman and A. Zichichi, Il Nuovo Cimento 104 A (1991) 1817.

[4] F. Anselmo, L. Cifarelli, A. Peterman and A. Zichichi, Il Nuovo Cimento 105 A (1992) 1025.

[5] F. Anselmo, L. Cifarelli, A. Peterman and A. Zichichi, Il Nuovo Cimento 105 A (1992) 581.

[6] F. Anselmo, L. Cifarelli, A. Peterman and A. Zichichi, Il Nuovo Cimento 105 A (1992) 1179.

[7] F. Anselmo, L. Cifarelli, A. Peterman and A. Zichichi, Il Nuovo Cimento 105 A (1992) 1201.

[8] F. Anselmo, L. Cifarelli and A. Zichichi,Il Nuovo Cimento 105 A (1992) 1335.

[9] F. Anselmo, L. Cifarelli and A. Zichichi,Il Nuovo Cimento 105 A (1992) 1357.

[10] For a review see A.B. Lahanas and D.V. Nanopoulos, Phys. Rep. 145 (1987) 1.

[11] J. Ellis, S. Kelley, D.V. Nanopoulos, Nuclear Physics B 373 (1992) 55.

[12] J.L. Lopez, D.V. Nanopoulos and A. Zichichi, preprint CERN-TH.6667/92, CTP-TAMU-68/92, ACT-20/92 (to be published in Nuclear Physics B).

[13] See "The Superworlds of SU(5) and SU(5)×U(1): A Critical Assessment and Overview" by J.L.Lopez, D.V. Nanopoulos and A. Zichichi, in this volume.

Figure Captions

Fig. 1 The wrong approach to understand SUSY [2].

Fig. 2 The Evolution of Gaugino Masses (the EGM effect) on the prediction for SUSY thresholds [5].

Fig. 3 An example of spectra for neutralinos ($\chi^0_{1,2,3,4}$), charginos ($\chi^{\pm}_{1,2}$), sleptons, squarks, gluinos and Higgs as predicted by a Supergravity model with inputs (m_t=130 GeV, $m_{\tilde{g}}$=300 GeV and $\mu<0$) [11].

Fig. 4 The correct approach to understand SUSY [3-11].

Fig. 5 The dependence of M_{GUT} from $\alpha_3(M_Z)$ and the ratio of the two crucial heavy threshold masses M_V/M_Σ, as described in the [8]. Note that the extreme value for M_{GUT} is above 10^{18} GeV.

Fig. 6 The evolution of the three gauge couplings (α_1,α_2,α_3) and their mutual correlation as described by a supergravity model. Note that the number of families (N_F=3) and the knowledge of α_3 are vital inputs.

Physics is not Euclidean geometry

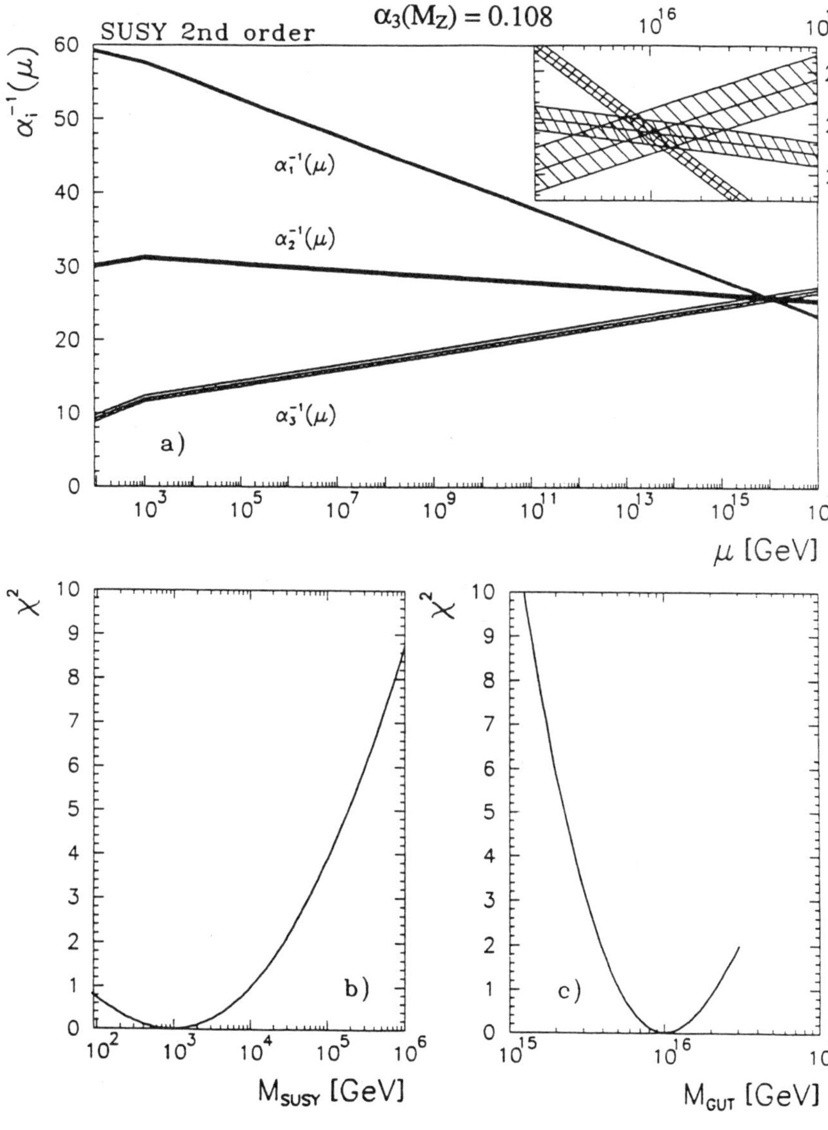

Figure 1

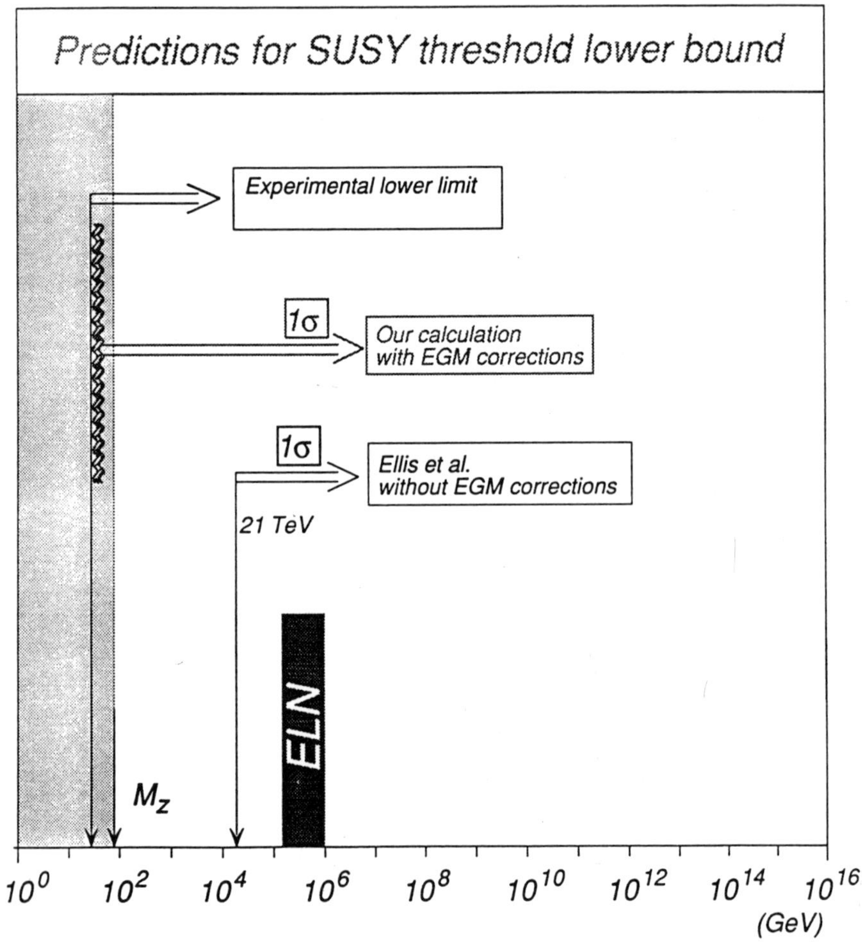

Predictions for SUSY threshold lower bound
Experimental lower limit
1σ
Our calculation
with EGM corrections
1σ
Ellis et al.
without EGM corrections
21 TeV
ELN
M_z
10^0
10^2
10^4
10^6
10^8
10^10
10^12
10^14
10^16
(GeV)

Figure 2

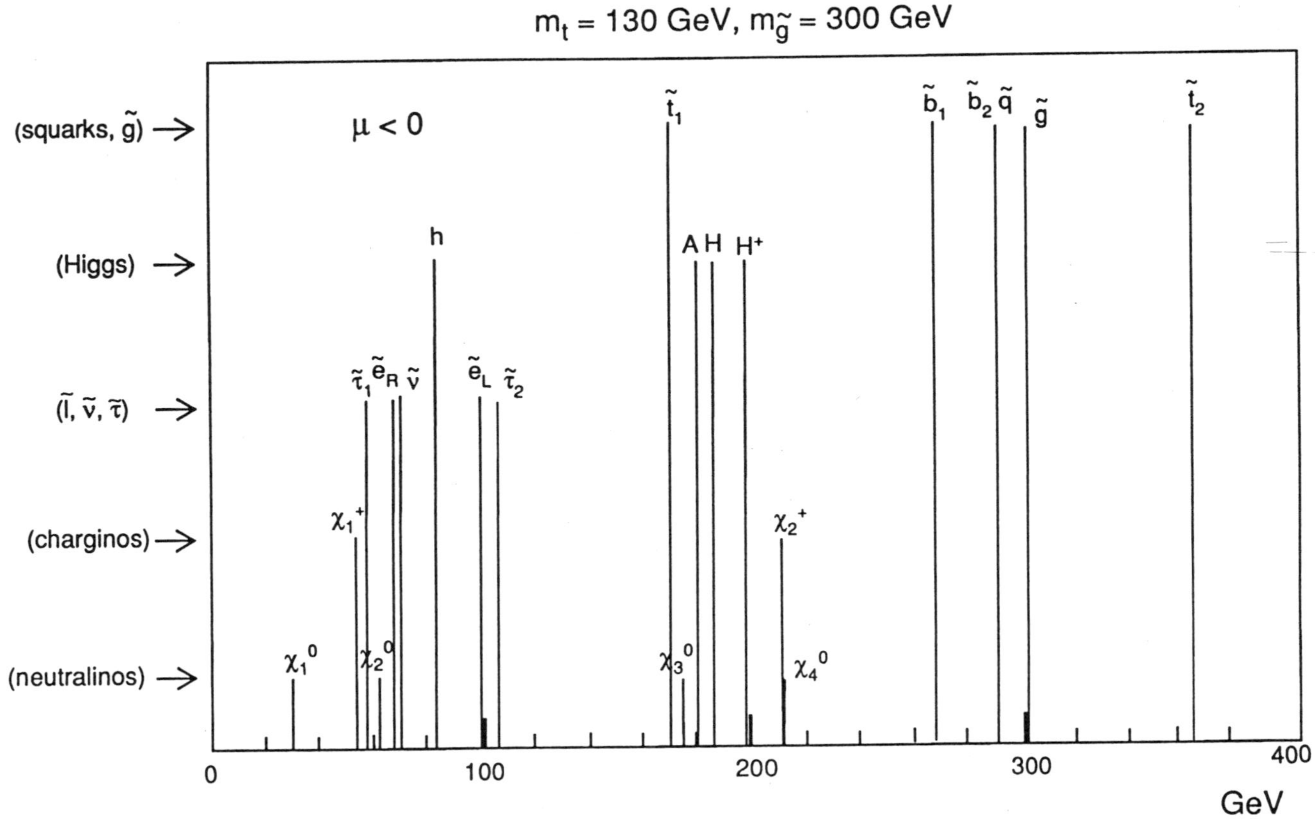
m_t = 130 GeV, $m_{\tilde{g}}$ = 300 GeV
μ < 0
(squarks, $\tilde{g}$)
(Higgs)
($\tilde{l}$, $\tilde{\nu}$, $\tilde{\tau}$)
(charginos)
(neutralinos)
$\tilde{t}_1$ $\tilde{b}_1$ $\tilde{b}_2$ $\tilde{q}$ $\tilde{g}$ $\tilde{t}_2$
h A H H^+
$\tilde{\tau}_1$ $\tilde{e}_R$ $\tilde{\nu}$ $\tilde{e}_L$ $\tilde{\tau}_2$
χ_1^+ χ_2^+
χ_1^0 χ_2^0 χ_3^0 χ_4^0
0
100
200
300
400
GeV

Figure 3

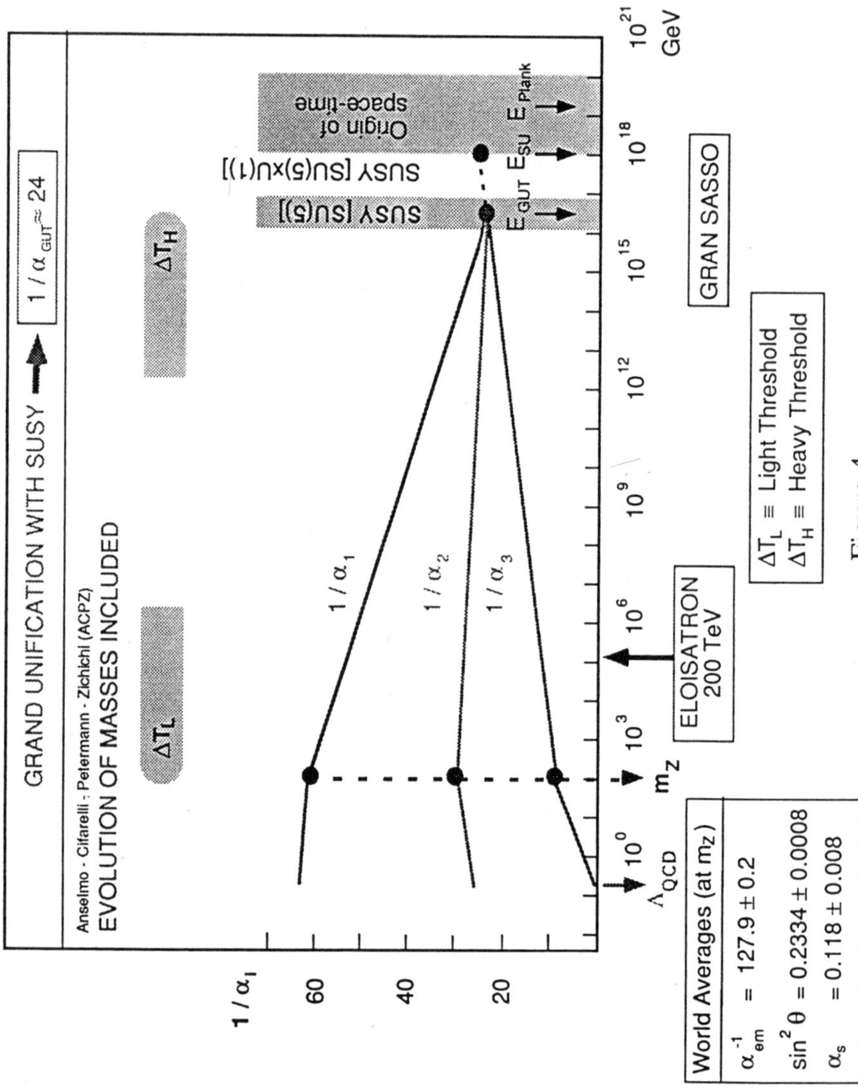
GRAND UNIFICATION WITH SUSY
1 / α_GUT ≈ 24
Anselmo - Cifarelli - Petermann - Zichichi (ACPZ)
EVOLUTION OF MASSES INCLUDED
ΔT_L
ΔT_H
SUSY [SU(5)]
SUSY [SU(5)xU(1)]
Origin of space-time
1 / α_1
1 / α_2
1 / α_3
1 / α_i
60
40
20
Λ_QCD
10^0
m_Z
10^3
10^6
10^9
10^12
10^15
10^18
10^21
GeV
E_GUT
E_SU
E_Plank
ELOISATRON 200 TeV
GRAN SASSO
World Averages (at m_Z)
α_em^-1 = 127.9 ± 0.2
sin^2 θ = 0.2334 ± 0.0008
α_s = 0.118 ± 0.008
ΔT_L ≡ Light Threshold
ΔT_H ≡ Heavy Threshold

Figure 4

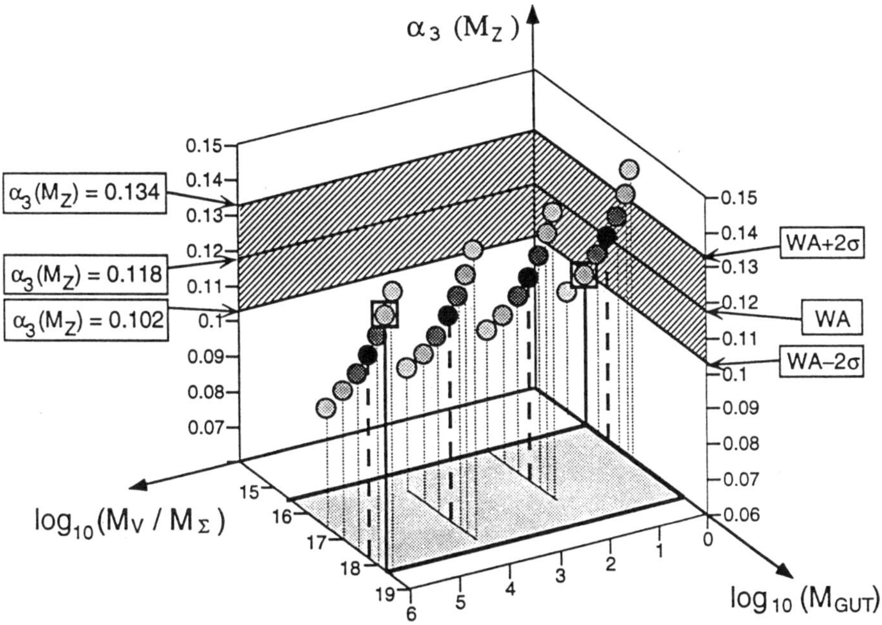

- World Average: WA
- WA ± 1σ
- WA ± 2σ
- WA ± 3σ

Extreme values

$$\begin{cases} WA - 2\sigma \;\; \& \;\; \dfrac{M_V}{M_\Sigma} = 10^0 \;\; \& \;\; \left(\dfrac{m_0}{m_{1/2}}\right)^2 = \left(\dfrac{m_4}{m_{1/2}}\right)^2 = 1 \\ WA + 2\sigma \;\; \& \;\; \dfrac{M_V}{M_\Sigma} = 10^6 \;\; \& \;\; \left(\dfrac{m_0}{m_{1/2}}\right)^2 = \left(\dfrac{m_4}{m_{1/2}}\right)^2 = 10^2 \end{cases}$$

Figure 5

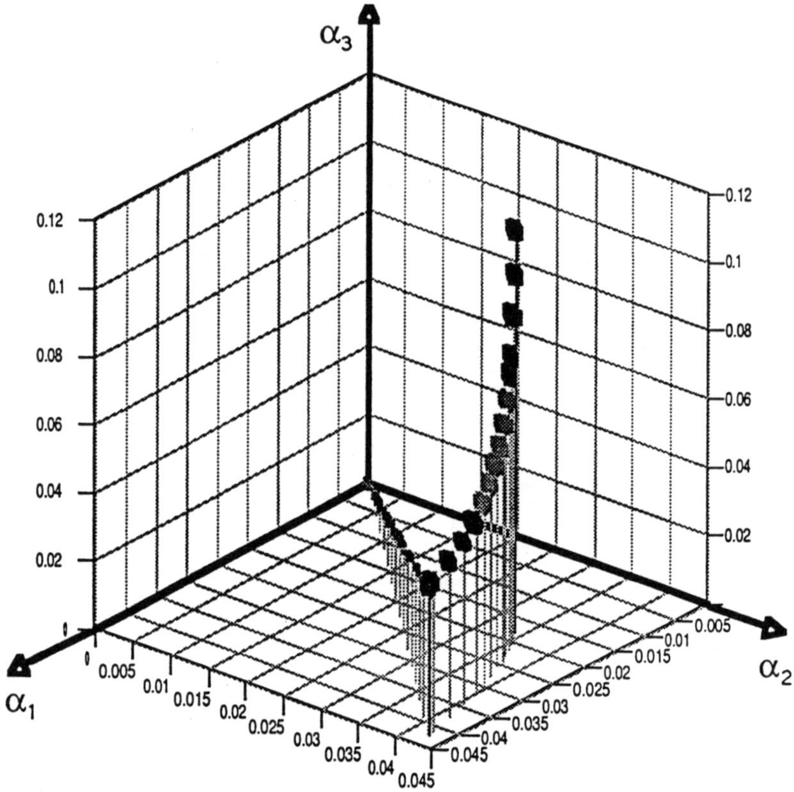

Supergravity Model

$N_F = 3$

$N_H = 2$

$M_{SUSY} = 10^2$ GeV

$E_{GUT} = 10^{16.2}$ GeV

$\alpha_3(M_Z) = 0.118 \pm 0.008$

$\sin^2\theta_w(M_Z) = 0.2334 \pm 0.0008$

Figure 6

CHAIRMAN: A. Zichichi

Scientific Secretaries: F. Block, E. Eskut, R. Malik

DISCUSSION

– *Malik:*

a) The trend emerging from your three approaches is:

$$\alpha_3(\mathrm{m_Z})(\uparrow) \equiv \text{ Theoretical Accuracy} \uparrow$$
$$\alpha_3(\mathrm{m_Z})(\uparrow) \equiv \mathrm{M_{GUT}} \uparrow$$
$$\alpha_3(\mathrm{m_Z})(\uparrow) \equiv \mathrm{M_{SUSY}} \downarrow$$

Can you explain the opposite trend of $\mathrm{M_{SUSY}}$ and $\mathrm{M_{GUT}}$?

b) The difference between $\mathrm{E_{SU}}$ and $\mathrm{E_{GUT}}$ is very large. What sort of physics can you predict in this energy range?

– *Zichichi:*

a) By one-loop and more precise two-loop calculations of our model we found that the value of $\alpha_3(\mathrm{m_Z})$, which fits with the measurements of $sin^2\theta_w$, should be higher than the values obtained from various experiments. The anticorrelation between $\mathrm{M_{SUSY}}$ and $\mathrm{M_{GUT}}$ is a result of our calculation. As emphasized in my talk the relevance of the lower bound of $\mathrm{M_{SUSY}}$ is around the LEP energy range and $\mathrm{M_{SUSY}}$ is not a mass. In addition, the important point is to know the expected spectrum if SUSY is really there. A spectrum of SUSY particles has been given in my talk. An example is given in Table 1.

b) If the unification point is not at 10^{16} but at 10^{18} GeV then SU(5) × U(1) (called flipped) is the best candidate for the right extension of our present knowledge (see Fig. 1). In this case the particles needed to fill the "Gap" are of two species. One group is at 10^6 GeV and the other at 10^{12} GeV.

– *Eskut:*

Which will be the LSP (Lightest Supersymmetric Particle)?

– *Zichichi:*

This depends on many details. If by LSP you mean "detectable" we have worked out many possibilities. In one case the $\tilde{\mathrm{w}}$ is the lightest, in another it is the $\tilde{\mathrm{h}}$. But if you refer to LSP assumimg R-parity conservation,the answer is χ_1^0 and this could be as light as 40 GeV or even less.

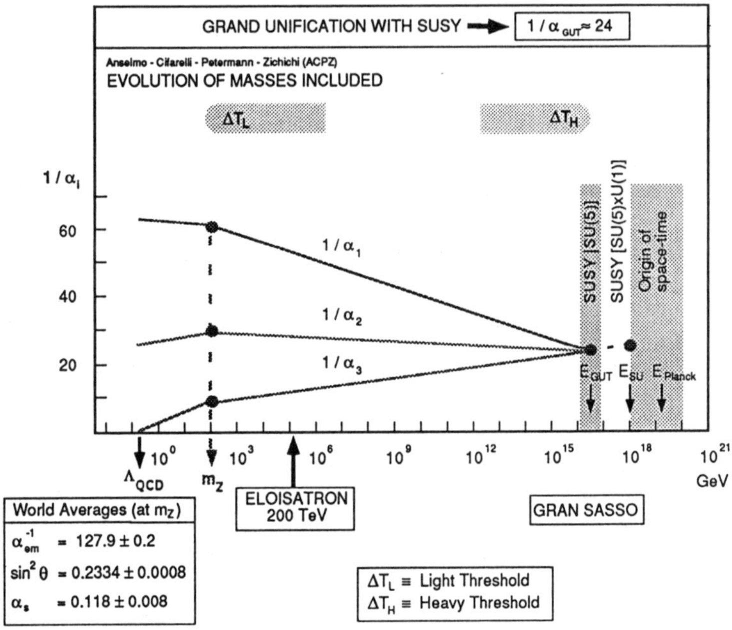

Figure 1

– *D. Nanopoulos and J. Ellis:*

Theoretically the Neutralino or Photino can be the LSP. But there is still no clear evidence as to which one it is.

– *do Couto e Silva:*

In order to detect sparticles one might need to upgrade current experiments or build new ones. Radiation may cause problems for detectors placed very close to the beam. How much does one need a microvertex detector in order to search for these particles?

– *Zichichi:*

The search for supersymmetric particles is very difficult. We need an almost perfect 4π detector to detect sparticles. If you want to discover a sparticle, you should design an experimental setup capable of detecting neutral objects, hadronic stuff, leading particles, momenta and vertices in order to look for anything missing.There is nothing special about the microvertex detector. The basic ingredients are hermeticity, energy, momentum and space resolution.

– *Ellis:*

The decays of sparticles can be complicated cascade decays, that means the amount of missing energy can be diluted. We need improvement of lepton detection and missing energy measurements and not only a microvertex detector.

– *Sehgal:*

You are making a linear extrapolation from 10^2 GeV to 10^{16} GeV. How safe do you think this is?

– *Zichichi:*

It is not an arbitrary extrapolation. It is controlled by the RGEs (Renormalization Group Equations). The extrapolation contains all the physics one can think of.

– *Sannino:*

Everyone believes in QCD as a fundamental theory of strong interactions. How do you think one should solve analytically and not numerically the problems of the low energy spectrum?

– *Zichichi:*

If by spectrum you mean the masses of the known particles, so far even lattice QCD has failed. No-one knows how to work out analytically the masses of the known hadrons. In any case, I must admit that this problem is of no interest to me. My interest lies in the search for new physics.

– *Liu:*

Do you think that baryon number violation should be taken seriously when you deal with unified theory? There are several proton decay experiments going on. What do you expect from them?

– *Zichichi:*

If you want a unified theory you must put baryonic and leptonic numbers in the same box. Therefore the proton must decay. It could be that the present experiments will detect proton decay. Concerning conservation laws let me tell you my point of view. I think that the conservation of "global" quantum numbers (baryon, lepton) will, sooner or later, be discovered to be violated. Only local conservation laws will resist.

– *Strogova:*

I would like to know about the possibility of observing $\tilde{w}$ or $\tilde{z}$ at LEP energy.

– *Zichichi:*

We have produced several spectra of sparticles. Let me show you an example where the $\tilde{w}$ is the lightest of all sparticles.

Table 1. SUSY PARTICLE SPECTRUM

$$M_{\tilde{w}} = 77 \text{ GeV}$$
$$M_{\tilde{g}} = 237 \text{ GeV}$$
$$M_{\tilde{t}_R} = 896 \text{ GeV}$$
$$M_{\tilde{t}_L} = 897 \text{ GeV}$$
$$M_{\tilde{t}_R} = 903 \text{ GeV}$$
$$M_{\tilde{t}_L} = 909 \text{ GeV}$$
$$M_{\tilde{q}} = 914 \text{ GeV}$$
$$M_{\tilde{h}} = 1208 \text{ GeV}$$
$$M_H = 1505 \text{ GeV} = \sqrt{m_{\tilde{h}}^2 + m_{\tilde{\ell}_L}^2}$$

– *Sopczak:*

You said that you prefer $\tan\beta$ to be much larger than 1. Why do you exclude $\tan\beta < 1$?

– *Zichichi:*

Because $m_{top} \gg m_b$.

– *Misiak:*

In your talk you mentioned that:

$$\Lambda_{QCD} = 260 \text{ MeV (world average)}$$
$$\Lambda_{QCD} = 390 \text{ MeV (from high energy data)}$$

Could you give:

i) the errors of these quantities;

ii) number of flavours corresponding to your Λ's;
iii) number of loops in the RGE you have used?

– *Zichichi:*

i) We wanted to study all α_s values obtained below LEP energies. We did our own analysis of the world data and found that the experimental results excluding the low energy data gave $\Lambda_{QCD} = (390 \pm 70)$ MeV. On the other hand, the inclusion of the low energy data gave $\Lambda_{QCD} = (260 \pm 40)$ MeV. These different values of Λ_{QCD} indicate sources of unquoted "systematic" errors.
ii) 5 flavours;
iii) 2 loops.

– *Leike:*

You have shown in your plot the meeting of running coupling constants of SU(3), SU(2) and U(1). How does this figure change with:
i) the errors of α_s;
ii) the uncertainty of the top quark mass;
iii) the uncertainty of the Higgs mass.

– *Zichichi:*

The effects of these uncertainties are studied in detail and can be found in the following references:

1. F. Anselmo, L. Cifarelli, A. Peterman and A. Zichichi, "*The Effective Experimental Constraints on M_{SUSY} and M_{GUT}*", preprint CERN-PPE/91-123,July 1991; Il Nuovo Cimento Vol. **104A**, No. 12 (1991) 1817;
2. F. Anselmo, L. Cifarelli, A. Peterman and A. Zichichi, "*The Convergence of the Gauge Couplings at E_{GUT} and Above: Consequences for $\alpha_3(M_Z)$ and SUSY Breaking*", preprint CERN/LAA/MSL/91-026,December 1991;
3. F. Anselmo, L. Cifarelli, A. Peterman and A. Zichichi, "*The Evolution of the Gaugino Masses and the SUSY Threshold*", preprint CERN-TH.6429/92, February 1992; Il Nuovo Cimento Vol. **105A** No. 4 (1992) 581;
4. F. Anselmo, L. Cifarelli, A. Peterman and A. Zichichi, "*The Simultaneous Evolution of Masses and Couplings: Consequences on Supersymmetry Spectra and Thresholds*", preprint CERN-PPE/92-103, June 1992;
5. F. Anselmo, L. Cifarelli, A. Peterman and A. Zichichi, "*Analytic Study of the Supersymmetry-Breaking Scale at Two Loops*", preprint CERN-TH. 6543/92, June 1992;
6. F. Anselmo, L. Cifarelli and A. Zichichi, "*A Study of the Various Approaches to M_{GUT} and α_{GUT}*", preprint CERN/LAA/MSL/92-011, July 1992;

7. F. Anselmo, L. Cifarelli and A. Zichichi, "*A χ^2-test to Study the $\alpha_1\alpha_2\alpha_3$ Convergence for High Precision LEP Data, Having in Mind the SUSY Threshold*", preprint CERN-PPE/92-122, July 1992.

Let me quote that our study is the most complete ever performed in this field.

– *Tao:*

The crucial point of SU(5) theory is that all couplings $\alpha_1, \alpha_2, \alpha_3$ will meet at one point, when we go to high energy. At one-loop level they meet at one point. Will they also meet at the same point when we go to the two-loop level?

– *Zichichi:*

It changes if we go from one-loop to the two-loops. If you want the details of our calculations I suggest you read our paper: F. Anselmo, L.Cifarelli, A. Peterman and A. Zichichi, "*The Convergence of the Gauge Couplings at E_{GUT} and Above: Consequences for $\alpha_3(M_Z)$ and SUSY Breaking*", preprint CERN/LAA/MSL/91-026, Il Nuovo Cimento Vol.**105A** No.7 (1992)1025.

ELECTROWEAK ASPECTS OF THE STANDARD MODEL

John Ellis

Theoretical Physics Division

CERN, Geneva, Switzerland

1. Constraints on particles within the Standard Model

1.1 Holdouts in the Standard Model

These are the undetected particles required to complete the Standard Model spectrum. One is the top quark t, an essential feature of the Standard Model that is required to complete the third electroweak doublet of quarks:

$$\begin{pmatrix} \mathrm{u} \\ \mathrm{d_c} \end{pmatrix} \begin{pmatrix} \mathrm{c} \\ \mathrm{s_c} \end{pmatrix} \begin{pmatrix} \mathrm{t} \\ \mathrm{b_c} \end{pmatrix} \tag{1.1}$$

LEP tells us that there are precisely 3 lepton doublets [1], and 3 quark doublets are therefore required in order to cancel their anomalous triangle diagrams so that the theory is renormalizable [2]. More experimentally, measurements including that of $\mathrm{Z} \rightarrow \bar{\mathrm{b}}\mathrm{b}$ decay at LEP tell us that the electroweak isospin of the left-handed bottom quark b is - 1/2 [3]:

$$\mathrm{I_3(b_2)} = -0.490^{+0.015}_{-0.012} \tag{1.2}$$

If the b has $\mathrm{I_3} = -1/2$, some other quark has to have $\mathrm{I_3} = +1/2$, namely the t. Another phenomenological argument is provided by the absence of flavour-changing b decays $\mathrm{b} \rightarrow (\mathrm{s\ or\ d}) + \ell^+\ell^-$ [4]. By a repetition of the arguments that motivated charm from the absence of strangeness-changing neutral currents [5], we infer that the t must exist. However, there is no model-independent prediction of

m_t: all we have are lower limits from LEP (46 GeV) [6] and $\bar{p}p$ collider experiments (91 GeV) [7].

Equally essential to the Standard Model is some sort of Higgs boson. One motivation [8] for it is to cancel the leading high-energy behaviours of the amplitudes for $\bar{f}f \to W^+W^-$ and $W^+W^- \to W^+W^-$.

Without a Higgs boson,

$$A(\bar{f}f \to W^+W^-) \sim g^2 \frac{m_f E}{m_W^2} \quad , \quad A(W^+W^- \to W^+W^-) \sim g^2 \frac{E^2}{m_W^2} \tag{1.3}$$

which seem fine for giving large cross-sections, but give rise to uncontrollable divergences in loop diagrams. To avoid these, one must [8] introduce some scalar field whose couplings to fermions (vector boson) must be:

$$g_{H\bar{f}f} = O(gm_f/m_W) \quad , \quad g_{HVV} = O(g) \tag{1.4}$$

respectively. Since the growths (1.3) apply at energies $E \leq M_H$, a unitarity upper bound on the tree-level $W^+W^- \to W^+W^-$ amplitude tells us that [9]:

$$M_H \leq 1 \text{ TeV} \tag{1.5}$$

which is strengthened in higher orders of perturbation theory [10]. The couplings of the Higgs boson are completely fixed in the Standard Model, where no other particles appear:

$$g_{H\bar{f}f} = \frac{g}{2} \frac{m_f}{m_W} \quad , \quad g_{HW^+W^-} = \frac{e}{\sin\theta_W} m_W \tag{1.6}$$

These imply that the dominant decays of the Standard Model Higgs boson are those into the heaviest particles kinematically accessible. Unfortunately, there is also no model-independent prediction for the mass of the Higgs boson, and the only experimental information we have is the lower limit of 60 GeV from LEP [11]. Experiments at LEP should increase this to about m_Z, and experiments at the LHC and SSC should be able to explore the rest of the mass range up to the limit (1.5).

1.2 Electroweak Renormalization

In order to interpret the precision LEP and other electroweak data, one must include one- and some two-loop quantum corrections. These have infinities in general, which can be renormalized by fixing selected physical quantities at their measured values. These are generally taken as the two historically most precisely measured quantities [12]:

$$\alpha_{em} = \frac{1}{137.03} \quad , \quad G_\mu = 1.166389(22) \times 10^{-5}\ \mathrm{GeV}^{-2} \tag{1.7}$$

and now also $m_Z = 91.187 \pm 0.007$ GeV [1] . It is convenient to compare different precision experiments in terms of an auxiliary variable [13]:

$$\sin^2\theta_W \equiv 1 - \frac{m_W^2}{m_Z^2} \tag{1.8}$$

in the on-shell scheme or $\sin^2\theta_W(m_Z)|_{\overline{MS}}$ in the modified minimal subtraction scheme [14]. The former is more physical, but the latter is more convenient for testing GUTs in combination with measurements of $\alpha_3(m_Z)$. Consistency of the different determinations of $\sin^2\theta_W$ can be used to constrain m_t and/or M_H.

In the on-shell prescription at the one-loop level [15]:

$$m_W^2\sin^2\theta_W = m_Z^2\cos^2\theta_W\sin_W^\theta = \frac{\pi\alpha}{\sqrt{2}s_\mu}\ (1 + \Delta r) \tag{1.9}$$

The radiative correction Δr can be decomposed as follows:

$$\Delta r = \Delta r_{oblique} + \Delta r_{direct} \tag{1.10}$$

where the first "oblique" term comprises the vacuum polarization diagrams and the second "direct" term includes vertex and box diagrams in μ decay. The former can in turn be decomposed into three terms:

$$\Delta r_{oblique} = \Delta\alpha - \cos^2\theta_W\Delta\rho + \widehat{\Delta}r \tag{1.11}$$

The first reflects the running of the fine structure constant between the Thompson limit (1.7) and m_Z, and its principal uncertainty is that associated with light quark loops. The second reflects the effective electroweak isospin violation which appears in the low-energy neutral to charged current cross-section ratio:

$$\Delta\rho = \frac{\Pi^{WW}(0)}{m_W^2} - \frac{1}{m_Z^2}\left(\Pi^{ZZ}(0) + 2\ \tan\theta_W\Pi^{ZA}(0)\right) \tag{1.12}$$

Its dominant contribution is:

$$\Delta\rho \ni \frac{3G_\mu m_t^2}{8\pi^2\sqrt{2}} \tag{1.13}$$

for $m_t \gg m_b$. The last term in (1.11) also depends on m_t:

$$\widehat{\Delta}r \ni \frac{\sqrt{2}G_\mu m_W^2}{8\pi^2}\left(\cot^2\theta_W - \frac{1}{3}\right)\ell n\ m_t^2/m_W^2 \tag{1.14}$$

for $m_t \gg m_W$, and Δr also depends logarithmically on M_H:

$$\Delta r \ni \frac{\sqrt{2} G_\mu M_W^2}{16\pi} \left\{ \frac{11}{3} (\ell n \; M_H^2/m_W^2 - \frac{5}{6}) \right\} \tag{1.15}$$

for large M_H. The dependences (1.13, 1.14, 1.15) on m_t and M_H mean that a measurement of Δr is not sufficient to determine either m_t or M_H: a trade-off is always possible, as seen in fig. 1. Fortunately LEP provides other measurements which, in principle, enable both m_t and M_H to be determined, as we shall see shortly. However, before we go on to the data and their interpretation, two more theoretical points are worth noting. One is that there is one important m_t-dependent vertex correction [15] :

$$\frac{\Delta r(Z \to \bar{b}b)}{r(Z \to \bar{b}b)} \ni \frac{-20}{13} \frac{\alpha}{\pi} \left[\frac{m_t^2}{m_W^2} + \frac{13}{6} \ell n \; m_t^2/m_W^2 \right] \tag{1.16}$$

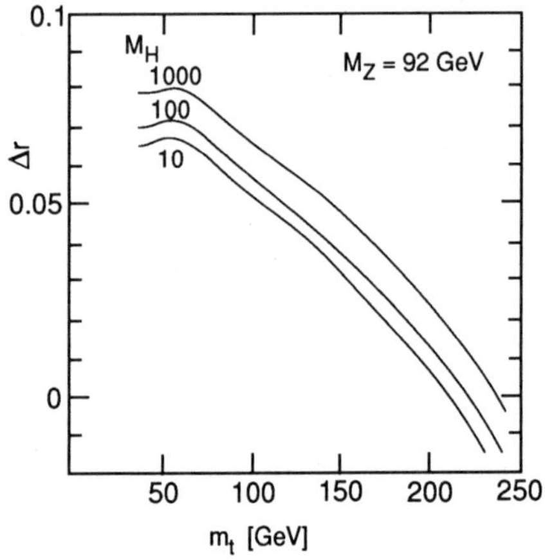

Fig. 1. Values of Δr in the Standard Model for different values of m_t, M_H.

The other is that the leading m_t-dependence is known at the two-loop level [16]: for example in equation (1.11) the factor $1 + \Delta r$ becomes $1/(1 - \Delta r)$ where:

$$\Delta r \ni \frac{3 G_\mu m_t^2}{8\pi^2 \sqrt{2}} \left[1 - (2\pi^2 - 19) \frac{G_\mu m_t^2}{8\pi^2 \sqrt{2}} + \ldots \right] \tag{1.17}$$

This is non-negligible for large m_t and included in the subsequent data analysis [17].

In addition to the LEP and $\bar{p}p$ collider data, these include low-energy neutral current data. Those on νq interactions are dominated by deep inelastic νN scattering, where the principal uncertainty is that in the charmed quark mass. The effective value of $\sin^2\theta_W$ extracted from these data [18]:

$$\sin^2\theta_W = 0.229 \pm 0.004 \ (\pm 0.004 \text{ due to } m_c) \tag{1.18}$$

is almost independent of m_t, but even these data by themselves do not allow a very large value of m_t. Data on eq interactions are now dominated by those on parity violation in atoms, principally in Cesium [19]. One extracts:

$$\sin^2\theta_W = 0.229 \pm 0.007 \text{ (exp) } \pm 0.008 \text{ (theory)} \tag{1.19}$$

for $m_t = 60$ GeV, but the extracted value decreases for larger m_t. Data on νe interactions [20] give:

$$\sin^2\theta_W = 0.232 \pm 0.008 \tag{1.20}$$

for $m_t = M_H = 100$ GeV, with an m_t-dependence similar to that of the eq sector. The combined low-energy data indicate by themselves that $m_t < 200$ GeV, as seen in fig. 2, but do not provide a lower bound on m_t. LEP measurements [1] (see Table 1) of the Z couplings are considerably more precise, but the agreement between these and the low-energy data is a most important consistency check on the Standard Model and constraint on any possible extension.

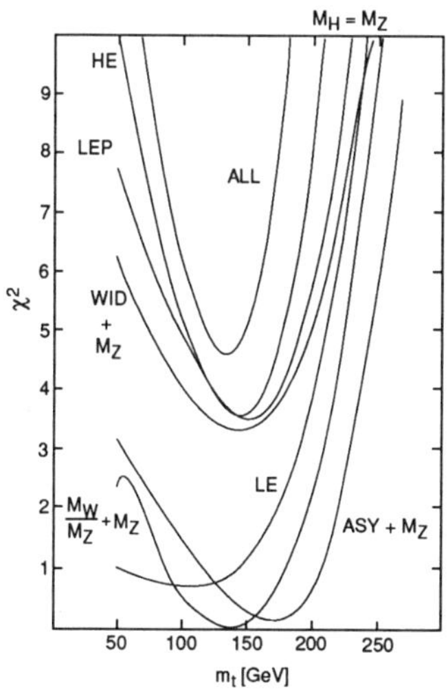

Fig. 2. The dependence on m_t of the contributions to the χ^2 function from different classes of experiments.

TABLE I

Compilation of LEP precision measurements

m_Z	=	91.187	$\pm$	0.007	GeV
Γ_Z	=	2.492	$\pm$	0.007	GeV
Γ_H	=	1.737	$\pm$	0.007	GeV
Γ_ℓ	=	83.33	$\pm$	0.030	MeV
g_a^ℓ	=	-0.4994	$\pm$	0.0009	
N_ν	=	3.04	$\pm$	0.04	
α_s	=	0.120	$\pm$	0.006	

1.3 Interpretation within the Standard Model

The comparisons between some of the LEP measurements and theoretical calculations for different values of m_t and M_H are shown in fig. 3. The measurement of Γ_Z is sensitive to both m_t and M_H, and also α_3, as seen in fig. 3a, whilst the measurement of Γ_ℓ (fig. 3b) is insensitive to α_3, but still sensitive to M_H and α_3. The measurement of the hadronic cross-section on the peak (fig. 3c) (fig. 3d)is almost independent of either m_t or M_H, but still sensitive to α_3, whilst the measurement of Γ_{inv}/Γ_L (see fig. 3d) is almost insensitive to any of these 3 parameters. Consistency between the various m_t-dependent measurements is only possible for a limited range of m_t. The most efficient way of combining globally all the experimental information on m_t is to determine the overall χ^2 function [17] .

Figure 4 shows the dependence of χ^2 on m_t for different choices of M_H. We see that it has a well-defined minimum at a value of m_t that increases slowly with M_H. Before the advent of very precise LEP data, the uncertainty in m_t associated with our ignorance of M_H was much smaller than other uncertainties, and fit values of m_t were often quoted for some fixed choice of M_H [21]. Nowadays, the uncertainty due to the correlation with M_H is no longer negligible. We prefer to quote constraints on m_t that are obtained with M_H free, that are obtained by taking the envelope of all the χ^2 curves in fig. 4, which now yields [17] :

$$m_t = 124 \pm^{26}_{28} \text{ GeV} \tag{1.21}$$

which is comfortably consistent with the direct CDF lower limit [7] . The contributions to the different sectors to the χ^2 function are shown in fig. 2. We see again that the low-energy data provide an upper bound on m_t, whilst the LEP $\bar{p}p$ collider measurements of m_W/m_Z are needed to provide a lower bound on m_t. The global consistency of the data, with a total minimum $\chi^2 = 4.3$, is very good.

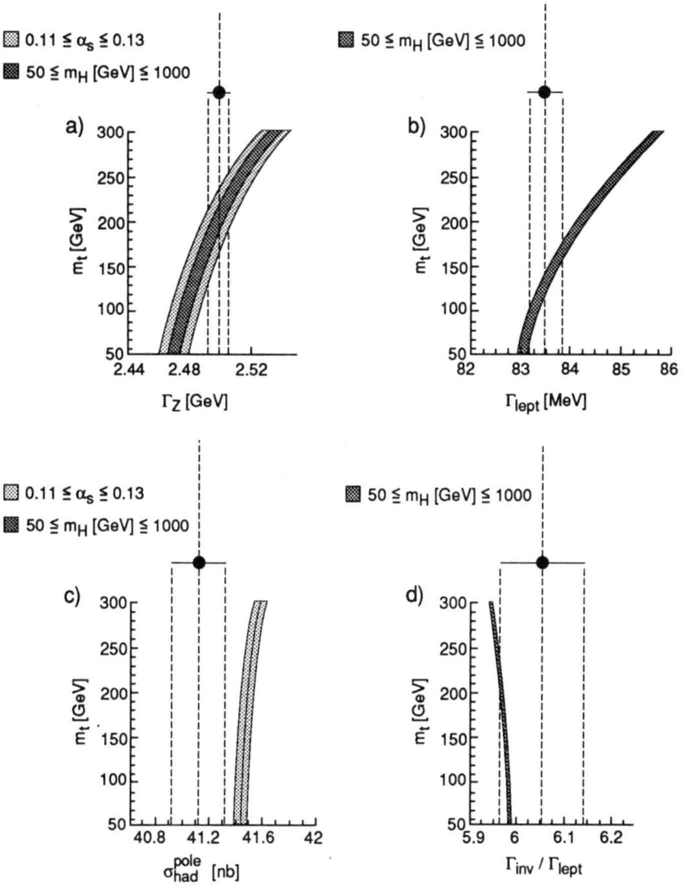

Fig. 3. Experimental determinations of different LEP observables, compared with calculations in the Standard Model.

We saw already in fig. 4 that the minimum value of χ^2 is sensitive to M_H, and indeed that lower values of M_H are slightly preferred. This feature of the data is shown in a different way in fig. 5, where the global χ^2 is shown as a function of M_H for different choices of m_t. The envelope of these χ^2 curves gives the global χ^2 as a function of m_t for M_H free, from which we infer [17] :

$$M_H = 25 \pm^{175}_{16} \text{ GeV} \tag{1.22}$$

Although the minimum of χ^2 is at a value now excluded by the LEP searches, the χ^2 price that one must pay for consistency is very small: $\Delta\chi^2 \simeq 0.2$. Therefore a constrained fit including the direct limit as a constraint would not relax very much the 68% confidence level upper value in (1.22). I am prepared to bet 2 lire against 1 lira that $M_H < 300$ GeV. The contributions of the different sectors to the χ^2 function for $m_t = 130$ GeV are shown in fig. 6. We see that the low-energy data by themselves are almost independent of M_H. However, they do contribute to the global bound on M_H indirectly, by bounding m_t which is correlated positively with M_H [17] .

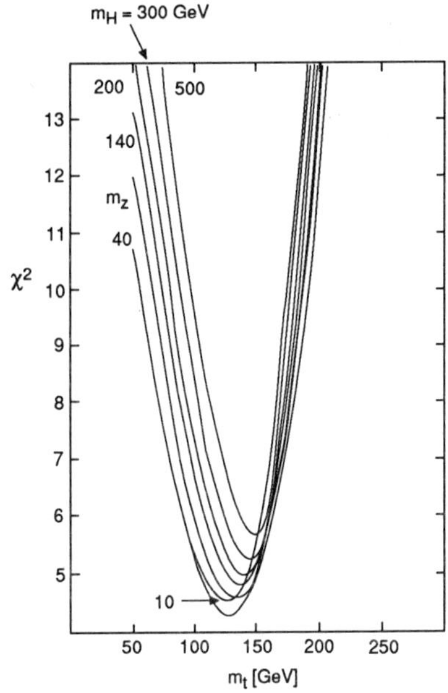

Fig. 4. The dependence of the χ^2 function on m_t for the indicated values of M_H, from a global electroweak fit.

This effect is seen clearly in fig. 7, which shows the $\Delta\chi^2 = 1$ and 2.7 contours in the (M_H, m_t) plane for the global fit, and in fig. 8 which shows the result of a fit to the high-energy LEP and $\bar{p}p$ collider data alone. We see that the lower bounds on m_t and M_H are almost the same in the two fits, whereas the high-energy fit allows larger values of m_t and M_H. Compare a recent estimate of m_t made by some irresponsible theorists [22]:

$$m_t = 131^{+22}_{-11} \text{ GeV} \tag{1.23}$$

who interpret a published CDF $e^+\mu^+$ missing p_T event [7] as an example of $\bar{t}t$ production. Is the consistency a coincidence?

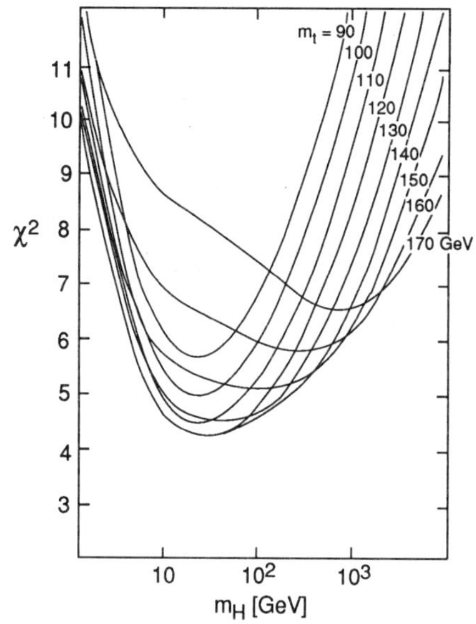

Fig. 5. The dependence of the χ^2 function on M_H for the indicated values of m_t, from a global electroweak fit.

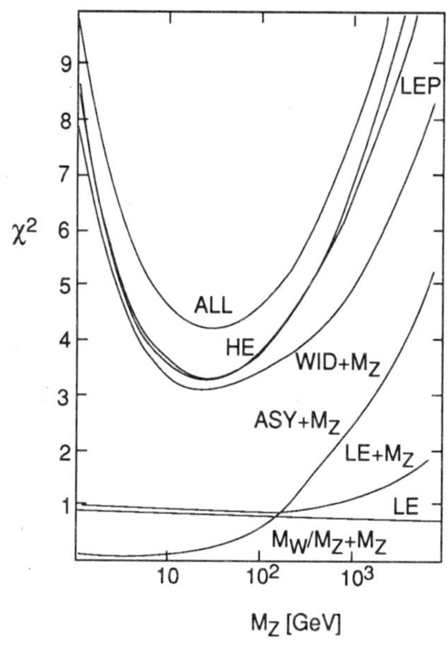

Fig. 6. The dependence on M_H of the contributions to the χ^2 function from different classes of experiments.

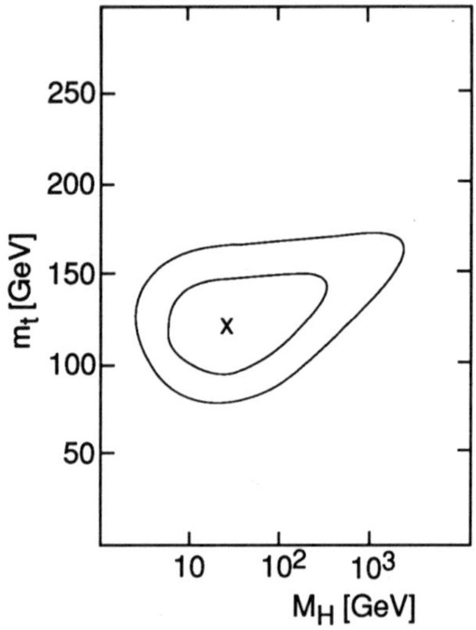

Fig. 7. Contours of the χ^2 function in the (M_h, m_t) plane for the global fit: the inner (outer) curves are for $\Delta\chi^2 = 1$ (2.7) relative to the minimum indicated by the cross.

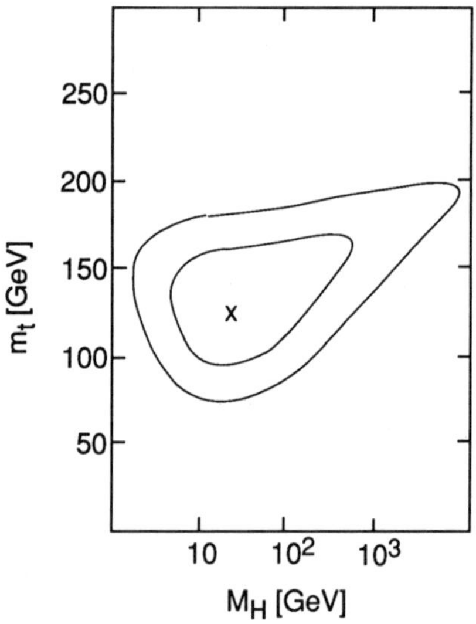

Fig. 8. As for fig. 7, for a fit to the LEP and $\bar{p}p$ collider data alone.

The range of M_H favoured by the LEP and other electroweak data is not merely below the unitarity limit, but is also below the more stringent bound:

$$M_H \leq 650 \text{ GeV} \tag{1.24}$$

quoted on the basis of non-perturbative lattice considerations [23]. As seen in fig. 9, it is also consistent with the upper bound:

$$M_H \leq 200 \text{ GeV} \tag{1.25}$$

that was obtained by requiring the Higgs self-coupling to remain finite at energies up to the Planck mass [24]. Encouraging for lovers of SUSY, it is also consistent with the range ($m_Z \pm 40$ GeV) of Higgs masses predicted in the minimal supersymmetric extension of the Standard Model [25]. The range (1.21) of m_t that we find is also comfortably consistent with the M_H-dependent cosmological bounds on m_t obtained by requiring our present electroweak vacuum to be stable, or even metastable with a lifetime in excess of 10^{10} y and not subject to stimulated vacuum decay in high-energy cosmic ray collisions, as also seen in fig. 9 [26].

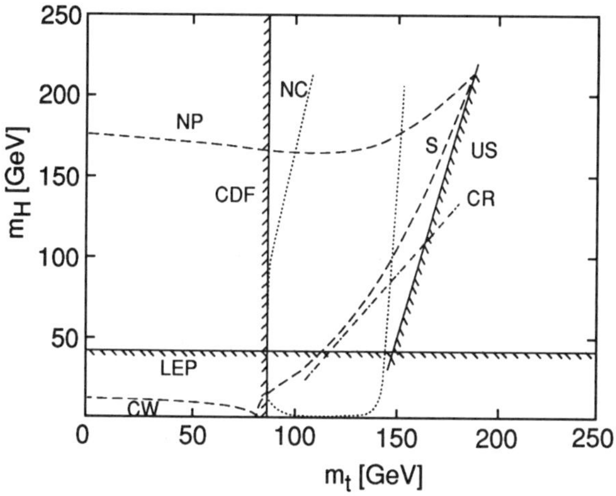

Fig. 9. Experimental, astrophysical and cosmological bounds on m_t and m_H. The line labelled LEP is the lower bound on m_H and the line labelled CDF is their bound on m_t experimental. The dotted line labelled NC encloses the region allowed at 1 standard deviation by precision electroweak data including radiative corrections. The line CW is a cosmological lower bound on m_H for small m_t and the line NP indicates where the Standard Model couplings become non-perturbative. Our present vacuum is stable to the left of the line S and has a spontaneous decay lifetime $< 10^{10}$ y to the right of the line US. The line CR bounds the region in which cosmic ray collisions might have triggered vacuum decay.

The consistency between theoretical, direct experimental, indirect experimental and cosmological constraints on m_t and M_H encourages us to think that we are on the right track. If the global electroweak fit is correct, both the top quark and the Higgs boson may lie "just around the corner". The direct coverage of different

Standard Model Higgs masses available with LEP2 and the LHC is shown in fig. 10. Good luck with the search!

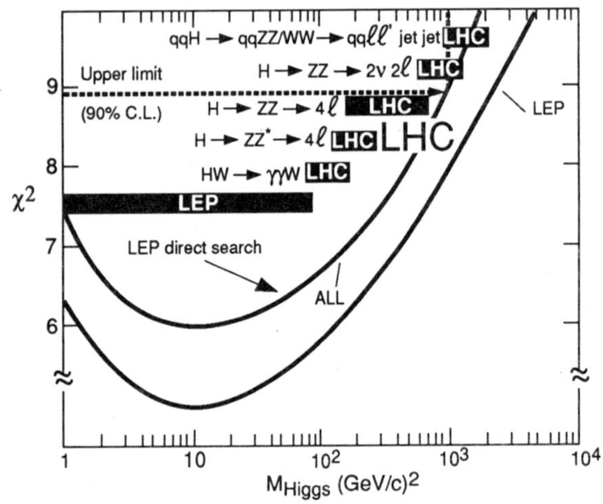

Fig. 10. The physics reaches of LEP and the LHC, looking for the Higgs boson in various different modes.

2. Constraints on Extensions of the Standard Model

2.1 Electroweak Symmetry Breaking

The generation of particle masses in the Standard Model requires the spontaneous breakdown of electroweak gauge symmetry by a vacuum expectation value for some field X with non-zero electroweak isospin. The measured value of m_W/m_Z and the existence of fermion masses tell us that much of X must have isospin 1/2. The next question is whether X is elementary or composite. The latter hypothesis strikes me as more conservative and boring: it would be a repetition of what happens in QCD, where $< 0|\bar{q}q|0 > \neq 0$, and in superconductivity with its Cooper pairs. Examples of composite models of symmetry breaking include the $\bar{t}t$ condensate [27] and technicolour [28] models that we will discuss shortly. The alternative hypothesis that electroweak symmetry is broken by an elementary Higgs field seems to me more radical, as no-one has yet seen an elementary scalar field, and its technical difficulties with quadratic divergences seem to require a new asymmetry principle, that of supersymmetry [29] as we will discuss later.

The basic idea [7] is to replace the Standard Model Lagrangian by one without a Higgs field but with higher-order multi-fermion interactions à la Nambu-Jona-Lasinio. In the simplest example where:

$$\mathcal{L} = \bar{\psi}\not{D}\psi^a - \frac{1}{2}G[(\bar{\psi}_a\psi^a)^2 - (\bar{\psi}_a\gamma^5\psi^a)^2] : a = 1,\ldots,N_c \qquad (2.1)$$

the physics is the same as that of the Standard Model with constraints [30], and one estimates [27] :

$$M_H \simeq (1 \text{ to } 2)m_t \ , \ m_t \simeq (200 \text{ to } 250) \text{ GeV} \tag{2.2}$$

Unfortunately, the first of these predictions runs foul of the precision electroweak data, which would charge [17] a price of $\Delta\chi^2 \geq 7$ for the prediction (2.2) of m_t. One can avoid this prediction by adding other higher-order terms to the Lagrangian (2.1), but then the model becomes equivalent to the Standard Model, but written using different parameters [30] , and loses its predictivity.

In technicolour models [28] one postulates a new set of strong interactions with characteristic mass-scale $\Lambda_{TC} \simeq 3000 \times \Lambda_{QCD}$, and the elementary Higgs boson is replaced by a technipion bound state of a doublet of technifermions T. These suffice to give masses to the W and Z, but not to the quarks and leptons, which require an extension of the minimal technicolour model. The conventional extension postulates [28] a complete technigeneration (N, L, U, D) analogous to a familiar generation (ν, ℓ, u, d). This model has long had problems with the experimental suppression of flavour-changing neutral interactions and the absence of light charged technipions, but these have recently become compounded by precision electroweak data.

These can be analyzed using a general parametrization of one-loop oblique radiative corrections in terms of 3 variables [31]:

$$S = \frac{4\sin^2\theta_W \epsilon_3}{\alpha} \ , \ T = \frac{\epsilon_1}{\alpha} = \frac{\Delta\rho}{\alpha} \ , \ U = \frac{-4\sin^2\theta_W \epsilon_2}{\alpha} \tag{2.3}$$

which are related to combinations of vacuum polarization functions. In the Standard Model [31] :

$$T = \frac{3}{16\pi}\frac{1}{\sin^2\theta_W \cos^2\theta_W}\frac{m_t^2}{m_Z^2} - \frac{3}{16\pi \cos^2\theta_W}\ell n(m_H^2/m_Z^2) + \ldots \tag{2.4a}$$

$$S = \frac{1}{12\pi}\ell n(M_H^2/m_Z^2) + \ldots \tag{2.4b}$$

and the previous bounds on (m_t, M_H) can be translated into bounds on (S, T, U). These then can be used to constrain extensions of the Standard Model with the same electroweak gauge group and extra matter fields, such as technicolour models. However, this type of analysis is necessarily approximate, because it does not take into account vertex and two-loop effects, the one-loop vacuum polarizations are treated in a linearized approximation, and the Standard Model contributions (2.4) must be treated as uncertain, with m_t and an effective M_H left free.

Fitting the usual electroweak measurements and assuming as default values $m_x = 130$ GeV, $M_H = M_Z$, we [17] find the $\delta\chi^2 = 1$ ellipsoid in the three-parameter space (2.3) that is shown in fig. 11 together with the two-dimensional submanifold spanned by the Standard Model for 90 GeV $< m_t <$ 250 GeV and 50 GeV $< M_H <$ 1 TeV. The data are quite consistent with the Standard Model, and give the following constraints:

$$\delta\epsilon_1 = (0.09 \pm 0.25) \times 10^{-2}, \delta\epsilon_2 = (0.09 \pm 0.72) \times 10^{-2}, \delta\epsilon_3 = -(0.24 \pm 0.38) \times 10^{-2} \tag{2.5}$$

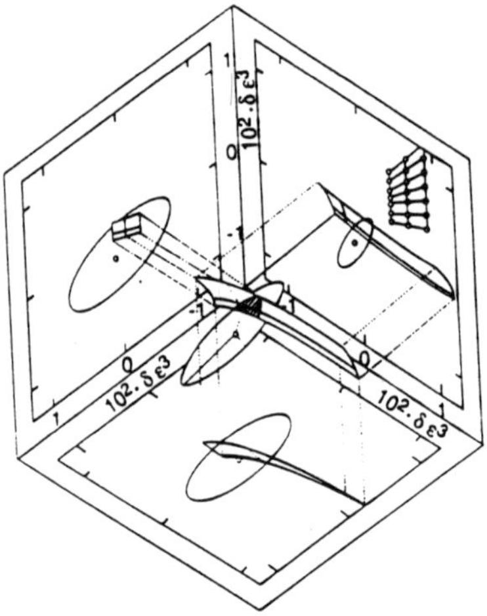

Fig. 11. Three-dimensional model-independent fit to the electroweak data (ellipsoid) compared with the surface allowed in the Standard Model and the predictions of available technicolour models.

on contributions from beyond the Standard Model for $m_t = 130$ GeV, $M_H = m_Z$. As seen in fig. 11, these seem to disfavour or even exclude a large class of technicolour models [28] , although we cannot exclude the possibility that some as-yet-unknown technicolour model might fit the data better.

2.2 Supersymmetry

As has already been mentioned, the radiative corrections to an elementary Higgs mass squared are quadratically divergent:

$$\delta m_H^2 \simeq g^2 \int^{\Lambda} \frac{d^4p}{(2\pi)^4} \frac{1}{p^2} = O\left(\frac{\alpha}{\pi}\right) \Lambda^2 \tag{2.6}$$

where Λ is a cutoff representing the appearance of new physics beyond the Standard Model. If $\Lambda \simeq m_{GUT} \simeq 10^{15}$ GeV or $\simeq m_P \simeq 10^{19}$ GeV, the correction would be much larger than the physical Higgs mass-squared. This is mathematically possible, but looks physically unnatural. It would require fine-tuning the bare mass-squared m_0^2:

$$m_H^2 = m_0^2 + \delta m_H^2 \ : \ m_0^2 = -\delta m_H^2 + 0(100 \text{ GeV})^2 \tag{2.7}$$

Theorists would prefer a "natural" theory in which $|\delta m_H^2| \leq m_H^2$, as in the case of corrections to fermion masses, which would require introducing some new physics at an energy scale below about 1 TeV. The idea of using supersymmetry [32] is that boson and fermion loops have opposite signs, and hence can be made to cancel up to mass corrections if one postulates identical number of fermions and bosons with identical couplings. Then:

$$\begin{aligned} \delta m_H^2 &= \sum_i \frac{g_i^2}{4\pi^2}\left[-(\Delta^2 + m_{F_i}^2) + (\Delta^2 + m_{B_i}^2)\right] \\ &= \sum_i O\left(\frac{\alpha_i}{\pi}\right)(m_{B_i}^2 - m_{F_i}^2) \end{aligned} \tag{2.8}$$

which is small enough to be "natural" if:

$$|m_B^2 - m_F^2| \leq 1 \text{ TeV}^2 \tag{2.9}$$

for supermultiplets with couplings to the Higgs that are $O(\alpha/\pi)$.

The minimal supersymmetric extension of the Standard Model (MSSM) has a Lagrangian containing a supersymmetric part that has the same gauge interactions as the Standard Model, and Yukawa interactions derived from a superpotential:

$$P = \sum \lambda_l L_L L_L^C H_1 + \sum_{q,p^c} \lambda_p Q_L p_L^c H_2 + \sum_{q,n^c} \lambda_n Q_L n_L^C H_1 + \mu H_1 H_2 \tag{2.10}$$

The first term gives masses to the charged leptons, the second to the charge 2/3 quarks, the third to the charge 1/3 quarks, and the last term mixes the two Higgs doublets H_1 and H_2 that are needed to cancel anomalies and give masses to all the quarks and leptons. The supersymmetric part of the MSSM Lagrangian also contains an effective scalar potential that is determined by the gauge and Yukawa couplings.

In addition to this supersymmetric part of the MSSM Lagrangian, there must also be soft supersymmetry-breaking terms including:

$$\mathcal{L}_{SUSYX} \ni -\sum_i m_{0_i}^2 |\phi^i|^2 - \sum_{a=1,2,3} M_a \widetilde{V}_a \widetilde{V}_a \tag{2.11}$$

It is often thought that the scalar masses m_{0_i} and the gaugino masses M_a originate from a supergravity or superstring theory, and that their bare values would be universal. These values would then be renormalized differently by the different gauge interactions, so that their physical values would be approximately [33]:

$$M^2_{\tilde{0}_i} \simeq m_0^2 + C_i m_{1/2}^2 \ : \ C_{\tilde{q}} \simeq 6 \ , \ C_{\tilde{\ell}_L} \simeq 0.5 \ , \ C_{\tilde{\ell}^c_L} \simeq 0.15 \qquad (2.12a)$$

$$M_a \simeq \frac{\alpha_a}{\alpha_{GUT}}\, m_{1/2} \ : \ M_3 (\equiv M_{\tilde{g}}) \simeq 3 m_{1/2} \ , \ M_{\tilde{\gamma}} \simeq \frac{1}{2}\, m_{1/2} \qquad (2.12b)$$

With these theoretical prejudices it is possible to compare the constraints from different experiments as in fig. 12 for the $(m_0, m_{1/2})$ plane.

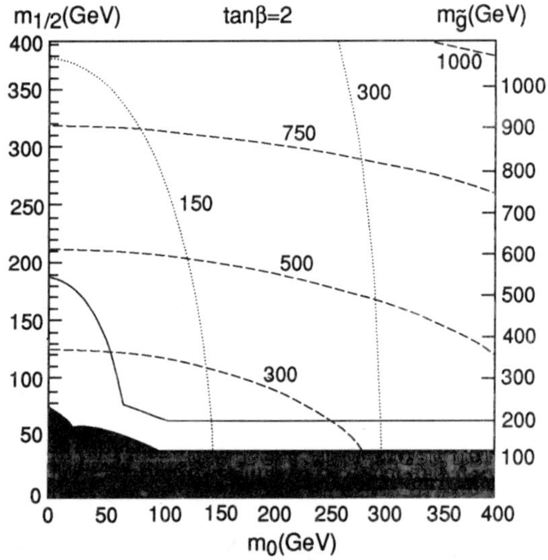

Fig. 12. Experimental constraints (solid region) in the $(m_0, m_{1/2})$ plane, together with the region accessible to LEP II and the Tevatron (solid line) and contours of $m_{\tilde{q}}$ (dashed lines) and $(m_{\tilde{e}})$ (dotted lines).

One can explore the effects of supersymmetric radiative corrections on the precision electroweak analysis. For example, there is a $(\tilde{t}, \tilde{b})$ contribution to $\Delta\rho$ that would be the same as the t quark contribution in the limit $m_{\tilde{q}} \ll m_{W,t}$, but is expected to be much smaller for realistic values of $m_{\tilde{q}}$ consistent with the constraints in fig. 12. However, it could reduce the inferred value of m_t by up to 10 GeV or so. Figure 13 shows the indirect constraints in the $(m_0, m_{1/2})$ plane that come from a global analysis assuming that $m_t = 130$ GeV [34]. They can make a significant improvement on the direct limits at intermediate values of $m_0/m_{1/2}$.

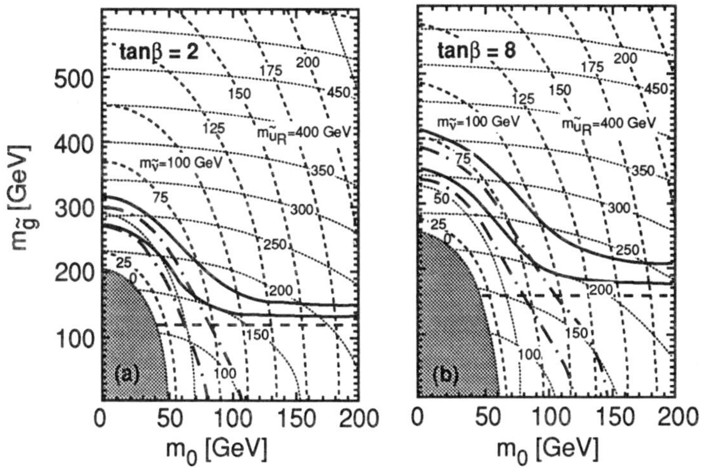

Fig. 13. The $\Delta\chi^2 = 1, 2.7$ countours in the $(M_0, m_{\tilde{g}})$ plane for $m_t = 130$ GeV, $\mu = -300$ GeV, $m_A = 1$ TeV and (a) $\tan\beta = 8$, shown as full heavy lines. The heavy dashed line is the limit due to the absence of the direct decay into the lightest chargino with mass $< M_Z/2$. Dash-dotted heavy lines indicate the limits when the chargino and neutralino effects are not included. Also shown is the excluded zone corresponding to $m^2_{\tilde{\nu}} < 0$, together with dashed curves for constants $m_{\tilde{\mu}}$ and dotted curves for constant $m_{\tilde{u}_R}$.

In order to understand the experimental constraints on electroweakly-interacting spin -1/2 sparticles, we must understand the structure of the chargino $(\widetilde{W}^{\pm}, \widetilde{H}^{\pm}_{1,2})$ mass squared matrix [35]:

$$M^2_{\widetilde{C}} = \begin{pmatrix} M_2^2 + g_2^2 v_1^2 & -M_2 g_2 v_2 + \mu g_2 v_1 \\ -M_2 g_2 v_2 + \mu g_2 v_1 & \mu^2 + g_2^2 v_2^2 \end{pmatrix} \tag{2.13}$$

and the neutralino $(\widetilde{W}^0/\widetilde{B}^0/\widetilde{H}^0_{1,2})$ mass matrix:

$$M_{\widetilde{N}} = \begin{pmatrix} M_2 & 0 & \frac{-g_2 v_2}{\sqrt{2}} & \frac{g_2 v_1}{\sqrt{2}} \\ 0 & M_1 & \frac{g' v_2}{\sqrt{2}} & \frac{-g' v_1}{\sqrt{2}} \\ \frac{-g_2 v_2}{\sqrt{2}} & \frac{g' v_2}{\sqrt{2}} & 0 & \mu \\ \frac{g_2 v_1}{\sqrt{2}} & \frac{-g' v_1}{\sqrt{2}} & \mu & 0 \end{pmatrix} \tag{2.14}$$

which depend on the parameters $m_{1/2}$ (via $M_{2,1}$), μ and $\tan\beta \equiv v_2/v_1$. The lightest neutralino eigenstate is particularly simple in the limit $M_2, M_1 \to 0$, where it becomes a pure photino $\tilde{\gamma}$, and in the limit $\mu \to 0$, where it becomes a pure Higgsino $\widetilde{H}^0$, but is quite a complicated mixture in general.

There are many constraints [36] on the parameters $(\mu, m_{1/2})$ coming from LEP searches for charginos and neutralinos, including the measurement of the invisible Z^0 decay width, and from searches for gluinos at $\bar{p}p$ colliders assuming the relation (2.12b). These are compiled [36] in fig. 14 for one particular choice of $\tan\beta$. The shaded regions are those where the cosmological relic density of the lightest neutralino could be critical for suitable values of the other model parameters [37].

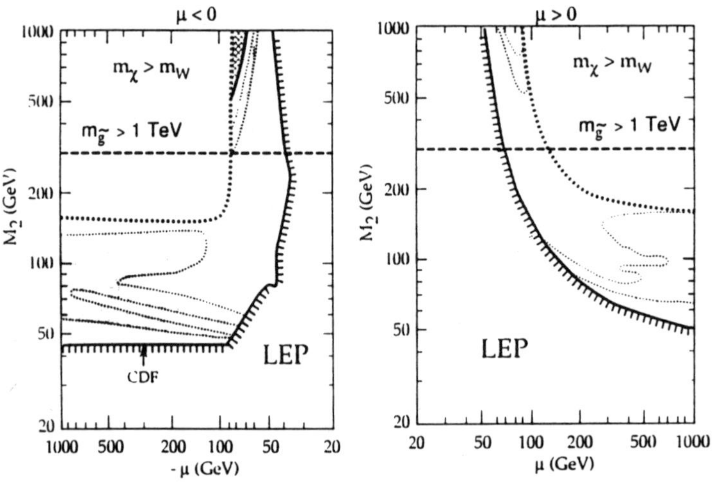

Fig. 14. Experimental constraints in the (μ, M_2) plane (hatched line), together with the line $m_x = m_W$ (thick dashed), and the stippled region where the relic LSP density can be cosmologically interesting.

The two Higgs multiplets in the MSSM give rise to 5 physical states [35] : two neutral scalars h, H, one neutral pseudoscalar A, and 2 charged bosons $H^\pm$. At the tree level, there are several formulae linking their masses:

$$\begin{gathered} m_h^2 + m_H^2 = m_A^2 + m_Z^2 \;,\; m_A^2 = m_{H^\pm}^2 - m_{W^\pm}^2 \\ m_{h,H}^2 = \frac{1}{2}\Big[m_A^2 + m_Z^2 \mp \sqrt{(m_A^2 + m_Z^2)^2 - 4m_Z^2 m_A^2 \cos^2 2\beta}\Big] \end{gathered} \tag{2.15}$$

which give, in particular, at least one state h weighing less than m_Z. However, radiative corrections to the supersymmetric Higgs masses are calculable and can be large [38]. For example:

$$\delta m_h^2 \simeq \frac{3g^2}{8\pi^2} \frac{m_t^4}{m_W^2} \, \ell n \, (m_{\tilde{q}}^2/m_t^2) + O(m_t^2) \tag{2.16}$$

which can increase m_h by tens of GeV, and make it heavier than the $\mathbf{Z^0}$ **and** hence beyond the planned reach of LEP 2. Figure 15 shows the results of a typical one-loop calculation of (m_h, m_A). The region where $m_A < m_h$ has by now probably been excluded by searches at LEP, leaving us with the expected range:

$$m_h = 80 \pm 50 \text{ GeV} \tag{2.17}$$

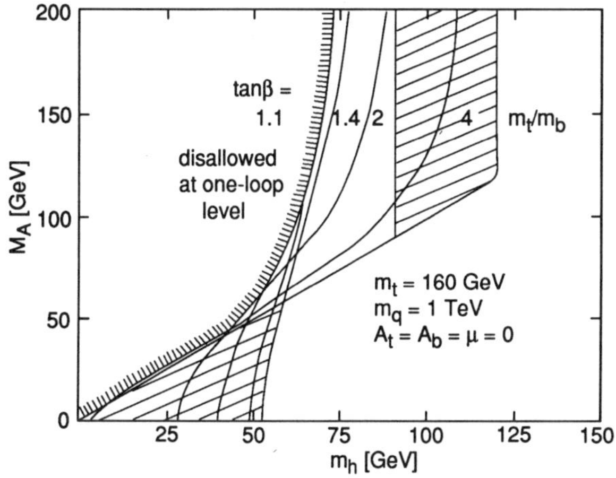

Fig. 15. The (m_H, M_A) plane for supersymmetric Higgses, indicating regions disallowed at the one-loop level, and disallowed at the tree level but allowed at the one-loop level (shaded), for the indicated values of tanβ, m_t, $m_{\tilde{q}}$, A_t, and μ.

which is consistent with the indications that we found previously from our precision LEP data analysis [17] .

We have now incorporated [34] the supersymmetric Higgs sector in a global fit to the electroweak data, assuming that the other sparticle masses are large. As seen in fig. 16, we find that small values of m_h and M_A are slightly preferred, although no values can be excluded if $m_t = 130$ GeV. The global minimum value of χ^2 is the same as in the Standard Model fit, but is located closer to the boundary of the region which is still allowed by direct searches for supersymmetric Higgses.

The production and detection of supersymmetric Higgses at LEP 2 and the LHC have been considered [39], with the conclusions summarized in fig. 17. Most of the supersymmetric Higgs parameter space can be explored, but there is still a hole for tan$\beta \simeq 10$ and $m_A \simeq 150$ GeV. This could be plugged by a higher-energy e^+e^- collider.

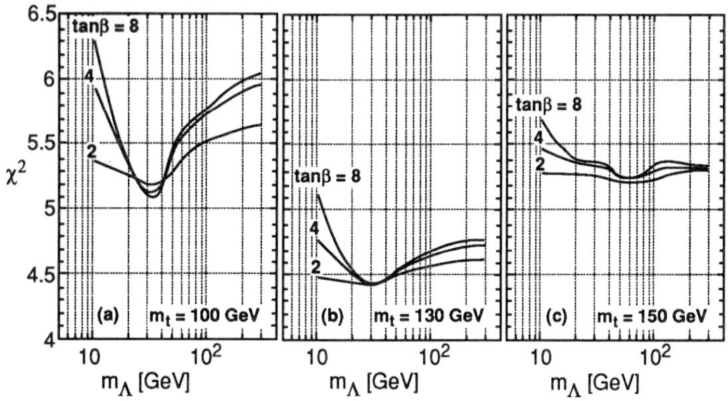

Fig. 16. The value of χ^2 as a function of m_A, for three choices of tanβ (2,4,8) and the values (a) m_t = 100 GeV, (b) m_t = 130 GeV, (c) m_t = 150 GeV. The statistical importance of the dependence on m_A is significant only for low m_t and large tanβ.

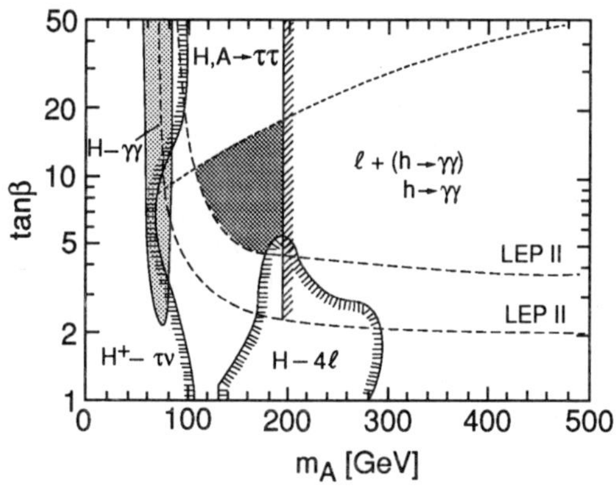

Fig. 17. Regions of the $(M_A, \tan\beta)$ plane for supersymmetric Higgses which should be detectable in pp and e^+e^- colliders.

2.3 Grand Unification

It is tempting to extrapolate the logarithmic decrease of the strong coupling observed in LEP and lower-energy experiments another 13 decades or so in energy until it becomes equal to the more slowly-varying SU(2) and U(1) couplings, where Grand Unification could occur. The simplest Grand Unified Theory (GUT) is SU(5) [40], but we will not worry about the details of the GUT model since the approach to the Grand Unification scale is determined by the light particle spectrum alone [41]. The following are the 2-loop equations for the renormalization group running of the SU(3), SU(2) and U(1) couplings in the Standard Model without supersymmetry:

$$Q\frac{\partial\alpha_i(Q)}{\partial Q} = -\frac{1}{2\pi}\left(b_i + \frac{b_{ij}}{4\pi}\,\alpha_j(Q)\right)[\alpha_i(Q)]^2 + \ldots \tag{2.18}$$

where the one-loop coefficients are:

$$b_i = \begin{pmatrix} 0 \\ -22/3 \\ -11 \end{pmatrix} + N_g \begin{pmatrix} 4/3 \\ 4/3 \\ 4/3 \end{pmatrix} + N_H \begin{pmatrix} 1/10 \\ 1/6 \\ 0 \end{pmatrix} \tag{2.19}$$

and the two–loop coefficients are:

$$b_{ij} = \begin{pmatrix} 0 & 0 & 0 \\ 0 & -\frac{136}{3} & 0 \\ 0 & 0 & -102 \end{pmatrix} + N_g \begin{pmatrix} \frac{19}{15} & \frac{3}{5} & \frac{44}{15} \\ \frac{1}{5} & \frac{49}{3} & 4 \\ \frac{11}{30} & \frac{3}{2} & \frac{76}{3} \end{pmatrix} + N_H \begin{pmatrix} \frac{9}{50} & \frac{9}{10} & 0 \\ \frac{3}{10} & \frac{13}{6} & 0 \\ 0 & 0 & 0 \end{pmatrix} \tag{2.20}$$

Using these, the experimental values of α_{em}, α_3, and the hypothesis that $\alpha_3 = \alpha_2 = \alpha_1$ at same scale m_{GUT}, one predicts [42]:

$$\sin\theta_W(m_Z)|_{\overline{MS}} = 0.208 + 0.006\ \ln\left(\frac{400\ \text{MeV}}{\Lambda_{\overline{MS}}(N_f = 4)}\right) \tag{2.21}$$

which disagrees significantly with the LEP value [17] [34] :

$$\sin^2\theta_W(m_Z)|_{\overline{MS}} = 0.2324 \pm 0.0006 \tag{2.22}$$

As has also been known for some time, the remedy is supersymmetry. The renormalization group equations for the MSSM are as in (2.18), but with [43]:

$$b_i = \begin{pmatrix} 0 \\ -6 \\ -9 \end{pmatrix} + N_g \begin{pmatrix} 2 \\ 2 \\ 2 \end{pmatrix} + N_H \begin{pmatrix} \frac{3}{10} \\ \frac{1}{2} \\ 0 \end{pmatrix} \tag{2.23}$$

and

$$b_{ij} = \begin{pmatrix} 0 & 0 & 0 \\ 0 & -24 & 0 \\ 0 & 0 & -54 \end{pmatrix} + N_g \begin{pmatrix} \frac{38}{15} & \frac{6}{5} & \frac{88}{15} \\ \frac{3}{5} & 14 & 8 \\ \frac{11}{15} & 3 & \frac{68}{3} \end{pmatrix} + N_g \begin{pmatrix} \frac{9}{50} & \frac{9}{10} & 0 \\ \frac{3}{10} & \frac{7}{2} & 0 \\ 0 & 0 & 0 \end{pmatrix} \tag{2.24}$$

again independently of the specific GUT model chosen. The resulting prediction [44] for $\sin^2\theta_W(m_Z)_{\overline{MS}}$ as a function of $\alpha_3(m_Z)$, and assuming that all the sparticles weigh m_Z, is shown in fig. 18. The agreement with the data is good, whilst the non-supersymmetric prediction is many standard deviations away. The question arises whether this agreement can be used to infer the sparticle mass scale with any precision. As seen in equations (2.12), the light sparticle thresholds are not expected to coincide, and it is not adequate to fit the data with a single universal sparticle mass scale [45]. In a general supersymmetric GUT there can be threshold effects around m_{GUT}. However, in the minimal supersymmetric SU(5) GUT it can be argued [46] that these are absent because of a lower limit on the Higgs colour triplet mass due to proton stability. Incorporating the low-mass sparticle thresholds, one finds [46] :

$$\begin{aligned}\sin^2\theta_W(M_{Z^0})\big|_{\overline{MS}} = 0.02029 + \frac{7\alpha_{em}}{15\alpha_3} &+ \frac{\alpha_{em}}{20\pi}\Big[-3\ell n\,(m_t/m_Z) \\ &+ \frac{28}{3}\,\ell n(m_{\tilde{g}}/m_Z) - \frac{32}{3}\,\ell n(m_{\widetilde{W}}/m_Z) - \ell n(m_H/m_Z) - 4\ell n(\mu/m_Z) \\ &+ \frac{3}{8}\,f(m_0^2/m_{1/2}^2\,,\ \tilde{m}^2/m_{1/2}^2)\end{aligned} \tag{2.25}$$

where:

$$\begin{aligned}f(y,w) \equiv \frac{15}{16}\,\ell n(6+y) - \frac{19}{96}\,\ell n(6+y+w) - \frac{35}{96}\,\ell n(6+y-w) \\ - \frac{9}{8}\,\ell n(0.5+y) + \frac{3}{4}\,\ell n(0.15+y)\end{aligned} \tag{2.26}$$

is associated with mixing of stop squarks and is numerically not very important.

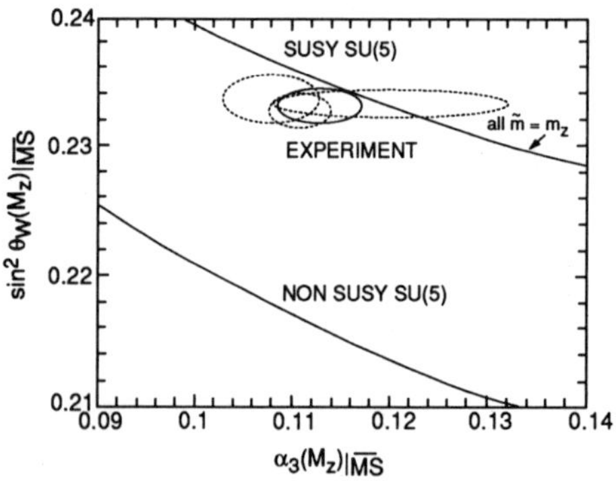

Fig. 18. Comparison of predictions of minimal non-supersymmetric SU(5) and a minimal supersymmetric GUT, assuming all sparticles weigh m_Z, with various different LEP values of $\alpha_3(m_Z)$ and $\sin^2\theta_W(m_Z)|_{\overline{MS}}$.

Inverting (2.25, 2.26), one finds [47] the prediction for $m_{1/2}$ shown in fig. 19. The uncertainty due to the lack of precision in our measurements of $\alpha_3(m_Z)$ is the largest: $1-\sigma$ variations in the other parameters correspond to $\Delta\alpha_3 \simeq \pm 0.0025$. Since experimental measurements of $\alpha_3(m_Z)$ vary all the way from below 0.110 to above 0.125 at the 1-σ level, it is clearly not possible to estimate $m_{1/2}$ with any reliability [48].

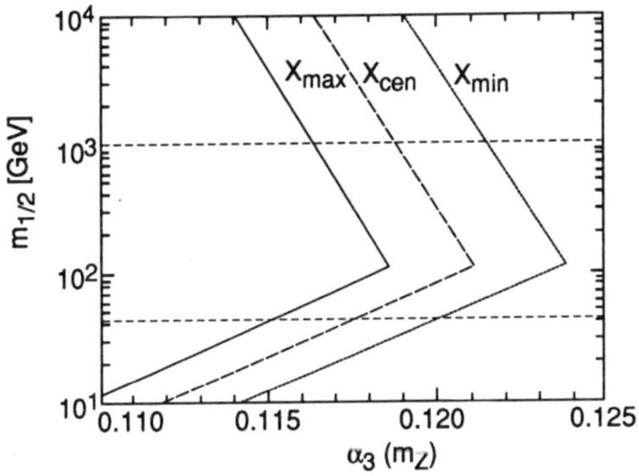

Fig. 19. Values of $m_{1/2}$ allowed with minimal supersymmetric GUTs by a given value of $\alpha_3(m_Z)$.

2.4 Is the TOE made of string?

The idea that what we now consider to be point particles are in part extended objects made of loops of string is very radical and has many far-reaching implications that we cannot discuss now [49]. It is now known that there are many consistent string models in which the original extra dimensions are replaced by internal degrees of freedom. The challenge to theorists is to make a definite prediction that can be proved or even better disproved by experiment. What I want to discuss here is the possibility that a string model might have observable implications for the gauge couplings measured at LEP and the unification programme discussed in the previous section.

The four-dimensional gauge groups in such models are often of the form $[SU(3)]^m \times [SU(2)]^n \times [U(1)]^p$. We have asked the question whether it is possible to obtain a GUT with a simple gauge group factor that contains the $SU(3) \times SU(2)$ of the Standard Model. The obstacle to this programme is that GUTs generally require adjoint Higgs representations for their breakdown to the Standard Model gauge group, whereas the class of string theories generally considered by model-builders does not allow for adjoint or larger Higgs representations.

A possible way around this obstacle is provided by the flipped $SU(5) \times U(1)$ GUT [50], whose gauge symmetry can be broken down to the Standard

Model's SU(3) × SU(2) × U(1) by $\underline{10}$ and $\overline{\underline{10}}$ representations of Higgs fields [51]. The particularity of this model is that the SU(2) doublet particle assignments are flipped compared with regular SU(5): $u \leftrightarrow d, \nu \leftrightarrow e$. This is possible because the Standard Model U(1) is partly inside the SU(5) group and partly the external U(1) factor, so that the charges in any given SU(5) multiplet do not have to sum to zero.

This model offers the possibility that the SU(3) and SU(2) gauge couplings might become equal to each other and to the SU(5) coupling at a mass-scale significantly below that at which the SU(5) and U(1) couplings become equal. In such a case, $\alpha_1 < \alpha_2 = \alpha_3$ at the GUT scale, and $\sin^2\theta_W(m_Z)|_{\overline{MS}}$ is lower than in the conventional SU(5) GUT. This possibility is certainly not excluded by the experimental data, as we see in fig. 18.

Moreover, it is possible to make in string theory a top-down calculation [52] of the unification scale, defined to be the scale at which any pair of low-energy couplings become equal. In general:

$$m_U = 3 \times 10^{17}\ \text{GeV} \times f\ (\text{moduli, matter}) \tag{2.27}$$

where the function f depends on the model considered. In the case of one version of the flipped SU(5) × U(1) GUT derived from string [53], the function $f \simeq 2$. In this case the string unification scale (2.27) appears to be significantly higher than m_{GUT} calculated by extrapolating upwards the values of the gauge couplings measured at LEP. Thus the possibility of a string discrepancy with the supersymmetric GUT value of $\sin^2\theta_W$ remains open.

References

[1] L. Rolandi, Rapporteur, Talk at the International Conference on High Energy Physics, Dallas 1992.

[2] C. Bouchiat, J. Iliopoulos and P. Meyer, *Phys. Lett.* **38B** (1972) 519.

[3] D. Schaile and P. Zerwas, *Phys. Rev.* **D45** (1992) 3262.

[4] M. Peskin and G.L. Kane, *Nucl. Phys.* **B195** (1982) 29.

[5] S.L. Glashow, J. Iliopoulos and L.Maiani, *Phys. Rev* **D2** (197) 1285.

[6] M. Davier, Proc. Joint International Lepton-Photon Symposium and Europhysics Conference on High Energy Physics, Geneva 1991, eds. S. Hegarty, K. Potter and E. Quercigh, (World Scientific, Singapore, 1992) p. 151.

[7] CDF Collaboration, F. Abe et al., *Phys. Rev. Lett.* **68** (1992) 447, and *Phys. Rev.* **D46** (1992) 147.

[8] C.H. Llewellyn Smith, *Phys. Lett.* **46B** (1973) 233;
J.S. Bell, *Nucl. Phys.* **B60** (1973) 427;
J. Cornwall, D. Levin and G. Tiktopoulos, *Phys. Rev.* **D16** (1974) 1145.

[9] B. Lee, C. Quigg and H. Thacker, *Phys. Rev.* **D16** (1977) 1519;
M. Veltman *Acta Phys. Pol.* **B8** (1977) 475.

[10] J.M. Johnson, L. Durand and J.L. Lopez, *Phys. Rev. Lett.* **64** (1990) 1215.

[11] E. Gross, CERN Preprint PPE 92-091 (1992).

[12] Particle Data Group, J.J. Hernandez et al., *Phys. Rev.* **D45** (1992).

[13] A. Sirlin, *Phys. Rev.* **D22** (1980) 971.

[14] A. Sirlin, *Phys. Lett.* **B232** (1989) 123.

[15] W.F.L. Hollik, *Fortschr. Phys.* **38** (1990) 165.

[16] J. Van der Bij and F. Hoogeveen, *Nucl. Phys.* **B283** (1987) 447.

[17] J. Ellis, G.L. Fogli and E. Lisi, CERN Preprint TH.6568/92 (1992).

[18] G.L. Fogli and D. Haidt, *Zeit für Phys.* **C40** (1988) 379.

[19] M.C. Noecker, B.P. Masterson and C.E. Wieman, *Phys. Rev. Lett.* **61** (1988) 310.

[20] CHARM II Collaboration, P. Vilain et al., *Phys. Lett.* **B281** (1992) 159.

[21] J. Ellis and G.L. Fogli *Phys. Lett.* **B232** (1989) 139; **B249** (1990) 543;
P. Langacker, *Phys. Rev. Lett.* **63** (1989) 1920.

[22] R. Dalitz and G.R. Goldstein, *Phys. Lett* **B287** (1992) 225.

[23] R. Dashen and H. Neuberger, *Phys. Lett.* **50** (1983) 1897.

[24] L. Maiani, G. Parisi and R. Petronzio, *Nucl. Phys.* **B136** (1978) 115;
N. Cabibbo, L. Maiani, G. Parisi and R. Petronzio, *Nucl. Phys.* **B158** (1979) 295.

[25] F. Zwirner, these proceedings.

[26] J. Ellis, A.D. Linde and M. Sher, *Phys. Lett.* **B252** (1990) 203.

[27] Y. Nambu, Proc. XI Int. Symp. on Elementary Particle Physics, eds. Z. Ajduk, S. Pokorski and A. Trautman (World Scientific, Singapore, 1989);
W.A. Bardeen, C.T. Hill and M. Lindner, *Phys. Rev.* **D41** (1990) 1647.

[28] E. Farhi and L. Susskind, *Physics Reports* **74C** (1981) 277.

[29] Yu.A. Gol'fand and E.P. Likhtman, *JETP Lett.* **13** (1971) 323;
D.V. Volkov and V.P. Akulov, *Phys. Lett.* **46B** (1973) 109;
J. Wess and B. Zumino, *Nucl. Phys.* **B70** (1970) 39.

[30] A. Hasenfratz et al., *Nucl. Phys.* **B365** (1991) 79.

[31] M. Peskin and T. Takeuchi, *Phys. Rev. Lett.* **65** (1990) 964;
G. Altarelli, R. Barbieri and S. Jadach, *Nucl. Phys.* **B369** (1992) 3.

[32] L. Maiani, Proc. Summer School on Particle Physics, Gif-sur Yvette;
E. Witten, *Nucl. Phys.* **B188** (1981) 513.

[33] A.B. Lahanas and D.V. Nanopoulos, *Physics Reports* **145C** (1987) 1.

[34] J. Ellis, G.L. Fogli and E. Lisi, CERN Preprint TH.6643/92 (1992).

[35] H.E. Haber and G.L. Kane, *Physics Reports* **117C** (1985) 75.

[36] J. Ellis, G. Ridolfi and F. Zwirner, *Phys. Lett.* **B237** (1990) 423.

[37] J. Ellis and L. Roszkowski, *Phys. Lett.* **B283** (1992) 252 and references therein.

[38] Y. Okada, M. Yamaguchi and T. Yanagida, *Progr. Theor. Phys.* **85** (1991)1;
J. Ellis, G. Ridolfi and F. Zwirner, *Phys. Lett.* **B257** (1991) 83;
H.E. Haber and R. Hempfling, *Phys. Rev. Lett.* **66** (1991) 1815;
R. Barbieri and M. Frigeni, *Phys. Lett.* **B258** (1991) 395.

[39] Z. Kunszt and F. Zwirner, CERN Preprint TH.6150/91 (1992).

[40] H. Georgi, S.L. Glashow, *Phys. Rev. Lett.* **32** (1974) 438.

[41] H. Georgi, H.R. Quinn and S. Weinberg, *Phys. Rev. Lett.* **33** (1974) 451.

[42] W. Marciano and A. Sirlin, *Phys. Rev. Lett.* **46** (1981) 163.

[43] M.B. Einhorn and D.R.T. Jones, *Nucl. Phys.* **B196** (1982) 475;
M.E. Machacek and M.T. Vaughn, *Nucl. Phys.* **B236** (1984) 221.

[44] J. Ellis, S. Kelley and D.V. Nanopoulos, *Phys. Lett.* **B249** (1990) 442 and **B260** (1991) 131.

[45] U. Amaldi, W. De Boer and H. Fürstenau, *Phys. Lett.* **B260** (1991) 447.

[46] J. Ellis, S. Kelley and D.V. Nanopoulos, *Nucl. Phys.* **B373** (1992) 55.

[47] F. Anselmo, L. Cifarelli, A Peterman and A. Zichichi, CERN Preprint TH.6429/92, LAA/MSL/92-003;
J. Ellis, S. Kelley and D.V. Nanopoulos, *Nucl. Phys.* **B287** (1992) 95.

[48] F. Anselmo, L. Cifarelli, A. Peterman and A. Zichichi, CERN Preprints PPE/91-123, LAA/MSL/91-015; PPE/92-103, LAA/MSL/92-008; TH.6543/92, LAA/MSL/92-009; PPE/92-122, LAA/MSL/92-012;
A. Zichichi, these Proceedings, CERN Preprint PPE/92-149, LAA/MSL/92-017.

[49] D.V. Nanopoulos, these Proceedings.

[50] S. Barr, *Phys. Lett.* **112B** (1982) 219;
J.-P. Derendinger, J. Kim and D.V. Nanopoulos, *Phys. Lett.* **B139** (1984) 170.

[51] I. Antoniadis, J. Ellis, J.S. Hagelin and D.V. Nanopoulos, *Phys. Lett.* **B194** (1987) 231, **B205** (1988) 459, **B208** (1988) 209 and **B231** (1989) 65.

[52] V. Kaplunovsky, *Nucl. Phys.* **B307** (1988) 145;
L.J. Dixon, V. Klapunovsky and J. Louis, *Nucl. Phys.* **B329** (1990) 27.

[53] I. Antoniadis, J. Ellis, R. Lacaze and D.V. Nanopoulos, *Phys. Lett.* **B268** (1991) 188.

CHAIRMAN: J. Ellis

Scientific Secretaries: A. Hanany, A. Leike

DISCUSSION

– *Shotkin:*

1) Could you please describe quantitatively how the plot of $m_{1/2}$ vs m_0 would change for other values of tanβ (shown for tan$\beta = 2$)?

2) Also please review why it is thought currently that tan$\beta \leq 30$ or so?

– *Ellis:*

1) Consider the plot of extended regions in the $(m_{1/2}, m_0)$ plane. I have here plots for tanβ = 2,4 that we can compare. For rising tanβ and small m_0 the slepton production can exclude higher $m_{1/2}$ masses relative to the strip excluded by wino searches. However, this increase is almost saturated when tan$\beta = 4$.

We expect that tan$\beta > 1$ because $m_t > m_b$. This has to do with the mechanism of dynamical symmetry breaking discussed in the lecture of D. Nanopoulos. Considering the evolution of effective Higgs masses as functions of the energy, we see that $\lambda_t > \lambda_b$ implies $v_t > v_b$. Therefore we can write:

$$30 \sim \frac{m_t}{m_b} = (\lambda_t V_t)/(\lambda_b V_b) > V_t/V_b$$

Finally we remember that tan$\beta = V_t/V_b$.

– *Do Couto e Silva:*

Let us say we have a real Z^0 that decays into a pair of dark matter objects, if s-neutrino is there it will affect the Z^0 line width. But at LEP people are looking for photinos.They are not supposed to affect the Z^0. Why is that?

– *Ellis:*

There is no coupling of the Z to photinos; i.e. $g_{Z\tilde{\gamma}\tilde{\gamma}} = 0$. So the decay $Z \to \tilde{\gamma}\tilde{\gamma}$ is not possible. However, the lightest supersymmetric particle (LSP) is, in general, not a photino but a mixture of the photino, zino and higgsino. For that mixture the coupling is in general not zero; i.e. $g_{Z(LSP)(LSP)} \neq 0$. So one can get constraints from LEP for regions of the $(m_0, m_{1/2})$ plane where the LSP is not a pure photino.

– *Malik:*

1) How do supersymmetric particles affect the width of the vector boson? Are they playing any role by affecting Γ_{inv}?

2) How would the gluino production events look like in a detector?

– *Ellis:*

1) The invisible decay width of the Z sets important lower limits on the mass of the sneutrino. So it is not a dark matter candidate and is excluded as a possible candidate for the lightest supersymmetric particle.

2) The decays of the gluino depend very much on the relation of its mass to the masses of the other supersymmetric particles, especially the left and right handed squarks. This leads to different decay cascades which could be quite complicated giving different signature in the detection from simple missing energy.

– *Benakli:*

What is known about the evolution of Newton's constant?

– *Ellis:*

It does not evolve: it is just a constant. Limits on a possible tiny evolution of Newton's constant come from big bang nuclear synthesis in the early universe. A second important bound on the variations comes from a natural nuclear reactor at Oklo in West Africa, see in the "Festschrift to Dirac" in 1978 an article by Freeman Dyson.

– *Benakli:*

What is the asymptotic character of the perturbation expansion?

– *Ellis:*

The accuracy of the perturbation expansion becomes better and better up to an order of $O(1/\alpha)$. If one calculates even higher orders the contributions start growing again. Since $1/\alpha$ is rather large we do not have this problem in present calculations.

– *Hambye:*

You spoke about the fact that supersymmetry is a good model to get rid of the quadratic divergence of the Higgs vector boson mass. Do you think it is by chance that some more simple models, where you simply constrain the top and Higgs mass to be such that you cancel these ones, (basing also on universality of quadratic divergences) agree with experimental constraints on m_t and m_H?

– *Ellis:*

With these models you cancel the quadratic divergences at one loop, not necessarily at all orders. So these assumptions are somewhat ad hoc. This is not the case for supersymmetry where you have a systematic way to cancel these to all orders. There is also considerable arbitrariness in the logarithmic divergences that are chosen to be cancelled. On the other hand, supersymmetry cancels systematically a large class of logarithmic divergences.

– *Baillie:*

Does the fact that γ vacuum polarization graphs are dependent on low-energy QCD dynamics significantly affect radiative correction predictions?

– *Ellis:*

To get $\alpha_{em}(M_Z^2)$ we have to make an evolution starting from the Thomson limit at low energies (m_e). In the evolution equation enter polarization diagrams (fermion loops). Lepton loops can be calculated exactly. Quark loops are not calculable perturbatively at low energies. They have to be replaced by the data. The experimental errors of these data, particularly those for 4 GeV $< E_{cm} <$ 10 GeV, is reflected in an uncertainty in $\alpha_{em}(M_Z^2)$. This uncertainty in $\alpha_{em}(M_Z^2)$ is important, for example in extracting $\sin^2\theta_W$ from measurements in m_Z.

– *Sannino:*

In SO(10) we need only one representation to classify all fermions and also it gives the right unification mass. Instead, in non-SUSY SU(5) we need two representations. So why do you prefer SU(5)?

– *Ellis:*

SU(5) is preferred for reasons of simplicity. Larger gauge groups contain, in general, more parameters. So it is hard to make some definite predictions, for instance for heavy mass thresholds, in such theories. But I would not exclude SO(10) a priori.

PHENOMENOLOGICAL SUPERSYMMETRY:

THE SUSY HIGGS SECTOR

Fabio Zwirner[1]

Theoretical Physics Division

CERN, Geneva, Switzerland

1) The Higgs sector of the Minimal Supersymmetric Standard Model

Despite the remarkable consistency of the Standard Model (SM) with all experimental data in particle physics, there is no experimental evidence yet that the minimal SM Higgs sector provides the correct description of electroweak symmetry breaking. One should then be prepared for alternatives: among the various possibilities, a particularly attractive one is the Higgs sector of the Minimal Supersymmetric Standard Model[2] (MSSM).

This possibility is theoretically well motivated: supersymmetry, effectively broken in the vicinity of the electroweak scale, is the only known theoretical framework that can solve the naturalness problem of the SM, while keeping elementary spin-0 fields in the theory. Also, in contrast with many other alternatives to the SM, the MSSM successfully survives all the stringent phenomenological tests coming from precision measurements: in most of its parameter space, the MSSM predictions for the electroweak observables are extremely close to the SM ones, evaluated for a relatively light SM Higgs. Moreover, the MSSM has a high degree of predictivity, since, as we shall see, all masses and couplings of the Higgs boson sector can be expressed, at the tree-level, in terms of only two parameters, and radiative corrections can be kept under control: in particular, cross-sections and branching ratios for the MSSM Higgs bosons can be reliably computed in perturbation theory.

[1]On leave from INFN, Sezione di Padova, Padua, Italy.

[2]Complementary aspects of phenomenological supersymmetry are discussed in the lectures by J. Ellis, S. Ferrara, P. Nath, G.G. Ross and A. Zichichi, these Proceedings. For further reading and references see, e.g., ref. [1].

Furthermore, it is intriguing that the idea of grand unification, which fails in its minimal non-supersymmetric implementation, can be successfully combined with low-energy supersymmetry: supersymmetric grand unification predicts a value of the electroweak mixing angle that is in agreement with the measured one within the theoretical uncertainties, and a value of the grand-unification mass which could explain why proton decay has escaped detection so far. Finally, in the MSSM the lightest supersymmetric particle, typically neutral and weakly interacting, is absolutely stable, and thus a natural candidate for dark matter.

The previous considerations indicate that the Higgs sector of the MSSM is certainly worth a detailed study. As we shall see, the phenomenology of supersymmetric Higgs bosons displays important differences with the SM case. To prepare the ground for the comparison, we begin by briefly recalling the essential features of the SM Higgs sector, which consists of a single complex $SU(2)$-doublet,

$$\phi \equiv \begin{pmatrix} \phi^+ \\ \phi^0 \end{pmatrix} \sim (1,2,+1/2), \tag{1.1}$$

where the numbers in brackets denote the transformation properties under $SU(3)_C \times SU(2)_L \times U(1)_Y$. For practical purposes, it is often convenient to consider the equivalent object $\phi^c \equiv i\sigma^2\phi^* \sim (1,2,-1/2)$. In the SM, the tree-level potential reads

$$V_{SM} = -\mu^2|\phi|^2 + \lambda|\phi|^4 \qquad (\lambda\,,\,\mu^2 > 0)\,, \tag{1.2}$$

and the part of the Lagrangian containing the Yukawa couplings is

$$\mathcal{L}_{Yuk} = h^U \overline{u_R} q_L \phi + h^D \overline{d_R} q_L \phi^c + h^E \overline{e_R} l_L \phi^c + \text{h.c.}\,, \tag{1.3}$$

where q_L, u_R, d_R, l_L, e_R are the different quark and lepton fields, and group and generation indices have been left implicit. Around the $SU(2)\times U(1)$ breaking minimum of V_{SM}, $\langle\phi^0\rangle = v/\sqrt{2}$, the Higgs field can be expanded as

$$\phi = \begin{pmatrix} \dfrac{G^1 + iG^2}{\sqrt{2}} \\ \dfrac{v + \varphi + iG^0}{\sqrt{2}} \end{pmatrix}, \tag{1.4}$$

where G^i $(i = 0,1,2)$ are the Goldstone bosons, φ is the physical neutral Higgs particle, and it is not restrictive to assume v real and positive. Minimization gives $v = \sqrt{\mu^2/\lambda}$, which sets the scale for all masses in the theory: $m_W^2 = g^2v^2/4$, $m_Z^2 = (g^2 + g'^2)v^2/4$, $m_f^2 = h_f^2 v^2/2$, $m_\varphi^2 = 2\lambda v^2$. In summary, the physical spectrum of the SM contains just one spin-0 state, φ, and the only free parameter in the Higgs sector can be taken to be the Higgs mass m_φ.

In supersymmetric extensions of the SM, the Higgs sector is quite different. Spin-0 particles come in chiral supermultiplets, whose physical degrees of freedom are a complex spin-0 boson and a two-component spin-1/2 fermion. If one tries to build a realistic supersymmetric extension of the SM with just one Higgs chiral supermultiplet, for example $\mathcal{H} \equiv (H, \tilde{H}) \sim (1,2,+1/2)$, where H has spin 0 and $\tilde{H}$ spin 1/2, one is immediately faced with two serious problems. First, in supersymmetric models the Yukawa couplings are determined by the superpotential $\mathcal{W}$, an analytic function of the spin-0 fields in the chiral supermultiplets: as a consequence, $\mathcal{W}$ can depend on H but not on H^*. This would allow us to write down the Yukawa couplings giving mass to the charge-2/3 quarks, but not those giving mass to the charge-1/3 quarks and to the charged leptons [see eq. (1.3)]. Furthermore, the spin-1/2 field $\tilde{H}$ carries non-trivial $SU(2)\times U(1)$ quantum numbers, so that it contributes to the chiral anomaly, whose cancellation is necessary for the quantum consistency of the theory. As a result, any realistic supersymmetric extension of the SM must contain at least two complex Higgs doublets,

$$H_1 \equiv \begin{pmatrix} H_1^0 \\ H_1^- \end{pmatrix} \sim (1,2,-1/2), \qquad H_2 \equiv \begin{pmatrix} H_2^+ \\ H_2^0 \end{pmatrix} \sim (1,2,+1/2). \tag{1.5}$$

The fact that H_1 and H_2 have opposite $SU(2)\times U(1)$ quantum numbers guarantees the solution of the two above-mentioned problems.

In the following, we shall concentrate on the MSSM, defined by: 1) minimal gauge group: $SU(3)_C \times SU(2)_L \times U(1)_Y$; 2) minimal particle content: three generations of quarks and leptons and two Higgs doublets, plus their superpartners; 3) an exact discrete R-parity, which guarantees baryon- and lepton-number conservation in renormalizable interactions: $R=+1$ for SM particles and Higgs bosons, $R=-1$ for their superpartners; 4) supersymmetry breaking parametrized by explicit but soft breaking terms: gaugino and scalar masses and trilinear scalar couplings. However, the possibility of non-minimal supersymmetric extension of the SM, and in particular of larger Higgs sectors, cannot be excluded. For example, the most attractive non-minimal SUSY-Higgs sector contains an extra singlet, N, in addition to the two doublets H_1 and H_2: a phenomenological discussion of the (H_1, H_2, N) Higgs sector can be found, for example, in ref. [2].

For a discussion of Higgs-boson masses and couplings in the MSSM, the obvious starting point is the tree-level Higgs potential, whose general form is

$$\begin{aligned} V_{MSSM} &= m_1^2 |H_1|^2 + m_2^2 |H_2|^2 + m_3^2 (H_1 H_2 + \text{h.c.}) \\ &+ \frac{1}{8} g^2 \left(H_2^\dagger \vec{\sigma} H_2 + H_1^\dagger \vec{\sigma} H_1 \right)^2 + \frac{1}{8} g'^2 \left(|H_2|^2 - |H_1|^2 \right)^2 , \end{aligned} \tag{1.6}$$

where m_1^2, m_2^2, m_3^2 are essentially arbitrary mass parameters, g and g' are the $SU(2)$ and $U(1)$ coupling constants, respectively, and $\vec{\sigma}$ are the Pauli matrices. The $SU(2)$ indices are left implicit and contracted in the obvious way. A crucial difference between V_{SM} and V_{MSSM} is now apparent: in V_{SM} the quartic scalar couplings are determined by the arbitrary parameter λ, related to the SM Higgs mass m_φ. In V_{MSSM}, despite the presence of two Higgs doublets, the quartic scalar couplings do not contain any arbitrary parameter, but are determined by the $SU(2)_L \times U(1)_Y$ gauge couplings: supersymmetry strongly constrains the most general form of the scalar potential.

It is not restrictive to consider the following expansion

$$H_1 = \begin{pmatrix} v_1 + \dfrac{S_1 + iP_1}{\sqrt{2}} \\ H_1^- \end{pmatrix}, \qquad H_2 = \begin{pmatrix} H_2^+ \\ v_2 + \dfrac{S_2 + iP_2}{\sqrt{2}} \end{pmatrix}, \tag{1.7}$$

and to choose m_3^2 real and negative, so that the vacuum expectation values $v_1 \equiv \langle H_1^0 \rangle$ and $v_2 \equiv \langle H_2^0 \rangle$ can be taken to be real and positive.

One can then read from the MSSM Lagrangian the different mass terms in the R-even sector of the theory. The weak boson masses are given by

$$m_W^2 = \frac{g^2}{2}(v_1^2 + v_2^2), \qquad m_Z^2 = \frac{g^2 + g'^2}{2}(v_1^2 + v_2^2), \tag{1.8}$$

and the fermion mass matrices by

$$m^U = h^U v_2 \,, \qquad m^D = h^D v_1 \,, \qquad m^E = h^E v_1 \,. \tag{1.9}$$

A physical constraint comes from the fact that the combination $(v_1^2 + v_2^2)$, which determines the W and Z boson masses, must reproduce their measured values. Once this constraint is imposed, at the classical level the MSSM Higgs sector contains only two independent parameters, which can be conveniently taken to be $\tan\beta \equiv v_2/v_1$ and one combination of the mass parameters (of the three original ones, two are removed by the minimization conditions).

The spin-0 boson mass matrices factorize into 2×2 blocks, corresponding to the charged, neutral CP-odd and neutral CP-even sectors. One can easily identify the Goldstone bosons $G^+ = -\cos\beta (H_1^-)^* + \sin\beta H_2^+$, $G^- = (G^+)^*$ and $G^0 = -\cos\beta P_1 + \sin\beta P_2$. The orthogonal combinations $H^+ = \sin\beta (H_1^-)^* + \cos\beta H_2^+$, $H^- = (H^+)^*$ and $A = \sin\beta P_1 + \cos\beta P_2$ correspond to physical particles, with masses

$$m_A^2 = -m_3^2 \left(\tan\beta + \frac{1}{\tan\beta} \right) , \tag{1.10}$$

and

$$m_{H^\pm}^2 = m_W^2 + m_A^2 \,. \tag{1.11}$$

A convenient choice, which will be adopted throughout this paper, is to take as independent parameters m_A and $\tan\beta$. The mass matrix of the neutral CP-even sector then reads

$$\left(\mathcal{M}_R^0\right)^2 = \left[\begin{pmatrix} \cot\beta & -1 \\ -1 & \tan\beta \end{pmatrix} \frac{m_Z^2}{2} + \begin{pmatrix} \tan\beta & -1 \\ -1 & \cot\beta \end{pmatrix} \frac{m_A^2}{2} \right] \sin 2\beta \,. \tag{1.12}$$

Defining the mass eigenstates as

$$h = -\sin\alpha\, S_1 + \cos\alpha\, S_2 \,, \qquad H = \cos\alpha\, S_1 + \sin\alpha\, S_2 \,, \tag{1.13}$$

one obtains

$$m_{h,H}^2 = \frac{1}{2} \left[m_A^2 + m_Z^2 \mp \sqrt{(m_A^2 + m_Z^2)^2 - 4 m_A^2 m_Z^2 \cos^2 2\beta} \right] , \tag{1.14}$$

and also celebrated inequalities such as $m_W, m_A < m_{H^\pm}$, $m_h < m_Z < m_H$, $m_h < m_A < m_H$. Similarly, one can easily compute all the Higgs boson couplings by observing that the mixing angle α, required to diagonalize the mass matrix (1.12), is given by

$$\cos 2\alpha = -\cos 2\beta \, \frac{m_A^2 - m_Z^2}{m_H^2 - m_h^2} \,, \qquad -\frac{\pi}{2} < \alpha \leq 0 . \tag{1.15}$$

For example, the couplings of the three neutral Higgs bosons to vector boson and fermion pairs are easily obtained from the SM Higgs couplings: it is sufficient to multiply the latter by the α- and β-dependent factors summarized in table 1. The remaining tree-level Higgs boson couplings in the MSSM can be easily computed and are summarized, for example, in ref. [3].

	$d\bar{d}, s\bar{s}, b\bar{b}$ $e^+e^-, \mu^+\mu^-, \tau^+\tau^-$	$u\bar{u}, c\bar{c}, t\bar{t}$	W^+W^-, ZZ
h	$-\sin\alpha/\cos\beta$	$\cos\alpha/\sin\beta$	$\sin(\beta-\alpha)$
H	$\cos\alpha/\cos\beta$	$\sin\alpha/\sin\beta$	$\cos(\beta-\alpha)$
A	$-i\gamma_5\tan\beta$	$-i\gamma_5\cot\beta$	0

Table 1: Correction factors for the couplings of the MSSM neutral Higgs bosons to fermion and vector boson pairs.

2) Radiative corrections to masses and couplings

It was recently pointed out [4] that, in realistic supersymmetric extensions of the SM, Higgs boson masses are subject to large, finite radiative corrections, dominated by loops involving the top quark and its supersymmetric partners. Several papers [5–10] have subsequently investigated various aspects of these corrections and their implications for experimental searches at LEP and at the LHC/SSC. We begin this section with a pedagogical example, the explicit calculation, in a particularly simple case, of the leading radiative correction to the neutral CP-even mass matrix. We then summarize the present theoretical status of radiative corrections in the Higgs boson sector of the MSSM.

Considering only the functional dependence on the fields $\varphi_i \equiv \mathrm{Re}\, H_i^0$ $(i=1,2)$, the classical potential of the MSSM can be written as

$$V_0 = m_1^2\varphi_1^2 + m_2^2\varphi_2^2 + 2m_3^2\varphi_1\varphi_2 + \frac{g^2+g'^2}{8}\left(\varphi_2^2-\varphi_1^2\right)^2 . \tag{2.1}$$

The standard way of describing quantum corrections to the classical potential is to consider the effective potential, which at the one-loop level can be written as $V_1 = V_0 + \Delta V$. Including only top and stop loops, working in the $\overline{DR}$ scheme and neglecting as usual field-independent terms, we find

$$\Delta V = \frac{3}{16\pi^2}\left[f(m_{\tilde{t}}^2) - f(m_t^2)\right] , \qquad f(m^2) = m^4\left(\log\frac{m^2}{Q^2} - \frac{3}{2}\right) , \tag{2.2}$$

where $m_t^2 = h_t^2\varphi_2^2$ and $m_{\tilde{t}}^2 = m_t^2 + m_{sq}^2$ are the field-dependent top and stop masses, and Q is the renormalization scale. For simplicity, we have neglected D-terms and mixing terms in the stop squark mass matrix, and we have assumed a common soft supersymmetry-breaking squark mass m_{sq}.

In analogy with the tree-level case, we can use the one-loop minimization conditions,

$$\left(\frac{\partial V_1}{\partial\varphi_i}\right)_{\varphi=v} = 0 , \qquad (i=1,2) , \tag{2.3}$$

to solve for the mass parameters m_1^2 and m_2^2. We can then identify the one-loop-corrected entries in the neutral CP-even mass matrix with

$$\left(\mathcal{M}_R^2\right)_{ij} \equiv \left(\mathcal{M}_R^0\right)^2_{ij} + \left(\Delta\mathcal{M}_R^2\right)_{ij} = \frac{1}{2}\left(\frac{\partial^2 V_1}{\partial\varphi_i\partial\varphi_j}\right)_{\varphi=v} . \tag{2.4}$$

Since in our approximation ΔV does not depend on φ_1, we can immediately write

$$\left(\Delta\mathcal{M}_R^2\right)_{11} = \left(\Delta\mathcal{M}_R^2\right)_{12} = 0 . \tag{2.5}$$

After some very simple algebra, we also obtain

$$\left(\Delta\mathcal{M}_R^2\right)_{22} = \frac{1}{2}\left[-\frac{1}{v_2}\left(\frac{\partial\Delta V}{\partial\varphi_2}\right)_{\varphi=v} + \left(\frac{\partial^2\Delta V}{\partial\varphi_2^2}\right)_{\varphi=v}\right] . \tag{2.6}$$

From eq. (2.2), and the expressions for m_t^2 and $m_{\tilde{t}}^2$, we get

$$\left(\frac{\partial^2\Delta V}{\partial\varphi_2^2}\right)_{\varphi=v} = \frac{1}{v_2}\left(\frac{\partial\Delta V}{\partial\varphi_2}\right)_{\varphi=v} + \frac{3}{16\pi^2}\left(\frac{\partial m_t^2}{\partial\varphi_2}\right)^2_{\varphi=v}\left[f''(m_{\tilde{t}}^2) - f''(m_t^2)\right]_{\varphi=v} , \tag{2.7}$$

and then, observing that $f''(m^2) = 2\log(m^2/Q^2)$,

$$\left(\Delta\mathcal{M}_R^2\right)_{22} = \frac{3}{8\pi^2}\frac{g^2 m_t^4}{m_W^2\sin^2\beta}\log\frac{m_{\tilde{t}}^2}{m_t^2} . \tag{2.8}$$

It is now a simple exercise to derive the one-loop-corrected eigenvalues m_h and m_H, as well as the mixing angle α associated with the one-loop-corrected mass matrix (2.4). The most striking fact in eq. (2.8) is that the correction $(\Delta\mathcal{M}_R^2)_{22}$ is proportional to (m_t^4/m_W^2). This implies that, for m_t in the presently allowed range, 91 GeV $< m_t \lesssim$ 180 GeV, the tree-level predictions for m_h and m_H can be badly violated, and so for the related inequalities. The other free parameter in eq. (2.8) is m_{sq}, but the dependence on it is much milder.

The above formulae have been generalized to arbitrary values of the parameters in the stop mass matrix, and the effects of other virtual particles in the loops have been included. Renormalization group methods have been used to resum the large logarithms that arise when the typical scale of supersymmetric particle masses, M_{SUSY}, is much larger than m_Z. Two-loop corrections have been computed in the leading logarithmic approximation and found to be small. After all these refinements, eq. (2.8) still gives the most important mass correction in the most plausible region of parameter space.

To illustrate the impact of eq. (2.8), we display in fig. 1 contours of the maximum value of m_h, reached for $m_A \gg m_Z$ and $\tan\beta \gg 1$, in the (m_t, m_{sq}) plane. The calculation has been performed in the effective potential approach, including top, bottom, stop, sbottom loops, and neglecting mixing in the stop and sbottom mass matrices. For very large values of m_t and m_{sq}, renormalization group improvement can lower the actual value of m_h^{max} by a non-negligible amount. In the following, when making numerical examples we shall choose the numerical values $m_t = 140$ GeV, $m_{sq} = 1$ TeV: for this parameter choice, the effect of renormalization group improvement is of the order of a few GeV, comparable with other residual theoretical uncertainties. Another useful way of looking at the mass corrections is displayed in fig. 2, which shows contour curves of constant $\tan\beta$ in the (m_h, m_A)

plane, for our standard parameter choice. The region of the (m_h, m_A) plane allowed at the classical level, $m_h < m_A$ and $m_h < m_Z$, is delimited by the dashed lines. One can see that quantum corrections induce a drastic deformation of the allowed region. For given m_A and $\tan\beta$, the shift in m_h can be as large as $O(50\text{ GeV})$, for $\tan\beta \sim 1$. The upper bound on m_h is shifted by an amount $O(20\text{ GeV})$. In particular, after radiative corrections one can have $m_h > m_Z$ and $m_h > m_A$. For small values of m_h it is even possible that $m_h > 2m_A$, in which case $h \to AA$, kinematically forbidden at tree level, can become the dominant h decay mode.

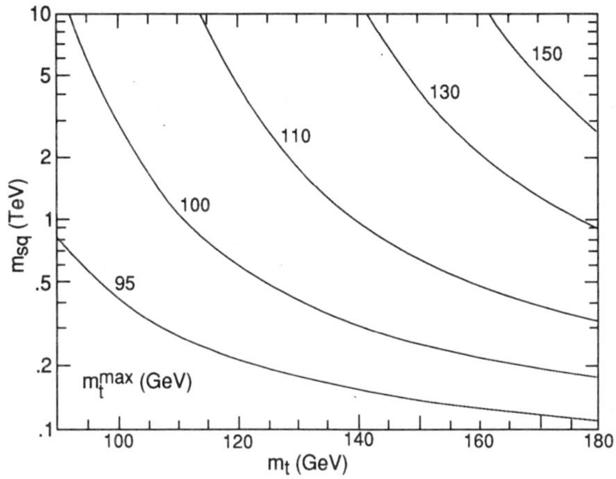

Fig.1: Contour curves of m_h^{max} (the maximum value of m_h, reached for $m_A \gg m_Z$ and $\tan\beta \gg 1$) in the (m_t, m_{sq}) plane.

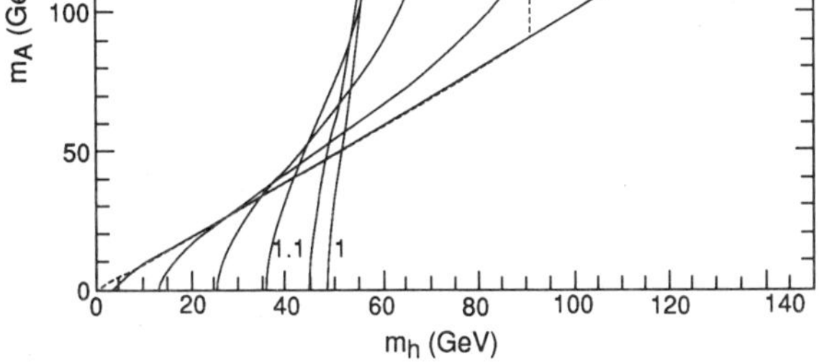

Fig.2: Contour curves of $\tan\beta$ in the (m_h, m_A) plane, for $m_t = 140$ GeV and $m_{sq} = 1$ TeV. The boundary of the region allowed at tree level ($m_h < m_Z$ and $m_h < m_A$) is also shown by the dashed lines.

The computation of radiative corrections can be extended to the other parameters of the MSSM Higgs sector. One-loop corrections to the charged Higgs mass have been computed and found to be small, at most a few GeV, for generic values of the parameters.

Whilst radiative corrections to Higgs boson masses are by now well under control, the study of radiative corrections to Higgs boson couplings is still at a less refined stage. In the case of Higgs boson self-couplings, which control decays such as $H \to hh$, $H \to AA$ and $h \to AA$ when they are kinematically allowed, radiative corrections can be numerically large. Detailed computations of these corrections have been performed by a variety of methods. For example, the leading radiative correction to the cubic Hhh coupling is $\lambda_{Hhh} = \lambda^0_{Hhh} + \Delta\lambda_{Hhh}$, where

$$\lambda^0_{Hhh} = -\frac{igm_Z}{2\cos\theta_W}[2\sin(\beta+\alpha)\sin 2\alpha - \cos(\beta+\alpha)\cos 2\alpha] \tag{2.9}$$

is the tree-level coupling, and

$$\Delta\lambda_{Hhh} = -\frac{igm_Z}{2\cos\theta_W}\frac{3g^2\cos^2\theta_W}{8\pi^2}\frac{\cos^2\alpha\sin\alpha}{\sin^3\beta}\frac{m_t^4}{m_W^4}\left(3\log\frac{m_{sq}^2+m_t^2}{m_t^2} - 2\frac{m_{sq}^2}{m_{sq}^2+m_t^2}\right). \tag{2.10}$$

Notice the explicit dependence on the ratio $(m_t/m_W)^4$.

Finally, one should consider Higgs couplings to vector boson and fermions. In most phenomenological studies, radiative corrections to these couplings have been taken into account only approximately, by improving the tree-level formulae with one-loop corrected values of the H–h mixing angle, α, and with running fermion masses, evaluated at the typical scale Q of the process under consideration. Residual corrections are expected to be numerically small in the experimentally interesting regions of parameter space.

3) SUSY-Higgs searches at e^+e^- and pp colliders

In this section, we briefly summarize the implications of the previous results on MSSM Higgs boson searches at LEP and at the LHC/SSC.

The relevant processes for MSSM Higgs boson searches at LEP I are $Z \to hZ^*$ and $Z \to hA$, which play a complementary role, since their rates are proportional to $\sin^2(\beta-\alpha)$ and $\cos^2(\beta-\alpha)$, respectively. We have seen that an important effect of radiative corrections is to render possible, for some values of the parameters, the decay $h \to AA$, which would be kinematically forbidden according to tree-level formulae. Experimental limits that take radiative corrections into account have by now been obtained by the four LEP collaborations [8], using different methods to present and analyse the data, and different ranges of parameters in the evaluation of radiative corrections. The presently excluded region of the $(m_A, \tan\beta)$ plane, for our standard parameter choice, is the one to the left of the solid line in fig. 3.

The situation in which the impact of radiative corrections is most dramatic is the search for MSSM Higgs bosons at LEP II. At the time when only tree-level formulae were available, there was hope that LEP could completely test the MSSM Higgs sector. According to tree-level formulae, in fact, there should always be a CP-even Higgs boson with mass smaller than (h) or very close to (H) m_Z, and significantly coupled to the Z boson. However, as should be clear from the previous section, this result can be completely upset by radiative corrections. A detailed evaluation of the LEP II discovery potential can be made only if crucial theoretical parameters, such as the top-quark mass and the various soft supersymmetry-breaking masses, and experimental parameters, such as the centre-of-mass energy, the luminosity, and the b-tagging efficiency, are specified. Taking for example $\sqrt{s} = 190$ GeV, and our standard values for m_t and m_{sq}, in the region of $\tan\beta$ significantly greater than 1, the associated production of a Z and a CP-even Higgs can be pushed beyond the kinematical limit. Associated hA production could be a useful complementary signal, but obviously only for $m_h + m_A < \sqrt{s}$. Associated HA production is typically negligible at these energies. To give a measure of the LEP II sensitivity, we plot in fig. 3 contours associated to two benchmark values of the total cross-section $\sigma(e^+e^- \to hZ, HZ, hA, HA)$. The dashed line corresponds to $\sigma = 0.2\,\mathrm{pb}$ at $\sqrt{s} = 175$ GeV, which could be seen as a rather conservative estimate of the LEP II sensitivity. The dash-dotted line corresponds to $\sigma = 0.05\,\mathrm{pb}$ at $\sqrt{s} = 190$ GeV, which could be seen as a rather optimistic estimate of the LEP II sensitivity. Of course, one should keep in mind that there is, at least in principle, the possibility of further extending the maximum LEP energy up to values as high as $\sqrt{s} \simeq 230$–240 GeV, at the price of introducing more (and more performing) superconducting cavities into the LEP tunnel. A detailed study of the discovery potential of e^+e^- colliders for the SM and MSSM neutral Higgs bosons has recently been performed in ref. [9].

Similar considerations can be made for charged Higgs searches at LEP II with $\sqrt{s} \lesssim 190$ GeV. In view of the β^3 threshold factor in $\sigma(e^+e^- \to H^+H^-)$, and of the large background from $e^+e^- \to W^+W^-$, it will be difficult to find the $H^\pm$ at LEP II unless $m_{H^\pm} \lesssim m_W$, and certainly impossible unless $m_{H^\pm} < \sqrt{s}/2$. We also know that for generic values of the parameters there are no large negative radiative corrections to the charged Higgs mass formula, eq. (1.11). Thus there is very little hope of finding the charged Higgs boson of the MSSM at LEP II (or, to put it differently, the discovery of a charged Higgs boson at LEP II would most probably rule out the MSSM).

The next question is then whether the LHC/SSC can explore the full parameter space of the MSSM Higgs bosons. A systematic study of this problem, including radiative corrections, has recently been started in ref. [10]. The analysis is complicated by the fact that the R-odd particles could play a role both in the production (via loop diagrams) and in the decay (via loop diagrams and as final states) of the MSSM Higgs bosons. For simplicity, one can concentrate on the most conservative case in which all R-odd particles are heavy enough not to play any significant role. Still, one has to perform a separate analysis for each $(m_A, \tan\beta)$ point, to include radiative corrections (depending on additional parameters such as m_t and m_{sq}), and to consider Higgs boson decays involving other Higgs bosons. We make here only a few general remarks on the LHC case, for our representative parameter choice, sending the interested reader to ref. [10] for a more complete discussion and recent simulation work.

Beginning with the neutral states, when h or H are in the intermediate mass range (80–130 GeV) and have approximately SM couplings, the best prospects for detection are offered, as in the SM, by their $\gamma\gamma$ decay mode. In general, however, $\sigma \cdot BR(h, H \to \gamma\gamma)$ is smaller than for a SM Higgs boson of the same mass. As a rather optimistic estimate of the possible LHC sensitivity, we display, in fig. 4, lines in the $(m_A, \tan\beta)$ plane corresponding to $\sigma \cdot BR(h, H \to \gamma\gamma) \sim 30$ fb. The contour line for h is shown only for $m_h \gtrsim 80$ GeV. Only for $m_A \gtrsim 200$ GeV, $\tan\beta \gtrsim 3$ (in the case of h) and in the shaded area (in the case of H), the $\gamma\gamma$ signal exceeds the chosen reference value. Very similar considerations can be made for the production of h or H, decaying into $\gamma\gamma$, in association with a W boson or with a $t\bar{t}$ pair, giving an extra isolated lepton in the final state. When H and A are heavy, in general one cannot rely on the $ZZ \to 4l^\pm$ ($l = e, \mu$) decay mode, which gives the 'gold-plated' Higgs signature in the SM case, since H and A couplings to vector-boson pairs are strongly suppressed: only for small $\tan\beta$ and $2m_Z \lesssim m_H \lesssim 2m_t$ might the decay mode $H \to ZZ \to 4l^\pm$ still be viable despite the suppressed branching ratio. Taking into account that with sufficient experimental resolution one could exploit the small H width, as an estimate of the possible LHC sensitivity we take $\sigma \cdot BR(H \to 4l^\pm) \sim 1$ fb ($l = e, \mu$). This contour defines the area in fig. 4 indicated by the label $H \to 4l$. For very large values of $\tan\beta$, and moderately large m_A, one could take advantage of the enhanced production cross-sections and of the unsuppressed decays into $\tau^+\tau^-$ to obtain a visible signal for one or more of the MSSM neutral Higgs bosons, and in particular for H and A, whose masses can be significantly larger than 100 GeV. The simulation work for this process is still at a rather early stage, so that no definite conclusion can be drawn yet. For reference, the dotted line in fig. 4 corresponds to a (somewhat arbitrary) interpolation of $\sigma \cdot BR(A, H \to \tau^+\tau^-) \sim 10$ pb at $m_{H,A} \sim 100$ GeV and $\sigma \cdot BR(A, H \to \tau^+\tau^-) \sim 1$ pb at $m_{H,A} \sim 400$ GeV. Finally, in the region of parameter space corresponding to $m_A \lesssim m_Z$, the charged Higgs could be discovered via the decay chain $t \to bH^+ \to b\tau^+\nu_\tau$, which competes with the standard channel $t \to bW^+ \to bl^\pm\nu_l$ ($l = e, \mu, \tau$). A convenient parameter is the ratio $R \equiv BR(t \to \tau^+\nu_\tau b)/BR(t \to \mu^+\nu_\mu b)$, which measures the violation of lepton universality in top decays. As an estimate of the LHC sensitivity, we take $R > 1.15$. The corresponding region of the $(m_A, \tan\beta)$ plane is indicated by the label $H^\pm \to \tau\nu$ in fig. 4. For all processes considered above, similar remarks apply also to the SSC.

In summary, a global look at figs. 3 and 4 shows that there is a high degree of complementarity between the regions of parameter space accessible to LEP II and to the LHC/SSC. However, for our representative choice of parameters, there is a non-negligible region of the $(m_A, \tan\beta)$ plane that is presumably beyond the reach of both LEP II and the LHC/SSC. This potential problem could be solved, as we said before, by a further increase of the LEP II energy beyond the reference value of $\sqrt{s} \lesssim 190$ GeV. Otherwise, one might need a higher energy e^+e^- collider, for a full exploration of the parameter space describing the MSSM Higgs sector. However, one should not forget that another important test of the MSSM will be provided by squark and gluino searches at the LHC/SSC, which should be sensitive to most of the theoretically motivated parameter space.

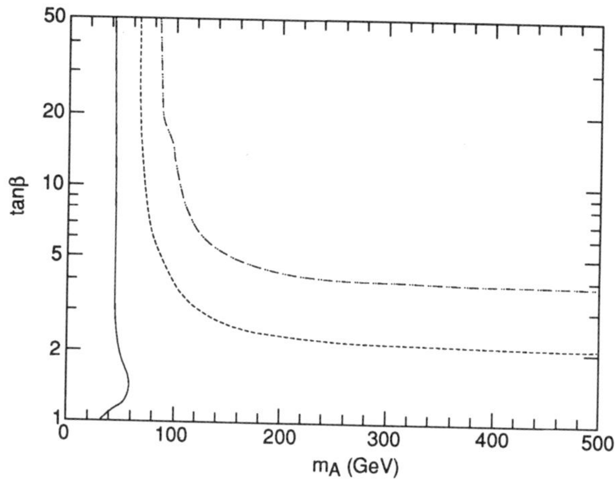

Fig.3: Schematic representation of the present LEP I limits and of the future LEP II sensitivity in the $(m_A, \tan\beta)$ plane, for $m_t = 140$ GeV and $m_{sq} = 1$ TeV. The solid line corresponds to the present LEP I limits. The dashed line corresponds to $\sigma(e^+e^- \to hZ, HZ, hA, HA) = 0.2\,\mathrm{pb}$ at $\sqrt{s} = 175$ GeV, which could be seen as a rather conservative estimate of the LEP II sensitivity. The dash-dotted line corresponds to $\sigma(e^+e^- \to hZ, HZ, hA, HA) = 0.05\,\mathrm{pb}$ at $\sqrt{s} =$ 190 GeV, which could be seen as a rather optimistic estimate of the LEP II sensitivity.

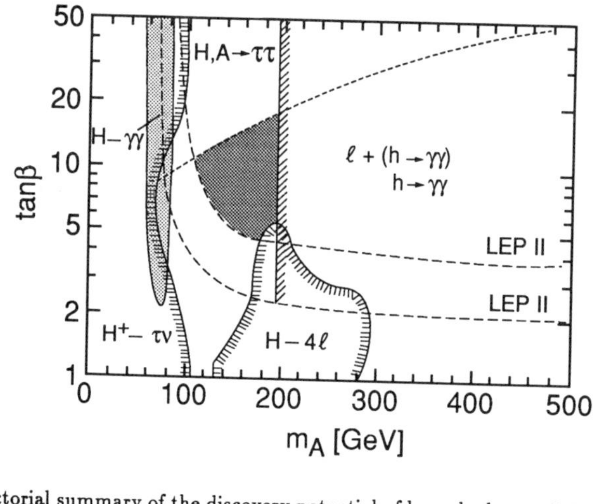

Fig.4: Pictorial summary of the discovery potential of large hadron colliders for $m_t =$ 140 GeV and $m_{sq} = 1$ TeV.

References

1. H.-P. Nilles, *Phys. Rep.* **110** (1984) 1;
   S. Ferrara, ed., *'Supersymmetry'* (North-Holland, Amsterdam, 1987);
   R. Barbieri, *Riv. Nuovo Cim.* **11** (1988) 1.

2. P. Fayet, *Nucl. Phys.* **B90** (1975) 104, *Phys. Lett.* **64B** (1976) 159 and **69B** (1977) 489;
   R.K. Kaul and P. Majumdar, *Nucl. Phys.* **B199** (1982) 36;
   R. Barbieri, S. Ferrara and C.A. Savoy, *Phys. Lett.* **B119** (1982) 36;
   H.P. Nilles, M. Srednicki and D. Wyler, *Phys. Lett.* **B120** (1983) 346;
   J.M. Frère, D.R.T. Jones and S. Raby, *Nucl. Phys.* **B222** (1983) 11;
   J.-P. Derendinger and C. Savoy, *Nucl. Phys.* **B237** (1984) 307;
   J. Ellis, J.F. Gunion, H.E. Haber, L. Roszkowski and F. Zwirner, *Phys. Rev.* **D39** (1989) 844;
   P. Binétruy and C.A. Savoy, *Phys. Lett.* **B277** (1992) 453;
   J.R. Espinosa and M. Quirós, *Phys. Lett.* **B279** (1992) 92 and Madrid preprint IEM-FT-60-92.

3. J.F. Gunion, H.E. Haber, G.L. Kane and S. Dawson, *The Higgs Hunter's Guide* (Addison-Wesley, New York, 1990).

4. Y. Okada, M. Yamaguchi and T. Yanagida, *Prog. Theor. Phys. Lett.* **85** (1991) 1;
   J. Ellis, G. Ridolfi and F. Zwirner, *Phys. Lett.* **B257** (1991) 83;
   H.E. Haber and R. Hempfling, *Phys. Rev. Lett.* **66** (1991) 1815.
   Large corrections to the neutral Higgs mass sum rule were found in:
   M. Berger, *Phys. Rev.* **D41** (1990) 225.

5. R. Barbieri, M. Frigeni and M. Caravaglios, *Phys. Lett.* **B258** (1991) 167;
   Y. Okada, M. Yamaguchi and T. Yanagida, *Phys. Lett.* **B262** (1991) 54;
   J.R. Espinosa and M. Quirós, *Phys. Lett.* **B266** (1991) 389;
   M.A. Diaz and H.E. Haber, *Phys. Rev.* **D45** (1992) 4246;
   K. Sasaki, M. Carena and C.E.M. Wagner, *Nucl. Phys.* **B381** (1992) 66.
   P.H. Chankowski, S. Pokorski and J. Rosiek, *Phys. Lett.* **B281** (1992) 100.

6. R. Barbieri and M. Frigeni, *Phys. Lett.* **B258** (1991) 395;
   J. Ellis, G. Ridolfi and F. Zwirner, *Phys. Lett.* **B262** (1991) 477;
   A. Brignole, J. Ellis, G. Ridolfi and F. Zwirner, *Phys. Lett.* **B271** (1991) 123 and (E) **B273** (1991) 550;
   M. Drees and N.M. Nojiri, *Phys. Rev.* **D45** (1992) 2482;
   D.M. Pierce, A. Papadopoulos and S.B. Johnson, *Phys. Rev. Lett.* **68** (1992) 3678;
   S. Kelley, J.L. Lopez, D.V. Nanopoulos, H. Pois and K. Yuan, *Phys. Lett.* **B285** (1992) 61;
   H.E. Haber, R. Hempfling and Y. Nir, *Phys. Rev.* **D46** (1992) 3015;
   H.E. Haber and R. Hempfling, Santa Cruz preprint SCIPP-91/33.

7. A. Yamada, *Phys. Lett.* **B263** (1991) 233;
   P.H. Chankowski, S. Pokorski and J. Rosiek, *Phys. Lett.* **B274** (1992) 191;
   A. Brignole, *Phys. Lett.* **B277** (1992) 313 and **B281** (1992) 284;

P.H. Chankowski, S. Pokorski and J. Rosiek, *Phys. Lett.* **B286** (1992) 307;
D. Pierce and A. Papadopoulos, Berkeley preprint UCB-PTH-92/23 and LBL-32498;
A. Brignole and F. Zwirner, preprint CERN-TH.6603/92.

8. G. Wormser, talk given at the XXVI Int. Conf. on High Energy Physics, Dallas, Texas, USA, 1992, to appear in the Proceedings, and references therein.

9. P. Janot, Orsay preprints LAL-91-61 and LAL-92-27.

10. Z. Kunszt and F. Zwirner, in Proceedings of the Large Hadron Collider Workshop, Aachen, 1990 (G. Jarlskog and D. Rein, eds.), Vol. II, p. 578 and Erratum; preprint CERN-TH.6150/91, ETH-TH/91-7, to appear in *Nucl. Phys.* **B**;
V. Barger, M.S. Berger, A.L. Stange and R.J.N. Phillips, *Phys. Rev.* **D45** (1992) 4128;
H. Baer, M. Bisset, C. Kao and X. Tata, *Phys. Rev.* bf D46 (1992) 1067;
J.F. Gunion and L. Orr, *Phys. Rev.* **D46** (1992) 2052;
J.F. Gunion, R. Bork, H.E. Haber and A. Seiden, *Phys. Rev.* **D46** (1992) 2040;
J.F. Gunion, H.E. Haber and C. Kao, *Phys. Rev.* **D46** (1992) 2907;
H. Baer, C. Kao and X. Tata, Florida preprint FSU-HEP-920717;
H. Baer, M. Bisset, D. Dicus, C. Kao and X. Tata, Florida preprint FSU-HEP-920724;
V. Barger, K. Cheung, R.J.N. Phillips and A.L. Stange, Madison preprint MAD-PH-696.
See also the Letters of Intent of the SDC, GEM, ATLAS, CMS and L3P Collaborations.

Figures

Fig.1: Contour curves of m_h^{max} (the maximum value of m_h, reached for $m_A \gg m_Z$ and $\tan\beta \gg 1$) in the (m_t, m_{sq}) plane.

Fig.2: Contour curves of $\tan\beta$ in the (m_h, m_A) plane, for $m_t = 140$ GeV and $m_{sq} =$ 1 TeV. The boundary of the region allowed at tree level ($m_h < m_Z$ and $m_h < m_A$) is also shown by the dashed lines.

Fig.3: Schematic representation of the present LEP I limits and of the future LEP II sensitivity in the $(m_A, \tan\beta)$ plane, for $m_t = 140$ GeV and $m_{sq} = 1$ TeV. The solid line corresponds to the present LEP I limits. The dashed line corresponds to $\sigma(e^+e^- \to hZ, HZ, hA, HA) = 0.2\,\mathrm{pb}$ at $\sqrt{s} = 175$ GeV, which could be seen as a rather conservative estimate of the LEP II sensitivity. The dash-dotted line corresponds to $\sigma(e^+e^- \to hZ, HZ, hA, HA) = 0.05\,\mathrm{pb}$ at $\sqrt{s} =$ 190 GeV, which could be seen as a rather optimistic estimate of the LEP II sensitivity.

Fig.4: Pictorial summary of the discovery potential of large hadron colliders for $m_t =$ 140 GeV and $m_{sq} = 1$ TeV.

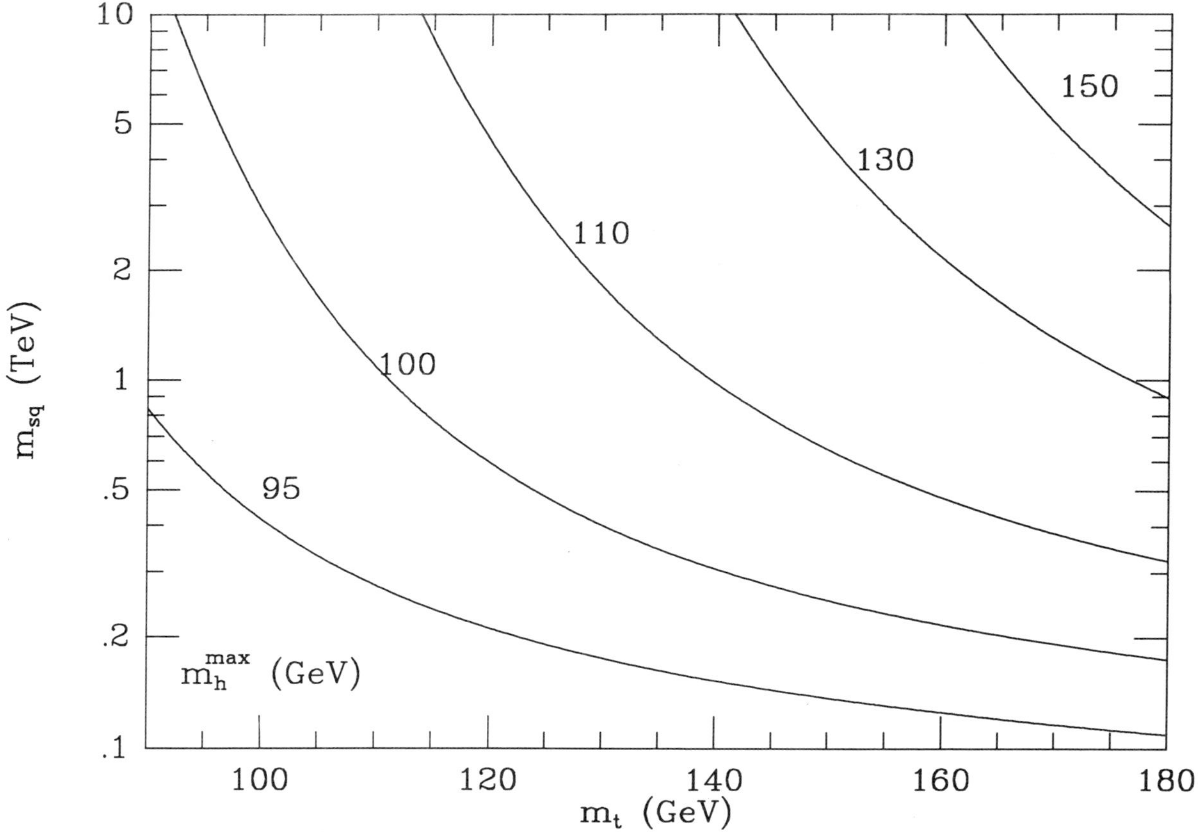
10
5
2
1
.5
.2
.1
m_{sq} (TeV)
100
120
140
160
180
m_t (GeV)
150
130
110
100
95
m_h^{max} (GeV)

Fig. 1

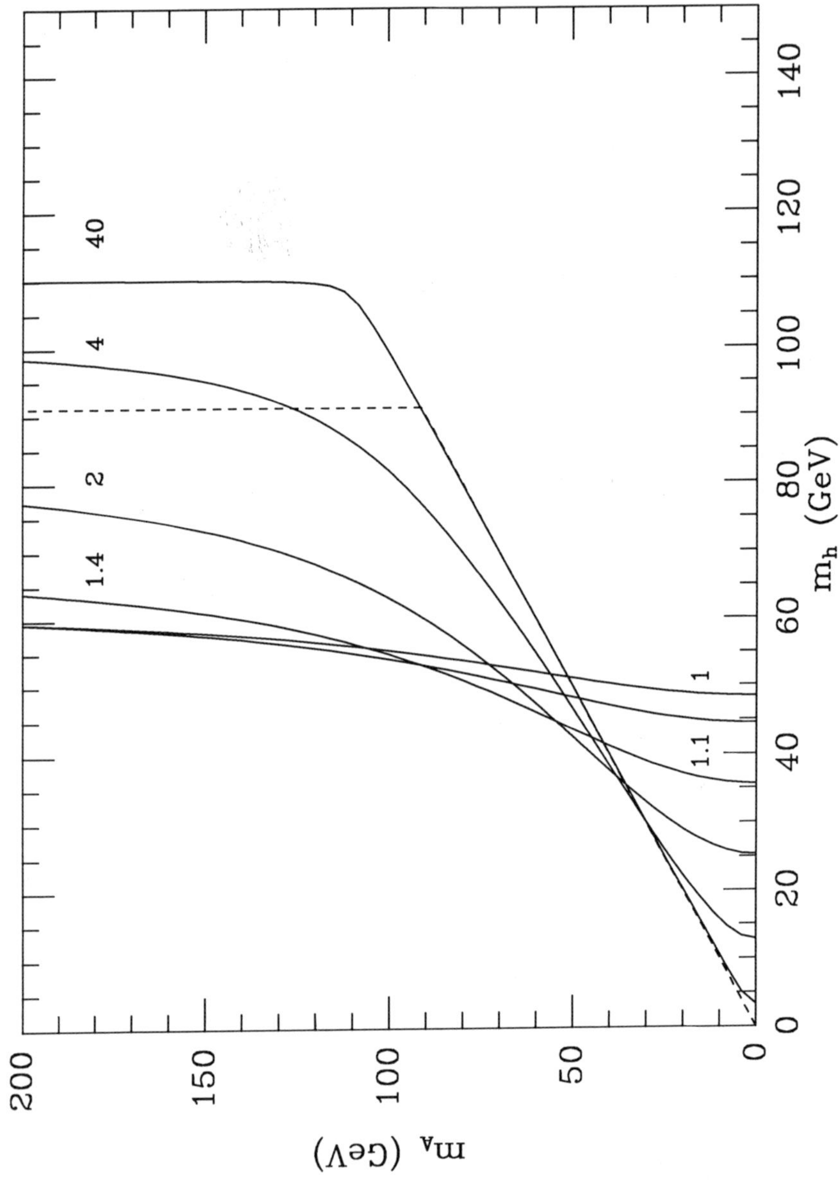

Fig. 2

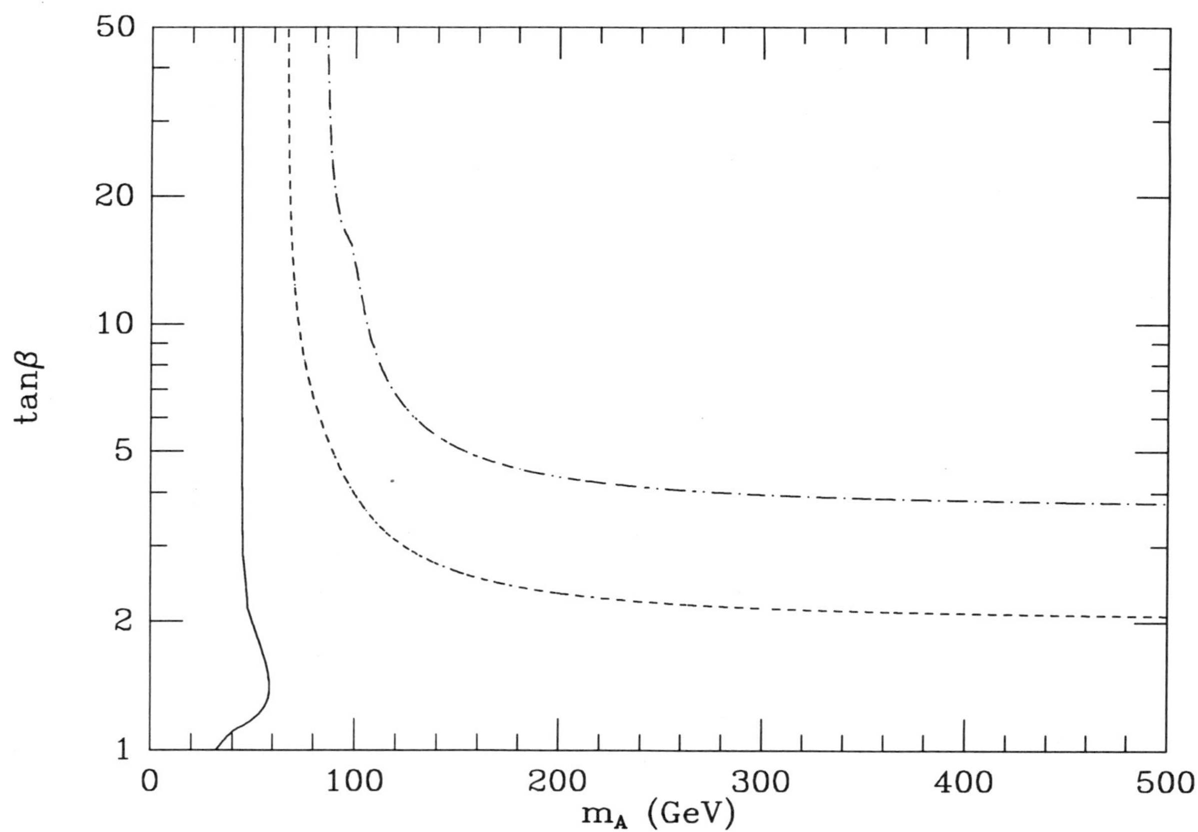

Fig. 3

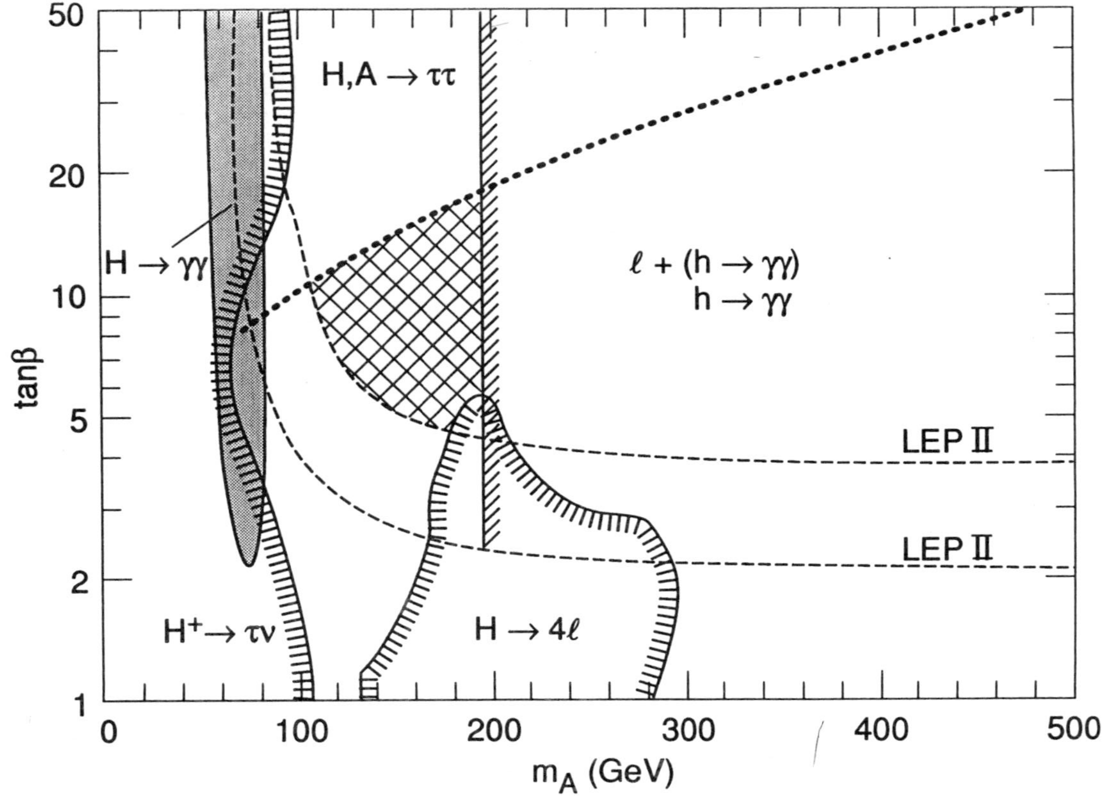
50
20
10
5
2
1
tanβ
0
100
200
300
400
500
m_A (GeV)
H,A → ττ
H → γγ
ℓ + (h → γγ)
h → γγ
LEP II
LEP II
H⁺→ τν
H → 4ℓ

Fig. 4

Chairman: F. Zwirner

Scientific Secretaries: H.I. Petrache, A. Sopczak

DISCUSSION

– *Shotkin:*

There are supersymmetric extensions of the Standard Model with broken R-parity. What is the most important implication for the Higgs sector?

– *Zwirner:*

If R-parity is broken explicitly by cubic superpotential terms, masses and couplings in the Higgs sector remain the same. The first important change which comes to mind is the signature of Higgs decays into supersymmetric particle pairs, if these are kinematically allowed. One can also build models with spontaneously broken R-parity: to be compatible with LEP data, however, one needs rather artifical constructions, and predictions become very model-dependent.

– *Baillie:*

Is there an intuitive way to understand why one-loop corrections to the neutral Higgs masses can be large, and two-loop corrections small?

– *Zwirner:*

A simple way to understand this point is provided by the renormalization group approach. Assume that $M_{SUSY} \gg m_Z$, and consider the effective theory for the degrees of freedom lighter than M_{SUSY}. The tree-level relations of the MSSM can be seen as boundary conditions on the quartic Higgs couplings at the scale $Q = M_{SUSY}$. The leading corrections to the Higgs masses can then be computed by solving (non-supersymmetric) renormalization group equations, to obtain running parameters down to the scale $Q = m_Z$. Since possible large logarithms are resummed by the renormalization group, the loop expansion is controlled by the size of the dimensionless couplings. If the top Yukawa coupling is small enough, as is the case in the preferred range of m_t and $\tan\beta$, one expects two-loop corrections to be small, as verified by explicit calculations. In some sense, a similar situation occurs when one considers corrections to the m_W/m_Z ratio in the SM, as an effect of the large top-bottom mass splitting.

– *Benakli:*

How do you deal with the renormalization-scale dependence of the derivatives of the effective potential with respect to the fields?

– *Zwirner:*

The one-loop effective potential is Q^2-independent up to higher-order corrections: the explicit Q^2-dependence is cancelled by the implicit Q^2-dependence of the renormalized parameters appearing in its tree-level part. However, derivatives of the effective potential with respect to the fields depend on Q^2, because of wave-function renormalization effects. If one correctly defines the physical masses as poles of the renormalized propagators, the one-loop-corrected Higgs masses are scale-independent. Since there are no Higgs wave-function renormalization effects of order h_t^4, at that level of approximation it is correct to identify the Higgs masses with the second derivatives of the effective potential at the minimum: in fact, the leading-order result of eq. (2.8) is Q^2-independent.

– *Sopczak:*

The exchange of a virtual charged Higgs boson could strongly enhance the semileptonic decay $b \to \tau\nu X$. Could this be used to explore values of the charged Higgs masses beyond the kinematical limit for pair production at LEP?

– *Zwirner:*

The possibility you mention could be realized in a generic model with two Higgs doublets, if the charged Higgs mass is just above the present limit from direct searches, and $\tan\beta$ is very large. This is not the case for the MSSM, since the mass formula (1.11) does not receive large negative corrections and $\tan\beta \lesssim m_t/m_b$.

– *Moulai:*

Since the Higgs can decay into two photons, is there any hope to find the Higgs in two-photon collisions?

– *Zwirner:*

This possibility has been discussed in relation with the physics of high-energy linear e^+e^- colliders. It is expected that photon beams of comparable energy and luminosity can be generated by backscattering of laser light. However, the smallness of the Higgs couplings to photons makes $\gamma\gamma$ collisions much less useful for discovery than e^+e^- collisions.

– *Summers:*

In how much of the MSSM parameter space can one distinguish the MSSM from the SM, just from the Higgs sector of the theory?

– *Zwirner:*

For all points of parameter space, such that one can discover one or more Higgs bosons via a process that is absent in the SM, the distinction is clear. If only a neutral Higgs can be discovered, via one or more processes allowed by the SM, one should compare the signal cross-sections and the Higgs width with the predictions of the two models: given the experimental limitations, this comparison can be rather difficult. In particular, in the limit $m_A \gg m_Z$ the states H, A and $H^\pm$ decouple from the low-energy effective theory, leaving behind a light Higgs h with SM-like properties. This is the most difficult case. Even if one or more non-standard Higgs bosons were found, however, with properties compatible with the MSSM, this would not be enough to establish low-energy supersymmetry. Conclusive positive evidence can only come from the discovery of an R-odd supersymmetric particle.

GAUGE COUPLING RUNNING IN SUPERSTRING THEORY

Sergio Ferrara

Theoretical Physics Division

CERN, Geneva, Switzerland

1) String Unification

One of the main developments in physical consequences of unifying quantum gravity with other forces in string dynamics [1] is the fact that unification of coupling constants is a necessary phenomenon. In fact, in 4D heterotic string theory [2], with gauge group $G = \prod_a G_a$, the tree level gauge couplings g_a are given by the relation [3]:

$$\frac{1}{g_a^2} = k_a/g^2_{STRING} \tag{1.1}$$

where k_a is the level of the Kac-Moody algebra for G_a and:

$$\frac{1}{g^2{}_{STRING}} = <\phi> \tag{1.2}$$

is the v.e.v. of the dilaton field which is the scalar partner of a chiral multiplet often called S [3] containing also an axion field a, dual to the usual antisymmetric tensor $b_{\mu\nu}$. The Newton constant is also related to g_{STRING} by the relation [4]:

$$G_N = \alpha' g^2_{STRING} \tag{1.3}$$

where α' is the inverse square of the string tension. Associated to α' there is a natural string mass-scale [5]:

$$M^2_{STRING} = a\ \alpha'^{-1} \text{ (a is a numerical constant)} \tag{1.4}$$

In renormalizable theories the running of gauge couplings $g_a^2(\mu)$ due to renormalization effects is given (at one loop) by the Georgi-Quinn-Weinberg equation [6]:

$$\frac{1}{g_a^2(\mu)} = k_a \frac{1}{g_{GUT}^2} + \frac{b_a}{16\pi^2} \log \frac{M_{GUT}^2}{\mu^2} + 0(1) \tag{1.5}$$

where a runs over the low-energy couplings (at scales $\mu < M_{GUT}$) and k_a are the tree level relation between these couplings. b_a, the one loop β function, contains the information of renormalization effects below M_{GUT} and, for supersymmetric GUT's:

$$b_a = -3T(G_A) + \sum_R T(R) \tag{1.6}$$

The 0(1) terms are threshold corrections Δ_a whose values depend on masses of heavy particles associated with the unification threshold.

String unification is a sort of extension of GUT's in the sense that the natural unification scale is M_{STRING} and Δ_a are stringy threshold effects [5] i.e. due to the integration of the stringy heavy modes, which are supposed to be integrated out in the low-energy field theory, describing the effective interactions of the light particles (with masses much lower than M_{STRING}).

One therefore expects a formula, modifying eq. (1.5) of the type:

$$\frac{1}{g_a^2(\mu)} = k_a \frac{1}{g_{STRING}^2} + \frac{b_a}{16\pi^2} \log \frac{M_{STRING}^2}{\mu^2} + \Delta_a \tag{1.7}$$

where eq. (1.1) and (1.4) have been used and the parameter a, which depends on the renormalization scheme, is given (in the $\overline{MS}$ scheme) by:

$$a = 2\, e^{1-\gamma_E}/\sqrt{27}\pi \tag{1.8}$$

and $\gamma_E = 0.57722\ldots$ is the Euler constant.

The interesting phenomenon which occurs in string theory is that eq. (1.7) can be obtained by computing an ultraviolet finite diagram but the logaritmic μ dependence is entirely due to an infrared singularity [5] [7] (of logaritmic nature in D=4 dimensions) which is responsible for the second term in eq. (1.7).

Indeed, as was shown [7] , eq. (1.7) can be obtained in a Kaluza-Klein theory by splitting the gauge-boson one loop self-energy diagram $\Pi_{\mu\nu}(Q)$ in a region $Q^2R^2 << 1$ and in a region $1/R^2 << Q^2 << M_{STRING}^2$. In a simple circle compactification of radius R it is easy to see that the dominant contribution in the region $Q^2R^2 << 1$ comes from the massless modes giving a contribution to the gauge coupling running of the form:

$$\beta \log Q^2R^2 \tag{1.9}$$

while the other region gives the threshold contribution of the Kaluza-Klein states in the large R-limit, of the form:

$$\Delta \sim \text{ const } Q^2R^2 \tag{1.10}$$

The contribution given by eq. (1.10) comes from an energy region where the theory is effectively higher-dimensional (5-dimensional in this case) and is non-singular, being related to having intergrated out the K-K massive states.

It does not come as a surprise that eq. (1.9) is the usual field theory result, with a dynamical reference scale given by $1/R^2$, while eq. (1.10) comes from the extension of the massless theory, by including the K-K states.

The previous argument shows that in the full fledged string theory we expect a similar result but with a more precise calculation of Δ_a, which takes into account both Kaluza-Klein and winding modes; which is the main difference between string theory and K-K theory.

The main difference between string theory and field theory is that both g^2_{STRING} and Δ_a are related to v.e.v. of dynamical fields, namely the dilaton and the moduli fields [2] [8] [9] [10] [11], which parametrize the shape of the internal manifold of compactification. These fields generalize the radial variable R of a simple circle compactification described earlier.

The simplest compactification preserving N=1 supersymmetry at D=4 is a six-dimensional torus modded out [12] by some Z_n symmetry (orbifold).

The complex modulus, which parametrizes the overall volume det $g_{i\bar{j}}$, corresponds to a torus modulus [13] , $T = R^2 + ia$, with an associated stringy symmetry $\Gamma = SL(2, \mathcal{Z})$ which combines together the radius inversion $R \to \alpha'/R$ and the Peccei-Quinn symmetry of the internal axion $a \to a + 1$.

This duality symmetry is entirely analogous to the O(D,D;$\mathcal{Z}$) Narain symmetry [14] of toroidal compactifications of superstrings compactified on a D-dimensional torus. For generic compactifications on smooth manifolds the target space duality group Γ is no longer $SL(2, \mathcal{Z})$ [15]. However for Calabi-Yau manifolds, or more generically (2,2) superstring vacua, the group Γ is linked to the monohomy group Γ_M of the Picard-Fuchs equations [15] - [22]. These are the linear differential equations which determine all low-energy couplings of the N=1 effective supergravity action.

For manifolds described by a Landau-Guizburg superpotential [23] W which also defines the Calabi-Yau manifold [24], the symmetry group of W, Γ_W is related to Γ and Γ_M by the relation [17] (valid at least for the one-modulus case):

$$\Gamma_W = \Gamma/\Gamma_M \tag{1.11}$$

In general Γ is a discrete infinite subgroup of SL(2,R).

The general outcome of the target-space duality symmetry [25] is that the threshold effects encoded in the function Δ_a are duality invariant expressions of the moduli [26]. Δ_a turns out to be the sum of a non-holomorphic function of the moduli, which can be computed purely in terms of the data of the massless theory and of the geometry of the moduli space, and an harmonic piece which comes by integrating out the massive stringy states and makes Δ_a Γ-invariant.

For the one-modulus case of orbifold compactifications Δ_a has typically the form [26] :

$$\Delta_a \sim C_a \log(T + \overline{T})|\eta(T)|^4 \tag{1.12}$$

where $\eta(T)$ is the Dedekind function: $\eta(T) = e^{-\pi T/12} \prod_{n=1}^{\infty}(1 - e^{2\pi n T})$.

The non-holomorphic term and the holomorphic term in eq. (1.12) reproduce, for T large, expressions given by eqs. (1.9) and (1.10).

In the next section it will be shown that, using target-space supersymmetry, the two terms in eq. (1.12) can be seen to be the supersymmetric partners of the Γ-anomaly [27] and of the local Wess-Zumino term which cancels the anomaly coming from the massless fermion loops.

2) Target Space Duality Anomalies and application to gauge-coupling running

The identification of points in the moduli space connected by the duality group Γ actually means that the moduli space of string compactifications is not a smooth manifold $\mathcal{M}$ but an orbifold $\mathcal{M}/\Gamma$ with fixed points giving rise to conical singularities. In the low-energy effective action for the massless modes of heterotic string, Γ appears as an exact symmetry and there are indications that this symmetry survives (higher genus) perturbation theory.

To discuss quantum target space duality it is important to realize that a transformation Γ on a generic moduli coordinate $\psi \to \Gamma\psi$ acts with a combined Kähler gauge transformation and coordinate transformation on matter fields as well as on Yukawa couplings and moduli and families metrics [26] [27] . The latter give the absolute normalization of physical fields and therefore is needed to compute the "invariant" Yukawa couplings.

Since fermions are coupled to gauge and gravitational fields, these transformations (being field-dependent) may induce mixed σ-model-gauge (or gravitational) anomalies through ordinary (one-loop) triangular graphs in which a σ-model U(1) connection couples to gauge or gravitational gauge fields [26] - [30]. We will give here the transformation properties, under the duality group Γ, of all relevant quantities appearing in the low-energy action for N=1 heterotic strings [2] compactified on a Calabi-Yau manifold with gauge group $E_6 \times E_8$.

The Γ transformations, since they act on different sections of a line bundle, will be accompanied by (holomorphic) Kähler gauge transformations. The holomorphic bundle in question is the line bundle L whose first Chezu-class coincide with the Kähler class on $\mathcal{M}$. This is a property of restricted Kähler manifolds [31].

The holomorphic gauge transformation can be split in a real (Kähler) transformation and in a U(1) phase transformation:

$$\Lambda(\varphi)\ (\overline{D}_2\Lambda = 0) \ = \ \frac{1}{2}\ (\Lambda + \overline{\Lambda}) + \frac{1}{2}\ (\Lambda - \overline{\Lambda}) \tag{2.1}$$

Kähler dependent quantities will transform under $(\Lambda + \overline{\Lambda})$ while Kähler independent quantities will only undergo the U(1) phase transformation $1/2(\Lambda - \overline{\Lambda})$. The U(1) connection is nothing but the imaginary part of the (complex) Kähler connection K_i and refers to a real U(1) bundle associated to the holomorphic bundle L.

Let's call ψ, φ the coordinates on $M_1 \times M_2$ of a generic moduli space of Calabi-Yau manifold.

The Kähler potential is [9] - [11] [32]:

$$K(\psi,\varphi) = K_1(\psi) + K_2(\varphi) \tag{2.2}$$

and under a coordinate transformation which is an isometry will have the property:

$$K(\Gamma\psi,\Gamma\varphi) = K(\psi,\varphi) - \Lambda_1(\psi) - \overline{\Lambda}_1(\bar{\psi}) - \Lambda_2(\varphi) - \overline{\Lambda}_2(\bar{\varphi}) \tag{2.3}$$

under the same action the Yukawa will transform as:

$$W^1_{i'j'k'}(\Gamma\psi) \Rightarrow e^{2\Lambda(\psi)}(\mathcal{M}^{i'}_i)^{-1}(\mathcal{M}^{j'}_j)^{-1}(\mathcal{M}^{k'}_k)^{-1}W^1_{ijk}(\psi) \tag{2.4}$$

with $\mathcal{M}^{i'}_i(\psi') = \partial\psi'^{i'}/\partial\psi^i$, $\psi' = \Gamma\psi$
analogously:

$$W^2_{\alpha'\beta'\gamma'}(\Gamma\varphi) \Rightarrow e^{2\Lambda_2(\varphi)}(\mathcal{N}^{\alpha'}_\alpha)^{-1}(\mathcal{N}^{\beta'}_\beta)^{-1}(\mathcal{N}^{\gamma'}_\gamma)^{-1}W^2_{\alpha\rho\gamma}(\varphi) \tag{2.5}$$

with $\mathcal{N}^{\alpha'}_\alpha(\varphi') = \partial\varphi'^{\alpha'}/\partial\varphi^\alpha$, $\varphi' = \Gamma\varphi$
Accordingly the scalar fields of 27, $\overline{27}$ families will transform as controvariant (vector) holomorphic sections [11] :

$$\phi^{i'}_{27} \Rightarrow e^{(\Lambda_2 - \Lambda_1)/3}\mathcal{M}^{i'}_i\,\phi^i_{27} \tag{2.6}$$

$$\phi_{\overline{27}}^{\alpha'} \Rightarrow e^{(\Lambda_1-\Lambda_2)/3}\mathcal{N}_{\alpha}^{\alpha'}\phi_{\overline{27}}^{\alpha} \tag{2.7}$$

The variation of the moduli and family metric is given by:

$$\begin{aligned}
G^1_{i'\bar{j}'} &\Rightarrow \left(\mathcal{M}_i^{i'}\right)^{-1}\left(\overline{\mathcal{M}}_{\bar{j}}^{\bar{j}'}\right)^{-1} G_{i\bar{j}} \\
G^2_{\alpha'\bar{\beta}'} &\Rightarrow \left(\mathcal{N}_{\alpha}^{\alpha'}\right)^{-1}\left(\overline{\mathcal{N}}_{\bar{\beta}}^{\bar{\beta}'}\right) G_{\alpha\bar{\beta}} \\
G^1_{27\ i'\bar{j}'} &\Rightarrow e^{(\Lambda_1+\overline{\Lambda}_1)/3-(\Lambda_2+\overline{\Lambda}_2)/3}\left(\mathcal{M}_i^{i'}\right)^{-1}\left(\overline{\mathcal{M}}_{\bar{j}}^{\bar{j}'}\right)^{-1} G'_{27\ i\bar{j}} \\
G^2_{\overline{27}\ \alpha'\bar{\beta}'} &\Rightarrow e^{(\Lambda_2+\overline{\Lambda}_2)/3-(\Lambda_1+\overline{\Lambda}_1)/3}\left(\mathcal{N}_{\alpha}^{\alpha'}\right)^{-1}\left(\overline{\mathcal{N}}_{\bar{\beta}}^{\bar{\beta}'}\right)^{-1} G^2_{27\ \alpha\bar{\beta}}
\end{aligned} \tag{2.8}$$

If we now consider the following holomorphic quantities:

$$\log \det \mathcal{M} = \Sigma^1\ ,\ \log \det \mathcal{N} = \Sigma^2 \tag{2.9}$$

we may rewrite:

$$\mathcal{M}^{-1} = \widehat{\mathcal{M}}^{-1} e^{\,-\Sigma^1}\ ,\ \mathcal{N}^{-1} = \widehat{\mathcal{N}}^{-1} e^{\,-\Sigma^2} \tag{2.10}$$

where $\widehat{\mathcal{M}}, \widehat{\mathcal{N}}$ are $SL(n,c)$ matrices ($n = h_{11}$ for $\mathcal{M}$ and $n = h_{12}$ for $\mathcal{N}$).

Let's now consider the supermultiplet whose first component is:

$$B = \log \det G \tag{2.11}$$

Then it immediately follows that its $(\theta\sigma_\mu\overline{\theta})$ vector component is the gauge field:

$$A_\mu = i(\partial_\mu\psi^i B_i - \partial_\mu\overline{\psi}^i B_{\bar{i}}) \tag{2.12}$$

Under the action of Γ, B, A_μ transform as:

$$B \to B - \Sigma - \overline{\Sigma} \tag{2.13}$$

$$A_\mu \to A_\mu + i(\overline{\Sigma} - \Sigma) \tag{2.14}$$

where A_μ is the $U(1)$ holonomy gauge field of the $H \subset U(n)$ holonomy group of the Kähler manifold $M(M^1$ or $M^2)$. On the other hand B is invariant under $\widehat{\mathcal{M}}, \widehat{\mathcal{N}}$ coordinate transformations on the moduli space.

The holomorphic transformations Λ and Σ are actually related one to another since charged matter fields may be required to be tensor densities under Γ, so that $\Lambda_i = \gamma_i\Sigma_i$ ($i = 1,2$).

The transformation laws of the fermions can be easily read from the bilinear fermionic terms in the effective action which are not chiral invariant (bilinear with an even number of γ-matrices) namely:

$$e^{k/2}(W^1\chi_{27}\chi_{27}\phi_{27} + W^2\chi_{\overline{27}}\chi_{\overline{27}}\phi_{\overline{27}}) \tag{2.15}$$

$$G^1_{27}\chi_{27}\lambda\bar{\phi}_{27} \ , \ G^2_{27}\chi_{\overline{27}}\lambda\bar{\phi}_{\overline{27}} \tag{2.16}$$

$$e^{k/2}\bar{\psi}_\mu\sigma^{\mu\nu}\psi_\nu W \ , \ \chi^i G_{i\bar{j}}\partial_\mu\psi^{\bar{j}}\gamma^\nu\gamma^\mu\psi_\nu \tag{2.17}$$

$$\lambda\sigma^{\mu\nu}F_{\mu\nu}\chi_s \tag{2.18}$$

In the above equations all fermions have (controvariant) vector indices. With respect to Σ, Λ transformations they transform as follows:

$$\chi_{27} \rightarrow \chi_{27}e^{\Sigma_1}e^{-5/12\,(\Lambda_1-\bar{\Lambda}_1)-1/12(\Lambda_2-\bar{\Lambda}_2)-1/6(\Lambda_1+\bar{\Lambda}_1)+1/6(\Lambda_2+\bar{\Lambda}_2)} \tag{2.19}$$

(the same for $\chi_{\overline{27}}$ with $1 \rightarrow 2$)

$$\chi_{M_1} \rightarrow \chi_{M_1}e^{\Sigma_1}e^{-1/4(\Lambda_1-\bar{\Lambda}_1)-1/4(\Lambda_2-\bar{\Lambda}_2)} \tag{2.20}$$

$$\begin{aligned} \lambda &\rightarrow \lambda\ e^{1/4(\Lambda_1-\bar{\Lambda}_1)+1/4(\Lambda_2-\bar{\Lambda}_2)} \\ \psi_\mu &\rightarrow \psi_\mu\ e^{1/4(\Lambda_1-\bar{\Lambda}_1)+\frac{1}{4}(\Lambda_2-\bar{\Lambda}_2)} \\ \chi_s &\rightarrow \chi_s\ e^{-1/4(\Lambda_1-\bar{\Lambda}_1)-1/4(\Lambda_2-\bar{\Lambda}_2)} \end{aligned} \tag{2.21}$$

Eqs. (2.20) (2.21) are the U(1) transformations [31] of restricted Kähler spaces. Eq. (2.19) differs because the heterotic fermions are not the partner of moduli fields but rather of charged scalar fields which are holomorphic sections on M. If we define flat (holonomy) indices for matter and moduli fermions by the substitution:

$$\widehat{\chi}_{27} \rightrightarrows e_{27}\chi_{27} \ , \ \widehat{\chi}_M \rightrightarrows e\chi_M \tag{2.22}$$

where e_{27}, e_M are (vierbein) one-forms such that:

$$e_{27} \otimes \bar{e}_{27} = G_{27} \ , \ e_M \otimes \bar{e}_M = G_M \tag{2.23}$$

then we have that under (Σ) coordinate and (Λ) Kähler transformations χ_{27} and χ transform only under a combined U(1) holonomy and U(1) (Kähler) transformation as follows:

$$\begin{aligned} \hat{\chi}_{27} &\to \hat{\chi}_{27} e^{\,1/2(\Sigma_1-\bar{\Sigma}_1)-5/12(\Lambda_1-\bar{\Lambda}_1)-1/12(\Lambda_2-\bar{\Lambda}_2)} \\ \widehat{\chi}_M &\to \chi_M e^{\,1/2(\Sigma_1-\bar{\Sigma}_1)} e^{\,-1/4(\Lambda_1-\bar{\Lambda}_1)-1/4(\Lambda_2-\bar{\Lambda}_2)} \end{aligned} \tag{2.24}$$

This follows from the fact that under Γ:

$$\begin{aligned} e_M &\to e^{-\Sigma}\widehat{\mathcal{M}}^{-1} e_M h_\Gamma \\ e_{27} &\to e^{-\Sigma+\ \Lambda_1-\Lambda_2/3}\widehat{\mathcal{M}}^{-1} e_{27} h_\Gamma \end{aligned} \tag{2.25}$$

An outcome of the previous analysis is that the discrete isometry group Γ may be anomalous at the quantum level because of the chiral U(1) transformation properties of Weyl fermions (2.19) (2.21) under Γ.

In fact the combination of target-space duality, gauge invariance and supersymmetry relates $F^2_{\mu\nu}$ and $F_{\mu\nu}\widetilde{F}_\mu$ terms in the effective action:

$$g_{ab}(\phi,\bar{\phi})F^a_{\mu\nu}F^b_{\mu\nu} + \theta_{ab}(\phi,\bar{\phi})F^a_{\mu\nu}F^b_{\rho\sigma}\ \epsilon^{\mu\nu\rho\sigma} \tag{2.26}$$

as follows [33] [27] :

$$g_{ab} = \mathrm{Re}\ K_{ab}\ ,\ \ \theta_{ab} = \mathrm{Im}\ K_{ab} \tag{2.27}$$

when K_{ab} is some (generally non-local) chiral superfield. If K_{ab} is local then it follows (up to higher derivative terms) that K is holomorphic:

$$K_{ab} = f_{ab}(\phi) = g_{ab}(\phi,\bar{\phi}) + \theta_{ab}(\phi,\bar{\phi}) \tag{2.28}$$

and therefore g and θ are harmonic [33] .

On the other hand if $g_{ab}(\phi,\bar{\phi})$ is non-harmonic it follows necessarily that K_{ab} is a non-local chiral superfield whose $\theta = 0$ has a local non-holomorphic piece in its real part.

The reason why this happens is because the chiral superfield K given by:

$$K = \frac{\overline{D}^2 D^2}{\Box}\ H \tag{2.29}$$

where H is a real superfield with expansion:

$$H = H_{\theta=0} + \ldots \theta\sigma\mu\bar{\theta}\ A_\mu + \ldots \theta^2\bar{\theta}^2 D \tag{2.30}$$

has a $\theta = 0$ component given by:

$$K_{\theta=0} = \left(H_{\theta=0} + \frac{D}{\Box}\right) + i\,\frac{\partial_\mu A_\mu}{\Box} \tag{2.31}$$

If $H_{\theta=0}$ is local we see that $K_{\theta=0}$ has a local term in spite of the fact that the imaginary part is non-local. By a further inspection of eq. (2.31) we see that the bosonic component of the superfield action [34]:

$$\int W^2 \frac{\overline{D}^2 D^2}{\Box} K \quad \text{(W is the gauge superfield strength)} \tag{2.32}$$

precisely contains a term of the form: $F^a_{\mu\nu}\widetilde{F}^a_{\rho\sigma} \ \in^{\mu\nu\rho\sigma} \ \partial^\rho A_\rho/\Box$ which is a typical mixed $U(1) \times G^2$ anomaly graph contribution coming by integrating out the massless fermions which are both charged under U(1) and G. Indeed the U(1) gauge variation of such non-local term is local and coincides with the anomolous variation of the fermion loop: $\Lambda F\widetilde{F}$ under $(A_\rho \to \partial_\rho \Lambda)$.

When the gauge group G has different gauge group factors like in Calabi-Yau compactification, the anomaly terms are usually different for different group factors since gauginos are only charged under E_6 but 27 and $\overline{27}$ fermions have different U(1) phase transformation.

Let us call K_A the generic non-holomorphic superfield which gives the anomaly diagram. The generic expression for K_A is [26] , [27] :

$$K_A = [C(G_A) - \sum_R T(R_A)]\,K + 2\sum_R T(R)\,\log\,\det G_{R_A} \tag{2.33}$$

In the case of Calabi-Yau manifolds we have the further relations [12] , [35]:

$$\begin{aligned} G^1_{27} &= e^{(K_2-K_1)/3} G^1 \\ G^2_{\overline{27}} &= e^{(K_1-K_2)/3} G^2 \end{aligned} \tag{2.34}$$

$$-e^{2k}\, W^1_{ijk}\, \overline{W}^1_{\bar{\imath}\bar{\jmath}\bar{k}}\, G^{1 j\bar{\jmath}} G^{k\bar{k}} = +\partial_i\partial_{\bar{j}} \log\,\det\, N^1_{IJ} = R^1_{i\bar{j}} \ + \ (n+1) G^1_{i\bar{j}}$$

where N^1_{IJ} is the imaginary part of the period matrix $F_{IJ} = \partial_I\partial_J F(I, J = 0, \ldots n)$.

Using eq. (2.31) this gives the following formula for the $E_6 \times E_8$ mixed U(1) anomalies [27] [36]:

$$\begin{aligned} K_A &= \sum_{a=1}^{2} C^i_A H_i \ (A = E_6) \\ C^1_A H_1 &= [T(G = E_6) - \frac{5}{3}\,T(R = 27) h_{11} - \frac{1}{3}\,T(\overline{R} = \overline{27}) h_{12}] K_1 \\ &\quad + \ 2T(R = 27)\,\log\,\det\,G^1 \end{aligned} \tag{2.35}$$

$C_A^2 H_2$ = the same with $1 \leftrightarrow 2$ $h_{11} \leftrightarrow h_{12}$ $R \leftrightarrow \overline{R}$.

$$\sum_{a=1}^{2} C_A^i H_i = (T(E_6) + 2T(27)(K_1 + K_2) + \frac{1}{6}\, T(27)\chi(K_1 - K_2)$$

$$+ 2T(27) \log\, (\det N^1 \det N^2)$$

with $\chi = 2(h_{11} - h_{12}), T(E_8) = 30,, T(E_6) = 12, T(27) = T(\overline{27}) = 3$.

$$K_A(A = E_8) = T(E_8)(K_1 + K_2) \tag{2.36}$$

The change of $\mathcal{L}_{ne}$ (induced by the anomaly):

$$\mathcal{L}_{ne} = \mathrm{Re} \sum_A \int d^2\theta\; W^{2A}\, \frac{\overline{D}^2 D^2}{\Box}\, K_A \tag{2.37}$$

is given by:

$$\delta(\mathcal{L}_{ne}) = \sum_a \mathrm{Re} \int d^2\theta\; W^{2A} \Lambda_\Gamma^A \tag{2.38}$$

where:

$$\begin{aligned} \Lambda_\Gamma^{E(6)} = &- \left[T(E_6) - T(27)\left(\frac{5}{3}\, h_{11} + \frac{1}{3}\, h_{21}\right)\right]\Lambda_\Gamma^1 \\ &- \left[T(E_6) - T(27)\left(\frac{5}{3}\, h_{21} + \frac{1}{3}\, h_{11}\right)\right]\Lambda_\Gamma^1 \\ &- 2T(27)(\Sigma_\Gamma^1 + \Sigma_\Gamma^2) \end{aligned} \tag{2.39}$$

$$\Lambda^{E(8)} = \frac{1}{4}\, T(E_8)(\Lambda_\Gamma^1 + \Lambda_\Gamma^2) \tag{2.40}$$

The anomaly may therefore be cancelled by appropriate Wess-Zumino terms:

$$\mathrm{Re} \sum_A \int d^2\theta\; W^{2A} \log f_A \tag{2.41}$$

where f_A are appropriate automorphic functions such that:

$$f_A \to e^{-\Lambda_\Gamma^A} f_A\, e^{i\varphi_\Gamma} \tag{2.42}$$

where f_A is holomorphic and φ_Γ is a real number.

Another possible anomaly cancelling term may be present for a gauge-independent part of the anomaly [27] - [30] . This is the 4D version of the

Green-Schwarz mechanism [37] and requires the dilaton to transform under duality transformation:

$$S \rightarrow S + \eta(\psi, \varphi) \tag{2.43}$$

and a modified dilaton Kähler potential:

$$-\log\,(S + \overline{S} + \Delta) \tag{2.44}$$

where Δ is an anomaly term:

$$\sum_{A} \mathrm{Re} \int d^2\theta W^{2A} \, \frac{\overline{D}^2 D^2}{\Box} \, \Delta \text{ such that } \Delta \rightarrow \Delta - \eta - \bar{\eta} \tag{2.45}$$

under target space duality transformations.

The effective gauge coupling constant, defined by eq. (2.24) are therefore given by:

$$\frac{1}{g^{2A}} = k^A(S + \overline{S}) - \frac{1}{16\pi^2}\,(K_A + \log f_A \bar{f}_A) + \gamma^A \tag{2.46}$$

(where, in eq. (2.44) we defined for convenience, a duality invariant dilaton chiral multiplet $\underset{\sim}{S}$). γ^A is a moduli independent quantity related to the N=1 β function as follows:

$$\gamma^A = \frac{b^A}{16\pi^2} \log M^2_{STRING}/\mu^2 \tag{2.47}$$

$b^A = -3T(G^A) + (h_{11} + h_{12})T(R^A)$.

By taking the overall volume variable T and by using the fact that:

$$K^1(T, \overline{T}) \rightarrow -3\log\,(T + \overline{T}) \text{ with } T + \overline{T} = R^2 M^2_{STR}, \quad \text{for } T \rightarrow \infty \tag{2.48}$$

one recovers the Kaluza-Klein result:

$$\frac{1}{g^{2A}_{(\mu)}} = k^A(S + \overline{S}) + \frac{1}{16\pi^2}\, b^A \log R^2\mu^2 + 0(R^2 M^2_{STR}) \tag{2.49}$$

valid for $0 < \mu^2 < 1/R^2$.

The $0(R^2)$ terms come from the Kaluza-Klein states which are relevant in the (infrared finite) regions $1/R^2 < \mu^2 < 1/\ell^2_{STRING}$.

From eq. (2.44) we can compute the unification scale M_x at which the E_6 and E_8 gauge couplings become equal [27] - [30] .

By defining $K_A = \ell n \, Y_A$ and by imposing:

$$\frac{1}{k_6 g_6^2} = \frac{1}{k_8 g_8^2} \text{ at } \mu = M_x \tag{2.50}$$

we obtain:

$$\frac{M_x^2}{M_{STRING}^2} = \left(\frac{Y_6|f_6|^2}{Y_8|f_8|^2}\right)^{1/b_8-b_6} \quad [b_8 - b_6 = -3(18 + h_1 + h_2)] \tag{2.51}$$

We observe that eq. (2.51) is duality invariant, independent of the Green-Schwarz term and justifies the definition of the moduli-dependent term in eq. (2.46) as string threshold effect.

The role of target-space duality is to make the moduli dependent thershold effect in eq. (2.46) duality invariant.

Similar results hold for gravitational anomalies [28] - [30] , which just add to a gauge factor another factor (Lorentz group) with now all chiral fermions of the theory (including the dilatino, moduli fermions and gravitino) contributing to the anomaly.

In principle one should also take care of pure σ-models holonomy anomalies [38] induced by the h_Γ transformation defined by eq.(2.25). In this case appropriate Wess-Zumino terms should be constructed (they correspond to higher-derivative terms in the moduli fields) which cancel the pure σ-model anomaly induced by the discrete isometry group Γ. It should be stressed that the existence of Wess-Zumino terms, related to automorphic functions f^A of Γ is a phenomenon which is possible only because the Γ group is discrete. The cancellation of anomalies related to continuous isometries is only possible through a Green-Schwarz mechanism and it requires the existence of new degrees of freedom (like the dilaton multiplet).

Conclusions

We have described in simple terms effects occuring in 4D string theory, responsible for the gauge coupling running.

The main features are the following:

1) The logaritmic running of the gauge couplings is an infrared phenomenon of 4D strings and is entirely due to the light particle spectrum of the thery.
2) Gauge coupling unification at the string scale is affected by loop-effects manifesting in threshold effects, depending on the geometry of the compactified space or more properly of the N=2 internal superconformal field theory. These threshold effects are severely constrained by target-space duality [27] - [30][39] and can be seen as the supersymmetric counterpart of the Γ-anomaly cancellation mechanism. We have not discussed here the more physically relevant evolution toward energies near the Fermi scale. This requires the discussion of supersymmetry breaking in string theory [40], an issue which can be phenomenologically discussed but which is not the subject of the present report. In particular it is an open question how one can get a realistic string model

[41] [42] where the low-energy gauge group is $SU(3) \times SU(2) \times U(1)$ and the threshold effects are such to successfully meet the requirements of the low-energy couplings which seem to indicate [43] - [45] a unification scale lower than the string scale.

References

[1] See i.e. M. Green, J. Schwarz and E. Witten in: *"Superstring Theory"* Cambridge University Press 1987;
A.N. Schellekens in: *"Superstring Constructions"* North-Holland Amsterdam, 1989.

[2] P. Candelas, G. Horowitz, A. Strominger and E. Witten, *Nucl. Phys.* **B258** (1985) 46.

[3] E. Witten, *Phys. Lett.* **B155** (1985) 185;
J.P. Derendinger, L.E. Ibanez, H.P. Nilles, *Nucl. Phys.* **B272** 365;
M. Dine, R. Rhom, N. Seiberg, E. Witten, *Phys. Rev. Lett.* **B156** (1986) 55.

[4] D. Gross, J. Harvey, E. Martinec and R. Rhom, *Phys. Rev. Lett.* **54** (1985) 502.

[5] V.S. Kaplunovsky, *Nucl. Phys.* **B307** (1988) 145.

[6] H. Georgi, H.R. Quinn, S. Weinberg, *Phys. Rev. Lett.* **33** (1974) 451.

[7] T.R. Taylor and G. Veneziano, *Phys. Lett.* **B212** (1988) 147.

[8] N. Seiberg, *Nucl. Phys.* **B303** (1988) 286.

[9] S. Cecotti, S. Ferrara and Girardello, *Int. J. Mod. Phys.* **A9** (1989) 2457.

[10] P. Candelas and X. de la Ossa, *Nucl. Phys.* **B385** (1991) 455.

[11] L. Dixon, V.S. Kaplunovsky and J. Louis, *Nucl. Phys.* **B329** (1990) 27.

[12] L. Dixon, J. Harvey, C. Vafa and E. Witten, *Nucl. Phys.* **B261** (1985) 671; **B274** (1986) 285.

[13] R. Dijkgraaf, E. Verlinde and H. Verlinde, *Com. Math. Phys.* **115** (1988) 649.

[14] K.S. Narain, *Phys. Lett.* **B169** (1986) 369;
K.S. Narain, M.H. Sarmadi and E. Witten, *Nucl. Phys.* **B289** (1987) 414.

[15] P. Candelas, X. de la Ossa, P. Green and L. Parkes, *Phys. Lett.* **B258** (1991) 118; *Nucl. Phys.* **B359** (1991) 21.

[16] A. Cadavid, S. Ferrara, *Phys. Lett.* **B267** (1991) 193.

[17] W. Lerche, D. Smit, N. Warner, *Nucl. Phys.* **B372** (1992) 87.

[18] D. Morrison, Preprint Duke DUK-M-91-14 (1991).

[19] A. Font, Preprint Caracas UCVFC/DF-1-92.

[20] A. Klemm, S. Theisen, Preprint Karlsruhe KA-THEP-03/92.

[21] S. Ferrara, J. Louis, *Phys. Lett.* **B278** (1992) 240.

[22] A. Ceresole, R. D'Auria, S. Ferrara, W. Lerche, J. Louis, *Int. J. Mod. Phys.***A6** (1993) 79.

[23] C. Vafa, N.P. Warner, *Phys. Lett.* **B218** (1989) 51;
E. Martinec, *Phys. Lett.* **B217** (1989) 431;
W. Lerche, C. Vafa, N.P. Warner, *Nucl. Phys.* **B324** (1989) 427.

[24] P. Candelas, *Nucl. Phys.* **B298** (1988) 458.

[25] S. Ferrara, D. Lüst, A. Shapere, S. Theisen, *Phys. Lett.* **B233** (1989) 147; for a Review see S. Ferrara in: *"From Superstrings to the Real Superworld"*, Erice Int. School of H.E.P. Erice (1992) Plenum Press, Ed. A. Zichichi.

[26] L. Dixon, VS. Klapunovsky, J. Louis, *Nucl. Phys.* **B355** (1991) 649;
I. Antoniadis, K. Narain, T. Taylor, *Phys. Lett.* **B267** (1991) 37.

[27] J.P. Derendiger, S. Ferrara, C. Kounnas, F. Zwirner, *Nucl. Phys* **B372** (1992);
G. Cardoso and B. Ovrut, *Nucl. Phys.* **B369** (1992) 351;
J. Louis, PASCOS 1991 Proc., World Scientific, Ed. P. Nath 1991;
S. Ferrara in: *"String and Symmetries"* (1991) (N. Berkovits et al. Ed.), World Scientific (1992), page 82.

[28] I. Antoniadis, E. Gava, K.S. Narain, *Phys. Lett* **B283** (1992) 209; *Nucl. Phys.* **B383** (1992) 93.

[29] G.L. Cardoso, B. Ovrut, Pennsylvania Report UPR-0502T (1992); *Nucl. Phys.* **B369** (1992) 351.

[30] L.E. Ibanez, D. Lüst, CERN-TH.6380 (1992).

[31] E. Witten, J. Bagger, *Phys. Lett.* **115B** (1982).

[32] S. Ferrara and A. Strominger in: *"Strings 89"* eds. R. Arnowitt et al., World Scientific (1989) p. 245;
A. Strominger, *Com. Math. Phys.* **133** (1990) 163;
L. Castellani, R. D'Auria , S. Ferrara, Class. Quantum Gravity **1** (1990) 1767;
R. D'Auria, S. Ferrara, P. Fré, *Nucl. Phys.* **B359** (1991) 205.

[33] E. Cremmer, S. Ferrara, L. Girardello, A. Van Proeyen, *Nucl. Phys.* **B212** (1983) 413.

[34] S.J. Gates, M.T. Grisaru, M. Roček, V. Siegel in: *"Superspace"*, (Benjamin Cummings Publ. Co., Reading, Mass. 1983).

[35] E. Cremmer et al., *Nucl. Phys.* **B250** (1985) 385.

[36] L. Dixon, talk given at the MRS Berkeley Workshop (1991).

[37] M. Green, J. Schwarz, *Phys. Lett.* **149B** (1984) 117.

[38] G. Cardoso, B. Ovrut, Preprint Pennsylvania UPR-05220T.

[39] M.K. Gaillard, R. Xiu, Preprint Berkeley LBL-32324.

[40] See i.e. J. Louis, SLAC-PUB-5645 and CERN-TH.6492/92.

[41] L. Ibanez, D. Lüst, G. Ross, *Phys. Lett.* **B272** (1991) 25.

[42] J. Ellis, S. Kelley and D.V. Nanopoulos, *Phys. Lett.* **B260** (1991) 441; **B260** (1991) 131;
I. Antoniadis, J. Ellis, S. Kelley and D.V. Nanopoulos, CERN-TH.6169 (1991).

[43] U. Amaldi, W. de Boer, H. Füsternau, *Phys. Lett.* **B260** (1991).

[44] P. Langacher and A.X. Luo, *Phys. Rev.* **D44**, 817 (1991).

[45] F. Anselmo, L. Cifarelli, A. Peterman and A. Zichichi, *Il Nuovo Cimento* **104A** (1991) 1817;
A. Zichichi, CERN-PPE/92-19; CERN/LAA/MSL/92-017 and reference therein.

CHAIRMAN: S. Ferrara

Scientific Secretaries: A. Hanany, R. Liu

DISCUSSION

– *Banakli:*

1) Can you explain what is the phase transformation of the fermions?
2) Can you tell us about this Picard-Fuchs equations?

– *Ferrara:*

$R \to \alpha'/R$ generalizes to $T = R^2 + ib$, as $T \to aT - ib/icT - d(ad - bc = 1, a, b, c, d, \in \mathcal{Z})$. This group of transformation is called $SL(2, \mathcal{Z})$ (T is a space-time scalar) and it acts on all other fields in the theory. This transformation acts on the fermions as a chiral local transformation with a potential violation of duality producing an anomaly in the quantum theory. But duality is not broken thanks to appropriate WZ terms which cancel this anomaly. Theories with N=1 SUSY are needed to solve the hierarchy problem and to have chiral fermions under $SU(3) \times SU(2) \times U(1)$. In this case the compactification space is more complicated than a toroidal compactification as I discussed before and the six dimensions must be curled up, for example, on a Calabi-Yau manifold. In this case, the cubic superpotential in the family-antifamily chiral multiplets looks like:

$$W(\phi, \psi, 27, \overline{27}) = W^1_{ijk}(\phi)(27)^i(27)^j(27)^k + \\ + W^2_{\alpha\beta\gamma}(\psi)(\overline{27})^\alpha(\overline{27})^\beta(\overline{27})^\gamma$$

with gauge group E_6 at the compactification scale. The Yukawa couplings can be calculated as follow: $W^1_{ijk}(\phi) = \partial_i\partial_j\partial_k\mathcal{F}^1(\phi)$ (integrable) and $\mathcal{F}^1(\phi)$ can be computed through differential equations derived by methods of algebraic geometry. These are the Picard-Fuchs equations.

– *Azimov:*

You emphasized in the lecture that the string theory is non-local. But locality is usually considered as one of basic points for the CPT-theorem. Therefore, does the CPT-theorem work for string? If yes, then why? Is it the general case, or, maybe the CPT-invariance restrict somehow the structure of the string?

– *Ferrara:*

I do not know of any work proving this but violation of locality does not mean violation of CPT. In field theory, in the days of current algebra there were attempts to write theories with infinite component fields which may violate some of the axioms of QFT. String theory is in some sense an infinite component field theory but it does not mean that it necessarily violates CPT.

– *Hanany:*

According to what you say the universe is invariant under $R \leftrightarrow 1/R, \vec{H} \leftrightarrow \vec{E}$, what about electric $\leftrightarrow$ magnetic monopoles in closed string? What about anomalies?

– *Ferrara:*

Space-time duality means that the theory is dual under σ-model coupling constant inversion $g^2 \rightarrow 1/g^2$. This is similar to a phenomenon in pure gauge theories where you replace the magnetic field with electric field. This duality is generally broken when you introduce sources. There are some special theories like N=4 Yang-Mills SUSY which has this duality property but in general many theories do not have this property.

– *Sannino:*

This morning you wrote the "string extension" of Heisenberg relation for Δp and Δx. Is there a similar extension for the other conjugated variables? In which way do you derive this? Which are the implications for the singularity (black hole and first instant of universe)?

– *Ferrara:*

The extension is due to α' corrections, namely: $\Delta t \sim 1/\Delta E + \alpha' \Delta E$.

– *Sanchez, comment:*

No, such work is known. The relation $\Delta x \sim 1/\Delta p + \alpha' \Delta p$ was suggested after a detailed calculation of graviton scattering (Gross, Veneziano). The coefficient α' was calculated by resummations of string diagrams but the coefficient is model dependent.

– *Raciti:*

Is there an agreement on the number of compactified dimensions in string theory?

– *Ferrara:*

In the case of Calabi-Yau compactification the 6 compactified dimensions have a geometric interpretation, but there are other 4D-strings in which the internal degrees of freedom have no interpretation as extra (compactified) dimensions.

– *Graudenz:*

Superstrings are supposed to be renormalizable. The sector describing gravity in supergravity suffers from non-renormalizable UV-divergences. Could you explain why renormalizability is spoiled in the infrared limit? Does this happen simply because supergravity is considered to be an effective theory?

– *Ferrara:*

Yes. String theory has an intrinsic UV cut-off which is the string tension and it is related to the length of the one dimensional extended object. The theory after integrating out the massive modes has infinite terms in the Einstein curvature $\sum_n (1/\alpha')^n R^n$. After integrating out the massive modes, when you restrict the theory to lower derivative terms, you get a theory which looks non-renormalizable but in fact if you include all the higher terms the theory is finite. A phenomenon like the 4 fermion interactions in the Fermi theory, happens in this case for gravity.

– *Zichichi:*

How is the Heisenberg Principle $\Delta x \cdot \Delta p \geq \hbar$ changed by the existence of a fundamental length L? Take $\hbar = 1$.

$$\Delta x_L \sim \frac{1}{\Delta \phi} + \underbrace{L^2 \Delta p}_{\Delta x'} = \frac{1}{\Delta p} + \underbrace{(10^{-33})}_{\alpha'} cm^2 \cdot \Delta p$$

$L^2 \Delta p$ can be written as $\Delta x'$ and this is simply another error and not a modification of the Heisenberg Principle.

SUPERSYMMETRIC PHENOMENOLOGY.

G. G. Ross
SERC Senior Research Fellow
Department of Physics
Theoretical Physics
University of Oxford
1 Keble Road
Oxford OX1 5DX

Abstract

An introduction to the construction of a supersymmetric version of the standard model is presented. The changes in phenomenology due to radiative corrections involving virtual supersymmetric states are discussed. The expectation for the SUSY spectrum following from the requirement of Unification of couplings and masses is presented. Finally the prospects for discovery of these new SUSY states is considered, both in the minimal supersymmetric standard model and in non-minimal extensions.

1 Introduction to Supersymmetry.

There are several reasons why supersymmetry has been widely suggested as a likely extension of the standard model. In the first place supersymmetry provides the only possible (finite) symmetry beyond those based on Lie groups. It is also a necessary ingredient in string theories which many people think is the best candidate for a "Theory of Everything" unifying all the fundamental forces including gravity. While these points may suggest we consider supersymmetry as an underlying symmetry of nature the only reason we have for expecting a *low-energy* realisation of the symmetry is the "hierarchy problem", namely the difficulty in field theory of keeping the electroweak scale small compared to the unification scale without having the interactions become strong.

In these lectures I will try to explain how supersymmetry explains the "hierarchy problem" and determine the constraints the solution places on the supersymmetric spectrum. I will also discuss what the phenomenology of the new supersymmetric states is likely to be and the prospects for finding experimental evidence for supersymmetry. First, however, I turn to a very brief introduction to supersymmetry.

The simplest possible supersymmetry algebra is given by the anticommutation relations involving spinorial generators $Q_\alpha, \overline{Q}_{\dot{\alpha}}$, two dimensional Weyl spinors [1].

$$\begin{aligned} \{Q_\alpha, Q_\beta\} &= \{\overline{Q}_{\dot{\alpha}}, \overline{Q}_{\dot{\beta}}\} = 0 \\ \{Q_\alpha, \overline{Q}_{\dot{\beta}}\} &= 2\sigma^\mu_{\alpha\dot{\beta}} P_\mu \end{aligned} \tag{1}$$

where σ^i are the 2×2 Pauli matrices and σ^0 is the unit matrix. The appearance of the momentum operator shows that the full supersymmetry algebra should include the generators of the Poincare group giving a super-Poincare algebra.

	Spin	Name	SM partner
$\tilde{g}$	$\frac{1}{2}$	gluino	gluon
$\tilde{W}^{+-}$	$\frac{1}{2}$	Wino	W boson
$\tilde{Z}$	$\frac{1}{2}$	Zino	Z boson
$\tilde{q}$	0	squark	quark
$\tilde{l}^{+-}$	0	slepton	lepton
$\tilde{\nu}$	0	sneutrino	neutrino
$\tilde{H}$	$\frac{1}{2}$	Higgsino	Higgs boson

Table 1: SUSY states in the MSSM.

The (massless) representations of supersymmetry are easily obtained from eq 1 and consist of just two states of helicity λ and $\lambda + 1/2$. This immediately follows from the fact that, from eq 1, only one product of generators is non-vanishing. When building supersymmetric (SUSY) models the only multiplets that lead to consistent theories are left handed "chiral" supermultiplets with helicity (0,1/2) pairs, vector supermultiplets with helicity (1/2,1) pairs and the gravitino supermultiplet with helicity (3/2,2) together with the conjugate fields.

The supersymmetry algebra of eq 1 may be extended by introducing more than one SUSY generator, $Q_\alpha^{a=1,..,N}$. These N-extended algebras are not used to construct low-energy supersymmetric theories because they have only non-chiral representations and thus lead to "mirror" families of quarks and leptons with the identical gauge couplings but opposite helicity to their standard model partners. For example the N=2 algebra has the representation with helicities (-1/2,0,1/2) (following because there are now *two* non-zero products of generators). Since there is no evidence for such mirror states essentially all work on low-energy SUSY has concentrated on the N=1 supersymmetry case. In the next section I will discuss how a N=1 SUSY version of the standard model is built.

2 A supersymmetric version of the standard model

If the standard model is to be incorporated in a field theory with N=1 supersymmetry, it is first necessary to assign the states of the standard model to N=1 supersymmetric representations [1]. Let us start with the SU(3) x SU(2) x U(1) vector bosons. The only super-multiplet available which contains vectors is the vector supermultiplet containing states of helicity 1 and $\frac{1}{2}$. Thus we must assign the vector bosons to sector multiplets transforming in the same way as the vectors under SU(3) x SU(2) x U(1); this is shown in Table 1. It is clear that this means we have to introduce gauginos, spin $\frac{1}{2}$ fermionic partners of the gauge bosons, which transform under the gauge group in exactly the same way as their vector partners.

Turning now to the assignment of the quarks and leptons to supermultiplets it appears that there is the option of assigning them to vector supermultiplets, with vector partners, or to chiral supermultiplets, with scalar partners. However the former possibility is not available, unless we increase the size of the gauge group, for the transformation properties of the quarks and leptons is different from the gauge bosons of the standard model (cf Table 1). Thus we cannot identify any of the gauginos of Table 1 with the quarks or leptons and we have to assign the latter to *chiral* supermultiplets with the transformation properties of the quarks and leptons. These introduce scalar partners to the quarks and leptons transforming in the same way as their fermionic partners as shown in Table 1.

Finally we must assign the Higgs bosons of the standard model to a supermultiplet. Having spin 0, they can only be assigned to a chiral supermultiplet and thus they must have fermionic partners called higgsinos. However the introduction of an SU(2) doublet of fermions of definite helicity introduces an anomaly to the SU(2) x U(1) gauge theory. This is unacceptable and to avoid it we are forced to introduce two Higgs chiral supermultiplets with *opposite* hypercharge (so that the higgsinos have cancelling contributions to the anomaly) as shown in Table 1. It turns out that both Higgs multiplets are needed in supersymmetry to give quarks and leptons a mass so this is indeed the minimal choice (i.e. attempts to identify the H_2 supermultiplet with a lepton supermultiplet without introducing an H_1 would have failed).

To summarise, the supersymmetric version of the standard model requires more than twice the number of particles; gauginos, partners for each of the gauge bosons, and squarks and sleptons, scalar partners for each of the quarks and leptons and two Higgs doublets with fermionic "Higgsino" partners.

2.1 Couplings in the supersymmetric model - the MSSM.

Since the gauge multiplet structure of the new supersymmetric states is the same as that of their standard model partners, the interaction of gauge bosons with these states is fixed. Supersymmetry relates these gauge couplings to new couplings involving two superpartners in place of the original states (for example a quark-quark-gluon vertex is related to a squark-quark-gluino vertex). Thus the gauge-related interactions are fixed given the multiplet structure of Table 1, and may be found, for example, in [1].

The remaining coupling that must be specified to complete the definition of the supersymmetric model correspond to the Yukawa couplings and those couplings related to them by supersymmetry. These are most conveniently given by the "F" term of the superpotential which is a polynomial in the chiral field ϕ (and does not involve their complex conjugates).

In terms of P the associated Yukawa interactions are given by $\Sigma_{i,j} \frac{\partial^2 P(\Phi)}{\partial \Phi_i \partial \Phi_j} |_B \psi_i \psi_j$, where the subscript B means the chiral fields Φ_i are to be replaced by their scalar components ϕ_i and ψ_i are the corresponding fermionic components. There are scalar couplings related to these Yukawa couplings which are given by $\Sigma_i \mid \frac{\partial P}{\partial \Phi_i} |_B^2$. The couplings needed to give quarks and leptons a mass are contained in the superpotential given by

$$P = h_{ijk} l_i h_{2j} \overline{e}_k, h'_{ijk} q_i h_{2j} \overline{d}_k, h''_{ijk} q_i h_{1j} \overline{u}_k \tag{2}$$

where l and e (q and $\overline{u}\overline{d}$) are the left-handed components of lepton doublet and antilepton singlet (quark doublet and antiquark singlet) chiral superfields respectively and $h_{1,2}$ are Higgs superfields.

The model with the minimal multiplet content of Table 1, with only these terms in the superpotential, is known as the minimal supersymmetric standard model (MSSM) (As we will see there are additional trilinear terms allowed by the gauge symmetry which give rise to non-standard supersymmetric version of the standard model.) Note that each term in eq 2 gives three separate Yukawa couplings. For example the first term gives

$$L_{Yukawa} = l_i h_{2j} \overline{e}_k + \tilde{l}_i \tilde{h}_{2j} \overline{\tilde{e}_k} + l_i \tilde{h}_{2j} \overline{\tilde{e}_k} \tag{3}$$

where we denote by a supertwiddle the scalar partners to the quarks and leptons namely the squarks and sleptons. The first term is the usual term in the standard model needed to give cha rged leptons a mass. The new couplings associated with the supersymmetric states related to the first term by the operation of the supersymmetry generator are given by the second and third terms.

Note that the last two terms involving the new supersymmetric states involves them in pairs. This is a general feature of the MSSM which preserves an R-parity, R =

$(-1)^{2s+3B+L)}$, where s is the spin quantum number. Under R-parity the new supersymmetric states are all odd while the standard model states are even [2]. This has a profound effect on the phenomenology of the MSSM for the new SUSY states may only be produced in pairs and the lightest supersymmetric state (the LSP) is stable.

There is one further coupling needed to complete the couplings of the minimal supersymmetric version of the standard model. In order to generate a mass for the Higgsinos associated with the Higgs doublets $H_{1,2}$ it is necessary to add a term to the superpotential given by

$$P' = \mu H_1 H_2 \tag{4}$$

In addition to giving a mass μ to the Higgsinos, this term plays an important role in determining the Higgs scalar potential and the pattern of electroweak symmetry breaking. The scalar term following from eq 4 aligns the vacuum expectation values (vevs) of the two Higgs fields so that the photon is left massless, obviously a crucial ingredient for a viable theory.

2.2 Unification Hints

Clearly the important question for SUSY searches is the mass of the supersymmetric states. Recently it has been shown that if one adds some assumptions about Grand Unification to the MSSM the scale of supersymmetry breaking and the associated superpartner masses is constrained [3, 4]. Since supersymmetry was introduced to allow for such unification it seems reasonable to treat seriously such predictions for the mass scale of supersymmetry.

The prediction follows from the Grand Unified prediction for the gauge couplings such as occurs in $SU(5)$ together with the form of the running strong, weak and electromagnetic couplings. In the standard model this was analyzed Georgi, Quinn and Weinberg [5] using the RG equations for $\alpha_i = g_i^2/(4\pi), i = 2,3$ and $\alpha_1 = \frac{5}{3}g_i^2/(4\pi)$ where g_i are the $SU(3), SU(2), U(1)$ couplings

$$\alpha_i^{-1}(q^2) = \alpha_i^{-1}(\mu^2) + \beta_{0i} ln(q^2/\mu^2) + .. \tag{5}$$

where the one-loop β function is simply evaluated knowing the light matter content

$$2\pi\beta_{0i} = \begin{pmatrix} 0 \\ \frac{-22}{3} \\ -11C^2 \end{pmatrix} + N_f \begin{pmatrix} \frac{4}{3} \\ \frac{4}{3} \\ \frac{4}{3}C^2 \end{pmatrix} + N_H \begin{pmatrix} \frac{1}{10} \\ \frac{1}{6} \\ 0 \end{pmatrix} \tag{6}$$

and N_f, N_H denote the number of families and Higgs doublets respectively. In SU(5) $C^2 = 1$ and $\alpha_i(M_X^2) = \alpha_G$; the same applies to many of the phenomenologically viable GUTs and superstring unification schemes as I will shortly discuss. Given these equations the low-energy couplings should evolve in energy to meet at the unification value α_G. However, the recent precision measurements of $\sin\theta_W$ are inconsistent with these predictions; the couplings fail to meet by more than six standard deviations [3, 4].

The situation is quite different in the MSSM due to the effects of the new supersymmetric states on the evolution of the couplings. Including this gives [6]

$$2\pi\beta_{0i} = \begin{pmatrix} 0 \\ -6 \\ -9C^2 \end{pmatrix} + N_f \begin{pmatrix} 2 \\ 2 \\ 2C^2 \end{pmatrix} + N_H \begin{pmatrix} \frac{3}{10} \\ \frac{1}{2} \\ 0 \end{pmatrix} \tag{7}$$

A description of the evolution of the couplings following from these equations (extended to include two-loop effects) has already been presented at this school by Professors Zichichi and Ellis. As they have discussed the SUSY unification predictions are consistent with the measured couplings provided the mass of the new supersymmetric states (assumed degenerate here) are low $M_{SUSY} \approx 10^{2.5\pm1}$. On the basis of these results it is tempting to argue that there is evidence both for new forms of (relatively) light supersymmetric matter *and* an underlying unified theory[1]. This conclusion is so dramatic that it needs to be evaluated with considerable caution. The most obvious reservation is that it relies on the extrapolation of the standard model, albeit extended to its supersymmetric version, some twelve orders of magnitude beyond the energy scale at which it has been tested. No theory has proved to be so robust in the past so it is understandable if one views this extrapolation with some caution. Even given the framework of Grand Unification there is considerable uncertainty for the relative values of the couplings may differ in different unification schemes. Within the (large) class of theories which give the $SU(5)$ predictions there are corrections coming from virtual states with mass $\geq O(M_X)$ which have not been included in the analysis and which can affect the predictions substantially. These predictions are even more sensitive to additional light states beyond those needed for the minimal supersymmetric standard model (Indeed it is possible to bring the predictions of non-supersymmetric $SU(5)$ into agreement with experiment if additional light Higgs doublets are added). Furthermore if the GUT should not break immediately to the standard model there will be several intermediate scales of breaking and again the results will change substantially.

Despite these caveats it is remarkable that the simplest possible extension of the standard model to include supersymmetry coupled with the simplest assumption about Grand Unification yields predictions in detailed agreement with experiment provided the new supersymmetric states are very light. The main objection to the analysis of the likely mass of the new supersymmetric states is that, due to large radiative corrections, the most reasonable supersymmetric spectrum is not degenerate. To discuss the likely effects of non-degenerate SUSY states I will use a spectrum motivated by most realistic supergravity/superstring models of supersymmetry breaking that have been extensively developed in recent years.

3 Supersymmetry breaking

So far the discussion of supersymmetric unification has relied only on the global supersymmetry aspects of the theory which involve the supermultiplet structure and the renormalisable couplings. However most workers in the field assume that it is the locally supersymmetric version that is realised [7]. There are several reasons for this. In the first place the local version of the supersymmetric algebra necessarily includes the local version of the Poincare group and hence includes a theory of gravity. The hope is that the ultimate T.O.E. will involve a unification of all the fundamental forces including gravity and so the local version seems desirable. Moreover the superstring theories which offer the best hope of such a T.O.E. naturally lead at low energies to local supersymmetric

[1]It is perhaps appropriate to comment about the meaning of the SUSY analysis, for the three gauge couplings are described in terms of three effective parameters M_X, α_X and the effective supersymmetry mass scale, meaning there will always be a fit and apparently no test of unification! However this is not quite fair for the resulting values of the parameters must be reasonable if the scheme is to make sense. Thus M_X should be less than the Planck scale but large enough to inhibit proton decay in Grand Unified theories. Also we would like α_X to be within the perturbative domain although it may be sensible to contemplate non-perturbative unification. Finally the supersymmetry mass scale must be large enough to explain why no supersymmetric states have been found and small enough to avoid the hierarchy problem (as we will see the latter gives a very strong constraint). As we will discuss these conditions are indeed satisfied.

theories. In addition the local version offers a very plausible explanation for the origin of the scale of supersymmetry breaking consistent with the requirements of a solution to the hierarchy problem.

3.1 The hidden sector.

In SUSY models it is normally assumed that supersymmetry is broken in a sector of the theory, the so-called "hidden" sector, which has only gravitational interactions with the states of the standard model. Such a structure emerges quite naturally in superstring theories [8]. For example the original heterotic string generates an $E_8 \otimes E_6$ structure in which the E_8 plays the role of the "hidden sector". In general 4D string models the E_8 group is broken at the compactification scale to some subgroup which plays the role of the hidden sector. Apart from triggering supersymmetry breaking the hidden sector plays no role in low energy phenomenology for the states of this sector are confined with mass of order the gaugino condensate scale ($\approx 10^{13} Gev$).

The most plausible origin for supersymmetry breaking in the hidden sector follows because the gauge interactions of the hidden sector are likely to be asymptotically free and become strong at the condensation scale Λ_c, below the compactification scale. Then, in analogy with what happens in QCD, it is likely that a fermion condensate, $< \lambda\lambda >$, forms, the fermions, λ, being the gauginos of the hidden sector [9, 10, 11, 12, 13, 14]. In local supersymmetry (but not in global supersymmetry) such a condensate breaks supersymmetry. The characteristic measure of this breaking is the mass, $m_{3/2}$, it generates for the (spin- 3/2) gravitino, supersymmetric partner of the graviton and is given by $m_{\frac{3}{2}} \propto \frac{<\lambda\lambda>}{M_P^2}$. Since the scale determining the gaugino condensate is given by the scale, Λ_c, at which the gaugino binding becomes nonperturbative we have $m_{\frac{3}{2}} \propto \frac{\Lambda_c^3}{M_P^2} = M_P exp(-\frac{3}{2b_0 g^2})$, where b_0 is the coefficient of the one loop β-function and the last equality follows from using the running of the gauge coupling from its value, g, at the Planck scale [11, 15, 16]. Thus the condensation scale can be much less than the Planck scale and hence the gravitino mass may easily be hierarchically smaller than the Planck scale.

This is the best explanation I know of in supersymmetric models for the magnitude of the hierarchy. As we will see the presence of supersymmetry then guarantees that the supersymmetry breaking effects in the visible sector are also small, of order the gravitino mass. It may be seen that the supersymmetric solution to the hierarchy problem has copied the essential feature of technicolour theories, namely symmetry breaking via a condensate. However, as we will also discuss, in contrast with the extended technicolour theories the existence of elementary scalar states in supersymmetry subsequently allow for a straightforward pattern of symmetry breaking in the visible sector and generation of quark and lepton masses.

3.2 Gaugino condensation, the hierarchy and the fine structure constant.

In string theories the possibility that gaugino condensation generates an hierarchy of mass scales requires further analysis because, apparently, there are no free parameters in addition to the Planck scale which could be responsible for determining the magnitude of the hierarchy [8]. Thus the form for the gravitino mass given above, namely $m_{\frac{3}{2}} \propto M_P \exp(\frac{-3}{2b_0 g^2})$, is incomplete for the gauge coupling, g, should not be treated as a constant because it is really given by $g^2 = 1/Re(< S >)$ where $< S >$ is the vev of a scalar field, a moduli field, associated with the dilaton. We see that in string theories, in order to determine the magnitude of the hierarchy of mass scales, it is also necessary to determine the gauge coupling (i.e. the fine structure constant). This is just a manifestation of the fact that *all* parameters are determined in string theories and, as we will now discuss, the determination of the hierarchy of scales forces us to come to grips with the problem of realising this predictive power.

In order to determine the vevs of the moduli which, in a string theory, specify the couplings in the theory and to determine the vev of the gaugino condensate which determines the magnitude of the mass hierarchy it is necessary to compute the effective potential for the moduli and the gaugino bilinear and to determine its minimum. The potential determining the moduli is intimately related to that determining the condensate for it is known that, in the absence of supersymmetry breaking, the potential for the moduli is exactly flat or degenerate.

There has been a considerable theoretical effort in recent years seeking to determine the form of the effective potential following from 4D string theories and appropriate at scales below the mass scale of string excitations [17, 18, 19, 20, 21, 22]. For simple orbifold examples this effective potential is known but for more complicated schemes (in which the curvature in the compactified dimensions is not concentrated in isolated points) the solutions are not known completely. The work so far done on gaugino condensation is in the context of these simple models. A further problem in the way of a complete understanding of gaugino condensation is that it is a non-perturbative phenomenon and hence requires techniques to deal with the strong coupling behaviour of the theory.

A model calculation starting with a 4D-superstring theory built using orbifold compactification and estimating the non- perturbative effects using Nambu-Jona-Lasinio techniques illustrates how the superstring can *generate* the hierarchy of mass scales as well as determine the fine structure constant [23]. In this model calculation an effective potential with a stable minimum for the gaugino condensate and the moduli vevs was found once the stringy effects of the Kaluza Klein and winding modes are included. The latter are needed to stabilise the potential for small and large values of the moduli determining the compactification radii and correspond to the inclusion of 4D-string states which normally have Planck scale masses and become massless only in the large and small radius limits. In this parameter-free orbifold string model it is the multiplet content that determines the fine structure constant and the hierarchy. This is because the value, ¡S¿, of S at the minimum of the potential (which determines the gauge coupling) is given approximately by $8\pi\sqrt{(2/n_g)}$ where n_g is the number of gauge bosons in the hidden sector. Similarly, as we saw above, the hierarchy is determined by $< S >$ and b_0, both of which are fixed by the multiplet content. It is easy to construct multiplet structures which lead to reasonable numbers. For example [23] a $SU(5)$ theory with 6 chiral supermultiplets in the vector representation gives S=2.3, corresponding to a reasonable prediction for the fine structure constant at the compactification scale. The resultant gravitino mass is $m_{\frac{3}{2}} = 500GeV$, showing that a large hierarchy of mass scales may easily develop.

3.3 The visible sector.

Since supersymmetry breaking occurs in the hidden sector it can only be communicated to the visible sector via gravitational, flavour blind, interactions suggesting that there will be generated a common mass, $m_{\frac{1}{2}}$, for the gauginos of the standard model and another common mass, m_o, for the scalars (the expectation is that these are both of order $m_{3/2}$)[2]. In addition there must be a new term $\mu H_1 H_2$ in the superpotential giving a common

[2]Recently it has been observed that in string theories this universality of masses may be broken if the fields have different modular weights. Although possible the necessity to avoid large flavour changing neutral currents strongly constrains the amount of such non-universality and suggests that, in a viable model, flavour blind masses at the unification scale is a good approximation [24].

Higgsino, Higgs mass at the unification scale. It is also found in specific supergravity models [25] that there are additional supersymmetry breaking terms given by $(A_0P_3 + B_0P_2)$ where A_0 and B_0 are masses of order $m_{\frac{3}{2}}$ and P_3 and P_2 are the trilinear and quadratic terms of the superpotential with the supermultiplets replaced by their scalar components. The expectation for A_0 and B_0 depends on the metric; in the case of the no-scale theories which descend from the string A_0 vanishes at the unification scale. The gauge bosons and fermions of the Standard Model do not acquire mass at this stage due to residual unbroken gauge and chiral symmetries.

The flavour independence of the supersymmetry breaking terms is broken by radiative corrections involving the gauge and Yukawa couplings of the standard model. These corrections may be calculated explicitly and are most conveniently included via the renormalisation group equations for the masses [26] and the A and B parameters using for initial values at M_X the common gaugino and scalar masses $m_{\frac{1}{2}}$ and m_o and the common A_0 and B_0 parameters.

The only Yukawa coupling large enough to give a significant contribution to this evolution in the standard model is likely to be the one responsible for the top quark mass and this is the only one kept in the subsequent analysis. In fig 1 the resultant spectrum for the superparticle masses is shown for a representative choice of the supersymmetry breaking parameters [27]. It may be seen that those states with the larger gauge coupling are systematically heavier; the gluinos are heavier than the Winos and Bino and the squarks are heavier than the sleptons.

4 Unification predictions.

4.1 Unification of gauge couplings.

What are the implications of unification for the superstring- inspired SUSY spectrum discussed in the last section? Using this spectrum to determine when to change from the non-supersymmetric to the supersymmetric β functions we may repeat the analysis of the unification predictions [27, 28]. In Fig 2 we show the contours of constant α_3 which lead to unification. It may be seen that, allowing for the range $\alpha_3(M_Z) = 0.102 - 0.12$, values for the supersymmetry mass scale M_{SUSY} in the range $(10^2 - 10^5)GeV$ are allowed (I define M_{SUSY} to be $Min(m_0, m_{\frac{1}{2}}, \mu)$). Thus the effect of the non- degenerate spectrum is to increase the anticipated mass of the SUSY states (the bounds given above were for $\alpha_3 = 0.108 \pm 0.005$). The value obtained for the unification scale is $M_X \approx 3.10^{16} GeV$.

4.2 Unification of masses.

We have shown that supersymmetry together with unification relations between gauge couplings gives a simple and entirely consistent picture. However this is not all that can be said about supersymmetric unification. One of the main arguments for a low-energy supersymmetry was the need to explain the hierarchy of masses and in particular the difference between the Planck scale and the electroweak breaking scale. It is therefore reasonable to ask how the supersymmetric unification we have constructed explains this difference. Remarkably the theory contains a mechanism for symmetry breaking which automatically selects the correct breaking pattern, namely $SU(3) \otimes SU(2) \otimes U(1) \rightarrow SU(3) \otimes U(1)$. To see how this comes about we note that the potential describing the Higgs fields is largely determined by supersymmetry having the form [29]

$$V(H_1, H_2) = m_1^2 \mid H_1^0 \mid^2 + m_1^2 \mid H_2^0 \mid^2 + \mu B H_1^0 H_2^0 + \frac{\frac{3}{5}g_1^2 + g_2^2}{8}(\mid H_1^0 \mid^2 - \mid H_2^0 \mid^2)^2 \quad (8)$$

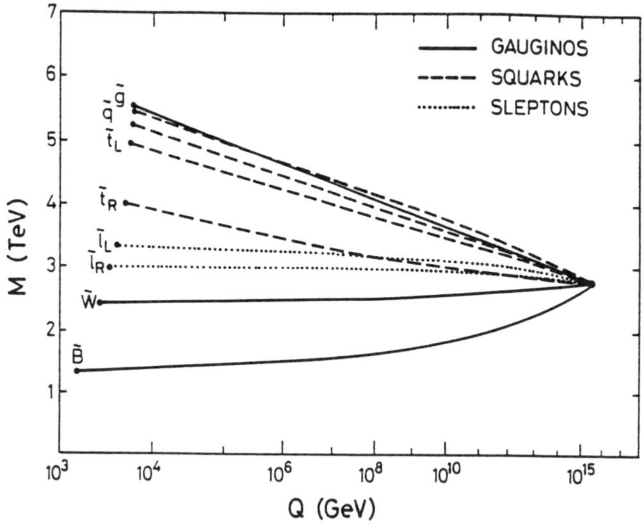

Figure 1: Running masses in the MSSM

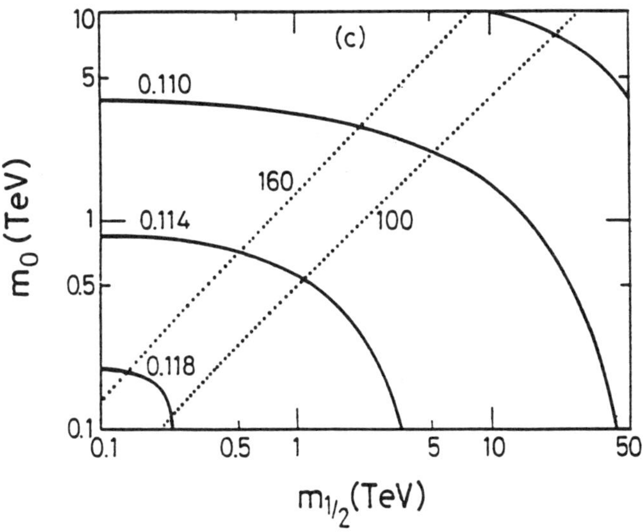

Figure 2: SUSY scale determination in minimal supersymmetric unification. The solid lines give the constant α_3 contours while the upper and lower dashed lines denote the $m_t = 160, 100 GeV$ contours respectively following from the requirement that M_Z be generated correctly.

Note that the coefficient of the quartic term is given in terms of gauge couplings by supersymmetry whereas it is an unknown parameter in the Standard Model. The masses of the two Higgs fields are given by

$$\begin{aligned} m_1^2 &= m_{H_1}^2 + \mu^2 \\ m_2^2 &= m_{H_2}^2 + \mu^2 \end{aligned} \tag{9}$$

with

$$m_{H_1}^2(M_X) = m_{H_2}^2(M_X) = m_0^2 \tag{10}$$

Using this we see that the Higgs mass terms are given via the RG equations in terms of the same mass parameters discussed above.The masses μ, m_1, m_2 evolve differently; for m_i we have [30]

$$\frac{dm_i^2}{dlog(Q)} = \frac{1}{8\pi^2}(-3M_2^2g_2^2 - \frac{3}{5}M_1^2g_1^2 + 3h^2\delta_{i2}(m_{\tilde{t}}^2 + m_{\tilde{t}^c}^2 + m_2^2 + A^2)) \tag{11}$$

where the M_i are the gaugino masses and the tilde denotes the supersymmetric state. Thus the difference between m_1, m_2 is due to the term involving the top Yukawa coupling. This term will drive m_2^2 negative if it is large enough, triggering electroweak breaking.

Now we can see why it is $SU(2)$ and not $SU(3)$ that is broken by this radiative breaking mechanism for, although the RG equations for both the Higgs scalars and the top (and bottom) squarks have destabilising terms due to the top Yukawa coupling only the squarks have large stabilising terms due to QCD interactions [31, 32, 33]. Thus it is the Higgs $(mass)^2$ that is driven negative, and it is $SU(2)$ that is broken by the resultant Higgs vev giving

$$\begin{aligned} \frac{v_2}{v_1} &= \omega = cot\theta \equiv tan\beta; (\beta = \frac{\pi}{2} - \theta) \\ m_3^2 &= B\mu \\ sin2\theta &= \frac{2B\mu}{m_1^2 + m_2^2} \end{aligned} \tag{12}$$

and the Z mass

$$M_Z^2 = \frac{(m_1^2 - m_2^2) - (m_1^2 + m_2^2)cos2\theta}{cos2\theta} \tag{13}$$

From eqs 11 and 13 we see that the initial value of h may be chosen to give the correct value of M_Z. Thus to each point of the solution plane of Fig 2 it is possible to assign a definite h (or equivalently m_t) needed to give the correct electroweak breaking scale. Using this the results of the analysis of electroweak breaking may be conveniently summarised by drawing contour plots of constant m_t in the $m_0, m_{\frac{1}{2}}$ plane [27]. These are shown by the dotted lines in Fig 2 for the case $m_t = 100, 160GeV$ from which it may be seen that part of the previously allowed region is excluded by the requirement of acceptable radiative electroweak breaking coupled with the LEP bounds on the top mass. In Fig 2 the origin of the variation with respect to m_0 is obvious for larger m_0 requires larger h to drive m_2^2 negative at the correct scale. The same applies to increasing the value of μ. It may be seen from Fig 2 that a heavy top quark leads to a very satisfactory explanation of both the existence of electroweak breaking and its magnitude for a wide range of supersymmetry breaking parameters.

This very pleasing self-consistency check of the MSSM shows how the electroweak breaking scale is generated from the unification scale, realising the original aim in introducing supersymmetry. It also allows us to investigate the hierarchy problem for, from eq 11, one may see the Higgs mass squared, and hence the electroweak breaking scale, is generated largely by the squark mass and has its natural value proportional to this mass. In this analysis this effect is manifested by the need to fine tune, for large squark masses, the top quark Yukawa coupling to make the constant of proportionality small enough to get the correct electroweak breaking scale.

4.3 The hierarchy problem and the SUSY spectrum.

The naturalness constraints may be made more precise by writing the relation between the W mass and the squark mass in a direct form. The renormalisation group equations are just a convenient method for summing the large logarithms $\propto log(M_x/M_W)$ that occur in calculating the radiative corrections to the Higgs mass. However these radiative corrections relate on-shell masses and so it is clear that the solution to the renormalisation group equations leads to relations relating on-shell quantities to on- shell quantities and free of uncertainties related to the definition of "running" masses. The result of calculating these radiative corrections to the Higgs mass and finally expressing the result in terms of the associated W mass is given in the equation

$$M_W^2 \approx \frac{3}{8\pi^2} h_t^2 (m_{\tilde{t}}^2 + m_{\tilde{t}^c}^2) \log(\frac{M_X^2}{m_{\tilde{t}}^2})[1 - \frac{8\pi^2}{6h_t^2 \log(\frac{M_X^2}{m_{\tilde{t}}^2})})] \tag{14}$$

No symmetry makes the term in square brackets small for, as may be seen, it involves supersymmetry *breaking* masses. Thus the expectation is that it is of order 1 and taking it to be much smaller corresponds to a fine-tuning". To quantify this we may rewrite eq 14 in the form

$$M_W^2 = c[1 - \frac{8\pi^2}{6h_t^2 \log(\frac{M_X^2}{m_{\tilde{t}}^2})})]M_W^2 \tag{15}$$

where

$$c = \frac{3h_t^2(m_{\tilde{t}}^2 + m_{\tilde{t}^c}^2)}{8\pi^2 M_W^2} \log(\frac{M_X^2}{m_{\tilde{t}}^2}) \tag{16}$$

In ref [34] "reasonable" scales for the supersymmetry thresholds were estimated by demanding that the sensitivity of the electroweak breaking scale to any of the parameters of the standard model should be less than some value. In eq 15 the sensitivity of M_W to the dominant supersymmetry breaking mass, the top squark mass, is determined by the quantity c and the equivalent statement is that $1/c$ should be less than some value.

No fine tuning would correspond to $c \approx 1$ but, more conservatively, we will use the value of $c = 10$ that was chosen by Barbieri and Giudice [34] as a measure of a reasonable theory. The requirement that the electroweak breaking scale be generated from the unification scale introduces more sensitivity to the top Yukawa coupling than is found in the MSSM without unification as may be seen from the appearance of the large logarithm in eq 14. As a result constraining $c \leq 10$ dramatically reduces the allowed region in Fig 2. This is shown in Fig 3 using, instead of the approximate form of eq 16, the value of c determined from the full renormalisation group equations.

We see that the condition $c \leq 10$ restricts us to the region $| \frac{\mu_0}{m_0} | \approx 1$, $\alpha_s = 0.118$ in the $m_0, m_{\frac{1}{2}}$ plane. The full SUSY spectra for each of the "extremes" labelled by Z and X in Fig(2) are listed in Table 2 including the (small) contributions to the masses coming from the spontaneous breaking of electoweak symmetry [1]. The neutralino and chargino masses are obtained by diagonalising the full mass matrices (cf. Section 5.3. The spectra are very strongly constrained and lie not far from the present experimental limits with all SUSY masses less than 1TeV and sleptons much less. of order 200GeV. Note that the

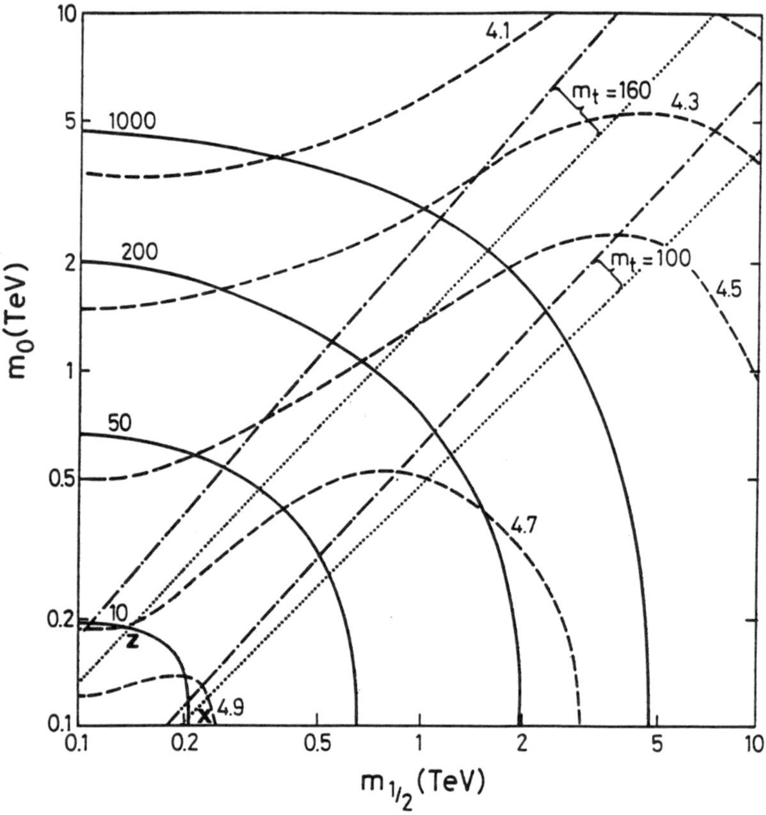

Figure 3: $m_0, m_{\frac{1}{2}}$ plot for $| \frac{\mu_0}{m_0} = 1 |$, $A_0 = 0$ in the low mass region, 100-300 Gev. The diagonal band (dotted) is the area constrained by $100 < m_t < 160GeV$ for $B_0 = 0$; the diagonal (dot-dashed) lines indicate how the band is modified for $| B_0 |= m_0$. Other contours (solid) indicate values of the fine tuning constant c (evaluated exactly) and (dashed) m_b (the bottom quark mass) using the full two-loop expressions for $\frac{m_b}{m_\tau}$. The contour c=10 here virtually coincides with the $\alpha_s = 0.118$ contour of Fig2.

	(a)	(b)
	Parameters	
$m_{1/2}$	140	230
m_0	190	120
μ_0	190	-120
m_t	160	100
$\tan\beta$	21	5
	Gauginos	
$\tilde{\gamma}$	57	83
$\tilde{Z}$; $\tilde{W}$	99; 99	120; 112
$\tilde{g}$	354	559
	Sleptons	
$\tilde{\ell}_L$	220	206
$\tilde{\ell}_R$	195	146
	Squarks	
$\tilde{u}_L, \tilde{c}_L$; $\tilde{d}_L, \tilde{s}_L$	365; 373	511; 517
$\tilde{u}^c_L, \tilde{c}^c_R$	359	495
$\tilde{d}^c_R, \tilde{s}^c_R, \tilde{b}_R$	358	491
$\tilde{t}_L$; $\tilde{b}_L$	325; 335	491; 497
$\tilde{t}^c_R$	273	452
	Higgs, Higgsinos	
H^0	91, 264	84,221
$H^\pm$; A^0	276; 264	232; 218
$\tilde{H}^0$	205, 225	139, 226
$\tilde{H}^\pm$	229	227

Table 2: Masses of the supersymmetric states for the two solutions (Z and X of Fig(4)) with $\alpha_s(M_z) = 0.118$, (a) with $m_t = 160$ GeV, (b) with $m_t = 100$ GeV. Here $|\mu_0/m_0|$=1 and $B_{\bar{0}} = A_0 = 0$. The soft supersymmetry breaking contributions have been corrected by including electroweak symmetry breaking terms. The masses (which are in GeV units) of the two solutions represent the range of values of our predictions. The light Higgs does not have the quartic corrections included.

fine tuning constraint is largely independent of the details of the unification at M_X and may be expected to give similar limits on the SUSY spectrum in any unification scheme having a very large M_X.

4.4 String Unification

In string theories the constraints of Unification are much more severe than in GUTs. A given four dimensional string theory has a definite multiplet structure and, moreover, the unification scale, M_{SU}, is *known* in terms of the Planck scale. The one loop running gauge coupling is given by [35, 36, 37, 38]

$$\alpha_i^{-1} = k_i \alpha_{SU}^{-1} + \beta_{0i} ln(\frac{M_{SU}^2}{\mu^2}) + \Delta_i \quad (17)$$

Here k_i are the Kac-Moody levels of the $U(1)$, $SU(2)$ and $SU(3)$ factors and Δ_i are threshold corrections from loops involving massive (Kaluza-Klein) modes with mass of order the compactification scale. The tree level gauge couplings are related by

$$g_1^2 k_1 = g_2^2 k_2 = g_3^2 k_3 = \frac{4\pi}{\alpha'} G_{Newton} \equiv g_{SU}^2 \quad (18)$$

and α' is the inverse of the string tension squared. The values $k_3 = k_2 = (3/5)k_1$ give the standard SU(5)- like predictions for the couplings (C=1 in eqs. 6 and 7). They arise in many 4D string models, both with Grand Unified gauge groups and in models which do not have a stage of Grand Unification. The string unification scale is (in the $\overline{MS}$ scheme is given by $M_{SU} = 0.7 g_{SU} 10^{18} GeV$. The simplest realisation of minimal string unification is for the gauge group after compactification to be just $SU(3) \otimes SU(2) \otimes U(1)$ with the minimal particle content *i.e.* there is no need for any additional heavy states. While no example of such a string theory has yet been constructed it seems likely that something quite close to it can be found. However the result found in the Section 4, $M_X \approx 3.10^{16} GeV$ is difficult to satisfy for a survey [37, 24] of all orbifold models shows that typically M_X is *increased* by the string threshold corrections Δ_i from the string unification scale M_{SU}. Only for very special assignments of the matter fields to (twisted) multiplets can the unification scale be reduced.

What then are the implications of coupling unification for string theories? The first possibility is that the threshold corrections *do* conspire to lower M_X. At least the existence of an orbifold example shows this is possible [37]. A second possibility is that the string gives a gauge group larger than the standard model which preserves the ratios of gauge couplings and which subsequently breaks at an intermediate scale $M_X \approx 10^{16} GeV$ to the standard model. This group need not be Grand Unified. My favourite example is the group $SU(3)^3$ which is known to occur in some promising compactification schemes. Provided there are equal numbers of quark and lepton multiplets the couplings renormalise in the same way between M_{SU} and M_X as required. Another possibility is to give up the minimal unification assumption and to add states beyond the minimal set. For example in flipped $SU(5)$ it is necessary to evolve the $SU(5) \otimes U(1)$ couplings from M_{SU} to M_X before using the $SU(3) \otimes SU(2) \otimes U(1)$ evolution. Unfortunately this spoils the agreement with the low-energy couplings in the original version of the theory but in non- minimal versions the effect of states with masses of order $3.10^{13} Gev$ can lead to acceptable results [36]. This example nicely illustrates both the additional predictive power of the string theory and the uncertainties introduced once non- minimal theories are introduced.

To summarise, the success of minimal unification predictions with $M_X \approx 3.10^{16} GeV$ can be understood within the framework of specific string models. However since these models do not correspond to minimal string unification some of the aesthetic appeal of simplicity is lost.

5 Phenomenology of the MSSM

As we have seen the spectrum of supersymmetric states is constrained by the hierarchy problem to be quite low and accessible to experimental detection. The effects of supersymmetry may be detected either by their virtual effects on processes involving just the standard model states, or by direct production of the new states. We start with a discussion of the most important virtual effects of SUSY that have been identified.

5.1 New flavour changing gauge interactions.

The discussion of gauge related interactions given in section 2.1 applied to the current eigenstates. Due to the non-degeneracy of quarks we know that the charged weak gauge interactions can change flavour when expressed in terms of quark mass eigenstates. There is an equivalent source of flavour-changing interactions in the new squark and slepton sector which we will now discuss [1].

The connection between the current quark (or lepton) basis and the mass eigenstate quark (or lepton) basis is given by unitary transformations, $\mathbf{V}^{u,d}_{L,R}$. In terms of these the Cabibbo, Kobayashi, Maskawa mixing matrix is $\mathbf{U}_{CKM} = \mathbf{V}^{u\dagger}_{L}\mathbf{V}^{d}_{L}$. In general we have to define *different* unitary rotations $\tilde{\mathbf{V}}^{u,d}_{L,R}$ diagonalising the squarks. Thus the interaction of the W bosons with squarks will be describes by a new matrix $\tilde{\mathbf{U}}_{CKM} = \tilde{\mathbf{V}}^{u\dagger}_{L}\tilde{\mathbf{V}}^{d}_{L}$. Perhaps more significant is the immediate implication that SUSY theories have flavour changing *neutral* currents (FCNC). For example the gluino-squark-quark vertex has interactions given by $\mathbf{U}^{d\tilde{d}}_{CKM} = \mathbf{V}^{d\dagger}_{L,R}\mathbf{V}^{\tilde{d}}_{L,R}$ and $\mathbf{U}^{u\tilde{u}}_{CKM} = \mathbf{V}^{u\dagger}_{L,R}\mathbf{V}^{\tilde{u}}_{L,R}$ for the down and the up sectors respectively. If SUSY were exact $\mathbf{U}^{d\tilde{d}}_{CKM} = \mathbf{U}^{u\tilde{u}}_{CKM}$ and the FCNC would vanish. However, SUSY is not exact and this equality is broken at some level. The expectation for the resulting mixing matrices depends sensitively on the structure of the model at high scales, up to the Grand Unified or Planck mass. In the minimal unification scheme the expectation is quite simple namely $\mathbf{V}^{\tilde{u}}_{L,R} = \mathbf{V}^{u}_{L,R}$, $\mathbf{V}^{\tilde{d}}_{R} = \mathbf{V}^{d}_{R}$ and $\mathbf{V}^{\tilde{d}}_{L} = \mathbf{V}^{u}_{L}$ [39]. The reason for this asymmetrical result is that, due to the large top Yukawa coupling, there are significant radiative corrections to the down squark mass matrix proportional to the up quark mass matrix. These dominate over the original (non-radiative) supersymmetric contribution which is just given by the down quark mass matrix. The radiative contribution is diagonalised by the same rotation that diagonalises the up-quark mass matrix and hence $\mathbf{V}^{\tilde{d}}_{L} = \mathbf{V}^{u}_{L}$. Using this gives $\mathbf{U}^{u\tilde{u}}_{CKM} = 1$ and $\mathbf{U}^{d\tilde{d}}_{CKM} = \mathbf{U}_{CKM}$. Non-minimal models may give both matrices non-vanishing but the expectation is still [40] that they have the same order of magnitude as $\mathbf{U}_{CKM}$.

Although these effects introduce new sources of FCNC, it turns out that for SUSY masses in the range expected these do not lead to unacceptably large effects. For example, the box diagram of Fig 4 gives a contribution to the $K_L - K_S$ mass difference due to gluino FCNC. Since the gaugino has QCD strength couplings it may be expected that this is the dominant contribution. However in the minimal unification scheme discussed in Section 4 the gluinos and squarks are the heaviest states due to their QCD interactions and their large mass suppresses this contribution. Using $\mathbf{U}^{d\tilde{d}}_{CKM} = \mathbf{U}_{CKM}$, the limit from the observed $K_L - K_S$ mass difference is [41], for the dominant contribution coming from the first two generations,

$$\frac{m^2_{\tilde{d}} - m^2_{\tilde{s}}}{(Max(m^2_{\tilde{u}}, m^2_{\tilde{g}}))^2} \le \frac{10^{-4}}{M^2_W} \tag{19}$$

Since the $\tilde{d}, \tilde{s}$ mass difference comes from the quark mass difference in the first two generations this bound is satisfied for squark, gluino masses of more than 30 Gev and lies

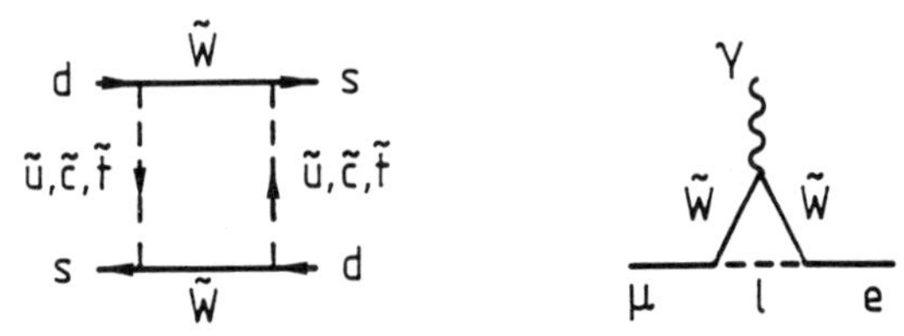

Figure 4: Graphs generating flavour changing processes in supersymmetry.

much below the observed value for the spectrum of Table 2. Similar conclusions follow for the Wino contribution of Fig 4.

Finally in the slepton sector there are new lepton number violating processes coming from the fact that the sneutrinos, unlike the neutrinos, are massive and need to be diagonalized when determining the structure of the charged currents. The graph of Fig 4 generates the process $\mu \to e\gamma$ and, imposing the experimental limits found for this process, gives the bonds [41] (assuming mixing angles of the same order as the CKM angles)

$$\frac{m_{\tilde{e}}^2 - m_{\tilde{\mu}}^2}{(Max(m_{\tilde{\mu}}^2, m_{\tilde{W}}^2))^2} \leq \frac{10^{-3}}{M_W^2} \tag{20}$$

This gives a similar bound to that found for the gluino but, given the minimal model expectation for the mass spectrum, is a more severe limit. However, the choice of mixing angle used in this bound is quite arbitrary so the only reasonable message to be drawn from the calculation is that lepton number violation in the MSSM may easily be within the experimental bounds.

5.2 New sources of CP violation in SUSY

In the standard model the source of CP violation is the complex nature of the CKM matrix giving the CKM phase $\delta = Arg det(U_{CKM})$. (The effect of T reversal on a constant, C, is $TCT_{-1} = C^*$, so complex couplings act as a source of T or CP violation.)

In the MSSM there are additional complex couplings which may give rise to CP violation [42]. With the minimal unification scheme of section 4 there are two further complex couplings giving rise to CP violating phases, namely $\phi_A = Arg(A m_{1/2}^*)$ and $\phi_B = Arg(B m_{1/2}^*)$ [42]. Finally there is the effective theta parameter $\overline{\theta}$ specifying the strong CP violation. In this case is given by $\overline{\theta} = \theta - Arg(det \lambda_u \lambda_d) - 3Arg(m_{1/2})$ where the λs are the matrices of Yukawa couplings.

These new sources of CP violation contribute to CP violation in various processes, the neutron and electron dipole electric moments, rare Kaon decays, rare B decays etc. The most sensitive of these to the SUSY CP violation effects turns out to be the dipole electric moments. The analysis of these moments starts with the $\Delta S = 0$ CP violating Hamiltonian, written in terms of the elementary quark and gluon degrees of freedom as

$$H_{CPV} = \Sigma_i C_I(\overline{g}, \mu) O_i(\mu) \tag{21}$$

where the O_i are the quark/gluon operators and $C_i(\overline{g}, \mu)$ the coefficients expresses in terms of the running strong coupling, which takes account of large radiative corrections, and the renormalisation scale, μ. It was pointed out by Weinberg [43] that a significant contribution comes from the gluonic operator

$$O_g = f^{abc} G_a G_b \tilde{G}_c \tag{22}$$

where $G_a^{\mu\nu}$ is the gluon field strength, $\tilde{G}_a^{\alpha\beta} = \epsilon^{\mu\nu\alpha\beta} G_{a,\mu\nu}$ and the Lorentz indices have been suppressed in eq 22. Subsequently two further operators were indentified as significant [44], the colour dipole operator

$$O_q = \overline{q} \sigma_{\mu\nu} q T^a G_a^{\mu\nu} \tag{23}$$

and the electric dipole operator

$$O_\gamma = \overline{q} \sigma_{\mu\nu} q F^{\mu\nu} \tag{24}$$

In SUSY these operators are all generated. For example the coefficient C_q of the colour dipole operator is

$$C_q = \frac{2 A m_u g_3^2}{4\pi m_{\tilde{g}}^2 sin(\phi)} \tag{25}$$

where ϕ gets contributions from both ϕ_A and ϕ_B. The net contribution is estimated to be [44, 45] $d_n^{Tot} \approx 10^{-22} sin(\phi) ecm$ which, when compared to the experimental bound $| d_n | < 10^{-25} ecm$ requires $sin(\phi) < 10^{-3}$. This is a strong constraint on the magnitude of $Arg(A m_{1/2}^*)$ and $Arg(B m_{1/2}^*)$ and the question arises how natural is such a value in the context of a SUSY model? The expectation for many supergravity/string models is that these phases are very small for A and B vanish at the Planck scale (If both A and B were related to trilinear terms in the superpotential they vanish in the so-called no-scale models). Although radiative corrections will still generate a non-zero value for $Arg(A m_{1/2}^*$ at low scales, an estimate of these effects shows it is a very small of $O(10^{-11})$ [47] and gives a neglible contribution to d_n. A similar result has been found in the case of the electron dipole electric moment [48]. The conclusion is that these dipole moments are quite sensitive to the new sources of CP violation in SUSY models. At least in an interesting subject of these models, however, these new sources are expected to be very small.

5.3 Superparticle searches in the MSSM

The spin-1/2 sector has gluinos with mass $m_{\tilde{g}} = m_{1/2}$ at the unification scale radiatively corrected as in Fig 2. The charginos $\tilde{W}^\pm, \tilde{H}^\pm$ mix via the 2×2 mass matrix

$$\begin{pmatrix} M_2 & \sqrt{2} M_W \sin\beta \\ \sqrt{2} M_W \cos\beta & \mu \end{pmatrix} \tag{26}$$

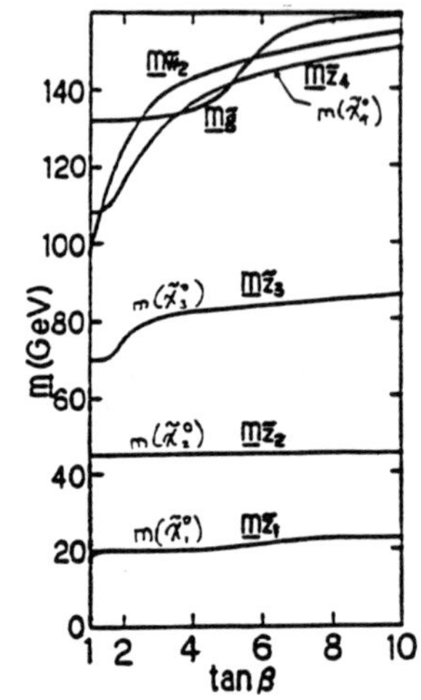

Figure 5: Bounds on neutralino masses.

and the neutralinos $(\tilde{B}, \tilde{W}_3, \tilde{H}_1^o, \tilde{H}_2^o)$ mix via the 4×4 matrix

$$\begin{pmatrix} M_1 & 0 & -M_Z \cos\beta \sin\theta_W & M_Z \sin\beta \sin\theta_W \\ 0 & M_2 & M_Z \cos\beta \cos\theta_W & -M_Z \sin\beta \cos\theta_W \\ -M_Z \cos\beta \sin\theta_W & M_Z \cos\beta \cos\theta_W & 0 & -\mu \\ M_Z \sin\beta \sin\theta_W & -M_Z \sin\beta \cos\theta_W & -\mu & 0 \end{pmatrix} \quad (27)$$

where θ_W is the weak angle, $M_{1,2}$ ($=m_{1/2}$ at the unification scale) are the triplet and singlet neutralino supersymmetry breaking masses (cf Fig 2) and $\tan\beta = v_2/v_1$, the ratio of the Higgs vevs.

In the MSSM the LSP, $\tilde{\Lambda}$. is stable and its character determines much of the phenomenology of supersymmetry. As John Ellis has discussed, a stable LSP should be neutral otherwise. given the expected initial abundance from cosmological origins. it should have been found in nuclear matter. Thus we expect the LSP to be either a neutralino or a sneutrino and. given the favoured spectrum discussed in section 4, it is more likely to be a neutralino. It is often assumed that the LSP following from diagonalizing the matrix 27 is the photino. $\tilde{\gamma}$, but this need not be the case. In the limit $M_1, M_2 \to 0$ (i.e.$m_{1/2}/m_0 << 1$) the photino is indeed the LSP but in the limit $\mu \to 0$ the LSP is the Higgsino. More generally one should allow for general mixing. denoting the mass eigenstates as χ_i^o.

5.3.1 Bounds on SUSY states.

- For weakly interacting sparticles the bounds are dominated by LEP (c.f. John Ellis' talk) following from the decay of the Z into a pair of sparticles. Thus $M_y \geq M_Z/2$ for $y = \tilde{l}, \tilde{H}, \tilde{\nu}, \tilde{q}, \tilde{q}, \tilde{W}$. For neutralinos one may still have lighter sparticles provided they couple less strongly to the Z. The situation is summarised in Figure 5 [49].

- For strongly interacting sparticles. $\tilde{g}$ and $\tilde{q}$; the bounds are dominated by the hadronic colliders and. at the $90 m_{\tilde{q}} > 150 GeV$. for $m_{\tilde{g}} > m_{\tilde{q}}$ or $m_{\tilde{g}} > 100 GeV$. both at the 90% confidence level.

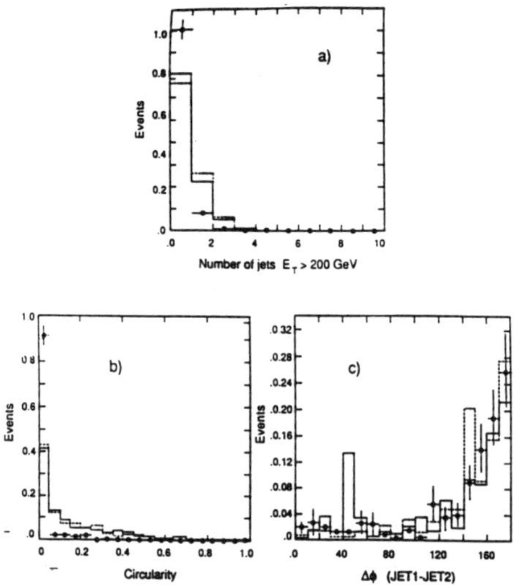

Figure 6: Distribution of a) $N_{jet} \geq 3$ for $E_T^{jet} > 200$ GeV, b) circularity C, and c) $\Delta\phi_{12}$ from ISAJET for gluino pair production of 300 GeV mass for tan $\beta = 2$ (solid histogram) and tan $\beta = 10$ (dashed histogram), compared with Standard Model background (points with error bars). $E_T^{miss} > 100$ GeV has been required and the distributions are normalized to 1.

To illustrate how these bounds are established and the prospects for finding strongly interacting sparticles at the SSC and the LHC I will discuss a case study done for the LHC of gluino pair production and decay [47]. The basic process is

$$\begin{aligned} pp &\rightarrow \tilde{g}\tilde{g} \\ &\rightarrow (q\bar{q}\tilde{\Lambda})(q\bar{q}\tilde{\Lambda}) \equiv (E_T)_{Missing} + n\ jet \end{aligned}$$

However if, as is expected from the radiatively corrected mass spectrum, Table 1, the gluinos are heavier than several other superparticles, cascade decays become relatively important compared to the direct decay to the LSP, shown in eq 28. Such decays soften the $(E_T)_{Missing}$ which is the characteristic signal of this process. Particularly at higher gluino masses, the cascade decays are very important. Using these various decay patterns the cross section has been determined via Monte Carlo simulations and the results are shown in Figs 6 for various cuts. It may be seen that the standard model background, coming principally from $t\bar{t}$ decays, is substantial. However cuts, which rely on the different dependence on azimuthal angle of jets or on transverse energy distribution are able to distinguish the signal from the background for gluinos with mass up to 1 TeV.

The cascade decays also open up the possibility of signals which do not rely on $(E_T)_{Missing}$. For example the process

$$\tilde{g}\tilde{g} \rightarrow Z + Z + \tilde{X} \rightarrow 2(l^+l^-) + \tilde{X} \tag{28}$$

was also studied [47]. Although the rates are quite small (some 30 events for a 750 Gev gluino after imposing an $((E_T)_{Missing}$ cut which eliminates the background) they provide a very clear signal, which would be a good second generation experiment of the $(E_T)_{Missing}$ search provides evidence for gluinos. The results of the Monte Carlo study of this process may be found in the LHC study [47].

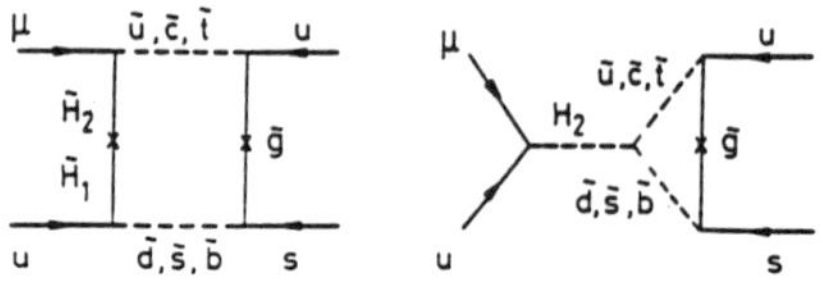

Figure 7: Graph giving proton decay in R-parity broken models

6 Not the MSSM

The direct supersymmetry searches rely largely on the existence of a stable LSP giving rise to $(E_T)_{Missing}$ signals. This in turn follows from R-parity. Given the importance of R-parity for phenomenology there is an obvious question - is R-parity inevitable?

The answer is no! R-parity may be broken spontaneously through a sneutrino acquiring a vacuum expectation value [48, 50]. If this happens there is a Majoron associated with the spontaneous breakdown of lepton number. LEP bounds on the Z width rule out a pure $I_W = \frac{1}{2}$ Majoron so viable models require spontaneous breakdown principally through "right handed" isosinglet sneutrinos [51][3].

A second possibility is that R-parity is explicitly broken. As mentioned in section 2.1 the couplings of eq 2 are not the only ones allowed by $SU(3)\otimes SU(2)\otimes U(1)$. The following terms are also allowed in the superpotential [52, 53]

$$[\lambda_{ijk} l_i l_j \bar{e}_k + \lambda'_{ijk} l_i q_j \bar{d}_k + \lambda''_{ijk} \bar{u}_i \bar{d}_j \bar{d}_k]_F \tag{29}$$

As before each of these terms give three Yukawa couplings; for example the first term gives

$$\lambda_{ijk}(l_i l_j \tilde{e}_k + l_i \tilde{l}_j \bar{e}_k + \tilde{l}_i l_j \bar{e}_k) \tag{30}$$

It may be seen that these involve a *single* SUSY state and thus break the R-parity of the standard model. I will discuss in the next sections whether such terms may be expected.

6.1 Explicit R-parity violation

The first point to note is that if the terms proportional to λ and λ' are simultaneously present the nucleon is unstable through the graph of Fig 7. Since the amplitude is proportional to $(\lambda\lambda')/m^2_{\tilde{d}}$, and the squark mass is at most in the TeV range, the decay rate is quite unacceptably fast. The cure is to forbid these new terms by a discrete symmetry, known as matter parity under which the quark and lepton superfields appearing in the superpotential change sign while the Higgs superfields are left invariant. Thus the last three terms of eq 29 change sign under this symmetry and are forbidden while the terms of eq 2 are invariant and allowed [54, 55].

Clearly this forbids all the terms of eq 29 and leaves just the couplings of the MSSM. Thus the matter parity leads to the R-parity of the MSSM. However, there are more possibilities to stabilize the proton than to forbid all the terms of eq 29 [52]. Provided the

[3] It is necessary that the minimum develops along the stau direction to avoid e- or μ- violation. It is not obvious why this should happen.

terms proportional to λ and λ' are not simultaneously present the graph of Fig 7 will be absent. It is possible to eliminate one or other of these operators by symmetries other than matter parity provided one allows for the possibility that quarks and leptons transform differently [56]. Although this is not possible if the theory is embedded in SU(5), in which the quarks and leptons transform under discrete symmetries in the same way, it is possible in other GUT's and also in string unification, which need not be embedded in a GUT. Indeed a study of all Z_N symmetries (N<5) shows that it is easy to obtain any of the following, all of which inhibit nucleon decay[58]

- Matter parity. $\tilde{\lambda} = \tilde{\lambda'} = \tilde{\lambda''} = 0$; $\Delta B = \Delta L = 0$
- Lepton "parity". $\tilde{\lambda'} = \tilde{\lambda''} = 0$; $\Delta B \neq 0, \Delta L = 0$
- Baryon "parity".$\tilde{\lambda} = 0$; $\Delta B = 0, \Delta L \neq 0$

It is clear there are many possibilities, indeed allowing for flavour dependence there are 45 different operators possible in eq 29. Obviously it would be useful to limit the possible new terms and various constraints have been considered. An interesting constraint comes from baryogenesis because the addition of any new interaction violating baryon number may, if this interaction is ever in thermal equilibrium, erase a pre-existing baryon asymmetry. A similar result follows if one adds one of the lepton number violating operators. In this case baryon number may be erased through a combination of sphaleron effects which convert baryon number to lepton number and the R- parity violating operator. Assuming all of the flavours of operator occur with the same strength it has been shown that their coefficients are less than 10^{-7} [59]. At this level the LSP will not decay within the apparatus and then the phenomenology will not be different from the MSSM. However if one allows for a significant flavour dependence of the couplings (as might be expected since they are like Yukawa couplings which show significant variation) the baryogenesis bounds are weakened and do not rule out R-parity violating terms with large couplings [60].

A second constraint on the possible magnitude of the R-parity violating couplings is via the radiative corrections to SM processes that would be induced by the new couplings. These constraints are summarised in Tabl 3. From this Table it may be seen that for most of the flavour structures the limits are not severe and that the addition of R-parity violating operators could change the phenomenology of the new supersymmetric states.

The third constraint on R-parity violating operators that has been considered follows from more theoretical considerations. It has been argued that the discrete symmetries which give rise to some allowed set of such operators must be "discrete gauge symmetries" if they are to survive large gravitational corrections. Discrete gauge symmetries are symmetries that remain after breaking a continuous gauge symmetry and, as I discuss in the next section, are constrained by the requirement that the original gauge symmetry be anomaly free. The constraints on discrete symmetry following from such anomaly cancellation appear also to be necessary in string theories even though the origin of such symmetries is not obviously a gauge symmetry. As such the constraints following from the requirement of discrete anomaly cancellation are more general than that following from a pattern of spontaneous breaking of a gauge symmetry. With this motivation I turn to a discussion of these conditions.

6.2 Discrete Gauge Symmetries

Remarkably one can show that many discrete symmetries cannot be discrete gauge symmetries from the transformation properties of the light states only [61]. These are the states which do not get a mass when the continuous gauge symmetry is broken leaving a discrete gauge factor. The origin of these constraints is simply illustrated from the $U(1)^3$ anomaly for the case a U(1) symmetry is broken to Z_N.

Consider the (LH) fermion fields, ψ_i, which have discrete charge q_i ($\psi_i \rightarrow \exp(i2\pi q_i/N)\psi_i$). The $U(1)^3$ anomaly vanishes provided

$$\Sigma_{i=Heavy+light}(q_i + m_i N)^3 = \Sigma_{i=Heavy+light} q_i^3 + m'N = 0 \tag{31}$$

where $m_i, m' \epsilon Z$.

The appearance of the $m_i N$ terms occurs because the charge under the continuous gauge group is only determined by the discrete charge modulo N. We have also summed over all fermions, both heavy and light.

Heavy states have constrained Z_N charges because the states may only acquire mass via Z_N invariant mass terms in the Lagrangian. For fermions ψ^i acquiring a Dirac mass via the term $M\psi_1^j\psi_2^j$, where the ψs are Weyl fermions, we have the constraint

$$Q_1^j + Q_2^j = p_j N, \; p_j \epsilon Z \tag{32}$$

which implies

$$\begin{aligned} (Q_1^j)^3 + (Q_2^j)^3 &= p_j N(3(Q_1^j)^2 - 3p_j N Q_1^j + p_j^2 N^2) \\ &\equiv p_j' N, \; p_j' \epsilon Z \end{aligned}$$

For a Majorana mass term $M\psi_1^k\psi_1^k$, we have the constraint

$$Q_1^j = \frac{N}{2} p_j'', \; p_j'' \epsilon Z \tag{33}$$

which implies

$$(Q_1^j)^3 = \frac{N^3}{8} p_j'''^3 \tag{34}$$

Using these results in eq(eq:u1) gives, for the *light* states only,

$$\Sigma_{i=Light}(q_i)^3 = mN + \eta n \frac{N^3}{8}, \; m, n \epsilon Z \tag{35}$$

where $\eta = 0, 1$ for N even, odd respectively.

This is the first (Z_N^3) discrete gauge anomaly cancellation condition. Similarly one may derive the mixed Z_N-gravitational anomaly cancellation condition

$$\Sigma_{i=Light} q_i = r'N + \eta s' \frac{N}{2} \tag{36}$$

and the $Z_N - SU(M) - SU(M)$ anomaly cancellation condition

$$\Sigma_{i=Light} T q_i = \frac{1}{2} r'' N \tag{37}$$

Given the light multiplet contant these conditions may be used to identify those discrete symmetries that may be discrete gauge symmetries [61].

Applying these constraints to the Z_N symmetries ($N \leq 4$) considered in Section 6.1 which are capable of inhibiting nucleon decay one may immediately determine which may be discrete gauge symmetries with the minimal light multiplet content of Table 1. Remarkably, of the large number of candidate symmetries, only two remain [58], namely

- Matter parity, Z_2 (R-parity)

- Baryon parity, Z_3 ($\lambda = 0, \Delta L \neq 0, \Delta B = 0$)

The first is the usual matter parity of the MSSM. The second acts on the quarks and leptons as follows

$$g(Q,u,d,L,e) = (1,\alpha^2,\alpha,\alpha^2,\alpha^2)$$
$$\psi_i \rightarrow g_i\psi_i;\alpha = exp(i\frac{2\pi}{3}) \tag{38}$$

Both these symmetries forbid the dangerous dimension-4 nucleon decay processes (cf. Fig 7). However at higher order there are further operators which can mediate proton decay. At dimension-5 the operators $[QQQL]_F$ and $[udde]_F$ give proton decay amplitudes suppressed only by a single power of M, where M is the scale of physics beyond the standard model. Even taking M to be the Planck scale these operators would generate too rapid proton decay so additional suppression is necessary. The simplest way to ensure this is to forbid these operators by a symmetry. It may be seen that matter parity allows these dangerous operators while the new baryon parity forbids these operators. For this reason I think that at the very least it should be considered on a par with the MSSM as a minimal supersymmetric extension of the standard model.

6.3 Phenomenology of broken R-parity.

In the direct search for supersymmetric particles the phenomenology is altered considerably when including R-parity-violating terms in the superpotential. In general both the production mechanisms and the decay patterns can change. Other than the standard supersymmetric pair production of particles there is now the possibility of production of an R-parity-odd final state as well. Also, if all supersymmetric particles decay in the detector, we will no longer have the standard $\not{p}_T$ signal and the decay patterns will all be altered, sometimes very dramatically.

Since we must allow for the most general family structure in the operators of Eq. (29) there are a very large number ($2^{45}-1$) of possible R-parity-violating schemes! In order to be able to make a systematic study it is necessary to introduce some notation and to make some reasonable simplifying assumptions [62]. We shall refer to the superpotential terms in Eq. (29) as operators and denote them by $\mathcal{O}$. Furthermore we shall denote the R-parity-even particle content of an operator by $\mathcal{O}_\mathcal{E}$. Thus, for example, for the operator $\mathcal{O} = [L_1L_2\overline{E}_3]_F$ we write

$$\mathcal{O}_\mathcal{E} = \{e^-\nu_\mu\tau^+, \nu_e\mu^-\tau^+\}, \quad \overline{\mathcal{O}}_\mathcal{E} = \{e^+\overline{\nu}_\mu\tau^-, \overline{\nu}_e\mu^+\tau^-\}. \tag{39}$$

The R-parity violating terms allow for decay of the LSP. For example the photino ($\tilde{\gamma}$) decay can be written as $\tilde{\gamma} \rightarrow \mathcal{O}_\mathcal{E}$, $\tilde{\gamma} \rightarrow \overline{\mathcal{O}}_\mathcal{E}$ respectively. Since the photino is a Majorana fermion it can decay with equal probability to both the operator $\mathcal{O}_\mathcal{E}$ and to the hermitean conjugate operator $\overline{\mathcal{O}}_\mathcal{E}$. As we will see below, this is very important in producing exotic signals of charged leptons.

A similar 3-body decay of the chargino via the same operator, for example $\tilde{\chi}^\pm \rightarrow l^\pm\tilde{\nu} \rightarrow l^\pm + \mathcal{O}_\mathcal{E}(\tilde{\nu})$, results in the final states $\tilde{\chi}^+ \rightarrow = \{e^+\mu^+\tau^-, \nu_e\nu_\mu\tau^+\}$, and $\tilde{\chi}^- \rightarrow = \{e^-\mu^-\tau^+, \overline{\nu}_e\overline{\nu}_\mu\tau^-\}$ We collectively denote these by $\mathcal{O}_\mathcal{W}$ to indicate that one of the final states of $\mathcal{O}_\mathcal{E}$ has been replaced by its $SU(2)$ doublet partner.

A scalar electron ($\tilde{e}$) can decay directly via the operator $\mathcal{O} = [L_1L_2\overline{E}_3]_F$: $\tilde{e}^- \rightarrow \tau^-\overline{\nu}_\mu$; we collectively denote such R-parity violating 2-body decays of a scalar electron as: $\tilde{e} \rightarrow \mathcal{O}_\mathcal{E}(\tilde{e})$, and analogously for other particles and operators.

The simplifying assumptions are:

$\mathbf{LL\overline{E}}$		$\mathbf{LQ\overline{D}}$		$\mathbf{\overline{UDD}}$	
ijk	$\lambda_{ijk} <$	ijk	$\lambda'_{ijk} <$	ijk	$\lambda''_{ijk} <$
121	$0.04^{(a)}$	111	$0.01^{(f)}$	112	$10^{-9(h)}$
122	$0.04^{(a)}$	112	$0.03^{(a)X}$		
123	$0.04^{(a)}$	113	$0.03^{(a)X}$		
131	$0.10^{(b)}$	121	$0.26^{(d)}$		
132	$0.10^{(b)(c)X}$	122	$0.45^{(c)}$		
133	$0.01^{(g)}$	123	$0.26^{(c)}$		
231	$0.09^{(b)X}$	131	$0.26^{(d)}$		
232	$0.09^{(b)X}$	133	$0.01^{(g)}$		
233	$0.09^{(b)X}$	211	$0.09^{(b)}$		
		212	$0.09^{(b)}$		
		213	$0.09^{(b)}$		
		221	$0.22^{(e)X}$		
		231	$0.22^{(e)X}$		

Table 3: Limits on the R-parity-violating parameters from the physical processes: (a) charged current universality; (b) e, μ, τ universality; (c) forward backward asymmetry; (d) atomic parity violation; (e) ν_μ deep inelastic scattering; (all from ref.) (f)Neutrinoless double beta decay ; (g) Neutrino masses ; (h)Heavy nuclei decay. The superscript X denotes that the corresponding number is at the 2σ level, and is excluded at the 1σ level. All numbers should be multiplied by $\left[\tilde{M}_f/100\,GeV\right]$.

- Of the allowed couplings consistent with no nucleon decay, only one dominates [65]. This is perhaps the most likely circumstance as the new couplings are of the same character as the MSSM Yukawa couplings responsible for quark and lepton masses and these show a vast difference in magnitude ($\frac{h_e}{h_{top}} \sim 5 \cdot 10^{-6}$).

- The operators related to the dominant operator by a flavor mixing are suppressed by a mixing angle. For example for a given operator, $\mathcal{O} = \lambda'' \cdot [\overline{TBS}]_F$, we have

$$\mathcal{O}_{mix} = \lambda''_{mix} \cdot [\overline{TBD}]_F \tag{40}$$

$$\lambda''_{mix} = \lambda'' \cdot \sin\theta. \tag{41}$$

  For the mixing angle θ we have assumed the typical values of the corresponding Kobayashi-Maskawa matrix elements.

- The LSP, denoted by $\tilde{\Lambda}$, is a neutralino, *i.e.* it is a mixed state consisting of $(\tilde{\gamma}, \tilde{Z}^0, \tilde{H}^0)$. This is a non-trivial assumption, as the standard bounds on *stable* charged or strongly interacting relics in the universe no longer apply, since the LSP can decay via an R-parity-violating operator.

- We will consider the case that the LSP decays within the detector. If not the LSP will give rise to the same $(E_T)_{Missing}$ signals as in the R-parityconserving scheme.

6.3.1 Experimental Bounds

The coefficients $\lambda, \lambda', \lambda''$ of the operators in Eq. (29) are limited in size by purely conventional (non-supersymmetric) processes, since the addition of a new interaction may affect these processes by the virtual exchange of an R-parity-odd particle ($\tilde{f}$, of mass $\tilde{M}_f$). In Table 3 we present a summary of these upper bounds on the λ's scaled by $\left[\frac{\tilde{M}_f}{100\,GeV}\right]$. These numbers compare with the electromagnetic coupling of $e \sim 0.30$, and $g_{weak} = e/\sin\theta_w \sim 0.62$. The only assumption made in compiling Table 3 is point 1. in the previous section. Note that there exist bounds on only 13 of the 27 operators $[L_iQ_j\overline{D}_k]_F$ and on only 1 of the 9 operators $[\overline{U}_i\overline{D}_j\overline{D}_k]_F$. The bounds on the purely leptonic operators $[L_iL_j\overline{E}_k]_F$ are already quite restrictive, but not yet beyond detection. As we will see below most of these bounds can be very much improved by direct searches for supersymmetric particles in R-parity-violating channels. Table 3 does not include any cosmological bounds [59] for, as discussed in [60], these are particularly sensitive to the detailed mechanisms responsible for baryogenesis and to the structure of the model above the electroweak breaking scale.

6.3.2 Collider Signals

We first consider the case that the LSP decays within the detector. Using the two forms of production discussed above and attaching the new R-parity-violating decays we arrive at two sets of signals:

$$\begin{array}{l}(1)(R-parityviolating\,\text{single susy production}) \oplus (R-parityviolating\ -\text{decay}),\\ (2)(\text{Standard susy production}) \oplus (R-parityviolating\ -\text{decay}).\end{array} \tag{42}$$

We have summarized the signals for (1) in Table 4 and for (2) in Table 5; the general situation is summarized below.

(1)R-parity violating Single Susy Production:

In both the $2 \to 1$ and the $2 \to 2$ single sparticle production there is an operator insertion. Thus the production cross sectionis suppressed by a factor of λ^2. For the $2 \to 2$ processes the additional phase space suppression results in a production rate too low to be visible [47]. It is thus only the $2 \to 1$ processes which are promising candidates for detection. In hadron colliders it is not possible to test the operators $[LL\overline{E}]_F$ in the production mechanism as they do not couple to the initial states. It is very difficult to test the operators $[\overline{UDD}]_F$, since in the dominant processes they usually result in multi-jet production, which suffers from a tremendous QCD background. (A promising case is the resonant bump at $M_{\tilde{q}}$ in the 2-jet production cross sectionfrom the process $q + q' \to \tilde{f} \to q + q'$, but even this is not detectable at the LHC [47]. As can be seen in Table 4 the situation for the $[\overline{UDD}]_F$ operators does not improve using the MSSM pair production of particles. Thus in order to test these operators one must consider cascade decays, which result in more visible signals. Although often not considered such cascade decays are likely to dominate for squarks and other heavier supersymmetric states. Using the high single squark production rate, signals can still be visible for small branching ratios. In [47] it was shown that the cascade decay of a squark through a gluino, as in the MSSM but without $\not{p}_T$, led to a visible signal.

(2)Standard Susy Pair Production:

In the second set of signals in Eq. (42) only gauge couplings are involved in the production mechanism and which is thus only suppressed by the masses of the supersymmetric particles. In Table 4 we see that the dominant decay modes for the operators $[\overline{UDD}]_F$ are not very promising as the LSP decays to jets. In this case cascade decays provide the only observable signals. The $[LL\overline{E}]_F$ operators however lead to very promising signals because the LSP decay leads to multi-leptons. (In [47] a case study was made for gluino

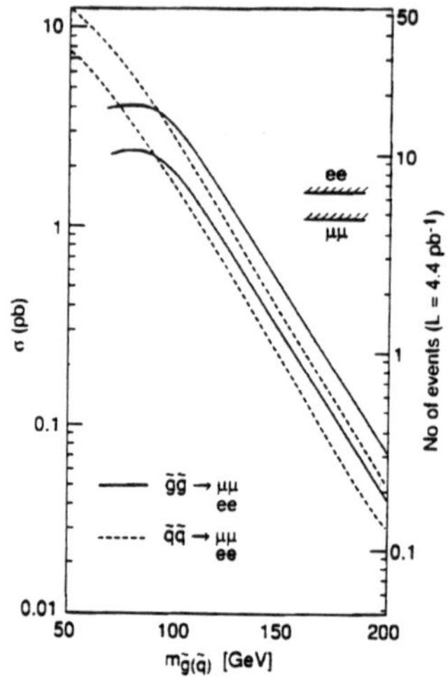

Figure 8: The visible dimuon and dielectron cross-sections as a function of the gluino (squark) mass[64].

pair production and decays via $\mathcal{O} = [L_1 L_2 \overline{E}_1]_F$.) As may be seen in Table 4 LSP decay converts the $\not{p}_T$ signals of the MSSM to multi-leptons which are highly visible. Similar results can be expected for the other cases of $[LL\overline{E}]_F$. For $[LQ\overline{D}]_F$ the standard signal is again like-sign dileptons.

For the operators $[\overline{UDD}]_F$ the $\not{p}_T$ signals of the MSSM are converted to hadronic jets through the decays of Table 2b. These give signals swamped by Standard Model QCD processes. However squarks (and gluinos) are expected to be the heaviest supersymmetric states and longer cascade decay chains may be expected to dominate. These can give rise to observable signals [47].

Recently there has been detailed studies of R-parity violating signals at HERA [63] and at the Tevatron [64]. As an example we show in Fig 8 the results of a study of dimuon and dielectron cross-sections expected at CDF together with the number of events expected for a luminosity of 4.4 pb^{-1} coming from R-parity violating lepton-number-violating operators via LSP decay. As may be seen there are visible signals for a range of SUSY masses.

6.4 Phenomenology of Spontaneous R-parity violation.

As mentioned in Section 4.2 it is also possible R- parity is spontaneously broken. Once again novel signals are generated. For example in Z decay there are two possibilities depending on whether the Majoron exists or is "eaten" by a new massive Z' boson. In both cases characteristic signals at LEP are: $Z \to \chi\tau$ occurring with BR $\approx 10^{-5}$ and $Z \to \chi\chi'$ with $\chi' \to \overline{f}f$. the so-called Zen events. Here χ is a chargino. I do not have the time adequately to discuss the phenomenology in this case and refer the reader to the literature [50].

PHYSICAL PROCESS	$LL\bar{E}$	$LQ\bar{D}$	$\overline{UDD}$
$\tilde{f} \rightarrow \mathcal{O}(\tilde{f})$	—	2 jets	2 jets
$\rightarrow f + \tilde{\Lambda}$	—	$2(e, \mu, \tau, \not{p}_T) + 2$ jets	4 jets
$f + \tilde{\chi}^0 \rightarrow f + \mathcal{O}_\mathcal{E}, \bar{\mathcal{O}}_\mathcal{E}$	—	$2(e^\pm, \mu^\pm, \tau^\pm, \not{p}_T) + 2$ jets	4 jets
$\rightarrow f + Z + \tilde{\Lambda}$	—	$Z + 2(e^\pm, \mu^\pm, \tau^\pm, \not{p}_T) + 2$ jets	$Z + 4$ jets
$f + \tilde{\chi}^\pm \rightarrow f + \mathcal{O}_W, \bar{\mathcal{O}}_W$	—	$2(e^\pm, \mu^\pm, \tau^\pm, \not{p}_T) + 2$ jets	4 jets
$\rightarrow f + W + \tilde{\Lambda}$	—	$W + 2(e^\pm, \mu^\pm, \tau^\pm, \not{p}_T) + 2$ jets	$W + 4$ jets
$f + \tilde{g} \rightarrow f + q + \bar{q} + \tilde{\Lambda}$	—	$2(e^\pm, \mu^\pm, \tau^\pm, \not{p}_T) + 4$ jets	6 jets
$\rightarrow f + \mathcal{O}_\mathcal{E}, \bar{\mathcal{O}}_\mathcal{E}$	—	$2(e^\pm, \mu^\pm, \tau^\pm, \not{p}_T) + 2$ jets	4 jets
$\tilde{f} + (g, W, Z, \gamma)$ $\rightarrow f + \tilde{\Lambda} + (g, W, Z, \gamma)$	—	$2(e, \mu, \tau, \not{p}_T) \otimes (g, W, Z, \gamma) + 2$ jets	$(g, W, Z, \gamma) + 4$ jets
$f + \tilde{f}' \rightarrow f + f' + \tilde{\Lambda}$	—	$2(e, \mu, \tau, \not{p}_T) + 3$ jets	5 jets

Table 4: Most visible signals in the single (R-parity violating) production of supersymmetric particles. The Table should be read as follows. In the first column we have listed the final states resulting from the $2 \rightarrow 2$ processes. This results in a final state containing an LSP and/or an operator ($\mathcal{O}_\mathcal{E}, \mathcal{O}_W$). The LSP is assumed to decay to $\mathcal{O}_\mathcal{E}$. The particle content of the operators depends on which operator is dominant. The various cases are contained in the following three columns. The notation $(e, \mu, \tau, \not{p}_T) \otimes (g, W, Z, \gamma)$ refers to all possible cross combinations of the particles. The process $q + q' \rightarrow f + gaugino$ can happen in t-channel as well as s- channel.

PHYSICAL PROCESS	$LL\overline{E}$	$LQ\overline{D}$	$\overline{UDD}$
$\tilde{l}^+\tilde{l}^- \longrightarrow l^+l^- + 2\tilde{\Lambda}$	$\not{p}_T + l^+l^- + 2(e,\mu,\tau)^\pm \otimes (e,\mu)^\mp$	$l^+l^- + 2(e,\mu,\tau)^\pm + 4$ jets	$l^+l^- + 6$ jets
$\longrightarrow l^+l^- + 2\tilde{\Lambda}$	—	$l^+l^- 2\gamma\nu$	—
$\tilde{g}\tilde{g} \longrightarrow 2q\bar{q} + 2\tilde{\Lambda}$	$2q\bar{q} + \not{p}_T + 2(e,\mu,\tau)^\pm \otimes (e,\mu)^\mp$	$2q\bar{q} + 2(e,\mu,\tau)^\pm + 4$ jets	$2q\bar{q} + 6$ jets
$\longrightarrow \mathcal{O}_\varepsilon\mathcal{O}_\varepsilon, \mathcal{O}_\varepsilon\overline{\mathcal{O}}_\varepsilon, \overline{\mathcal{O}}_\varepsilon\overline{\mathcal{O}}_\varepsilon$	—	$2(e,\mu,\tau)^\pm + 4$ jets	6 jets
$\tilde{q}\tilde{q}' \longrightarrow qq' + 2\tilde{\Lambda}$	$qq' + \not{p}_T + 2(e,\mu,\tau)^\pm \otimes (e,\mu)^\mp$	$qq' + 2(e,\mu,\tau)^\pm + 4$ jets	$qq' + 6$ jets
$\longrightarrow \mathcal{O}(\tilde{q})\mathcal{O}(\tilde{q}')$	—	$(l,\nu) \otimes (l,\nu) + 2$ jets	4 jets
$\tilde{\chi}^+\tilde{\chi}^o \longrightarrow \left\{ \begin{array}{c} W^+ + Z^o + 2\tilde{\Lambda} \\ l'^+l^+l^- + 2\tilde{\Lambda} \end{array} \right\}$	$\not{p}_T + l^+l^-l'^+ + 2(e,\mu,\tau)^\pm \otimes (e,\mu)^\mp$	$l^+l^-l'^+ + 2(e,\mu,\tau)^\pm + 4$ jets	$l^+l^-l'^+ + 6$ jets
$\longrightarrow \mathcal{O}_\varepsilon + \left\{ \begin{array}{c} u + \mathcal{O}(\tilde{d}) \\ d + \mathcal{O}(\tilde{u}) \end{array} \right\}$	—	$l^+l^- + 4$ jets	6 jets
$\longrightarrow \mathcal{O}_\varepsilon + \left\{ \begin{array}{c} \nu + \mathcal{O}(\tilde{e}) \\ e + \mathcal{O}(\tilde{\nu}) \end{array} \right\}$	$\not{p}_T + 5l^\pm$	$l^+l^- + 4$ jets	—

Table 5: The most visible signals for pair (associated) production of supersymmetric particles. $(e,\mu,\tau) \otimes (e,\mu)$ refers to the possible signals: ee, $e\mu$, $e\tau$, $\mu\mu$, and $\mu\tau$.

7 Summary.

There is a good case to be made for a stage of low-energy supersymmetry to explain the hierarchy of mass scales. It is quite surprising how successful the resulting unified theory is in fitting the observed couplings and masses, provided the spectrum of SUSY states is low. The phenomenology of the resultant low energy SUSY is consistent with present observations but the new SUSY states should be accessible to the new accelerators as studies of the MSSM suggest that squarks and gluinos should be visible up to a mass of 1TeV. In the light of our experience building unified models based on string theories it is reasonable to think that R-parity-violating versions of supersymmetry deserve to be considered on an equal footing with the MSSM. The phenomenology of such models can differ significantly from the MSSM and so it is essential to look for R-parity violatingsignals as well if a complete SUSY search is to be performed. A survey of the characteristic signals shows that, due largely to the important rôle that cascade decays are likely to play, there are signals visible at the LHC (and presumably at the Tevatron) in all R-parity violating variants. However it is essential, if supersymmetry is to be discovered in these variant models, that a search is made of channels other than those relying on conventional missing energy signals for the LSP decays converting missing into visible energy.isible as in the MSSM.

References

[1] For reviews see:
G. G. Ross, Grand Unified Theories, Benjamin/Cummings (1984)
H.-P. Nilles, Phys. Rep. 110 (1984) 1
H.E.Haber and G.L.Kane, Phys. Rep. 117 (1985) 75
A.B.Lahanas and D.V.Nanopoulos, Phys. Rep. 145 (1987) 1

[2] P. Fayet, Phys. Lett. 69B(1977)489; ibid. 84B(1979)416; ibid. 78B(1978)417;
G. Farrar and P. Fayet, ibid. 79B(1978)442; ibid 89B(1980)191.

[3] J.Ellis, S.Kelley, D.V.Nanopoulos,Phys.Lett.260B(1991)131;
U.Amaldi,W.deBoer,H.Furstenau,Phys.Lett.260B(1991)447;
P.Langacker,M.Luo,Univ. of Pennsylvania preprint,UPR- 0466T(1991).

[4] See J. Ellis and A. Zichichi, these proceedings and references therein.

[5] H.Georgi,H.R.Quinn,S.Weinberg,Phys.Rev.Lett.33(1974)451.

[6] L.E.Ibanez,G.G.Ross,Phys.Lett.105B(1982)439;
M.Einhorn,D.R.T.Jones,Nucl.Phys.B196(1982)475.
S.Dimopoulos,S.Raby,F.Wilczek,Phys.Rev.D24(1981)1681;

[7] E. Cremmer, S. Ferrara, L. Girardello and A. Van Proeyen, Nucl. Phys. B212 (1983) 413.

[8] For a review of 4D string theories see M. Green, J. Schwarz and E. Witten, Superstring Theory, Cambridge University Press, 1987.

[9] For a review see D. Amati, K. Konishi, Y. Meurice, G. Rossi and G. Veneziano, Phys. Rep 162 (1988) 169; H. P. Nilles, Int. J. Mod. Phys. A5 (1990) 4199 ; J. Louis, SLAC-PUB- 5645 (1991)

[10] H.P Nilles, Phys. Lett. 115B (1982) 193;
S. Ferrara, L. Girardello and H.P. Nilles, Phys. Lett. 125B (1983) 457;
J.P. Derendinger, L.E. Ibanez and H.P. Nilles, Phys. Lett. 155B (1985) 65;
J.P. Derendinger, L.E. Ibanez and H.P. Nilles, Nucl. Phys. B267 (1986) 365

[11] M. Dine, R. Rohm, N. Seiberg and E. Witten, Phys. Lett. 156B (1985) 55

[12] J.A. Casas, Z. Lalak, C. Munoz and G.G. Ross, Nucl. Phys. B347 (1990) 243;
L. Dixon, SLAC preprint 5229 (1990).

[13] A. Font, L.E. Ibanez, D. Lust and F. Quevedo, Phys. Lett. B245 (1990) 401.

[14] S. Ferrara, N. Magnoli, T.R. Taylor and G. Veneziano, Phys. Lett. B245 1990 409;
H.P. Nilles and M. Olechowski, Phys. Lett. B248 (1990) 268;
P. Binetruy and M.K. Gaillard, Phys. Lett. B253 (1991) 119;
J. Louis, SLAC-PUB-5645 (1991);
S. Kalara, J. Lopez and D. Nanopoulos, Texas A & M preprint CTP- TAMU-69/91.

[15] C. Kounnas and M. Porrati, Phys. Lett. 191B (1987)

[16] I. Affleck, M. Dine and N. Seiberg, Nucl. Phys. B241 (1984) 493; B256 (1985) 557

[17] S. Ferrara, D. Lust, A. Shapere and S. Theisen, Phys. Lett. 225B (1989) 363;
S. Ferrara, D. Lust and S. Theisen, Phys. Lett. 233B (1989) 147

[18] M. Cvetic, A. Font, L. Ibanez, D. Lust and F. Quevedo, Nucl. Phys. B361 (1991) 194;
H.P. Nilles and M. Olechowsky, Phys. Lett. B248 (1990) 268;
P. Binetruy and M.K. Gaillard, Phys. Lett. B253 (1991) 119;
D. Lust and T.R. Taylor, Phys. Lett. B253 (1991) 335

[19] D. Lust,and T.R. Taylor, Phys. Lett. B253 (1991) 335;
B. de Carlos, J. A. Casas and C. Munoz, Phys. Lett. 263B (1911) 248

[20] M. A. Shifman and A. I. Vainstein, Nucl. Phys. B277(1986)649; Preprint TPI-Minn 91/4-T.

[21] V. Kaplunovsky, preprint UTTG-15-91;
J. Louis, SLAC preprint SLAC-PUB 5527;
V. Kaplunovsky and J. Louis, SLAC-PUB 5638.

[22] A. Font, L. Ibanez, D. Lust and F. Quevedo, Phys. Lett. 245B (1990) 401;

[23] A. de la Macorra and G.G.Ross, Oxford University preprint, OUTP-92-14P (1992)

[24] L.Ibanez and D.Lust,CERN preprint, CERN-TH.6380/92.

[25] L. Girardello and M. Grisaru, Nucl. Phys. B194(1982)65.

[26] L. E. Ibanez, G. G. Ross, Phys. Lett., 110B (1982) 227
K. Inoue et al., Prog. Theor. Phys. 68, 927 (1982)
L. Alvarez-Gaume, J.Polchinski, M. Wise, Nucl. Phys. B221, 495 (1983)
J. Ellis, J. Hagelin, D.V. Nanopoulos, K. Tamvakis, Phys. Lett., B125, 275 (1983)
L. Ibanez, C. Lopez, Phys. Lett. B126, 54 (1983); Nucl. Phys. B233, 545 (1984)
C. Kounnas, A. Lahanas, D.V. Nanopoulos, M. Quiros, Phys. Lett. B132, 95 (1983)

[27] G.G.Ross and R.G.Roberts, Nucl. Phys. B377(1992)571.
See also rapporteur talk presented at 1991 Joint International Lepton-Photon Symposium and EPS conference,..

[28] F.Anselmo, L. Ciafarelli, A. Peterman and A. Zichichi, Nuovo Cim. 104A (1991) 1817 and CERN preprint CERN- TH.6429/92.

[29] For a recent review see L.E.Ibanez and G.G. Ross, CERN preprint CERN-TH.6412/92, to appear in "Perspectives in Higgs Physics", G. Kane, editor.

[30] L.E. Ibáñez and C. López, Phys. Lett. 126B (1983) 54; Nucl. Phys. B233 (1984) 511.

[31] L.E. Ibáñez and G.G. Ross, Phys. Lett. 110B (1982) 215

[32] K. Inoue et al. Prof. Theor. Phys. 568 (1982) 927

[33] L. Alvarez-Gaumé, J. Polchinsky and M. Wise, Nucl. Phys. B221 (1983)

[34] R. Barbieri, G. Giudice, Nucl. Phys. , B296, 75(1988)

[35] V. Kaplunovsky, Nucl. Phys. B 307 (1988) 145

[36] I. Antoniadis, J. Ellis, R. Lacaze, D. V. Nanopoulos, CERN preprint CERN-TH. 6136/91 (1991);
S.Kalara, J.L.Lopez, and D.V.Nanopoulos, CERN preprint CERN- TH-6168/91(1991)

[37] L.E.Ibanez, D. Lust and G. G. Ross, CERN preprint CERN-TH-6241/91(1991).

[38] I. Antoniadis, J. Ellis, S. Kelley, D. V. Nanopoulos, CERN preprint, CERN-TH. 6169/91 (1991)

[39] R. Barbieri and R. Gatto, Phys. Lett. 110B (1982) 211;
J. Ellis and D.V.Nanopoulos, Phys. Lett. 110B (1982)44.;
T. Inami and C.S.Lim, Nucl. Phys. B207 (1982) 533.

[40] G.G.Ross and S.Wilkinson, Nucl. Phys. B282(1987)382.

[41] S. Bertolini and A. Masiero, Phys. Lett. B174(1986)343.

[42] J. Polchinski and M.B.Wise, Phys. Lett. 125B (1983) 393;
F. del Aguila, M. Gavela, J. Grifols and A. Mendez, Phys. Lett. 126B (1983) 71;
W. Buchmuller and D. Wyler, Phys. Lett 121B (1983) 321;
M. Dugan, B. Grinstein and L. Hall, Nucl. Phys. B255 (1985) 413.

[43] S. Weinberg, Phys. Rev. Lett. 63 (1989) 2333.

[44] J.Dai and H.Dykstra, Phys. Lett. B237 (1990) 256;
E. Braaten, C.S.Li and T.C.Yuan, Phys. Rev. Lett. 64 (1990) 1709;
J. Dai, H. Dykstra, R.G.Leigh, S.Paban and D.A.Discus, Phys. Lett. B237 (1990) 216;
R.Arnowitt, J.L.Lopez and D.V.Nanopoulos, Phys. Rev. D42 (1990) 2423;
A.de Rujula, M.B.Gavela, O.Pene and F.J.Vegas, Phys. Lett. B245 (1990) 640.

[45] R.Arnowitt, M.J.Duff and K.S.Stelle, Phys. Rev. D43 (1991) 3085.

[46] P.Nath, Phys. Rev. Lett. 66 (1991) 2565.

[47] *et al*, ECFA Large Hadron Collider (LHC) Workshop, Aachen, 1990, Vol. II.

[48] C. Aulakh and R. Mohapatra, Phys. Lett B119 (1983) 136; G.G. Ross and J.W. F.Valle, Phys. Lett. B151 (1985) 375; J.Ellis *et al* Phys. Lett B150 (1985)142; R.Barbieri *et al.*, Phys. Lett B238 (1990)86;

[49] K. Hidaka, contributed paper to the 1991 joint lepton-photon and Europhysics conference on High Energy physics.

[50] A.Santamaria and J.F.W.Valle, Phys. Lett. B195 (1987)423; Phys.Rev. Lett.60 (1988)397; Phys.Rev.D39 (1989)1780;
M.C.Gonzalez-Garcia and J.F.W. Valle, CERN preprint, CERN- TH.5861/90;
P.Nogueira, J.C.Romao and J.W.F.Valle, Phys. Lett. B251(1990)142;
M.C.Gonzalez-Garcia, J.C.Romao and J.W.F.Valle, Nucl.Phys. B(1992);
J.C.Romao, N.Ruis and J.W.F.Valle, Nucl. Phys. B363(1991)369.

[51] A. Masiero and J.W.F.Valle, Phys. Lett. B251(1990)273.

[52] L. Hall and M. Suzuki, Nucl. Phys. B231 (1984) 419;
I. Lee, Nucl. Phys. B246 (1984) 120.

[53] F. Zwirner, Phys. Lett. 132B (1983) 103;
R. Barbieri and A. Masiero, Nucl. Phys. B267 (1986) 679.

[54] S. Dimopoulos and H. Georgi, Nucl. Phys. b193(1981)150.

[55] S.Weinberg, Phys. Rev. D26(1982)287;
N. Sakai and T. Yanagida, Nucl. Phys. B197(1982)533.

[56] L.J.Hall and R. Barbieri, LBL preprint (1991)

[57] M.Bento, L.J.Hall and G.G.Ross, Nucl.Phys. B292 (1987)400.

[58] L.E.Ibanez and G.G.Ross, Nucl.Phys. B368(1992)3.

[59] B.A.Campbell, S.Davidson, J.Ellis and K.Olive, Phys. Lett. B256(1991)457;
W.Fischler, G.Giudice, R.G.Leigh and S.Paban, Phys.Lett. B258(1991)45.

[60] H.Dreiner and G.G.Ross, Oxford University preprint OUTP-92-08P(1992).

[61] L.E.Ibanez and G.G.Ross, Phys. Lett. 260B(1991)291.

[62] H.Dreiner and G.G.Ross, Oxford University preprint OUTP-91-15P(1991)to be published in Nucl. Phys.

[63] J.Butterworth and H. Dreiner, Oxford University preprint OUNP-92-01 (1992).

[64] D.P.Roy,CERN preprint CERN-TH.6339/91(1992).

[65] S.Dimopoulos and L.J.Hall, Phys.Lett. B196(1987)135.

CHAIRMAN: G.G. Ross

Scientific Secretaries: N. Moulai, N. Yunus

DISCUSSION

– *Ren:*

You mentioned the possibility of observing the 4 lepton decay channel of $\tilde{g}\tilde{g}$ pair at LHC. Can you explain what the contents of $\tilde{X}$ in $\tilde{g}\tilde{g} \to 4\ell + \tilde{X}$ are?

– *Ross:*

This $\tilde{X}$ is producing a neutralino and a gluon. Then the gluon has the potential of producing jets in addition to electron signal. This means that there will be $4\bar{e}$, 2 gluons and missing energy.

– *Sopczak:*

Great progress has been made in astrophysics by observing an anisotropy in the microwave background radiation. What are the possible implications for the LSP and the expected Higgs structure?

– *Ross:*

For the case when the LSP is a stable state, it is a candidate for dark matter. One then has to analyze its couplings to determine what proportion of the closure density will be provided by the LSP. An analysis by Roskovsky suggests that for a small proportion of the parameter space, consistent with fine tuning of less than 10, the LSP may contribute a sizeable proportion of the closure density. The anisotropy of the universe may be explained by inflationary cosmology. In this case the fluctuations in microwave background arise as quantum fluctuations which are blown up by inflation. A lot of study has been done to find the magnitude of the fluctuations. This is related to the detailed shape of the potential which is driving inflation. The (new or chaotic) inflationary model requires a very flat potential and the slope of that potential and its height determines how big the fluctuations are.

– *Cho:*

1) Can SUSY and Technicolour be profitably synthesized?

2) Has $K - \overline{K}$ mixing via the two box diagrams been considered in light of the very heavy top mass?

– *Ross:*

1) SUSY helps to cure some Technicolour diseases associated with the generation of fermion masses. But SUSY itself does not suffer from these diseases, so it has no need for Technicolour. I am doubtful if supersymmetric Technicolour has any life in it.

2) The heavier the top the larger the potential contribution. But compensating this is the fact that the mixing angle of the light quark to the heavy top is very small, and seems to have an inverse relationship to the top mass. Yes, people have gone back to these calculations in the light of heavy top and done rather precise calculations. It still remains consistent.

– *Baillie:*

What are the implications of the MSSM for cosmology, in particular baryogenesis and nucleosynthesis?

– *Ross:*

Baryogenesis requires baryon number violation, CP violation and a departure from thermal equilibrium during its production. SUSY GUTs can satisfy these conditions and acceptable SUSY models for baryogenesis have been analysed. Physicists have recently started to consider the possibility of a baryon number excess at low energies through SM processes which violate the baryon number. However the situation is still unclear. The question of departure from thermal equilibrium is sensitively connected to whether it is a first order or second order transition at the electroweak phase transition and the calculations cannot be done reliably. However it is known that CP violation in the SM is insufficient to give acceptable baryogenesis. In the MSSM the new sources of CP violation are thought to be capable of rectifying this problem. Nuclear synthesis occurs around 1 MeV. By that time the heavy SUSY particles would have decayed and thus they play no significant role in nucleo-synthesis.

– *Vassilevskaya:*

I would like to know about the new upper limit of the branching ratio of the decay $\mu \rightarrow e\gamma$.

– *Zichichi:*

It is very small!

– *Vassilevskaya:*

Do you know the models in which the processes with lepton number violation are possible without lepton mixing?

– *Ross:*

Yes, SUSY is an example in which the mixing is in the slepton channel rather than the lepton channel.

– *Vassilevskaya:*

I would like to know if there is already experimental data concerning the mixing matrix of leptons.

– *Ross:*

In the Standard Model the mixing matrix for leptons exists only if we have neutrino masses. As we do not yet have any firm evidence of neutrino masses, we do not have firm evidence for the mixing angles yet.

– *Hanany:*

Can we probe discrete symmetries at low energies?

– *Ross:*

Yes, at low energies the form of the superpotential and equivalent resulting interactions are restricted by the need to be invariant under these symmetries. So, if one can identify a definite structure of couplings, then one can infer from it what the discrete symmetries are.

– *Misiak:*

Is the parameter arg $(\mathrm{Am}^*_{1/2})$ equal to zero in the simplest SU(5) SUSY GUT inspired by strings (SUGRA)?

– *Ross:*

The magnitude of A depends upon the detail of the string theory. There is a class of string theories called "no scale theories" which would have $A = 0$ at compactification scale. In such theories, although A is recreated at low energies, arg $(\mathrm{Am}^*_{1/2})$ would be zero at the compactification scale.

CHAIRMAN: G.G. Ross

Scientific Secretaries: K. Benakli, M.T. Dova

DISCUSSION

– Benakli:

You calculated the gaugino condensate in the hidden sector, then you used the gravitation sector of the supergravity theory to trigger the supersymmetric breaking to the observable sector. There you used the tree level approximation. Why do you still think it is valid for the spin 3/2 sector?

– Ross:

In supergravity theory, the one loop correction is estimated to be of order $1/4\pi^2 < \mathrm{F}\phi >^2 / \mathrm{m}_\mathrm{p}^2$ which comes with a factor $1/4\pi^2$ smaller than the tree level term.

– Liu:

Regarding proton decay, what is the lifetime you predict? What is the most important decay mode?

– Ross:

My philosophy in this talk has been a bottom-up approach, namely, I did not specify what the Grand Unified theory was, but tried to determine the low energy parameters. As such, I do not know, a priori, what the proton decay processes are. There are a lot of different possibilities. You may have, as in Nath's lecture, proton decay through dimension five operators which leads to a proton lifetime $\sim 1/\mathrm{M}_\mathrm{x}^2$, but you can avoid this, for example, by replacing by $\mathcal{Z}_2$ matter parity symmetry by the $\mathcal{Z}_3$, baryon parity R that forbids these operators. So I cannot give you an upper limit because I cannot do it in a model independent way.

– Liu:

Is there any new physics in the desert between 100 GeV and 10^{16} GeV?

– Ross:

Of course, before we meet the desert in the supersymmetric version, which I believe to be the only viable self-consistent version of the Grand Unified theories, we must find more than twice the particles known at present in the standard model. As far as filling the desert is concerned, obviously you can add states with intermediate masses. The trouble is that there is nothing to constrain your imagination doing this. My philosophy in using the desert scenario is to press the theory as far as you can to obtain its predictions that, if shown to be uniform, will require some new structure. But until there are some further constraints on this new structure, it seems to me just an empty speculation to add intermediate scale multiplets without particular motivation. Of course, various Grand Unified Theories do have such intermediate scale states.

– *Malik:*

The breaking of supersymmetry generates a large number of particles. Can they all fall in the gap between the lowest limit of M_{SUSY} and M_{GUT}? If not, can you please give the upper limit for M_{SUSY}?

– *Ross:*

The limit is based on the hierarchy problem. Once you give me your preferred value of "c" in the fine tuning formula I can tell you the upper limit on supersymmetry states. If supersymmetry is not discovered at such low energy it would not solve the hierarchy problem, so I would give it up and look to composite theories as a solution to the hierarchy problem.

– *Cho:*

Does the ordinary SU(5) GUT prediction $m_b = m_\tau$ persist in SUSY SU(5)?

– *Ross:*

The prediction at the unification scale is the same; what differs are the radiative corrections. So, when you come to the detailed predictions and look for a precise comparison of m_b/m_τ at low energies, you find the running that comes from SUSY works well with the unification scale of 10^{16} GeV. However the running without including SUSY is not entirely consistent with similar observations of the improvement of $\sin^2\theta_W$.

– *Cho:*

What progress, if any, has been made in incorporating non-trivial family structure in GUT's?

– *Ross:*

Five years ago there were several attempts to understand the family structure via enlarging the gauge group. In my opinion these attempts were not very successful and there was no progress in our understanding of the structure of the mass matrices. In contrast, I believe there has been considerable progress in string theories in the last five years. String theories automatically lead to a level structure. After compactification one finds families of quarks and leptons with identical quantum number under the standard model gauge group $SU(3)\times SU(2)\times U(1)$, but with different quantum numbers under the internal (compactified) string symmetries. Given a choice of string vacuum one can predict the number of families and the Yukawa couplings, which are in general different for different families. Thus one can predict the fermion mass matrices. In particular the zeros in the mass matrices at the GUT scale are direct descendents of the symmetries of the string. In summary I believe there have been considerable advances in our understanding of the fermion family structure in string theory.

– *Sehgal:*

You have partially answered my question while answering Malik. However just for clarity I would like to know what energy limit you would put after which, if none of the supersymmetric particles are found, would mean a major modification of the theory?

– *Ross:*

The upper limit constraint comes from the hierarchy problem. We have:

$$M_W^2 = c\left[1 - \frac{8\pi^2}{6h_t^2}\,\frac{1}{\log\frac{m_x^2}{m_{\tilde{q}}{}^2}}\right] M_W^2 \ .$$

When you give me a value for "c" which corresponds to the fine tuning you are happy with, I will give you the upper mass bounds. My personal taste is contained in the two spectra I showed in my lecture. The only terms not included in my analysis are the sources of corrections that Fabio Zwirner cited in his lecture. These are terms proportional to m_t^4 in the Higgs mass. I was informed last week that when this is included this analysis gets slightly weaker bounds, by a factor 2.

– *Sehgal:*

What number would you quote?

– *Ross:*

10 is an extreme value of the fine tuning for me, I would prefer 5. So the spectra I gave still give an estimate of the upper bounds on the masses.

– *Leike:*

You got nice predictions for the quark mixing matrix. Is there room in your theory for possible neutrino mixing?

– *Ross:*

I gave an example where there are predictions for mixing angles involving slepton number violation. But I do not have predictions for neutrino masses. Dimopoulos, Hall and Raby have another good candidate mass matrix which they have extended to give predictions for neutrino masses.

– *Hanany:*

What about the high temperature phase structure of this theory?

– *Ross:*

The phase structure of this theory is just like the phase structure of the supersymmetric standard model. At high temperature ($\simeq 10^{10}$ GeV) we expect supersymmetry restoration. Below this temperature some supersymmetry is broken and below approximately 10^2 GeV it will trigger a vacuum expectation value for the Higgs and thereafter the quarks and leptons will get masses as well as the W.

– *Benakli:*

How do you think the disagreement between the string predicted value $E_{SU} \sim 3.73 \times 10^{17}$ GeV and the MSSM value $M_{GUT} \sim 10^6$ GeV can be reconciled?

– *Ross:*

The question is: is your prediction true for all the string theories? There are some orbifolds in which this scale can go down. Much more plausible for me is the idea that we have a string theory which at the compactification scale gives you some group, say $[SU(3)]^3$. This group could persist down to the 10^{16} GeV. Other people have suggestions of additional multiplets at the intermediate scales which affect the unification scale and push it to 10^{18} GeV.

QCD – part of the Theory of Everything

W.J. STIRLING
Departments of Mathematical Sciences and Physics, University of Durham, Durham DH1 3LE, England

Abstract

Quantum Chromodynamics is one of the components of the Standard Model. In these Lectures we outline some of the fundamental properties of the theory, focussing in particular on the running coupling and on applications in high energy e^+e^-, lepton-hadron and hadron-hadron collider physics. The recent interest in the value of the strong coupling is motivated and discussed.

1. Introduction

Quantum Chromodynamics (QCD), the gauge field theory which describes the interactions of coloured quarks and gluons, is one of the components of the $SU(3) \times SU(2) \times U(1)$ Standard Model. It has the property of *asymptotic freedom* – at short distances, equivalently high energies, the effective coupling is small and the theory can be studied using perturbative techniques. The advent of high energy lepton and hadron colliders has allowed these short distance properties of QCD to be studied in great detail. Some of the most direct information on QCD comes from studying the short-distance structure of hadrons using electroweak gauge boson probes – deep inelastic scattering. Here we see the asymptotic property directly, as the gauge bosons scatter incoherently off the weakly interacting quarks and gluons. From such experiments we learn how the partons (i.e. the quarks and gluons) share the momentum and quantum numbers of the hadron. By studying how the 'structure functions' vary with the momentum transferred by the probe, precision measurements of the short-distance coupling can be made. Jet production in high-energy e^+e^- and $p\bar{p}$ collisions is another fertile testing ground of the theory. Here, the fundamental $q\bar{q}g$ and ggg interactions of the theory can be studied and, again, the coupling can be measured.

The strong coupling α_s is a fundamental parameter of the Standard Model, and this in itself justifies its measurement to as high a degree of precision as possible. However, in recent years it has become clear that α_s could also provide clues to 'Theories of Everything' which may lie beyond. First, the three gauge couplings are now known so accurately that simple 'grand unification' at some large energy scale can be ruled out (Fig.1). If supersymmetry at some intermediate scale $M_{\rm SUSY}$ is invoked to rescue unification,[1,2] then $M_{\rm SUSY}$ is strongly dependent on the boundary value $\alpha_s(M_Z^2)$. It is intriguing (Fig.1) that a (too) naive implementation of supersymmetry restoration at a single scale leads to values of this scale in the TeV region, possibly within reach of

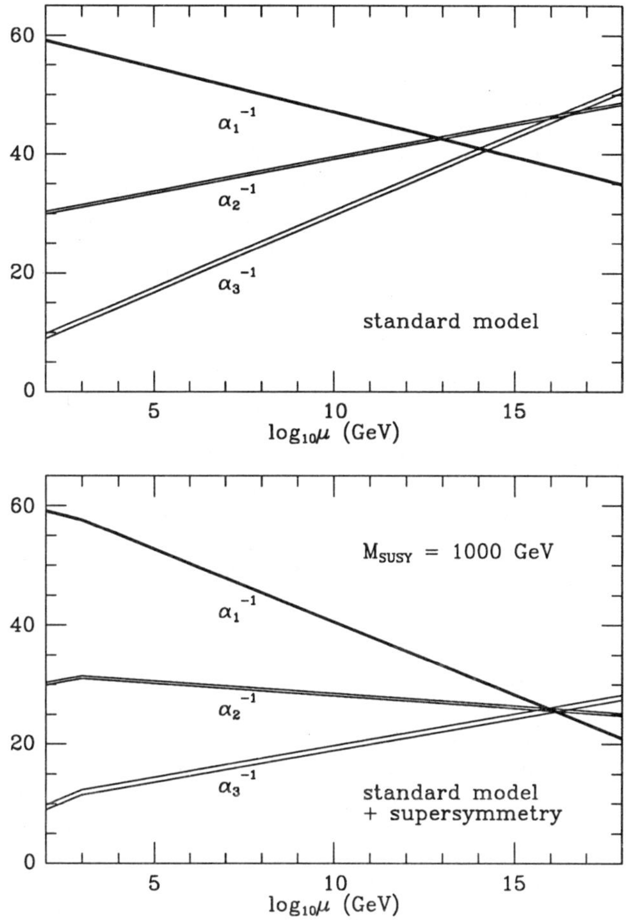

Figure 1: Unification of the three couplings of the Standard Model. In the upper figure the couplings are evolved using the standard β-functions of the theory, while in the lower figure the complete set of supersymmetric partners are introduced at a single scale of 1 TeV. The changes in the β-functions thus induced allow the couplings to meet at a single point.

the next generation of hadron-hadron colliders. A complete discussion can be found in the lectures of Zichichi.[1]

A second hint of possible new physics comes from comparing $\alpha_s(Q^2)$ values measured at LEP ($Q \sim M_Z$) with those evolved from lower energy measurements, for example from fixed-target deep inelastic scattering at $Q \sim 10$ GeV. Fig.2 shows a recent compilation[3] of $\alpha_s(Q^2)$ measurements, plotted at the 'typical' energy scale Q of the particular process. From this compilation, an interesting point emerges: there is a *hint* of a disagreement between α_s values measured at low energies and those at high energies. The solid line in Fig.2 corresponds to $\alpha_s(Q^2)$ evaluated and evolved at next-to-leading order in the $\overline{\text{MS}}$ scheme, with $\Lambda^{(4)}_{\overline{\text{MS}}} = 230$ MeV, a value consistent with all fixed-target deep inelastic experiments and related processes.[4] We see that when extrapolated to higher Q values, the coupling tends to lie below the high-energy measurements. This slight incompatibility is not, however, statistically significant. This notwithstanding, it has recently been argued[5–8] that a possible explanation of the mismatch in the evolved and measured high-energy couplings is that at some intermediate scale a new coloured degree of freedom is being excited, whose effect is to slow the running of α_s. A light supersymmetric gluino ($\tilde{g}$) has been suggested as a possible candidate. As the (Majorana) gluino mass threshold is crossed the β-function coefficients change:

$$\beta_i \rightarrow \beta_i + \Delta\beta_i\, \theta(Q - 2m_{\tilde{g}}) , \tag{1}$$

where $\Delta\beta_0 = -2$ and $\Delta\beta_1 = -48$. With $m_{\tilde{g}} = 5$ GeV, the coupling evolves as the dashed line in Fig.2, and consistency with the high-energy measurements is restored. It is not our purpose here to discuss in detail the theoretical and experimental consistency or otherwise of this hypothesis: discussions can be found in the recent literature.[7] The point to emphasize is the importance of accurate and reliable strong coupling measurements, from different processes at different energies.

These Lectures will focus on the fundamental properties of short-distance QCD. Our aim is to explain the basic theoretical ideas which underpin modern strong interaction phenomenology. Only a brief overview of these ideas can be given here, but more details can of course be found in the literature.[9,10] In the next section the running coupling and the Λ parameter are introduced. Section 3 describes the most straightforward application of these ideas, to the process $e^+e^- \rightarrow$ hadrons. The other cornerstone of modern QCD phenomenology, deep inelastic scattering, is described in Section 4. Finally, Section 5 describes how perturbative QCD provides a framework for calculating hard scattering cross sections in high energy hadron-hadron collisions, the basic physics of the LHC and SSC.

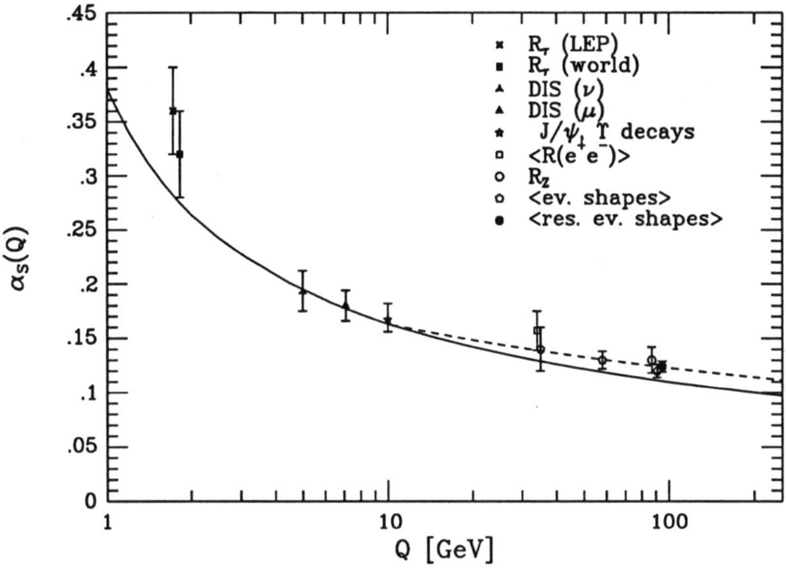

Figure 2: A compilation of α_s measurements, taken from the review by Bethke[3]. The solid line is the standard next-to-leading order QCD coupling for $\Lambda^{(4)}_{\overline{\rm MS}} = 230$ MeV. The dashed line corresponds to a gluino with $m_{\tilde{g}} = 5$ GeV and the same value of $\Lambda^{(4)}_{\overline{\rm MS}}$.

2. Basics of Perturbative QCD

2.1. *The QCD Lagrangian*

The QCD Lagrangian is, up to gauge-fixing terms,

$$\begin{aligned}
\mathcal{L}_{QCD} &= -\frac{1}{4}F^{(a)}_{\mu\nu}F^{(a)\mu\nu} + \sum_q \bar{\psi}^q_i \left(i\gamma^\mu (D_\mu)_{ij} - m_q\delta_{ij}\right)\psi^q_j \\
F^{(a)}_{\mu\nu} &= \partial_\mu A^a_\nu - \partial_\nu A^a_\mu + g_s f_{abc} A^b_\mu A^c_\nu \\
(D_\mu)_{ij} &= \delta_{ij}\partial_\mu - i g_s T^a_{ij} A^a_\mu
\end{aligned} \tag{2}$$

where g_s is the QCD coupling constant, T^a_{ij} and f_{abc} are the SU(3) colour matrices and structure constants respectively, the $\psi^q_i(x)$ are the 4-component Dirac spinors associated with each quark field of colour i and flavour q, and the $A^a_\mu(x)$ are the eight Yang-Mills gluon fields. From this Lagrangian, the Feynman rules can be derived in the usual way – the rules for a covariant gauge are summarized in Table 1.

Explicit forms for the SU(3) colour matrices and structure constants can be found, for example, in the *Review of Particle Properties.*[11] The following are some useful identities:

$$\begin{aligned}
[T^a, T^b] &= i f^{abc} T^c \\
\{T^a, T^b\} &= d^{abc} T^c + \frac{1}{3}\delta^{ab} \\
f^{acd} f^{bcd} &= C_A \delta^{ab} \\
(T^a T^a)_{ij} = T^a_{ik} T^a_{kj} &= C_F \delta_{ij} \\
\mathrm{Tr}(T^a T^b) = T^a_{ij} T^b_{ji} &= T_F \delta^{ab} \\
C_A &= N_c = 3 \\
C_F &= \frac{N_c^2 - 1}{2N_c} = \frac{4}{3} \\
T_F &= \frac{1}{2} \\
\mathrm{Tr}(T^a T^b T^c) &= \frac{i}{4} f^{abc} + \frac{1}{4} d^{abc} \\
f^{abc} f^{abc} &= 24 \\
d^{abc} d^{abc} &= \frac{40}{3}
\end{aligned} \tag{3}$$

where summation over repeated indices is understood.

2.2. *The QCD Coupling Constant*

Quantum Chromodynamics is an asymptotically free gauge field theory, that is, the strength of the interaction between the quarks and gluons becomes weaker in the short-distance limit. In QCD the renormalized coupling can be defined in a

Table 1: Feynman rules for QCD in a covariant gauge.

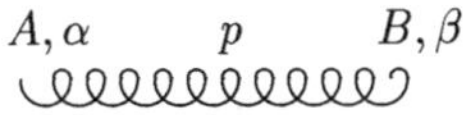

$$\delta^{AB}\left[-g^{\alpha\beta}+(1-\lambda)\frac{p^{\alpha}p^{\beta}}{p^2+i\varepsilon}\right]\frac{i}{p^2+i\varepsilon}$$

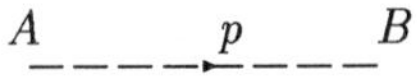

$$\delta^{AB}\frac{i}{p^2+i\varepsilon}$$

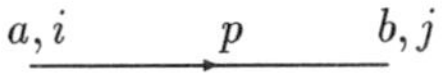

$$\delta^{ab}\frac{i}{(\hat{p}-m+i\varepsilon)_{ji}}$$

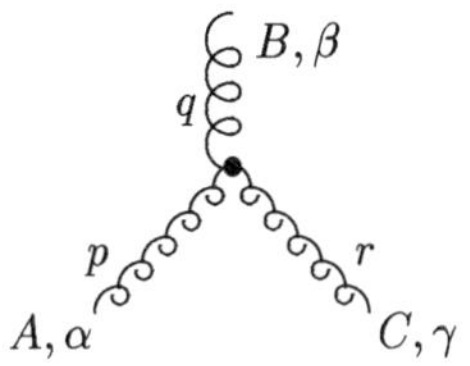

$$-gf^{ABC}\left[g^{\alpha\beta}(p-q)^{\gamma}+g^{\beta\gamma}(q-r)^{\alpha}+g^{\gamma\alpha}(r-p)^{\beta}\right]$$

(all momenta incoming)

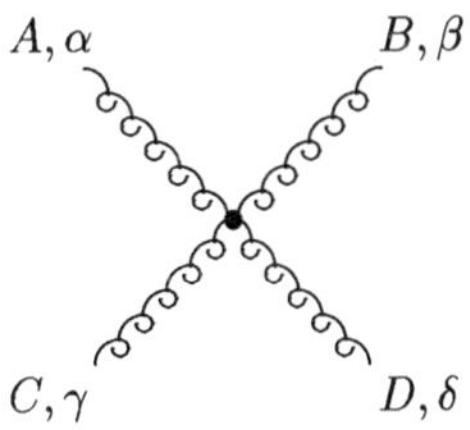

$$\begin{gathered}-ig^2 f^{XAC}f^{XBD}\left(g_{\alpha\beta}g_{\gamma\delta}-g_{\alpha\delta}g_{\beta\gamma}\right)\\ -ig^2 f^{XAD}f^{XBC}\left(g_{\alpha\beta}g_{\gamma\delta}-g_{\alpha\gamma}g_{\beta\delta}\right)\\ -ig^2 f^{XAB}f^{XCD}\left(g_{\alpha\gamma}g_{\beta\delta}-g_{\alpha\delta}g_{\beta\gamma}\right)\end{gathered}$$

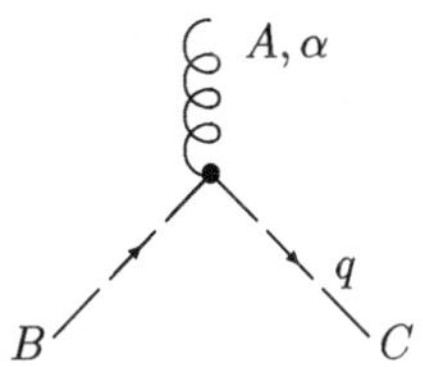

$$gf^{ABC}q^{\alpha}$$

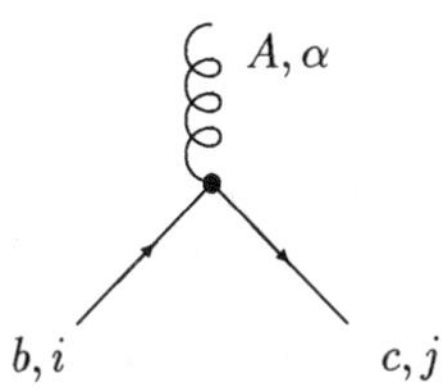

$$-ig\left(t^{A}\right)_{cb}(\gamma^{\alpha})_{ji}$$

variety of ways, for example from the 'dressed' qqg or ggg vertices. Renormalization of the coupling necessitates the introduction of a scale μ – effectively the scale at which the ultra-violet loop divergences are subtracted off. A dimensionless physical quantity R which depends on some energy scale Q will depend also on μ both *explicitly* and *implicitly* through the renormalized coupling, i.e. $R = R(\mu/Q, \alpha_s(\mu))$. The fact that such a quantity should not depend on the arbitrary scale μ (when calculated to all orders in perturbation theory) leads to an equation for the μ dependence of the renormalized coupling:

$$\begin{aligned}
\frac{\mu^2}{\alpha_s(\mu^2)}\frac{\partial\alpha_s(\mu^2)}{\partial\mu^2} &= -\frac{\alpha_s(\mu^2)}{4\pi}\beta_0 - \left(\frac{\alpha_s(\mu^2)}{4\pi}\right)^2\beta_1 - \left(\frac{\alpha_s(\mu^2)}{4\pi}\right)^3\beta_2 + \ldots \\
\beta_0 &= 11 - \frac{2}{3}n_f \\
\beta_1 &= 102 - \frac{38}{3}n_f \\
\beta_2(\overline{\mathrm{MS}}) &= \frac{2857}{2} - \frac{5033}{18}n_f + \frac{325}{54}n_f^2.
\end{aligned} \tag{4}$$

Retaining only the first two terms on the right hand side and solving the differential equation for $\alpha_s(\mu^2)$ gives

$$\frac{1}{\alpha_s} + b_1 \log\left(\frac{b_1\alpha_s}{1 + b_1\alpha_s}\right) = b_0 \log\frac{\mu^2}{\Lambda^2}, \tag{5}$$

with

$$b_0 = \frac{\beta_0}{4\pi}, \quad b_1 = \frac{\beta_1}{4\pi\beta_0}. \tag{6}$$

Note that a constant of integration in the form of a dimensionful parameter Λ has been introduced – replacing Λ by $c\Lambda$ also gives a solution to the differential equation. The convention chosen here is such that the left-hand-side vanishes when $\mu = \Lambda$. This is the standard definition of the 'two-loop' coupling constant as a function of the scale μ and the fundamental QCD scale parameter Λ. It is adequate for 'next-to-leading order' phenomenology. The above expression for α_s can be generalized to include also the β_2 term.[12]

An explicit form for α_s can be obtained by expanding (5) in inverse powers of $\log(\mu^2/\Lambda^2)$:

$$\alpha_s(\mu^2) = \frac{12\pi}{(33 - 2n_f)\log(\mu^2/\Lambda^2)}\left[1 - \frac{6(153 - 19n_f)}{(33 - 2n_f)^2}\frac{\log\log(\mu^2/\Lambda^2)}{\log(\mu^2/\Lambda^2)} + \ldots\right], \tag{7}$$

which illustrates the characteristic 'asymptotic freedom' property – the coupling decreases monotonically as μ^2 increases. Note however that this expansion corresponds to a slightly different definition of Λ from the implicit expression (5) for α_s, the expansion of which would contain a term $\sim$ const./$\log^2$. The freedom to multiply Λ by a constant can be used to remove this term. There is a $\mathcal{O}(15\%)$ difference in the Λ's defined by (5) and (7). Since in practice it is usually α_s which is measured experimentally, it is important when comparing Λ values to check that the same equation has been used to determine Λ from the coupling constant.

A second difficulty with the above definitions is that Λ depends on the number of active flavours. Values of Λ for different numbers of flavours can be defined by imposing the continuity of α_s at the scale $\mu = m$, where m is the mass of the heavy quark. For example, for the b-quark threshold: $\alpha_s(m_b^2, 4) = \alpha_s(m_b^2, 5)$. Using the next-to-leading order form (5) for $\alpha_s(Q^2)$ one can show that

$$\Lambda(4) \approx \Lambda(5) \left(\frac{m_b}{\Lambda(5)}\right)^{\frac{2}{25}} \left[\ln\left(\frac{m_b^2}{\Lambda(5)^2}\right)\right]^{\frac{963}{14375}} . \tag{8}$$

In practice, most higher order QCD corrections are carried out using the *modified minimal subtraction* ($\overline{\text{MS}}$) regularization scheme. To be consistent, then, one uses the above results for $\alpha_s(\mu^2)$ with $\Lambda \equiv \Lambda_{\overline{\text{MS}}}$.

Some recent α_s measurements, taken from the compilation by Bethke,[3] are shown in Fig.3. The lines indicate different values of $\Lambda^{(5)}_{\overline{\text{MS}}}$. Caution should be exercised when comparing the precision of the various measurements, as the errors have different meanings in different processes. A central value of $\Lambda^{(5)}_{\overline{\text{MS}}} \simeq 150$ MeV is indicated although, as pointed out in the Introduction, the LEP values on their own would suggest a higher value than this.

Historically, the dimensionful parameter Λ (in some particular renormalization scheme, e.g. $\overline{\text{MS}}$) has been taken as *the* fundamental parameter of QCD, to be measured by experiment. However, it is just as natural to take instead the value of the renormalized coupling at some reference mass scale. The reason this was not done was that there was no obvious 'standard' scale, high enough so that the coupling there was perturbatively small. The situation has changed with the advent of high precision α_s measurements at LEP. It now seems sensible to adopt $\alpha_s^{\overline{\text{MS}}}(M_Z^2)$ as the fundamental parameter, rather than $\Lambda_{\overline{\text{MS}}}$. (The small uncertainty in M_Z is completely negligible compared to the uncertainty in the coupling itself). Of course using the above equations it is straightforward to convert between the two. Table 2 gives the conversion between $\Lambda^{(5)}_{\overline{\text{MS}}}$ and $\alpha_s(M_Z^2)$ using the definition given in (5).

3. QCD in $e^+e^- \to$ hadrons

Many of the basic ideas and properties of perturbative QCD can be illustrated by considering the process $e^+e^- \to$ hadrons. We begin by discussing how the order α_s corrections to the total hadronic cross section are calculated, and how renormalization scheme dependence enters at order α_s^2. This cross section also provides one of the most precise measurements of the strong coupling, see Fig.3.

Perturbative QCD also predicts a rich 'jet' structure for the final state hadrons. We show how jet cross sections can be defined, and how the predictions compare with experiment.

Table 2: $\alpha_s(M_Z^2)$ for various $\Lambda_{\overline{MS}}^{(5)}$.

$\Lambda_{\overline{MS}}^{(5)}$ (MeV)	$\alpha_s(M_Z^2)$
50	0.0970
100	0.1060
150	0.1122
200	0.1170
250	0.1210
300	0.1245
350	0.1277
400	0.1305
450	0.1332
500	0.1356
550	0.1379
600	0.1401

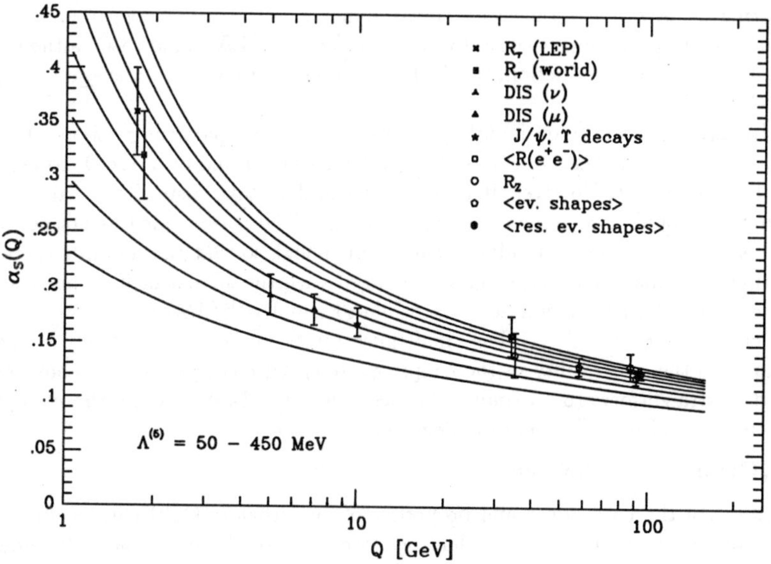

Figure 3: A compilation[3] of α_s measurements from different processes. The lines correspond to 50 MeV intervals of $\Lambda_{\overline{MS}}^{(5)}$.

3.1. *The total cross section for $e^+e^- \to$ hadrons*

One of the theoretically cleanest predictions of perturbative QCD is $R^{e^+e^-}$, the ratio of the total e^+e^- hadronic cross section to the muon pair production cross section. On the Z pole, as for example at LEP and SLC, the analogous quantity is the ratio of the partial decay widths of the Z to hadrons and to muon pairs. We begin by considering the high energy $2 \to 2$ process $e^+e^- \to f\bar{f}$ with f a light charged fermion, $f \neq e$. In lowest order the process is mediated by either a virtual photon or a Z^0 in the s–channel. With θ the centre-of-mass scattering angle of the final state pair, the differential cross section is:

$$\begin{aligned}\frac{d\sigma}{d\cos\theta} &= \frac{\pi\alpha^2}{2s}\Big[(1+\cos^2\theta)(Q_f^2 - 2Q_f v_e v_f \chi_1(s) + (a_e^2+v_e^2)(a_f^2+v_f^2)\chi_2(s)) \\ &\quad + \cos\theta(-4Q_f a_e a_f \chi_1(s) + 8a_e v_e a_f v_f \chi_2(s))\Big] \end{aligned} \tag{9}$$

where

$$\begin{aligned}\chi_1(s) &= \kappa \frac{s(s-M_Z^2)}{(s-M_Z^2)^2+\Gamma_Z^2 M_Z^2} \\ \chi_2(s) &= \kappa^2 \frac{s^2}{(s-M_Z^2)^2+\Gamma_Z^2 M_Z^2} \\ \kappa &= \left(\frac{\sqrt{2}G_F M_Z^2}{16\pi\alpha}\right) \end{aligned} \tag{10}$$

$$\tag{11}$$

and (v_f, a_f) are the vector and axial couplings of the fermions to the Z. The χ_2 term comes from the square of the Z-exchange amplitude and the χ_1 term from the photon-Z interference. Now at centre-of-mass scattering energies $\sqrt{s}$ far below the Z peak, the ratio s/M_Z^2 is small and so $1 \gg \chi_1 \gg \chi_2$. This means that the weak effects – manifest in the terms involving the vector and axial couplings – are quite small and can be neglected. Eqn.(9) then reduces to

$$\frac{d\sigma}{d\cos\theta} = \frac{\pi\alpha^2 Q_f^2}{2s}(1+\cos^2\theta). \tag{12}$$

Integrating over θ and setting $Q_f = -1$ gives the total cross section for $e^+e^- \to \mu^+\mu^-$:

$$\sigma_0 = \frac{4\pi\alpha^2}{3s}, \tag{13}$$

where $\sqrt{s}$ is the total centre-of-mass energy. On the Z pole, $\sqrt{s} = M_Z$, the χ_2 term in (9) dominates and the (peak) cross section is

$$\sigma_0 = \frac{4\pi\alpha^2\kappa^2}{3\Gamma_Z^2}(a_e^2+v_e^2)^2. \tag{14}$$

When an electron and a positron annihilate they can also produce hadrons in the final state. The formation of the observed final state hadrons is not governed by perturbation theory. Why then would one expect perturbation theory to give an accurate description of the hadronic production cross section? The answer can be understood by visualising the event in space-time. The electron and positron form a photon of virtuality $Q = \sqrt{s}$ which fluctuates into a quark and an antiquark. By the uncertainty principle this fluctuation occurs in a space time volume $1/Q$, and if Q is large the production rate should be predicted in perturbation theory. Subsequently the quarks and gluons form themselves into hadrons. This happens at a later time scale characterised by the scale $1/\Lambda$, where Λ is the typical mass scale of the strong interactions. The interactions which change quarks and gluons into hadrons modify the outgoing state, but they occur too late to modify the probability for an event to happen.

In leading order perturbation theory, therefore, the total hadronic cross section is obtained by simply summing over all kinematically accessible flavours and colours of quarks. Ignoring the Z exchange contributions (i.e. assuming $\sqrt{s} \ll M_Z$) we have

$$R^{QPM} = \frac{\sum_q \sigma(e^+e^- \to q\bar{q})}{\sigma(e^+e^- \to \mu^+\mu^-)} = 3\sum_q Q_q^2 \tag{15}$$

With $q = u, ..., b$ we obtain $R^{QPM} = 11/3 = 3.67$. At $\sqrt{s} = 34$ GeV the measured value is about 3.9. Even allowing for the Z contribution ($\Delta R_Z \simeq 0.05$), this result is some 5% higher than the lowest order prediction. It turns out that the difference is due to higher order QCD corrections, and in fact the comparison between theory and experiment gives one of the most precise determinations of the strong coupling constant.

The $O(\alpha_s)$ corrections to the total hadronic cross section are calculated from the real and virtual gluon diagrams shown in Fig.4. For the real gluon emission diagrams it is convenient to write the three-body phase space integration as

$$\mathrm{d}\Phi_3 \sim \mathrm{d}\alpha \mathrm{d}\beta \mathrm{d}\gamma \mathrm{d}x_1 \mathrm{d}x_2 \tag{16}$$

where α, β, γ are Euler angles, and $x_1 = 2E_q/\sqrt{s}$ and $x_2 = 2E_{\bar{q}}/\sqrt{s}$ are the energy fractions of the final state quark and antiquark. Integrating out the Euler angles gives a matrix element which depends only on x_1 and x_2 and the contribution to the total cross section is

$$\sigma^{q\bar{q}g} = \sigma_0 \, 3\sum_q Q_q^2 \int \mathrm{d}x_1 \mathrm{d}x_2 \, \frac{2\alpha_s}{3\pi} \, \frac{x_1^2 + x_2^2}{(1-x_1)(1-x_2)} \tag{17}$$

where the integration region is: $0 \leq x_1, x_2 \leq 1$, $x_1 + x_2 \geq 1$. Unfortunately, we see that the integrals are divergent at $x_i = 1$! These singularities come from regions of phase space where the gluon is *collinear* with either quark, $\theta_{qg} \to 0$, or where the gluon is *soft*, $E_g \to 0$. Evidently we require some sort of regularization procedure – to render the integrals finite – before the calculation can be completed. A variety of methods

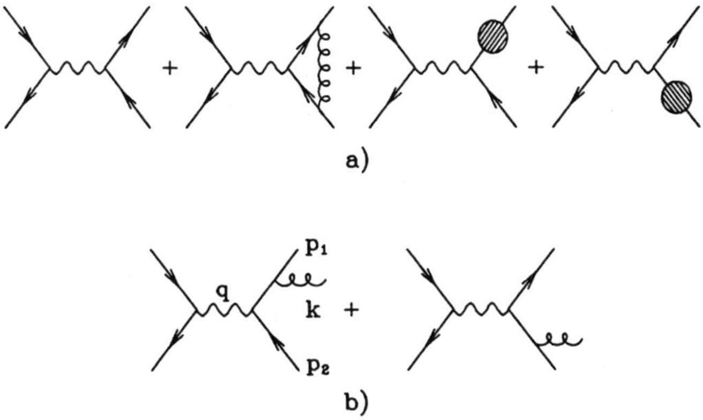

Figure 4: Feynman diagrams for the $O(\alpha_s)$ corrections to the total hadronic cross section in e^+e^- annihilation.

are suitable. One can give the gluon a small mass, or take the final state quark and antiquark off-mass-shell by a small amount. In each case the singularities are then manifest as logarithms of the regulating mass.

A more elegant procedure is to use dimensional regularization, with the number of space-time dimensions now $n > 4$. With the three-body phase space integrals now cast in N dimensions, the soft and collinear singularities appear as *poles* at $n = 4$. Details of how the calculation proceeds can be found for example in.[10] The result is that the cross section of Eqn.(17) becomes

$$\sigma^{q\bar{q}g} = \sigma_0\, 3\sum_q Q_q^2\, \frac{2\alpha_s}{3\pi}\, H(\epsilon)\left[\frac{2}{\epsilon^2} - \frac{3}{\epsilon} + \frac{19}{2} + O(\epsilon)\right], \tag{18}$$

where $H(\epsilon) = 1 + O(\epsilon)$.

The virtual gluon contribution can be calculated in a similar fashion, with dimensional regularization again used to render finite the infra-red divergences in the loops. The result is

$$\sigma^{q\bar{q}(g)} = \sigma_0\, 3\sum_q Q_q^2\, \frac{2\alpha_s}{3\pi}\, H(\epsilon)\left[-\frac{2}{\epsilon^2} + \frac{3}{\epsilon} - 8 + O(\epsilon)\right]. \tag{19}$$

When the two contributions Eqns.(18) and (19) are added together the poles exactly cancel and the result is *finite* in the limit $\epsilon \to 0$:

$$R^{e^+e^-} = 3\sum_q Q_q^2 \left\{1 + \frac{\alpha_s}{\pi} + O(\alpha_s^2)\right\}. \tag{20}$$

Note that the next-to-leading order correction is positive, and with a value for α_s of about 0.15, can accommodate the experimental measurement at $\sqrt{s} = 34$ GeV. In contrast, the corresponding correction is negative for a scalar gluon.

The cancellation of the soft and collinear singularities between the real and virtual gluon diagrams is not accidental. Indeed, there are theorems – the Bloch, Nordsieck[13] and Kinoshita, Lee, Nauenberg[14] theorems – which state that suitably defined inclusive quantities will be free of singularities in the massless limit. The total hadronic cross section is an example of such a quantity, whereas the cross section for the exclusive $q\bar{q}$ final state, i.e. $\sigma(e^+e^- \to q\bar{q})$ is not.

The $O(\alpha_s^2)$ and $O(\alpha_s^3)$ corrections to $R^{e^+e^-}$ are also known. At these higher orders we encounter the ultra-violet divergences associated with the renormalization of the strong coupling. Writing

$$\begin{aligned} \sigma_{tot} &= \frac{4\pi\alpha^2}{3s} R, \\ R &= K_{QCD}\, 3\sum_q Q_q^2, \\ K_{QCD} &= 1 + \sum_{n\geq 1} C_n \left(\frac{\alpha_s}{\pi}\right)^n, \end{aligned} \tag{21}$$

the coefficients C_1, C_2 and C_3 are (in the $\overline{\rm MS}$ scheme with the renormalization scale choice $\mu = \sqrt{s}$):

$$\begin{aligned}
C_1 &= 1 \\
C_2 &= \left(\frac{2}{3}\zeta(3) - \frac{11}{12}\right)n_f + \left(\frac{365}{24} - 11\zeta(3)\right) \\
&\simeq 1.986 - 0.115 n_f \\
C_3 &= \left(\frac{87029}{288} - \frac{1103}{4}\zeta(3) + \frac{275}{6}\zeta(5)\right) \\
&\quad -\left(\frac{7847}{216} - \frac{262}{9}\zeta(3) + \frac{25}{9}\zeta(5)\right)n_f \\
&\quad +\left(\frac{151}{162} - \frac{19}{27}\zeta(3)\right)n_f^2 \\
&\quad -\frac{\pi^2}{432}(33 - 2n_f)^2 + \eta\left(\frac{55}{72} - \frac{5}{3}\zeta(3)\right) \\
&\simeq -6.637 - 1.200 n_f - 0.005 n_f^2 - 1.240\eta,
\end{aligned} \tag{22}$$

where $\eta = (\sum_q Q_q)^2/3\sum_q Q_q^2$ and the sum extends over the (n_f) quarks which are effectively massless at the energy scale $\sqrt{s}$. The result for C_3 is from Samuel and Surguladze.[15] Apart from the η term, the result for the QCD corrections K is the same for the ratio of hadronic to leptonic Z decay widths: $R_Z = \Gamma_h/\Gamma_\mu$. In practice, quark masses (particularly m_b and m_t) have a small but non-negligible effect[16] and must be taken into account in precision fits to data.

Through $\mathcal{O}(\alpha_s^3)$ the μ dependence is restored by the replacements:

$$\begin{aligned}
\alpha_s &\to \alpha_s(\mu^2) \\
C_2 &\to C_2 - C_1\frac{\beta_0}{4}\log\frac{s}{\mu^2} \\
C_3 &\to C_3 + C_1\left(\frac{\beta_0}{4}\right)^2\log^2\frac{s}{\mu^2} - (C_1\frac{\beta_1}{16} + C_2\frac{\beta_0}{2})\log\frac{s}{\mu^2}.
\end{aligned} \tag{23}$$

where β_0 and β_1 are defined in Section 2.2. Note that the μ-dependence of the second order coefficient is exactly as specified by the renormalization group equation, i.e. the coefficient of $\log(s/\mu^2)$ is proportional to the β function coefficient defined in (4).

In general the coefficients of any QCD perturbative expansion depend on the choice made for the renormalization scale μ in such a way that as μ is varied, the change in the coefficients exactly compensates the change in the coupling $\alpha_s(\mu^2)$. However this μ-independence breaks down whenever the series is *truncated*. One can show in fact that changing the scale in a physical quantity such as $R^{e^+e^-}$ – which has been calculated to $O(\alpha_s^n)$ – induces changes of $O(\alpha_s^{n+1})$. This is illustrated in Fig.5, which shows $K_{QCD} = 1 + \delta$ for R_Z as a function of μ, as the higher order terms are added in. As expected, the inclusion of higher order terms leads to a more definite prediction. In the absence of higher order corrections, one can try to guess the 'best' choice of scale,

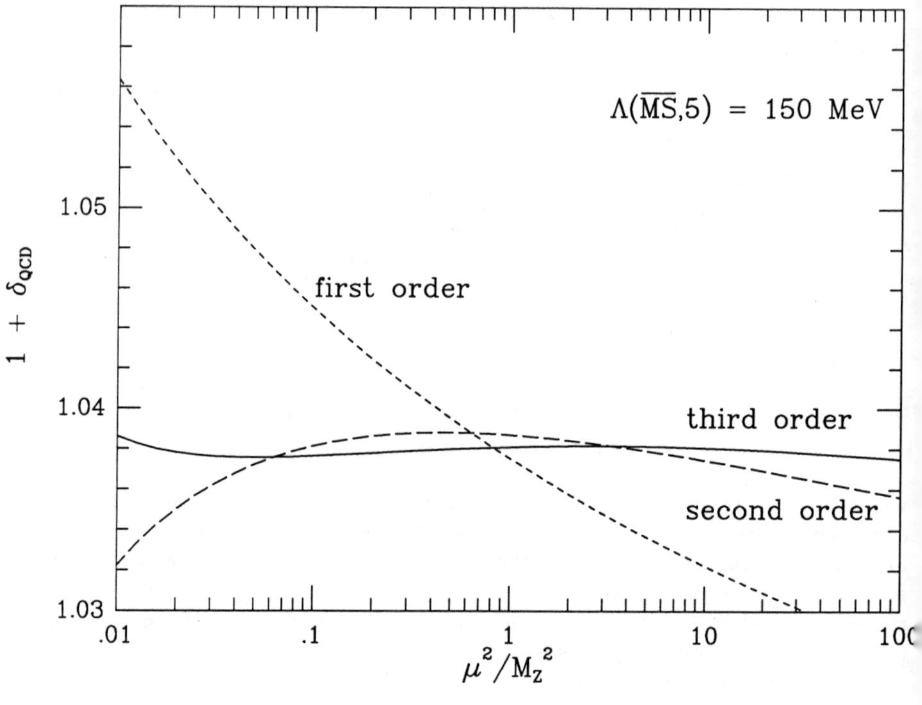

Figure 5: The effect of higher order QCD corrections to R_Z, as a function of the renormalization scale μ.

defined as the scale which makes the truncated and all-orders predictions equal. In the literature, two such choices have been advocated in particular. In the *fastest apparent convergence* approach,[17] one chooses the scale $\mu = \mu_{FAC}$, where

$$R^{(1)}(\mu_{FAC}) = R^{(2)}(\mu_{FAC}). \tag{24}$$

On the other hand, the *principle of minimal sensitivity*[18] suggests a scale choice $\mu = \mu_{PMS}$, where

$$\mu \frac{\mathrm{d}}{\mathrm{d}\mu} R^{(2)}(\mu)\Big|_{\mu_{PMS}} = 0. \tag{25}$$

These two special scales can be identified in Fig.5. it is important to remember that there are no theorems that prove that any of these schemes are correct. All one can say is that the theoretical error on a quantity calculated to $O(\alpha_s^n)$ is $O(\alpha_s^{n+1})$. Varying the scale is simply one way of quantifying this uncertainty.

3.2. Jet cross sections

The expression given for the total hadronic cross section in the previous section is very concise, but it tells us nothing about the kinematic *distribution* of hadrons in the final state. If the hadronic fragments of a fast moving quark have limited transverse momentum relative to the quark momentum, then the lowest order contribution – $e^+e^- \to q\bar{q}$ – can naively be interpreted as the production of two back-to-back jets. In this section we investigate how higher order perturbative corrections modify this picture.

Consider first the next-to-leading process $e^+e^- \to q\bar{q}g$. From the previous section (Eqn.(17)), we have

$$\frac{1}{\sigma}\frac{\mathrm{d}^2\sigma}{\mathrm{d}x_1\mathrm{d}x_2} = \frac{2\alpha_s}{3\pi}\frac{x_1^2 + x_2^2}{(1-x_1)(1-x_2)} \tag{26}$$

Recall that the cross section becomes infinitely large when either (a) the gluon is collinear with one of the outgoing quarks, or (b) the gluon momentum goes to zero. This corresponds to (a) only one and (b) both of the x_i approaching 1 respectively. In other words the gluon prefers to be soft and/or collinear with the quarks. If the gluon is *required* to be well-separated in phase space from the quarks – a configuration corresponding to a 'three jet event' – then the cross section is suppressed relative to lowest order by one power of α_s. It would appear, therefore, that the two-jet nature of the final state is maintained to next-to-leading order, since both the preferred configurations give a final state indistinguishable (after parton fragmentation to hadrons) from that at lowest order. This qualitative result holds in fact to *all* orders of perturbation theory. Multigluon emission leads to a final state which is predominantly 'two-jet-like', with a smaller probability (determined by α_s) for three or more distinguishable jets. A more complete discussion is given by Kramer.[19]

To quantify this statement we need to introduce the concept of a *jet measure*, i.e. a procedure for classifying a final state of hadrons (experimentally) or quarks and gluons (theoretically) according to the number of jets. To be useful, a jet measure should be free of soft and collinear singularities when calculated in perturbative QCD, and should also be relatively insensitive to the non-perturbative fragmentation of quarks and gluons into hadrons.

One of the most widely used jet measures is the 'minimum invariant mass' or JADE algorithm.[20] Consider a $q\bar{q}g$ final state. A three-jet event is one in which the invariant masses of the parton pairs are all larger than some fixed fraction y of the overall centre-of-mass energy:

$$(p_i + p_j)^2 > ys, \qquad i,j = q,\bar{q},g. \tag{27}$$

It is immediately clear that this region of phase space avoids the soft and collinear singularities of the matrix element. In fact in terms of the energy fractions, Eqn.(27) is equivalent to

$$0 < x_1, x_2 < 1 - y, \qquad x_1 + x_2 > 1 + y. \tag{28}$$

If we define f_2 and f_3 as the two and three jet fractions defined this way then to $O(\alpha_s)$ we obtain

$$\begin{aligned} f_3 &= \frac{2\alpha_s}{3\pi}\left[(3-6y)\log\left(\frac{y}{1-2y}\right) + 2\log^2\left(\frac{y}{1-y}\right) + \frac{5}{2} - 6y - \frac{9}{2}y^2 \right. \\ &\left. +4\ \mathrm{Li}_2\left(\frac{y}{1-y}\right) - \frac{\pi^2}{3}\right], \quad \mathrm{Li}_2(y) = -\int_0^y \frac{\mathrm{d}z}{1-z}\log z\ , \\ f_2 &= 1 - f_3 \end{aligned} \tag{29}$$

Note that the soft and collinear singularities reappear as large logarithms in the limit $y \to 0$. Clearly the result only makes sense for y values large enough such that $f_2 \gg f_3$, so that the $O(\alpha_s)$ correction to f_2 is perturbatively small.

The generalisation to multi-jet fractions is straightforward. Starting from an n-parton final state, identify the pair with the minimum invariant mass squared. If this is greater then ys then the number of jets is n. If not, combine the minimum pair into a single 'cluster'. Then repeat for the $(n-1)$-parton/cluster final state, and so on until all parton/clusters have a relative invariant mass squared greater than ys. The number of clusters remaining is then the number of jets in the final state. Note that an n-parton final state can give any number of jets between n (all partons well-separated) and 2 (for example, two hard quarks accompanied by soft and collinear gluons).

Since a soft or collinear gluon emitted from a quark line does not change the multiplicity of jets, the cancellation of soft and collinear singularities that was evident in the total cross section calculation can still take place, and the jet fractions defined this way are free of such singularities to all orders in perturbation theory.

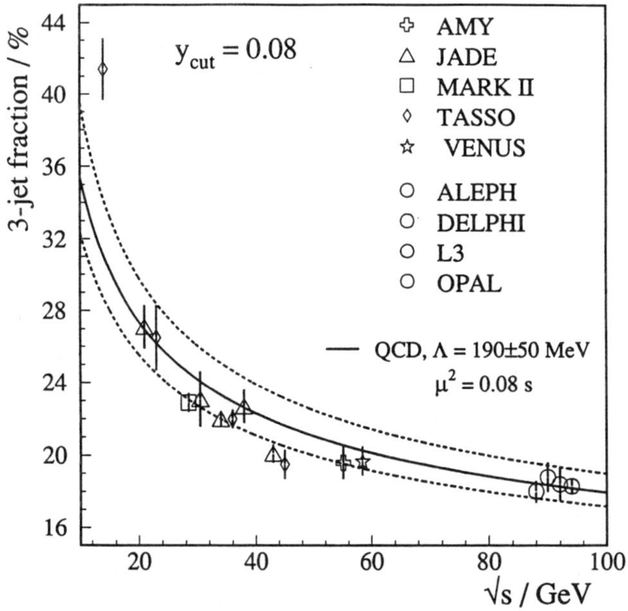

Figure 6: A compilation of three-jet fractions at different e^+e^- annihilation energies, from the L3 collaboration[26].

Now in general we have

$$\begin{aligned} f_{n+2}(\sqrt{s},y) &= \Big(\frac{\alpha_s(s)}{\pi}\Big)^n \sum_{j=0}^{\infty} C_{nj}(y)\Big(\frac{\alpha_s(s)}{\pi}\Big)^j, \quad n \geq 0, \\ \sum_{n=2}^{\infty} f_n &= 1\,. \end{aligned} \tag{30}$$

Note that since the jet parameter y is dimensionless, all the energy dependence of the jet fractions is contained in the coupling $\alpha_s(s)$. One can therefore exhibit, at least in principle, the *running* of the strong coupling by measuring a decrease in f_3 as $\sqrt{s}$ increases, see Fig.6. Note that experimentally the algorithm is applied to final state *hadrons* rather than *partons*. However studies using parton shower/fragmentation Monte Carlos have shown that – at least at high energy – the fragmentation corrections are small and therefore the QCD parton-level predictions can be reliably compared with the experimental data. A quantitative discussion can be found in Bethke et al.[21]

The next-to-leading order corrections to f_3 have been calculated.[25] Because the

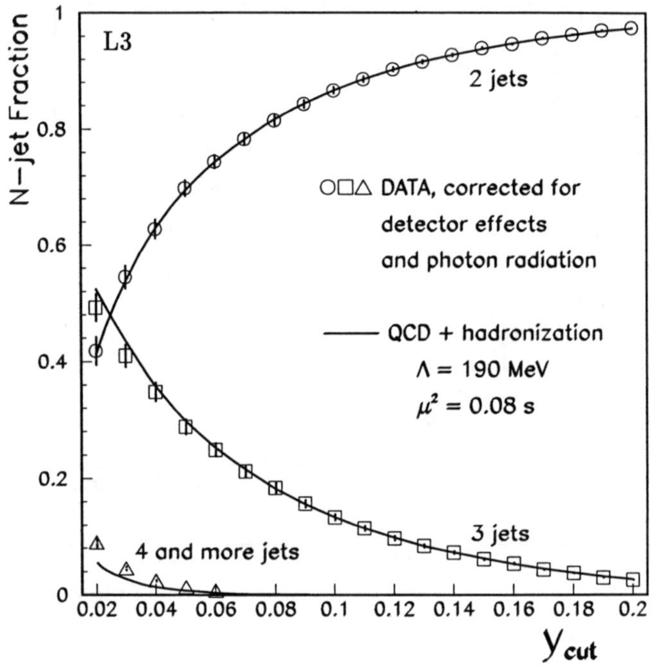

Figure 7: QCD fits to the jet rates at LEP as measured by the L3 collaboration[26].

hadronization corrections to f_3 are small, the three-jet rate provides one of the most precise measurements of α_s at LEP and SLC. A typical fit is shown in Fig.7.[26]

While the above definition is well suited to experimental jet measurements, it is not quite optimum from a theoretical point of view. The reason is that when y becomes small (as happens in practice), the large logarithms of y explicit in (29) begin to dominate the theoretical predictions. It is straightforward to show[22] that higher order corrections to jet fractions such as f_2 and f_3 will contain terms like $\alpha_s^n \log^{2n} y$. When y is small enough that $\alpha_s \log^2 y \sim 1$, then these terms must be resummed to obtain a reliable prediction. Unfortunately, the JADE algorithm is not well-suited to this type of resummation,[22] and so a variant – the 'Durham' or k_T algorithm – has been proposed.[23,24] In this new algorithm, the invariant mass measure of two partons (hadrons) given in (27) is replaced by the minimum of the relative transverse momenta:

$$\min k_{Tij}^2 = \min(E_i^2, E_j^2) \sin\theta_{ij}^2 > ys, \quad i,j = q,\bar{q},g \tag{31}$$

in the e^+e^- centre-of-mass frame with massless quarks and gluons. With this new definition of the jet measure, resummation of all large logarithms can be performed.[23] Finite, next-to-leading order corrections have also been calculated[21] and comparisons of theory and experiment have been performed,[3] as for the JADE algorithm. It is interesting that the (presumably) more reliable α_s values from resummed perturbative jet measures tend to be slightly larger than those obtained without resummation, see for example Fig.3.

4. Deep Inelastic Scattering

The original, and still the most powerful, test of perturbative QCD is the breaking of Bjorken scaling in deep inelastic lepton-hadron scattering. Nowadays, structure function analyses not only provide some of the most precise tests of the theory but also determine the momentum distributions of partons in hadrons for use as input in predicting cross-sections in high energy hadron-hadron collisions. In this section we discuss the parton structure of hadrons and its relation to perturbative Quantum Chromodynamics. We introduce the parton model of short-distance hadron structure and discuss how the basic 'scaling property' is modified by perturbative QCD corrections. Lack of space precludes an in-depth treatment of most of these issues, but further details can, as always, be found in the literature (see for example Roberts[27]).

4.1. The Parton Model

Consider the deep inelastic process $lp \to lX$. Label the incoming and outgoing lepton four-momenta by k^μ and k'^μ respectively, the incoming proton momentum by p^μ ($p^2 = M^2$) and the momentum transfer by $q^\mu = k^\mu - k'^\mu$. The standard deep inelastic variables are defined by:

$$\begin{aligned} Q^2 &= -q^2 \qquad p^2 = M^2 \\ x &= \frac{Q^2}{2p\cdot q} = \frac{Q^2}{2M(E-E')} \\ y &= \frac{q\cdot p}{k\cdot p} = 1 - E'/E \\ s &= (k+p)^2 = M^2 + \frac{Q^2}{xy}, \end{aligned} \tag{32}$$

where the energies are defined in the rest frame of the target. The structure functions $F_i(x, Q^2)$ are then defined in terms of the lepton scattering cross sections. For charged lepton scattering, $lp \to lX$,

$$\begin{aligned} \frac{\mathrm{d}^2\sigma^{em}}{\mathrm{d}x\mathrm{d}y} &= \frac{4\pi\alpha^2(s-M^2)}{Q^4}\left[\left(\frac{1+(1-y)^2}{2}\right)2xF_1^{em}\right. \\ &\left. +(1-y)(F_2^{em} - 2xF_1^{em}) - \frac{M^2}{s-M^2}xyF_2^{em}\right], \end{aligned} \tag{33}$$

and for neutrino (antineutrino) scattering, $\nu(\bar\nu)p \to lX$,

$$\begin{aligned} \frac{\mathrm{d}^2\sigma^{\nu(\bar\nu)}}{\mathrm{d}x\mathrm{d}y} &= \frac{G_F^2(s-M^2)}{2\pi}\left[\left(1-y-\frac{M^2}{s-M^2}xy\right)F_2^{\nu(\bar\nu)}\right. \\ &\left. +y^2xF_1^{\nu(\bar\nu)} + (-)y(1-y/2)xF_3^{\nu(\bar\nu)}\right]. \end{aligned} \tag{34}$$

In the quark-parton model, these structure functions are related to the quark 'distribution functions' or 'densities' $q(x, Q^2)$, where $q(x, Q^2)\mathrm{d}x$ is the probability that a parton carries a momentum fraction x of the target nucleon's momentum when probed (by a gauge boson γ^*, W or Z) at energy scale Q. Thus

$$\begin{aligned} F_2^{\nu} &= 2x[d + s + \bar{u} + \bar{c}] \\ xF_3^{\nu} &= 2x[d + s - \bar{u} - \bar{c}] \\ F_2^{\bar{\nu}} &= 2x[u + c + \bar{d} + \bar{s}] \\ xF_3^{\bar{\nu}} &= 2x[u + c - \bar{d} - \bar{s}] \\ F_2^{em} &= x[\frac{4}{9}(u + u + c + \bar{c}) + \frac{1}{9}(d + \bar{d} + s + \bar{s})] \\ 2xF_1 &= F_2. \end{aligned} \tag{35}$$

Note that when the nature of the target is unambiguous the notation $q(x, Q^2)$ and $g(x, Q^2)$ for the quark and gluon densities can be used, otherwise a general notation is $f_{a/A}(x, Q^2)$, where a = u, d, ... g and A = p, n, Fe, Cu, etc. In the 'naive' parton model the structure functions *scale*, i.e. $F(x, Q^2) \to F(x)$ in the asymptotic (Bjorken) limit: $Q^2 \to \infty$, x fixed. In fact, it was the observation of scaling in the original SLAC experiments that provided the first evidence of parton structure in the hadron. To a first approximation, therefore, one can take the parton distributions to be functions of x only: $q(x, Q^2) \to q(x)$. We shall see below how perturbative QCD induces logarithmic deviations from scaling, exactly in line with more recent high-precision experimental measurements.

The picture which emerges is of a proton consisting of three valence quarks (uud) and a 'sea' of $q\bar{q}$ pairs and gluons. In the most simple version of this parton model the sea would be (three) flavour symmetric and hence the net quark distributions would be given by $u = u_V + S$, $d = d_V + S$ and $s = \bar{s} = \bar{u} = \bar{d} = S$, with the sum rules

$$\int_0^1 u_V(x)\mathrm{d}x = 2\int_0^1 d_V(x)\mathrm{d}x = 2 \tag{36}$$

$$\int_0^1 x[u_V(x) + d_V(x) + 6S(x)]\mathrm{d}x = 1\,. \tag{37}$$

These represent conservation of proton quantum numbers and momentum fraction respectively.

Fig.8 shows some representative modern quark and gluon distributions (the MRS-D_0' distributions of Martin et al.[4]) extracted from deep inelastic scattering and other processes. The scale is $Q^2 = 20\ \mathrm{GeV}^2$, a typical value for fixed-target deep inelastic experiments. At this scale the charm quark distribution is very small and is not shown.

4.2. Scaling Violations – the Altarelli-Parisi Equations

In QCD, Bjorken scaling is broken by logarithms of Q^2. Physically, a quark in the proton can emit a gluon, with probability determined by α_s, and lose momentum as

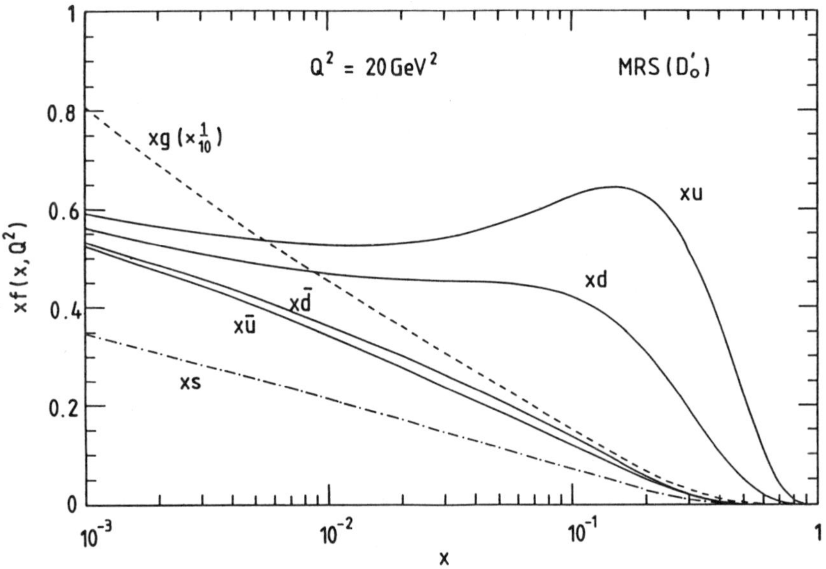

Figure 8: Quark and gluon distributions at $Q^2 = 20$ GeV^2 from Martin et al[4].

a result. Since the higher the Q^2 the more phase space is available for gluon emission, the expectation is that parton distributions should shrink to small x as Q^2 increases, with the rate of shrinkage being controlled by α_s.

In describing quantitatively the way in which scaling is violated it is convenient to define singlet and non-singlet quark distributions:

$$F^{NS} = q_i - q_j, \qquad F^S = \sum_i (q_i + \bar{q}_i). \tag{38}$$

The non-singlet structure functions have non-zero values of flavour quantum numbers such as isospin or baryon number. The variation with Q^2 of these functions is described by the so-called Altarelli-Parisi equations[28]:

$$\begin{aligned} Q^2 \frac{\partial F^{NS}}{\partial Q^2} &= \frac{\alpha_s(Q^2)}{2\pi} P^{qq} * F^{NS} \\ Q^2 \frac{\partial F^S}{\partial Q^2} &= \frac{\alpha_s(Q^2)}{2\pi} \left(P^{qq} * F^S + 2n_f P^{qg} * g \right) \\ Q^2 \frac{\partial g}{\partial Q^2} &= \frac{\alpha_s(Q^2)}{2\pi} \left(P^{gq} * F^S + P^{gg} * g \right), \end{aligned} \tag{39}$$

where $*$ denotes a convolution integral:

$$f * g = \int_x^1 \frac{\mathrm{d}y}{y} f(y) g\left(\frac{x}{y}\right). \tag{40}$$

In leading order the Altarelli-Parisi splitting functions are

$$\begin{aligned} P^{qq} &= \frac{4}{3}\left(\frac{1+x^2}{1-x}\right)_+ \\ P^{qg} &= \frac{1}{2}(x^2 + (1-x)^2) \\ P^{gq} &= \frac{4}{3}\left(\frac{1+(1-x)^2}{x}\right) \\ P^{gg} &= 6\left(\frac{1-x}{x} + x(1-x) + \left(\frac{x}{1-x}\right)_+\right) \\ &\quad - \left(\frac{1}{2} + \frac{n_f}{3}\right)\delta(1-x). \end{aligned} \tag{41}$$

Note the 'plus prescription' for those functions which are singular as $x \to 1$:

$$\int_0^1 \mathrm{d}x f(x) \left(g(x)\right)_+ = \int_0^1 \mathrm{d}x (f(x) - f(1)) g(x). \tag{42}$$

The Altarelli-Parisi equations can be solved analytically by defining *moments* (formally, the Mellin transforms) of the structure functions, $M_n^{NS} = \langle F^{NS} \rangle_n \equiv \int_0^1 \mathrm{d}x x^{n-1} F^{NS}$ etc.

The convolution integral then becomes a simple product. Introducing the leading order expression for the QCD coupling constant (see Section 2.2),

$$\alpha_s(Q^2) = \frac{4\pi}{\beta_0 \log(Q^2/\Lambda^2)}, \tag{43}$$

one obtains, for the non-singlet solution,

$$M_n^{NS}(Q^2) = M_n^{NS}(Q_0^2)\left(\frac{\alpha_s(Q^2)}{\alpha_s(Q_0^2)}\right)^{-d_n}, \tag{44}$$

where $d_n = 2\langle P^{qq}\rangle_n/\beta_0$. Note that $d_1 = 0$ and that $d_n < 0$ for $n \geq 2$, which implies that the parton distributions decrease and increase with increasing Q^2 at large and small x respectively. Solutions for the singlet and gluon moments can be found in a similar way, by first diagonalizing the coupled equations.

The precision of contemporary deep inelastic data demands that the QCD predictions are calculated to next-to-leading order. This amounts to the replacements (shown schematically):

$$\begin{aligned} P(x) &\rightarrow P^{(0)}(x) + \frac{\alpha_s}{2\pi}P^{(1)}(x) \\ F = \sum q &\rightarrow F^{(1)} = \sum C * q, \qquad C = \delta(1-x) + O(\alpha_s). \end{aligned} \tag{45}$$

Beyond leading order, the definition of parton distributions (like α_s) becomes scheme dependent. It is conventional nowadays to work in the $\overline{\rm MS}$ scheme, where for example

$$\begin{aligned} F_2^{\mu p}(x,Q^2) &= \sum_q e_q^2 x q^{\overline{\rm MS}}(x,Q^2) + \alpha_s^{\overline{\rm MS}}(Q^2)\sum_q x\int_x^1 \frac{dz}{z}c_q(z)q^{\overline{\rm MS}}\left(\frac{x}{z},Q^2\right) \\ &\quad + x\int_x^1 \frac{dz}{z}c_g(z)g^{\overline{\rm MS}}\left(\frac{x}{z},Q^2\right) . \end{aligned} \tag{46}$$

An example of a next-to-leading order QCD fit[4] to recent high-precision data on $F_2^{\mu p}$ from the BCDMS[29] and NMC[30] collaborations is shown in Fig.9.

A more complete discussion of how detailed QCD fits are performed can be found in the review by Roberts.[27] In summary, the observed scaling violations are exactly in line with perturbative QCD expectations and lead to some of the most precise measurements of the strong coupling, at a scale $\mu \sim 10$ GeV (see Fig.3).

4.3. Parton Distributions in Hadrons

As we have seen in the previous sections, distributions of the quarks (and gluons) in the proton are determined by values of the structure functions $F_i(x,Q^2)$ measured in muon (or electron)-nucleon (or nucleus) and neutrino-nucleus deep inelastic scattering experiments. It is relevant to ask why we should devote so much effort to the study of the distributions of partons in the proton. There are two main reasons – one

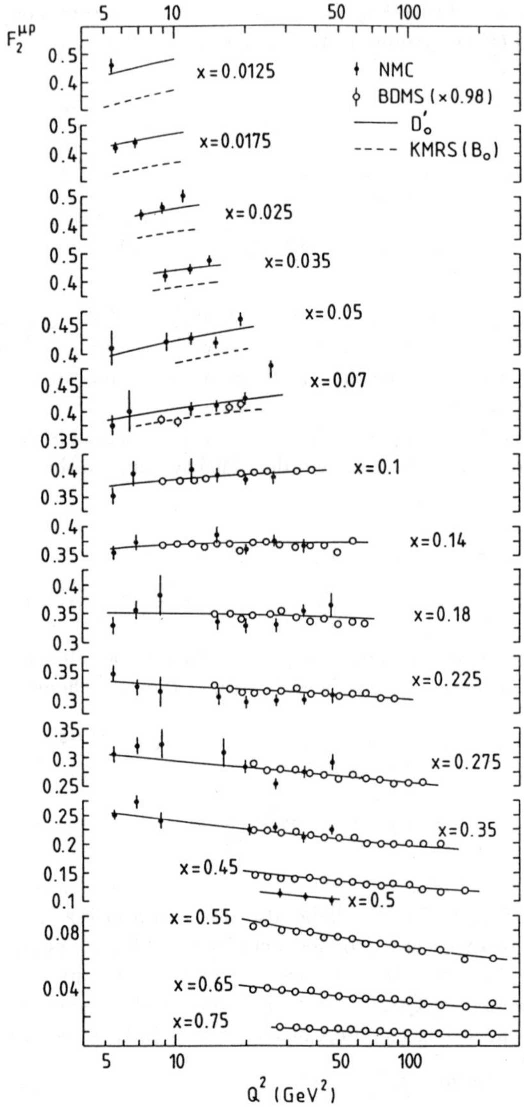

Figure 9: The BCDMS[29] and NMC[30] measurements of $F_2^{\mu p}$, together with the fits by KMRS[31] and MRS[4]. The NMC data were not available for KMRS.

experimental and one theoretical. First, a detailed knowledge of parton distributions is an essential ingredient in all 'hard' interactions involving protons, and so they are needed to estimate the production rates of the various 'hard' processes that may occur at current and forthcoming colliders (see Section 5). However the parton structure is interesting in its own right. In particular, novel perturbative QCD effects are expected to become apparent at small x (i.e. $x \lesssim 10^{-3}$). The reason is that at small x the sum over soft gluon emission off the incoming parton leads to a power series in $\alpha_s \log(1/x)$, which on resummation, via the Lipatov (or BFKL) equation,[32] suggests that the gluon behaves as

$$xg \sim x^{-\lambda} \tag{47}$$

as $x \to 0$, with λ predicted, typically, to be about 0.5. Such an increase in $xg(x,Q^2)$, as x decreases, cannot go on indefinitely. If the density of gluons becomes too large they can no longer be treated as free partons, and the effects of recombination or shadowing must be included. These small-x effects are reviewed in detail by Kwiecinski.[33]

There is a long history of determining parton distributions from deep-inelastic data. The recent MRS analysis[4] shows that a wide range of data are well described by remarkably simple parametrizations of parton distributions at $Q_0^2 = 4$ GeV^2; in total only about 15 parameters are required. An example of the quality of the fit has already been shown in Fig.9. In order to provide an indication of the possible range of extrapolations to smaller values of x, beyond the range of current fixed-target data, two sets of MRS partons are presented which describe the data equally well – one set omitting and one set including the Lipatov behaviour (47) of the gluon and sea quark distributions at small x. Namely

$$\begin{aligned} xg(x,Q_0^2) &\sim \text{constant} && \text{for the } \mathrm{D}_0' \text{ set,} \\ xg(x,Q_0^2) &\sim x^{-\frac{1}{2}} && \text{for the } \mathrm{D}_-' \text{ set,} \end{aligned} \tag{48}$$

as $x \to 0$; and similarly for $x\bar{q}(x,Q_0^2)$ since the sea is driven at small x by $g \to q\bar{q}$.

The parton sets MRS-D_0' and MRS-D_-' [4] are essentially identical for $x \gtrsim 0.02$ but (by construction) are dramatically different at small x. In particular they lead to the very different predictions for F_2^{ep} in the HERA regime, as shown by the continuous curves in Fig.10. Also shown are (i) the CTEQ1M and CTEQ1MS predictions[34] and (ii) the predictions from the 'dynamical quark' model of Glück, Reya and Vogt (GRV).[35] (Subsequent to the School, preliminary H1 data on F_2^{ep} were announced. These are shown in Fig.11, and indicate a strong preference for the singular D_-' and GRV predictions.)

5. Hard Processes in Hadronic Collisions

5.1. *Introduction*

It was first pointed out by Drell and Yan[37] that parton model ideas developed for deep inelastic scattering could be extended to certain processes in hadron-hadron

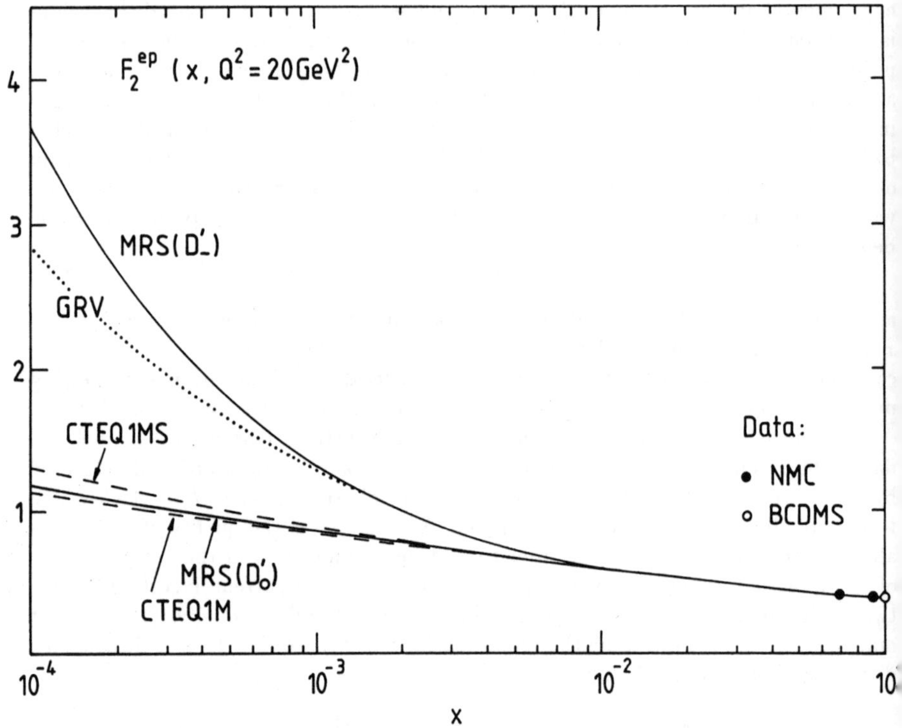

Figure 10: Predictions for F_2^{ep} at $Q^2 = 20$ GeV^2 obtained from the parton sets from MRS[4], CTEQ[34] and GRV[35].

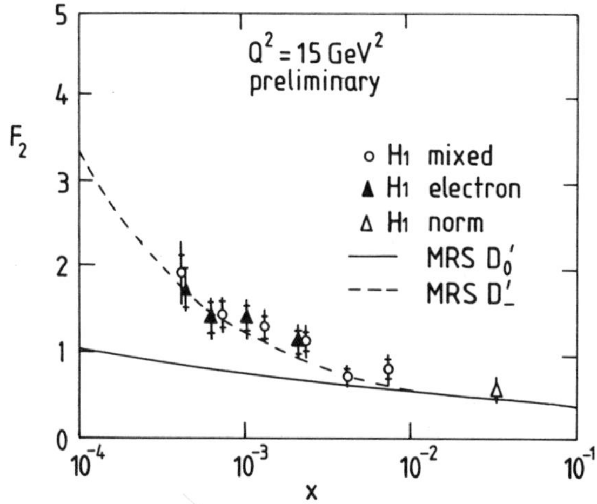

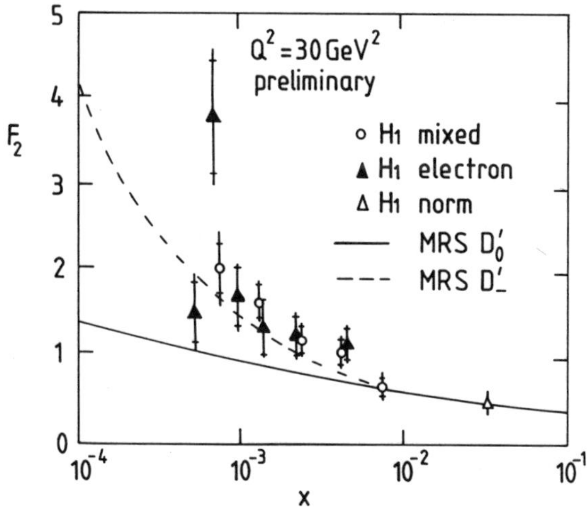

Figure 11: Preliminary H1 data[36] on the F_2 structure function at HERA together with the extrapolations resulting from MRS parton sets D_0' and D_-' [4]. The data are subject to an additional overall normalization uncertainty of $\pm 12\%$.

collisions. The paradigm process was the production of a massive lepton pair by quark-antiquark annihilation – the Drell-Yan process – and the hadronic cross section σ was to be obtained by weighting the subprocess cross section $\hat{\sigma}$ for $q\bar{q} \to \mu^+\mu^-$ with the parton distribution functions $f_{q/A}(x)$ extracted from deep inelastic scattering:

$$\sigma_{AB} = \int \mathrm{d}x_a \mathrm{d}x_b \; f_{a/A}(x_a) f_{b/B}(x_b) \; \hat{\sigma}_{ab\to X} \; . \tag{49}$$

The domain of validity is the asymptotic 'scaling' limit $M^2_{l^+l^-}, s \to \infty$, $\tau = M^2_{l^+l^-}/s$ fixed. The good agreement between theoretical predictions and the measured cross sections provided confirmation of the parton model formalism, and allowed for the first time a rigorous, quantitative treatment of hadronic cross sections. Studies were extended to other 'hard scattering' processes, for example the production of hadrons with large transverse momentum, with equally successful results. Problems, however, appeared to arise when perturbative corrections from real and virtual gluon emission were calculated. Large logarithms from gluons emitted collinear with the incoming quarks appeared to spoil the convergence of the perturbative expansion. It was subsequently realised that these logarithms were the same as those which arise in deep inelastic scattering structure function calculations, and could therefore be absorbed in the definition of the parton distributions, giving rise to logarithmic violations of scaling. The key point was that *all* logarithms appearing in the Drell-Yan corrections could be factored into renormalized parton distributions in this way, and *factorization theorems* which showed that this was a general feature of hard scattering processes were derived.[38] Taking into account the leading logarithm corrections, (49) simply becomes:

$$\sigma_{AB} = \int \mathrm{d}x_a \mathrm{d}x_b \; f_{a/A}(x_a, Q^2) f_{b/B}(x_b, Q^2) \; \hat{\sigma}_{ab\to X} \; , \tag{50}$$

corresponding to the structure depicted in Fig.12. The Q^2 which appears in the distribution functions is a large momentum scale which characterises the hard scattering, e.g. $M^2_{l^+l^-}$, p_T^2, ... Changes to the Q^2 scale of $\mathcal{O}(1)$ are equivalent in this leading logarithm approximation.

The final step in the story was the recognition that the *finite* corrections left behind after the logarithms had been factored were not universal and had to be calculated separately for each process, giving rise to $\mathcal{O}(\alpha_s)$ corrections to the leading logarithm (or leading order (LO)) cross section of (50). Schematically

$$\sigma_{AB} \; = \; \int \mathrm{d}x_a \mathrm{d}x_b \; f_{a/A}(x_a, M^2) \; f_{b/B}(x_b, M^2) \; \times \; [\, \hat{\sigma}_0 \; + \; \alpha_s(\mu^2) \, \hat{\sigma}_1 \; + \; ... \,]_{ab\to X} \; . \tag{51}$$

Here M^2 is the *factorization scale* and μ is the renormalization scale for the QCD running coupling. Formally, the perturbation series is invariant under changes in these parameters, the M and μ dependence of the coefficients, e.g. $\hat{\sigma}_1$, exactly compensating the explicit dependence of the parton distributions and the coupling constant. This compensation becomes more exact as more terms are included in the perturbation

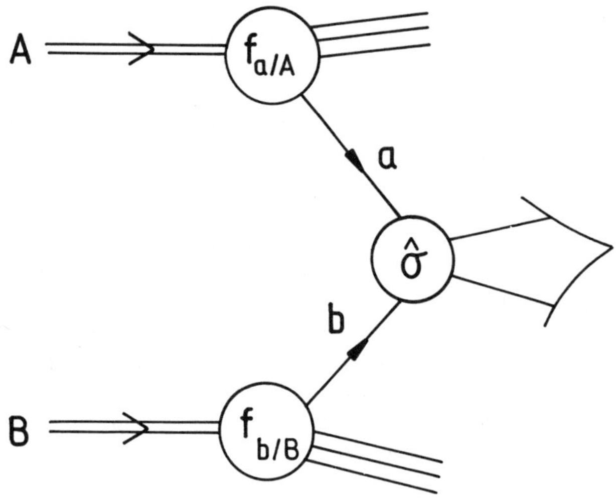

Figure 12: The basic structure of hard scattering processes in perturbative QCD.

series. To avoid unnaturally large logarithms reappearing in the perturbation series it is sensible to choose M and μ values of order the typical momentum scales of the hard scattering process, and $M = \mu$ is also often assumed.

In general, all the important hadronic processes have now been calculated to next-to-leading order (NLO), i.e. up to and including the $\hat{\sigma}_1$ terms. One process – the Drell-Yan process – is even calculated to one order higher (see below). As will be illustrated in the following section, this allows a very high degree of precision in a wide variety of processes. In many cases, the residual renormalization and factorization scale dependence is weak, and the precision of the theoretical prediction is limited only by uncertainties in the knowledge of the parton distributions, in particular the relatively poorly known gluon distribution.

What, then, are the most important applications of this formalism? Nowadays one can, for example, (i) attempt to measure α_s, particularly from those processes involving α_s at leading order, i.e. in $\hat{\sigma}_0$; (ii) obtain information on parton distributions complementary to that from deep inelastic scattering; (iii) attempt to extend the domain of applicability to other kinematic regions, in particular to small x, where additional complications arise[33]; (iv) perform precision analyses of signals and backgrounds for *new physics*, e.g. top, Higgs, supersymmetric particle production etc. There are many examples of situations where the ability to detect a signal for new particle production depends crucially on the accuracy of the Standard Model background estimate.

5.2. Higher Order Corrections to Inclusive Processes

As discussed in Section 5.1, many hard scattering cross sections in hadronic collisions have now been calculated to next-to-leading order. Some of the most important are:

1. The Drell-Yan process (γ^*, $W^\pm$, Z^0 production).[39]
2. Large p_T prompt photon production.[40]
3. Heavy quark pair production.[41]
4. Electroweak gauge boson pair production $(Z^0Z^0, W^+W^-, W^\pm\gamma, ...)$.[42]
5. Higgs production by gg fusion.[43]
6. Inclusive large p_T jet production.[44]

As an example of the impact of the higher order corrections on the cross section, it is instructive to consider the Drell-Yan process in more detail. As already noted, this has recently been calculated to next-to-next-to-leading order (i.e. $O(\alpha_s^2)$) by Hamberg et al.[45] Fig.13 shows the cross section for $p\bar{p} \to W^\pm X$ as a function of the collider energy $\sqrt{s}$. The curves are calculated using the results of Van Neerven et al.[45] (in the $\overline{\rm MS}$ scheme) with MRS(S0) parton distributions[48] and the renormalization and factorization scales $M = \mu = M_W$. The data points are from the UA2[46] and CDF[47]

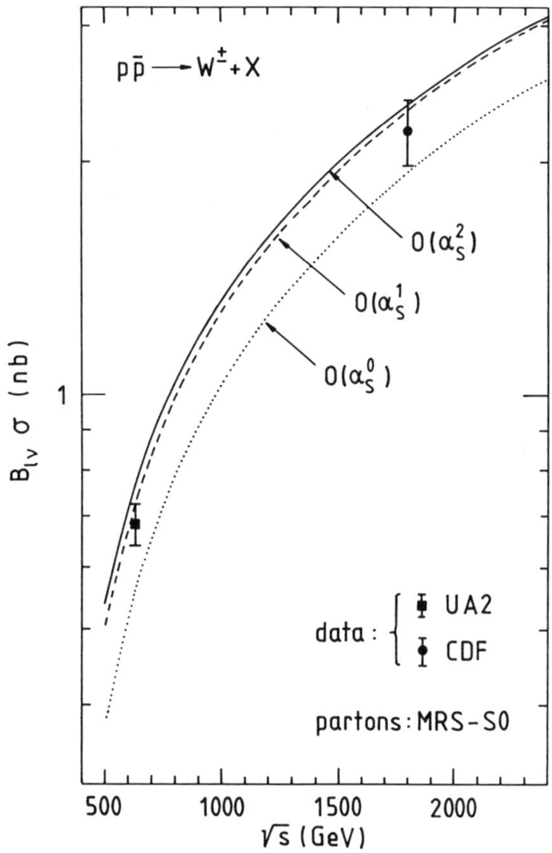

Figure 13: Theoretical predictions for the $W^{\pm}$ cross section in $p\bar{p}$ collisions, as a function of $\sqrt{s}$, with data from UA2[46] and CDF[47].

collaborations, at $\sqrt{s} = 630$ GeV and 1.8 TeV respectively. Notice how both the NLO and NNLO corrections increase the lowest order cross section, the net effect being of order +30%. The NLLO correction is significantly smaller than the NLO correction, due to a partial cancellation between the positive second order corrections involving the $q\bar{q}$ initial state and the negative corrections from the qg initial state. Perhaps the most important point to note from Fig.13 is that, aside from unknown (and presumably small) $O(\alpha_s^3)$ corrections, there is virtually no theoretical uncertainty associated with the predictions – the parton distributions are being probed in a range of $x \sim M_W/\sqrt{s}$ where they are constrained from deep inelastic scattering[48] and the scale dependence is weak.[45] The agreement with experiment, therefore, provides a powerful test of the whole theoretical edifice which goes into the calculation.

I am most grateful to Professor Zichichi for organizing such an excellent School. I would also like to thank all my colleagues, in particular Keith Ellis, Alan Martin and Dick Roberts,who helped in providing material for these Lectures.

References

1. A. Zichichi, these Proceedings.
2. J. Ellis, S. Kelley and D.V. Nanopoulos, Phys. Lett. **260B** (1991) 131.
   U. Amaldi, W. de Boer and H. Furstenaü, Phys. Lett. **260B** (1991) 447.
   P. Langacker and M. Luo, Phys. Rev. **D44** (1991) 817.
   R.G. Roberts and G.G. Ross, Nucl. Phys. **B377** (1992) 571.
   F. Anselmo et al., preprints CERN-PPE-92-103,122,145, CERN-TH-6429,6543-92 (1992).
3. S. Bethke, review talk 'Tests of QCD' at 26th Int. Conf. on HEP, Dallas 1992, Heidelberg preprint HD-PY-92-13 (1992).
4. A.D. Martin, W.J. Stirling and R.G. Roberts, Phys. Lett. **306B** (1993) 145.
5. I. Antoniadis, J. Ellis and D.V. Nanopoulos, Phys. Lett. **262B** (1991) 109.
6. M. Jezabek and J.H. Kühn, Phys. Lett. **301B** (1993) 121.
7. J. Ellis, D.V. Nanopoulos and D.A. Ross, Phys. Lett. **305B** (1993) 375.
8. L. Clavelli, Phys. Rev. **D45** (1992) 3276; **D46** (1992) 2112.
   L. Clavelli, P.W. Coulter and K. Yuan, University of Alabama preprint UAHEP-924 (1992).
   L. Clavelli, P.H. Cox and K. Yuan, University of Alabama preprint UAHEP-9211 (1992).
9. *Quantum Chromodynamics*, F. Yndurain, Springer-Verlag (1983).
10. *Foundations of Quantum Chromodynamics*, T. Muta, World Scientific (1987).
11. *Review of Particle Properties*, Phys. Lett. **239B** (1990) 1.
12. W.J. Marciano, Phys. Rev. **D29** (1984) 580.
13. F. Bloch and A. Nordsieck, Phys. Rev. **52** (1937) 54.
14. T. Kinoshita, J. Math. Phys. **3** (1962) 650.
    T.D. Lee and M. Nauenberg, Phys. Rev. **B133** (1964) 1549.

15. M. A. Samuel and L.R. Surguladze, Phys. Rev. Lett. **66** (1991) 560; **66** (1991) 2416(e).
S.G. Gorishny, A.L. Kataev and S.A. Larin, Phys. Lett. **259B** (1991) 144.
16. B.A. Kniehl and J.H. Kuhn, Nucl. Phys. **B329** (1990) 547.
17. G. Grunberg, Phys. Lett. **95B** (1980) 70.
18. P.M. Stevenson, Nucl. Phys. **B150** (1979) 357.
19. *Theory of jets in electron-positron annihilation*, G. Kramer, Springer Tracts in Modern Physics, Berlin (1984).
20. JADE collaboration: S. Bethke *et al.*, Phys. Lett. **213B** (1988) 235.
21. S. Bethke, Z. Kunszt, D.E. Soper and W.J. Stirling, Nucl. Phys. **B370** (1992) 310.
22. N. Brown and W.J. Stirling, Phys. Lett. **252B** (1990) 657.
23. S. Catani, Yu.L. Dokshitzer, M. Olsson, G. Turnock and B.R. Webber, Phys. Lett. **269B** (1991) 432.
24. N. Brown and W.J. Stirling, Z. Phys. **C53** (1992) 629.
25. G. Kramer and B. Lampe, Fortschr. Phys. **37** (1989) 161.
26. L3 collaboration: B. Adeva et al., Phys. Lett. **248B** (1990) 464.
27. *The Structure of the Proton*, R.G. Roberts, Cambridge University Press (1990).
28. G. Altarelli and G. Parisi, Nucl. Phys. **B126** (1977) 298.
29. BCDMS collaboration: A.C. Benvenuti et al., Phys. Lett. **223B** (1989) 485.
30. NMC: P. Amaudruz et al., Phys. Lett. **295B** (1992) 159.
31. J. Kwiecinski, A.D. Martin, R.G. Roberts and W.J. Stirling, Phys. Rev. **D42** (1990) 3645.
32. E.A. Kuraev, L.N. Lipatov and V.S. Fadin, Sov. Phys. JETP **45** (1977) 199.
Ya.Ya. Balitsky and L.N. Lipatov, Sov. J. Nucl. Phys. **28** (1978) 822.
33. J. Kwiecinski, Proc. of Durham Workshop on 'HERA – the new frontier for QCD', J. Phys. **G** (1993).
34. CTEQ collaboration: J. Botts, J. Morfin, J. Owens, J. Qiu, W.-K. Tung and H. Weerts, Phys. Lett. **304B** (1993) 159.
35. M. Glück, E. Reya and A. Vogt, Z. Phys. **C48** (1990) 471, **C53** (1992) 127; Dortmund preprint DO-TH-93-02 (1993), Phys. Lett. (in press).
36. H1 collaboration: A. De Roeck, Proc. of the Durham Workshop on 'HERA – the new frontier for QCD', to be published in J. Phys. **G** (1993).
37. S.D. Drell and T.M. Yan, Ann. Phys. **66** (1971) 578.
38. See for example: J.C. Collins and D.E. Soper, Ann. Rev. Nucl. Part. Sci. **37** (1987) 383, and references therein.
39. G. Altarelli, R.K. Ellis and G. Martinelli, Nucl. Phys. **B143** (1978) 521; **B146** (1978) 544(e); **B147** (1979) 461.
J. Kubar-Andre and F.E. Paige, Phys. Rev. **D19** (1979) 221.
J. Kubar-Andre, M. LeBellac, J.L. Meunier and G. Plaut, Nucl. Phys. **B175** (1980) 251;
40. P. Aurenche, R. Baier, M. Fontannaz and D. Schiff, Nucl. Phys. **B286** (1987) 509; **B286** (1987) 553; **B297** (1988) 661.

41. P. Nason, S. Dawson and R.K. Ellis, Nucl. Phys. **B312** (1989) 607.
W. Beenakker, H. Kuijf, W.L. van Neerven and J. Smith, Phys. Rev. **D40** (1989) 54.
42. J. Smith, D. Thomas and W.L. van Neerven, Z. Phys. **C44** (1989) 267.
B. Mele, P. Nason and G. Ridolfi, Nucl. Phys. **B357** (1991) 409.
J. Ohnemus and J.F. Owens, Phys. Rev. **D43** (1991) 3626.
B. Bailey, J. Ohnemus and J.F. Owens, Phys. Rev. **D46** (1992) 2018.
43. D. Graudenz, M. Spira and P.M. Zerwas, preprint DESY-92-149 (1992).
A. Djouadi, M. Spira and P.M. Zerwas, Phys. Lett. **264B** (1991) 440.
44. S.D. Ellis, Z. Kunszt and D.E. Soper, Phys. Rev. Lett. **64** (1990) 2121; Proc. Lepton-Photon Symp. and Int. Europhysics Conf. on High Energy Physics, Geneva, Switzerland (1991), eds. S. Hegarty, K. Potter and E. Quercigh, World Scientific (1992), Vol.1, p.417.
F. Aversa, M. Greco, P. Chiapetta and J.Ph. Guillet, Phys. Rev. Lett. **65** (1990) 401.
45. R. Hamberg, T. Matsuura and W.L. van Neerven, Nucl. Phys. **B345** (1990) 331; Nucl. Phys. **B359** (1991) 343.
W.L. van Neerven and E.B. Zijlstra, Nucl. Phys. **B382** (1992) 11.
46. UA2 collaboration: J. Alitti et al., Phys. Lett. **276B** (1992) 365.
47. CDF collaboration: F. Abe et al., Phys. Rev. **D44** (1991) 29.
48. A.D. Martin, W.J. Stirling and R.G. Roberts, Phys. Rev. **D47** (1993) 867.

CHAIRMAN: W.J. Stirling

Scientific Secretaries: O. Rind, J. Shukla

DISCUSSION

– *Rind:*

There has been some ongoing interest in the possible discrepancy between the values of α_s measured from R and those measured from event shape variables. Can you comment on the possible implications of a real discrepancy, specifically within the context of SUSY?

– *Stirling:*

As this is a large part of my lecture tomorrow, I will make a few general comments and leave the details until tomorrow. All these values of α_s are extracted by comparing experiment to theoretical prediction from truncated perturbation series, so we must hope that the effect of truncation is not large. No one really understands why coefficients have particular sizes in particular perturbative expansions and it is likely that some quantities have series which converge worse than others. It may be that there are large higher-order terms, past the next-to-leading order, which could contribute to a shift in α_s. There is one process calculated to NNLO, giving the measurement of α_s from the Z^0 hadronic cross-section, which is most out of line. Interestingly, of all of these measurements it is the one whose errors will most certainly come down over the next years. The issue of whether this measurement is anomalously high will be settled, so I do not worry too much about it.

– *Dova (comment):*

Could you comment on a new paper where it is said that it is possible that the difference between the measurements of α_s from the hadronic branching ratio of Z^0 and the world average is due to the decay of the Z^0 into squarks and gluinos.

– *Stirling:*

Yes, it is possible that the hadronic width of the Z^0 is larger than we think because of extra particles contributing to its decay. The measurement is made by counting the events with hadrons in the final state. It is possible that there are new particles hidden within, for example, the 4-jet sample, which you cannot distinguish because, as in the case of SUSY particles, they decay by cascades and the amount of missing energy is small. Thus, it could be that part of the hadronic states are actually due to new particles. I would guess that most of the possible new particles which might be causing this anomaly would be seen from the specific event topologies. But your point is correct, when we compute the total Z^0 hadronic width, we just add up the contributions from the udscb quarks, put in the corrections, and that's it. There could be something extra there.

– *Cho:*

What are the advantages and disadvantages of summing up leading log QCD corrections using the renormalization group vs. keeping finite $O(\alpha_s)$ or $O(\alpha_s^2)$ terms.

– *Stirling:*

The question has no answer which holds in all cases. One should try to identify the particular regions of phase-space that are responsible for generating large logarithms. As in QED, large logarithms associated with soft gluon emission can be generated in perturbative QCD computations. In such instances, the logarithms should be resummed to all orders using the renormalization group. On the other hand when the differences between two scales is not very large, the logarithms are not so important and one can try to extract finite contributions in a scheme-independent fashion.

– *Hanany:*

Can you please comment on the recent calculation of West claiming that we can calculate the large order behaviour of the current-current correlation function?

– *Stirling:*

West has written papers based on arguments that the perturbation series is an asymptotic series, and we can estimate when this behaviour will set in. Typically, these series have coefficients that will grow like m! One can do a toy-model calculation and convince oneself that this m! growth will not happen until safely way out in the series and will not matter. Third order coefficients, when first published, looked very large and could have been a sign of this asymptotic behaviour. Any previous estimate of the order when the asymptotic behaviour would occur was much higher than third order. Fortunately, the calculation was redone and the coefficients were found to be much smaller. For present day phenomenology and the sort of phenomenology that we will be doing with the series in the next 5-10 years, this behaviour is irrelevant.

– *Wadhwa:*

Could you say something about the experimental proof for three gluon vertices?

– *Stirling:*

This is an important question which I will discuss tomorrow. Experimentalists at LEP have a strong claim of evidence for the triple gluon vertex. At hadron colliders, it is clear that the large p_t jets observed come from things like quark and gluon scattering. Experimentalists at hadron colliders claim to have evidence for the triple gluon vertex from the computation of the scattering cross-sections; specifically the contributions of the triple gluon vertex to the scattering cross-sections are needed for the theory to agree with the data. This morning we saw evidence of this from CDF data. Tomorrow we will discuss the methods used to investigate this problem at LEP.

– *Baillie:*

Are uncertainties in quark masses significant in running α_s from M_Z down to lower scales? Also, are threshold effects important in the running of α_s?

– *Stirling:*

There is a prescription to go over a threshold, but it requires one to put in the values of the quark masses. Guido Altarelli will talk about it next week in more detail. What he has done is put in different values for the b mass in the evolution equation and this results in an uncertainty of 0.001. In general, it is better to measure α_s at low energy and run the value to higher energy because you gain an improvement in error. Some people have argued that the best way to measure α_s is from τ hadronic decays because of this effect and because of the large statistics obtainable. However, this measurement has other problems to deal with such as the thresholds and resonances you mentioned, the large value of α_s at this energy scale (~ 0.3) and quark mass effects can be important issues.

– *Ball:*

How can one consistently define quark masses in QCD?

– *Stirling:*

You must take into account the running of the quark masses just as you must take into account the running of the couplings. The situations are analogous. If you want to investigate this, it is best to start off at high enough mass scales where α_s

is small enough. This is why people have focussed on the running of the b quark mass. A nice example of this mass running comes from Higgs phenomenology. Because the Yukawa coupling of the Higgs to $b\bar{b}$ is proportional to $(m_b)^2$, it is necessary to include this effect if one calculates the branching ratio of $H^0 \to b\bar{b}$. At the scale of the Higgs mass, the b mass has run significantly down from 5 GeV, thus changing the predicted decay width by about a factor of 2!

– *do Couto e Silva:*

At the beginning of your talk, you said that even though Λ_{QCD} has dimensions, it can be taken as a "fundamental parameter." Could you explain this again?

– *Stirling:*

The theory has in it a coupling. Nature gives us the Lagrangians but not the coupling and we have to measure it. This coupling varies with scale and we are free to choose a scale; the value of the coupling at that scale is our fundamental parameter. Λ_{QCD} is just a function of the boundary condition at that reference scale. In this definition, the value of the reference scale (called μ_0) actually appears and so Λ_{QCD} has dimensions of μ_0.

– *Malik:*

The fact that the photon is massless makes the electromagnetic interaction one of long range. If the gluon is also massless, why is the strong colour interaction not also of long range?

– *Stirling:*

This is due to the fact that gluons have self-interactions and, therefore, carry colour charge. At long range, you find an "anti-screening" effect from the gluons which is of opposite sign to the screening effect in QED. This is a key feature of the non-abelian nature of QCD.

– *Sopczak:*

Low energy experiments favour smaller Λ_{QCD} than high energy experiments. Suppose this effect is not a statistical fluctuation, could it be due to the fact that higher order terms in the RGE are not known, or could this be an indication of physics in the intermediate energy range we do not understand yet?

– *Stirling:*

In the plots you refer to, we have used the so-called two-loop evolution of the coupling. One can use a 3-loop evolution, which should be more accurate, but the effect is extremely small. Thus we believe that the evolution of the couplings in this mass range is under control. There could be an effect from crossing a flavour threshold, resulting in a flattening of the slope, but this would also be a very small effect on this plot. A more interesting possibility is that one could construct a theory including some new colored particles, different from quarks and gluons, which would contribute to β_0 and, therefore, to the running of α_s. This is a bit tricky because the contributions of the new particles to β_0 depend on how you include them in the group. This type of thing has been looked at before, but I have a feeling that such new particles would tend to show up in event topologies at LEP.

– *Misiak:*

The RGE for α_{QCD} is used to avoid large logarithms like $\ell n Q^2/\mu^2$ in the perturbation series. But what should we do if we have two extremely different invariants in a process e.g.: $\ell n(s/t) \gg \Lambda$ (s,t: Mandelstam invariants)? Can we do something better than choosing μ somewhere between $\sqrt{s}$ and $\sqrt{t}$?

– *Stirling:*

One identifies what is causing the large logarithms; for example, soft gluons and collinear gluons can cause large logarithms. These logarithms are multiplied by a small coupling and we can resum them to all orders. There are no known cases which are exceptions to this.

– *Genovese:*

How can the threshold of heavy quarks be treated in a rigorous way not only using the requirement of the continuity of α_s changing simultaneously n_f and $\Lambda_{QCD}^{(n_f)}$?

– *Stirling:*

To understand this let us take a look at gluon emission with a loop in it. Near the threshold we have new terms in the series for the cross-section. We can either put them in α_s as a vertex correction or leave it in as a term in the cross-section, but whatever convention we choose we must stick to it and the final result is independent of it.

– *Hanany:*

Please comment on the scheme in which $\beta(\alpha_s) = b_0\alpha_s + b_1\alpha_s^2$.

– *Stirling:*

In general, scheme dependence can change things at next-to-leading order, but as I commented previously, β_0 and β_1 are scheme independent. Higher orders of $\beta(\beta_2, \beta_3 \ldots)$ are scheme dependent and people realized that choosing a value of $\beta_2, \beta_3 \ldots$ defines the scheme. This seems miraculous, since as you say, one can define a scheme in which all higher orders of β are zero. The catch is that when you actually try to calculate a cross-section you find that you have gained no advantage from choosing this scheme.

CHAIRMAN: W.J. Stirling

Scientific Secretaries: E. do Couto e Silva, Z. Tao

DISCUSSION

– *Leike:*

a) Suppose a b-factory is built. What are the most urgent QCD calculations which should be done in this case?

2) Would it be necessary to calculate the final states QCD corrections for massive particles, or do you think these contributions are small?

– *Stirling:*

a) A b-factory is important because there one can study b-decays and extract from these information about CP violation and CKM values. Of course, there are perturbative calculations that can be done, but the most important are the non-perturbative lattice calculations of the hadronic matrix elements; b) the corrections of order α_s in b-decays should be computed, but they are not a major part of the uncertainty in this game.

– *Martin:*

There is a problem of principle or maybe worse than that, on the one hand we say that the dependence of predictions on the scale of α_s is eliminated by trying the whole series, on the other hand we know that this series does not converge. This is not new because it was already the case in QED as shown by Dyson, but α_{QED} is so small that it is not a serious worry. in QCD we know that the series does not converge and, as shown by t'Hooft, it is not even Borel summable, at best, as G. West hopes, it is asymptotic, and one should stop at the term of minimum size, if we look at R we see that the 3rd order over 2nd order is 1/3 to 1/4.

– *Stirling:*

It is vital not just to calculate the corrections but to understand what the series is doing. One sees rises of coefficients because the number of Feynman diagrams is increasing. Yes this is a problem. One cannot anticipate the behaviour of the coefficients. It is necessary to find some pattern for higher orders and a comparison between QED and QCD could give us some help.

– *Benakli:*

When one wants to look at the infrared divergence it will be below the confinement scale. How do people working on the subject deal with all the bound states that appear?

– *Stirling:*

There was something I did not mention: infrared safety. Choose a quantity, for instance, a total cross-section or a thrust distribution where you have a regularization scheme for the soft part or infrared part. Then, by the time you have added all the Feynman diagrams together, you are insensitive to the infrared part and you can take your cutoff to zero. One has to restrict oneselves to quantities which receive contributions only from the pieces of the phase-space which are well away from the non-perturbative regions.

– *Cho:*

1) How are gluon graphs with loops that are both ultraviolet and infrared divergent regulated? Can dimensional regularization accommodate both types of infinities?

2) Are logarithmic infrared divergences independent of the regularization scheme (e.g. dimensional regularization, Feynman parameter cutoff, fictious gluon mass, etc.)?

– *Stirling:*

1) The same regularization scheme takes care of both UV and IR regions. You have gamma functions which exhibit poles at $\epsilon = 0$, and if you are required to come in from $+\epsilon$ you might refer to it as ultraviolet, and if you come in from $-\epsilon$ it will be infrared. A way to work this out is to split the regions of integration of all your integrals into separate ultraviolet and infrared pieces, for instance: integrate from 0 to 1 for dimensions bigger than 4 and from 1 to ∞ for dimensions less than 4. In this way the poles can be cancelled.

2) One could regulate an infrared singularity by this dimensional regularization to get a pole $1/\epsilon$, or one could impose a cut-off on the energy for the gluon and the coefficients obtained will have the same logarithmic dependence as the ϵ poles.

– *Sopczak:*

a) You pointed out the large scale dependence of the event shape variables. Can this theoretical uncertainty be removed by the next order calculations?

2) Are these calculations under way and is it necessary to do these calculations for each shape variable separately?

– *Stirling:*

a) Yes, this is the most important calculation to be done, because we need to reduce the scale dependence and more work has to go into higher order calculations. As far as I know, nobody has done it or is doing it.

2) For each shape variable one used to have to do a different calculation, but in the last year or two, techniques have been developed for setting up a procedure to calculate and to handle both real and virtual gluons, which allows you to impose whatever variable you want to get the correction. There is a nice discussion in the LEP yellow book on how theorists have learned how to make a generic calculation, so you do not have to do the same amount of work each time when you go from e.g. thrust to energy correlations.

– *Hanany:*

Can you put back the plot $1+\delta_{\rm QCD}$ vs. μ^2/μ_Z^2 for $\Lambda(\overline{\rm MS},5) = 150$ MeV.

1) Do you have a way to get the right scale from inspection of the 1,2,3 loop results?

2) Why do you have big errors near $\mu = 0$?

3) How do you calculate the $\bar{q}q$ final states what about calculating the cohomology at the BAST operator in perturbation theory?

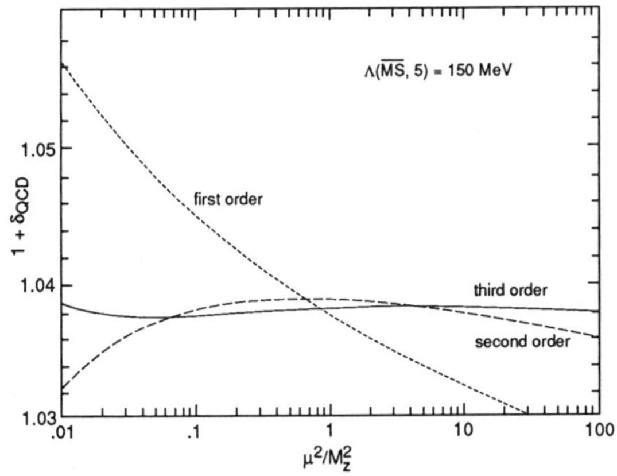

– *Stirling:*

1) I do not think anybody understands the way the coefficients behave therefore one can only guess at "right scales".

2) At $\mu = 0$, α_s becomes very large and one expects the series to go wrong because it is an expansion in terms of a large parameter.

3) Nobody really knows how to identify quarks and gluons in final states on an event by event basis. On the BAST method I have no comments.

– *Gosdzinsky:*

Could you please explain how the scale μ appears in dimensional regularization and how it does when regularizing with a cutoff?

– *Stirling:*

Using the cutoff method I put in μ as an external virtuality of the renormalized vertex. In dimensional regularization the scale μ is introduced to keep the dimensions correct, then I can actually do calculations for massless on-shell particles.

– *Dreiner:*

Is it still reasonable to consider an abelian model when studying the triple gluon vertex? The abelian model has been killed many times (running of α_s, confinement etc.). I agree that the triple gluon vertex is a good test of new physics beyond SU(3). What are the newest LEP limits on such new physics?

– *Stirling:*

I agree with your point about the abelian groups. Up to the scales of the order of the Z mass there is QCD and nothing else. There must be limits but I am not aware of them in detail.

– *Zamora:*

a) In the lecture Jets and Cross-Section you mentioned the matrix element-model for jet reconstruction; can you talk about the parton shower model?

b) Is there any difference in using the matrix element and parton shower models?

– *Stirling:*

a) The parton shower model will attempt to simulate the emission of many gluons (this actually cannot be done exactly), so if I keep just the largest collinear soft logarithms I can approximate the matrix element in such a way that I can compute the probability of any number of gluon emissions. I will make a mistake when the gluons are not collinear or soft, but nonetheless as far as the general properties of the events are concerned, I will get them right. Parton shower Monte Carlos are based on the emission of many gluons in a probabilistic way. The hadronisation is put in the end . This is an approximation to the field theory, to the extent that the partons that are emitted in the shower are soft and collinear. They are extremely useful tools for simulating the general properties of the event.

b) Matrix elements are something one computes with perturbative QCD at the level of quarks and gluons. When people say "matrix element model" they should mean: put the matrix elements inside a bigger black box which includes a hadronisation model.

QCD AND EXPERIMENT

Guido Altarelli

Theoretical Physics Division

CERN, Geneva, Switzerland

Content

PART I. MEASUREMENTS OF α_s

I 1. Introduction

This is a concise review of the present status of the experimental determination of the QCD coupling α_s [1]. In 1989 I summarised the situation at that time [2]. Since then many important developments have taken place, both in theory and experiment. In particular, LEP was commissioned and LEP experiments have much contributed to precise QCD tests [3]. In other experimental areas, the picture has become considerably clearer, for example in deep inelastic scattering. Many relevant theoretical calculations, often of unprecedented complexity, have been completed. Some erroneous results were corrected. As an outcome of 3 years of really remarkable progress, our confidence in QCD has been further consolidated and the determination of α_s is now more precise and a lot of additional checks from many different processes have become possible. In this presentation I will mostly concentrate on new developments and much of the general material of Ref. [2] is not repeated here.

Because of the asymptotic freedom property of QCD a perturbative approach to hard processes is possible and consequently the expansion parameter, i.e. the running coupling $\alpha_s(Q)$, can be measured, Q being of the order of the typical energy scale of the measured hard process. The experimental results on $\alpha_s(Q)$ are often expressed in terms of $\Lambda_{\overline{MS}}$ which is useful in order to compare $\alpha_s(Q)$ at different values of Q. However, it is important to keep in mind that what is directly measured is $\alpha_s(Q)$. For example, for a given precision in $\alpha_s(Q)$, the accuracy on $\Lambda_{\overline{MS}}$ decreases with Q. Thus, sometimes experiments at small Q may erroneously appear to be more sensitive to $\Lambda_{\overline{MS}}$ or to $\alpha_s(m_Z)$, i.e. the value of α_s at a much larger value of Q. But, of course, at small values of Q, the relation between $\alpha_s(Q)$ and what is actually measured is in general much less clear. In most cases, this apparent sensitivity is abundantly overcompensated if theoretical ambiguities are correctly evaluated. For measuring α_s, the real good experiments are those that directly lead to precise values of $\alpha_s(Q)$ at large Q.

At the present level of accuracy, when a value of $\Lambda_{\overline{MS}}$ is quoted it is extremely important to completely specify its relation with $\alpha_s(Q)$. In the following we shall always refer to the expression [1]:

$$\alpha_s(Q) \;=\; \frac{1}{b_f \log Q^2/\Lambda^2}\left[1 - \frac{b_f' \log \log Q^2/\Lambda^2}{b_f \log Q^2/\Lambda^2}\right] \tag{1}$$

with $\Lambda = \Lambda_{\overline{MS}}$ and $b_f = (33-2f)/12\pi, b_f = (153-19f)/(2\pi(33-2f))$ where f is the number of light (approximately massless) quarks with $m \ll Q$. Note that in order to relate α_s at different values of the scale one can also use the direct solution of the renormalisation group equation. To the same accuracy as in (1) one has (see Ref. [1], eq.3-27):

$$\frac{\alpha_s(\mu)}{\alpha_s(Q)} = 1 + b_f\alpha_s(\mu)t + b'_f\alpha_s(\mu)\log[1 + b_f\alpha_s(\mu)t] \tag{2}$$

where $t = \log Q^2/\mu^2$. Small numerical differences between (1) and (2) are due to truncation of the perturbative series and are part of the theoretical uncertainties. A more substancial source of theoretical error is encountered when comparing $\alpha_s(Q)$ at values of Q where different numbers for f are to be taken. For example, on the one hand, the results of experiments on deep inelastic scattering, on Υ-decays etc. are usually expressed giving the value of $\Lambda^{(4)}$, defined by eq.1 written in terms of b_4 and b'_4. On the other hand, experiments at PEP, Petra, Tristan and LEP naturally express their results by using $\Lambda^{(5)}$, defined by eq.1 written in terms of b_5 and b'_5. In the threshold region the running of $\alpha_s(Q)$ is deformed by non perturbative effects. As the physical running coupling is continuous, the matching between the perturbative domains below and above threshold is obtained by assuming a smooth extrapolation of the beta function. In practice, one relates $\Lambda^{(4)}$ and $\Lambda^{(5)}$ by imposing continuity of $\alpha_s(Q)$ at some threshold value of Q, $Q_{th} \cong 0(m_b) : \alpha_s(Q_{th})_{f=4} = \alpha_s(Q_{th})_{f=5}$. Slightly different procedures can also be envisaged (for an example, see Ref. [2]).The resulting difference between $\Lambda^{(4)}$ and $\Lambda^{(5)}$ is quite important. Indeed, for realistic values of Λ, it amounts to approximately:

$$\Lambda^{(5)} \cong 0.65\ \Lambda^{(4)} \tag{3}$$

Clearly, some theoretical error is involved when the results at low energy, for, say, Q = 3-5 GeV, are compared with LEP results at $Q = m_Z$. By varying Q_{th} in the range (0.75-2.5) m_b, with $m_b = 4.8$ GeV and using eq.1 or eq.2, we estimate an error:

$$\Delta\alpha_s(m_Z) \cong \pm 0.0015 \tag{4}$$

associated with translating a value of $\alpha_s(Q)$ measured below the beauty threshold (but above charm) into a corresponding figure for $\alpha_s(m_Z)$. A larger error on $\Delta\alpha_s(m_Z)$ would be implied if one would try to go from $Q \cong 1$ GeV, i.e. below charm threshold, up to $Q = m_Z$.

We now consider the various methods to determine $\alpha_s(Q)$, starting from the totally inclusive processes.

I 2. Deep Inelastic Scattering

Leptoproduction at large momentum transfer is in principle a very solid and powerful method. It is completely inclusive and the theory of scaling violations via the renormalisation group and the light cone operator expansion is particularly rigorous and clear [1]. Moreover several structure functions can be measured, with different beams and targets. The Q^2 dependence can be tested at each value of x. Thus deep inelastic scattering provides a whole laboratory for quantitative tests of QCD, for determining α_s and for measuring the quark and gluon densities in nucleons.

For each structure function $F(x, Q^2)$, perturbative QCD predicts the Q^2 evolution but not the x shape. However, the structure of the evolution equations is such that Q^2 evolution and x shape are interrelated. The maximum of emphasis on specifically testing the pattern of scaling violations predicted by QCD is obtained by focusing on the logarithmic slopes. For each x-bin, one derives the best fit for the slope of a straigth line through the data on $\log F(x, Q^2)$ plotted versus $\log Q^2$. The Q^2 span and precision of the data are not sensitive to the curvature, so that for each x-bin one has one number for $\mathrm{dlog}F(x, Q^2)/\mathrm{dlog}Q^2$. The QCD evolution equations relate, for each given x, the slopes to $\alpha_s(Q)$. In the simplest non singlet case one has:

$$\frac{\mathrm{dlog}F(x, Q^2)}{\mathrm{dlog}Q^2} = \frac{\alpha_s(Q)}{2\pi} \int_x^1 \frac{dy}{y} \frac{F(y, Q^2)}{F(x, Q^2)} \left[P_{qq}\left(\frac{x}{y}\right) + \text{two} - \text{loops}\right] \qquad (5)$$

A single value of Λ should fit the slopes at all values of x. The x shape of the structure function is obtained in terms of a suitable parametrisation by fitting the whole of the data.

As already mentioned, now after about 25 years of experimental work on scaling violations in deep inelastic scattering a clear pattern is finally emerging. In short, the important progress over the situation of a couple of years ago is listed in the following. On leptoproduction from muon (or electron) beams, the famous discrepancy between the results of BCDMS and EMC on H and D targets at low x has been to a great extent clarified by a reanalysis of the EMC data, by comparison with the SLAC data and by new data from the NMC collaboration. The results obtained on H, D, C and Fe targets by the various experiments are quite consistent among them. On neutrino experiments, the new data on the structure function F_3 on Fe by CCFR contradict previous results by CDHSW, which showed poor agreement with the QCD predictions at low x. The analysis of scaling violations in the CCFR data lead to values of $\alpha_s(Q)$ in perfect agreement with those obtained from charged lepton beams. Also, the discussion of theoretical errors in the determination of $\alpha_s(Q)$ from deep inelastic scattering has been considerably improved.

The BCDMS collaboration [4] has measured F_2 with muon beams on C, H and D targets. This experiment has the largest statistics at large Q^2. The C data are for $Q^2 > 25$ GeV^2 in the range $0.275 < x < 0.75$, while for H and D, $Q^2 > 8$ GeV^2 for $0.06 < x < 0.16, Q^2 > 14$ GeV^2 for $0.16 < x < 0.25$ and $Q^2 > 20$ GeV^2 for $0.25 < x < 0.80$. The QCD analysis of the BCDMS data is very solid and consistent results are obtained from singlet and non singlet structure functions on C, H and D targets.

There are no news on the BCDMS analysis of C data [4] . These data, at $x > 0.275$ where the gluon density is negligible, were analysed in the non singlet approximation. The results obtained for $\alpha_s(Q)$ are reported in Table 1 (only experimental errors are included in Table 1).

More recently a combined analysis of the BCDMS [4] and the SLAC [5] data on H and D targets has been performed [6]. The SLAC data were collected in the period 1970-85. They were recently reanalysed with a better set of radiative corrections and a more refined measurement of R (the ratio of longitudinal to transverse structure functions). The Q^2 and x range of the SLAC data is $Q^2 = 0.5 - 30$ GeV^2 for $x = 0.07 - 0.75$. The SLAC data are important to check the normalisation and to model higher twist effects that cause the main systematic error at large x. The minimum value of Q^2 is too low but, for the determination of $\alpha_s(Q)$, only the data at $Q^2 > 5 - 20$ GeV^2 were used. The average value of Q^2 is 75 GeV^2 for the BCDMS sample and 50 GeV^2 for the combined SLAC-BCDMS set of data used for $\alpha_s(Q)$. The higher twist contribution is sizeable at low values of Q^2. The higher twist effect on F_2 is parametrised according to $F_2(x, Q^2) = F_2^{LT}(x, Q^2)[1 + C(x)/Q^2]$, where F_2^{LT} is the lower twist structure function (with target mass corrections included) and C(x) is a function fitted to the data. The measured pattern of higher twist corrections is reasonable [6]. C(x) is of order 1 in GeV^2 at $x > 0.6$ and compatible with zero within errors at small x. The result of the QCD analysis of the measured logarithmic slopes, combined for H and D data (which lead to quite consistent results), is shown in fig. 1 [6]. The data extend down to very low values of x, so that a singlet analysis was performed, with a gluon density also fitted to the data. The uncertainties associated with varying the gluon density within errors and the effect of the higher twist correction are also shown in fig. 1. The corresponding results on $\alpha_s(Q)$ are displayed in Table 1. As seen, the determinations of $\alpha_s(Q)$ from C and H,D targets agree very well.

The famous discrepancy between the BCDMS and the EMC [7] data on H has now been clarified to a large extent. First, the EMC data were recently reanalysed [8]. One important difference is in the treatment of $R = \sigma_L/\sigma_T$. In the original analysis the position R=0 was made, while in the new analysis $R = R_{QCD}$ was taken (as was done by BCDMS as well) and the normalisation was consequently recomputed. The logarithmic slopes obtained by the new EMC analysis and their QCD fit are shown in fig. 2 [8]. The corresponding results for $\alpha_s(Q)$ are shown in Table 1. Second, the new NMC (New Muon Collaboration) [9] data on H and D targets are in good agreement with the BCDMS results but still show some disagreement with the reanalysed EMC data (see fig. 3). A similar indication in favour of BCDMS had already been obtained by comparison with the SLAC data. However in that case the comparison was only based on an extrapolation, because there is no overlap in x and Q^2 between the SLAC and the BCDMS data.

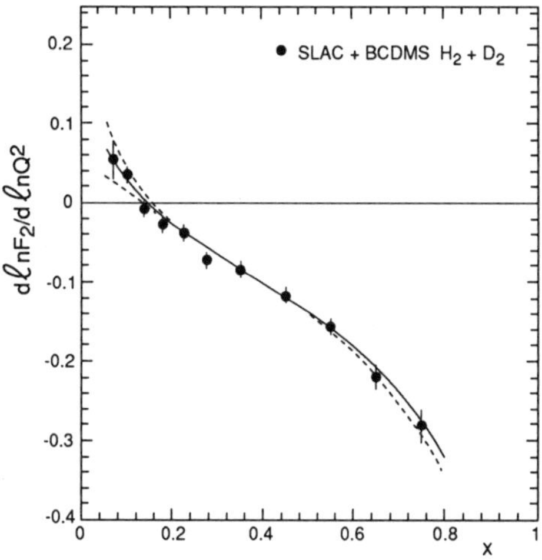

Fig. 1. Logarithmic slopes from the combined SLAC-BCDMS data on H and D. The QCD fit, corresponding to the value of $\Lambda^{(4)}$ shown in Table 1, is also displayed. The uncertainties from varying the fitted gluon density within the errors (at small x) and the effect of higher twist terms (at large x) are also presented.

The new NMC data cover a wide kinematic range, which extends down to very small values of x: $0.008 < x < 0.7$ and $0.8 < Q^2 < 50$ GeV^2. The preliminary analysis of the logarithmic slopes is presented in fig. 4 [9]. We see that the gluon density effect is quite substancial, because the data are at particularly low values of x. The results on $\alpha_s(Q)$ are reported in Table 1.

We now consider deep inelastic scattering from neutrino or antineutrino beams. Here the news are represented by the measurement of xF_3 on Fe by the CCFR collaboration [10] at $< Q^2 >= 25$ GeV^2. At low x, the CCFR data are in disagreement with the previous measurements of the same structure function on the same target by CDHSW [11]. The logarithmic slopes measured by CCFR are in good agreement with the QCD predictions (which was not the case for the CDHSW results), as can be seen from fig. 5, with results on $\alpha_s(Q)$ in perfect agreement with those obtained from muon beams and also,but with larger errors, with the old CHARM data [12] (see Table 1).

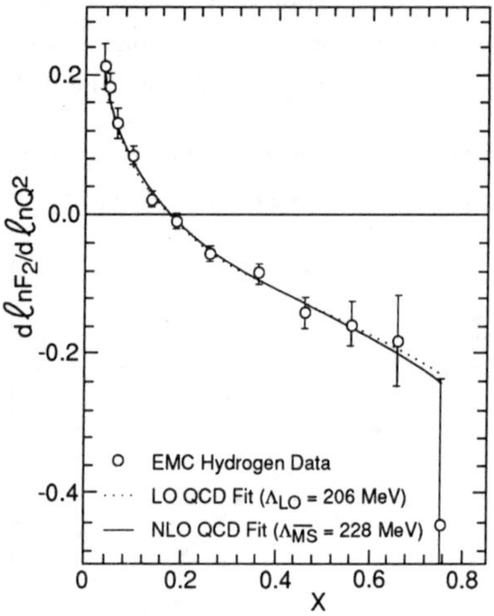

Fig. 2. Logarithmic slopes from the new analysis of the EMC data on H and their QCD fit (see Table 1).

Table 1

μ -beams		
Experiment	$\Lambda^{(4)}$ (MeV)	$\alpha_s(m_Z)$
BCDMS C	230 ± 63	0.110 ± 0.005
SLAC-BCDMS H,D	263 ± 42	0.113 ± 0.003
EMC H	$211 \pm^{117}_{108}$	0.108 ± 0.009
NMC D	307 ± 72	0.117 ± 0.005
ν-beams		
CCFR Fe	210 ± 50	0.111 ± 0.004
CHARM $CaCO_3$	310 ± 157	0.114 ± 0.009

Table 1. Results on $\Lambda_{\overline{MS}}$ for f = 4 and the corresponding values of $\alpha_s(m_Z)$ obtained by the various experiments. The indicated arrors are purely experimental, with statistical and systematic errors added in quadrature.

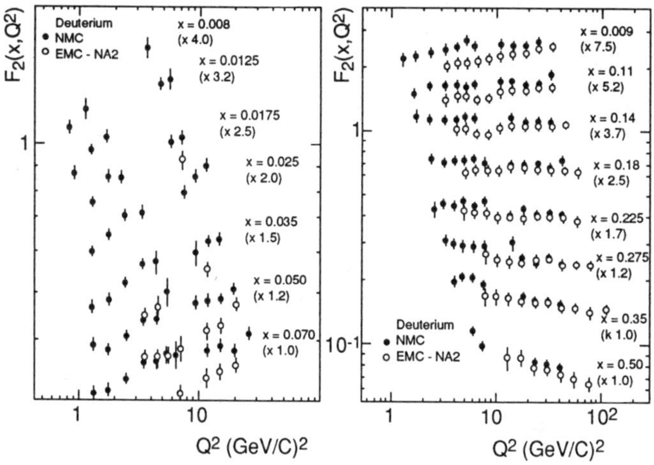

Fig. 3. Comparison between the NMC and the reanalised EMC data on D.

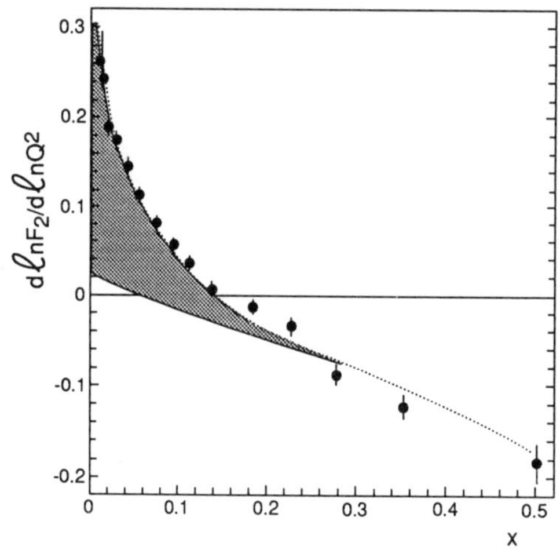

Fig. 4. Logarithmic slopes from the NMC data on D and their QCD fit (see Table 1). The shaded area is the difference due to the gluon density.

We now consider the important issue of estimating the theoretical error associated with the measurements of $\alpha_s(m_Z)$ in Table 1[1]. There are many possible

[1] For theoretical errors we must decide what we mean: a maximum error, similar to 3σ for experimental errors, or a 68% c.l. error similar to 1σ error. In the following we talk of 1σ errors, although there is no real way of quantifying a confidence level for theoretical errors.

sources of errors in the relation between what is actually measured and $\alpha_s(Q)$.

First, there are terms down by powers of Q^2 (e.g. ambiguities on target mass corrections and from higher twist terms), and uncertainties on the gluon density and on the x-parametrisation of structure functions. In Ref. [6] these sources of errors are included in the quoted experimental error. We have already described the treatment of $1/Q^2$ corrections given in Ref. [6] , which was precisely aimed at obtaining an estimate of the corresponding error directly from the combined SLAC-BCDMS data. The resulting effects are presented in fig. 1. The error from the gluon parametrisation is also shown in fig. 1. For the other experiments mentioned in our previous summary, the discussion of such errors is less transparent.

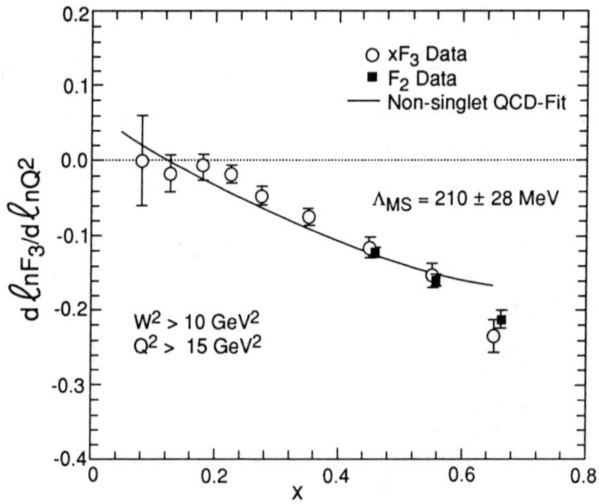

Fig. 5. Logarithmic slopes from the CCFR data on F_3 (which is a purely non-singlet structure function) and their QCD fit (see Table 1).

Second, there are theoretical errors associated with neglected higher order terms in $\alpha_s(Q)$ in the QCD evolution equations. The most usual way to obtain an estimate of these higher order terms is by varying the renormalisation scale μ (i.e. express the perturbative expansion in terms of $\alpha_s(\mu)$) in some fiducial interval around the natural scale of the hard process (Q in deep inelastic scattering) and observe the change in the value of Λ or of $\alpha_s(m_Z)$. More in general some other reparametrisation of the perturbative series can be tried, for example a change of the factorisation scale (the scale of parton densities). When the scale is changed in the leading term in $\alpha_s(Q)$ the subleading terms are correspondingly modified in order to lead to the same result, whithin the accuracy of the known terms in the expansion. The compensation is not exact because the series is truncated. This is a reasonable method, provided that it is correctly interpreted. Of course, the result depends on the range of μ which is chosen, so that some subjective element is immediately evident. In general, μ can be varied in a range such that $\alpha_s(Q)\log Q^2/\mu^2$ remains a small quantity (so that no new large logs are generated). In principle, μ could depend on the parameters which are fixed when Q^2 varies, e.g. on x in deep inelastic scattering, but it is usually taken as a constant. The method only provides an indication of a lower bound on the error from higher order terms: such error cannot be smaller than what is obtained in this way because otherwise a contradiction would be manifest. One obvious remark on this method is that it extrapolates the behaviour of the unknown terms in the series from those that are known: the scale dependence is considered small if the first corrective term is small. But, clearly there is no theorem that guarantees that higher order terms are small when the first correction is small. The scale dependence in the determination of $\alpha_s(m_Z)$ from the actual experiments of deep inelastic scattering was studied [6] , [13]. The result is something like $\Delta\alpha_s(m_Z) = \pm 0.004$. The total theoretical error which is usually quoted is close to this estimate [3] [6] [13] : $\Delta\alpha_s(m_Z) = \pm 0.005$, when the error from translating $\alpha_s(Q)$ measured at Q = 5-10 GeV into $\alpha_s(m_Z)$ is also included (see eq. 4).

In my opinion, the real theoretical error is probably somewhat larger, not only because of the previous words of caution about the scale dependence, or because of similar reservations on the estimate of higher twist terms, but also because other sources of errors can be imagined. For example, at realistic energies, when moving Q^2 at fixed x, one often goes through the threshold for production of heavy quarks in the final state, either charm or beauty, depending on the experiment and the Q^2 dependence of the structure function is altered in a non perturbative way. Thus I would be a little bit more conservative and assign a theoretical error $\Delta\alpha_s(m_Z) = \pm 0.007$ to $\alpha_s(m_Z)$ measured in deep inelastic scattering.

In conclusion, the situation of deep inelastic scattering is much clearer now. The very fundamental test of QCD based on measuring the small logarithmic deviations from scaling and comparing them with the quantitative theoretical predictions has been finally accomplished by several experiments to a reasonable degree of accuracy. The measured values of $\alpha_s(Q)$ can be expressed in the form:

$$\begin{aligned} \alpha_s(m_Z) &= 0.112 \pm 0.002\ (\text{exp})\ \pm 0.007\ (\text{th}) = \\ &= 0.112 \pm 0.007 \end{aligned} \tag{6}$$

where the central value and the experimental error are obtained by combining the data in Table1. Note that the theoretical error is dominant.

I 3. Total Hadronic Production in e^+e^- Annihilation and Z Decay

The total hadronic cross-section in e^+e^- annihilation, and its modern variant, the Z total hadronic width, are processes with the same gold-plated standing as deep inelastic scattering, as far as QCD tests and the measurement of $\alpha_s(Q)$ are concerned. Both are totally inclusive, both can be treated by the light-cone operator expansion plus renormalisation group approach. As already observed, deep inelastic scattering is wider in scope (many structure functions with Q^2 and x dependence), and the QCD scaling violations are relatively prominent (enhanced by logs). The total hadronic cross-section in e^+e^- annihilation is just one function of Q^2 and the QCD effect is a small correction. But there are some advantages. This process is experimentally simpler and the theoretical calculations were possible up to the order $\alpha_s(Q)^3$. Actually an error in the first computation of the next-next to leading correction has now been corrected [14]. The new result for the third order coefficient is numerically moderate so that the expansion looks quite reassuring. For f = 5 and pure photon exchange one has, at $Q = \sqrt{s}$:

$$R_{e^+e^-} = \frac{11}{3}\left\{1+\left(1+\frac{3}{11}\mu\right)\frac{\alpha_s(Q)}{\pi}+1.411\left(\frac{\alpha_s(Q)}{\pi}\right)^2-12.76\left(\frac{\alpha_s(Q)}{\pi}\right)^3+\ldots\right\} \tag{7}$$

with $\mu = 4m_b^2/Q^2$ describing, in linear approximation, the effect of the finite b-quark mass [15].

The corresponding formula for the ratio Γ_h/Γ_ℓ measured at LEP is given by [16]:

$$\frac{\Gamma_h}{\Gamma_\ell} = \left(\frac{\Gamma_h}{\Gamma_\ell}\right)_{EW}\left\{1+1.05\,\frac{\alpha_s(m_Z)}{\pi}+(0.9\pm0.1)\left(\frac{\alpha_s(m_Z)}{\pi}\right)^2-13\left(\frac{\alpha_s(m_Z)}{\pi}\right)^3+\ldots\right\} \tag{8}$$

where the b-quark mass correction (which is particularly large on the axial contribution) was taken into account in the first order coefficient. The second order coefficient is different for the vector and axial contributions to the width due to the effect on the axial part of the top-bottom mass splitting from diagrams as the one in fig. 6 [17]. The corresponding error is induced by our ignorance of m_t.

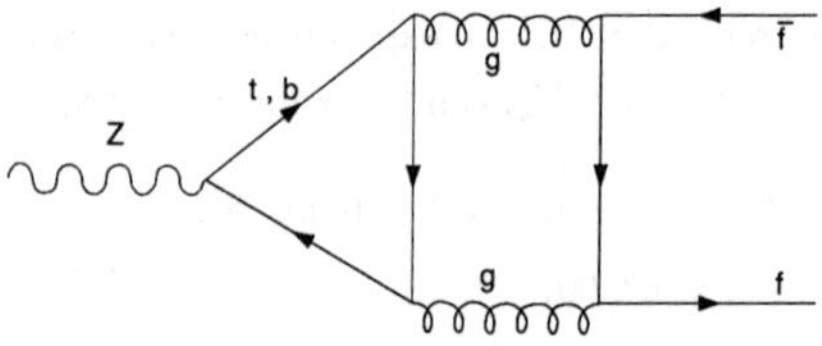

Fig. 6. A typical diagram that makes the difference between the QCD corrections for vector and axial couplings.

By now the LEP measurement of Γ_h/Γ_ℓ is the best experiment of this kind for the determination of $\alpha_s(m_Z)$: the energy is largest, the LEP detectors, being aimed at the very precise electroweak physics, have a better control of the systematics (acceptance, efficiency...) in comparison to previous e^+e^- experiments. Also a better treatment of radiative corrections has been implemented in the LEP esperiments. The most recent experimental value of Γ_h/Γ_ℓ, combined over the four LEP experiments, is given by [18]:

$$R_{exp} = \left(\frac{\Gamma_h}{\Gamma_\ell}\right)_{exp} = 20.86 \pm 0.08 \tag{9}$$

where a common systematic error is also taken into account. Combining this experimental result with the theoretical prediction for $\alpha_s(m_Z) = 0$ [16]:

$$R_{EW} = 19.97 \pm 0.03 \tag{10}$$

where the error is due to our ignorance of m_t and m_H, one obtains:

$$\alpha_s(m_Z) = 0.132 \pm 0.012 \tag{11}$$

The separate results of the four LEP experiments are:

$$\begin{aligned} \alpha_s(m_Z) &= 0.118 \pm 0.020 \text{ ALEPH} \\ &= 0.160 \pm 0.027 \text{ DELPHI} \\ &= 0.135 \pm 0.026 \text{ L3} \\ &= 0.127 \pm 0.026 \text{ OPAL} \\ &= 0.132 \pm 0.012 \text{ Average} \end{aligned} \tag{12}$$

In this case the theoretical error is negligible with respect to the experimental error. In fact, first, the experimental error is large. Second, the contribution of terms down by powers of $1/Q^2$ is quite small and the scale dependence is also very limited because $\alpha_s(m_Z)$ is small and the perturbative series extends up to the third order term. The statistical error is still a sizeable component of the error so that further improvement of this measurement is possible.

Starting with the works of Ref. [19] all available e^+e^- data at high energies have been analysed in order to extract values of $\alpha_s(Q)$. We consider a recent study by Branchina et al. [20]. These authors use an updated set of data from Spear up to Tristan energies, fit $\alpha_s(Q)$ using the LEP values for $m_Z, \sin^2\theta_W$ and so on and make use of the new corrected formula eq. 7 for the QCD correction. The results are given in Table 2 for two different fits, one for $\sqrt{s} > 5$ GeV and one for $\sqrt{s} > 14$ GeV.

Table 2

$\sqrt{s}$ range GeV	5-61.4	14-61.4
$\Lambda^{(5)}$ (MeV)	$397 \pm^{256}_{193}$	$990 \pm^{608}_{485}$
$\alpha_s(m_Z)$	0.129 ± 0.013	0.151 ± 0.017

Table 2. Values of $\Lambda^{(5)}$ or, equivalently $\alpha_s(m_Z)$, obtained from the fits of Ref. [20] in the two indicated ranges of energy.

I think that the error that appears in Table 2, and in other similar analyses, is not the whole error. To me is in fact a bit surprising that an error almost as small as the one from the combined LEP measurement of $\alpha_s(m_Z)$, given in eq. 11, could be obtained from a compilation of low energy experiments. In fact the errors in Table 2 are obtained by assuming uncorrelated systematic errors among the various experiments. But it is not difficult to imagine important correlated systematics. For example, most old experiments used first order electromagnetic radiative corrections for initial state photon emission. Since the effect of initial state radiation is large, the contribution of multiphoton radiation is still sizeable. By neglecting multiphoton initial state radiation the value of $R_{e^+e^-}$ is overestimated. In fact when a hard photon is emitted, the invariant mass of the hadronic system can fall below the corresponding cut. The second order correction reduces the probability for hard photon emission. Thus correcting for the cut on the basis of the lowest order estimate leads to a larger value of $R_{e^+e^-}$. The magnitude of this effect was studied in Ref. [21]. It was found that it can lead to a 1% change in $R_{e^+e^-}$. Similarly many corrections are implemented using the same methods and, often, the same computer codes. We also see that the central value of $\alpha_s(m_Z)$ depends sensitively on the range of energies which are included in the fit.

In conclusion, I think that the low energy experiments are less reliable and precise than the LEP experiments. With the further progression of LEP this statement will become more and more true. The errors on $\alpha_s(Q)$ obtained from the assumption of uncorrelated systematics are probably optimistic. From $R_{e^+e^-}$ at energies below LEP I would quote as a result something like:

$$\alpha_s(m_Z) = 0.14 \pm 0.02 \tag{13}$$

I 4. τ Decay

The possibility of measuring as from τ-decay has been extensively studied in a series of interesting papers [22], in particular by Braaten, Narison and Pich. The relevant quantity is $R_\tau = \Gamma(\tau \to \nu_\tau +$ hadrons$)/\Gamma(\tau \to \nu_\tau + \ell\nu)$, with $\ell = e, \mu$. It is claimed [23] that the present experimental value:

$$(R_\tau)_{exp} = 3.53 \pm 0.04 \tag{14}$$

implies $\alpha_s(m_\tau) = 0.32 \pm 0.04$ and, based on this result it is argued that $\alpha_s(m_Z) = 0.118 \pm^{0.004}_{0.006}$.

Given that m_τ is so small this determination of $\alpha_s(m_Z)$ appears a bit too precise! In defence of this method one can certainly point out that R_τ has several combined advantages. First, R_τ is even more inclusive than $R_{e^+e^-}$ (s). Dropping some inessential complications R_τ is an integral over a spectral function, analogue to $R_{e^+e^-}$ (s):

$$R_\tau \sim \int_0^{m_\tau^2} \frac{ds}{m_\tau^2}\left(1 - \frac{s}{m_\tau^2}\right)^2 \mathrm{Im}\,\Pi_\tau(s) \tag{15}$$

One expects that the asymptotic regime is more precocious for more inclusive quantities. Second, one can use analyticity in order to transform the above integral into an integral over the circle $|s| = m_\tau^2$:

$$R_\tau \sim \frac{1}{2\pi i} \int_{|s|=m_\tau^2} \frac{ds}{m_\tau^2}\left(1 - \frac{s}{m_\tau^2}\right)^2 \Pi_\tau(s) \tag{16}$$

This not only gives some confidence that the relevant scale of energy for the evaluation of $\Pi_\tau(s)$ is of order m_τ, but also shows that the integration over the low energy domain helps very much in smearing out the complicated behaviour in the resonance region. Also important is the phase space factor in front of $\Pi_\tau(s)$ in eqs. 15, 16 that kills the sensitivity near Re $s = m_\tau^2$, where there is a gap. Note that one could naively think of gaining even more (in terms of increasing the effective energy scale) by going from the circle to some other curve in the complex plane with $|s| > m_\tau^2$ except at the tip where Re $s = m_\tau^2$. But, as observed in Ref. [24], one would not obtain an advantage because the integrand at large $|s|$ becomes much larger than the result, so that there must be important cancellations and the precision of any asymptotic approximation on the leftover would not improve. This argument is used [24] to show that the choice of the circle is actually not conclusive on what the effective energy scale is, and that it can well be smaller than m_τ. Third, $\Pi_\tau(s)$ can be evaluated at $|s| \sim m_\tau^2$ by controlling non perturbative corrections by using the operator product expansion and some estimate (either experimental or by some model) of the dominant condensates, in the spirit of the QCD sum rules [25]. Finally the perturbative component of $\Pi_\tau(s)$ is known up to terms of order $\alpha_s(m_\tau)^3$.

This series of virtues of R_τ is indeed real but would not be sufficient in itself to justify the precision on $\alpha_s(m_\tau)$ which is claimed. The real point is another one which I now discuss. By applying the previous strategy eventually one arrives at an expression for R_τ of the following form (schematically) [22]:

$$\begin{aligned} R_\tau = R_\tau^0 \Big\{ 1 + \frac{\alpha_s}{\pi} + 5.202 \left(\frac{\alpha_s}{\pi}\right)^2 + 26.37 \left(\frac{\alpha_s}{\pi}\right)^3 + \ldots \\ + c\,\frac{m^2}{m_\tau^2} + d\,\frac{< m\bar{q}q >}{m_\tau^4} + \ldots + g\,\frac{< O_6 >}{m_\tau^6} + \ldots \Big\} \end{aligned} \tag{17}$$

where the perturbative coefficients are appropriate for f=3 and $\alpha_s = \alpha_s(m_\tau)$, c,d and g are known coefficients, m is indicative of light quark masses and $< O >$ stands for the condensate (i.e. the vacuum expectation value) of the operator O. The crucial ingredient that allows a precise determination of $\alpha_s(m_\tau)$ is that one replaces m by the current masses of u and d quarks (s quarks are essentially irrelevant, because the corresponding final state is Cabibbo suppressed in τ decay). With $m \sim 5 - 10$ MeV it is clear that the dangerous $1/m_\tau^2$ corrections are completely negligible. This replacement of m by the current masses is based on the assumption, usually made in the QCD sum rule approach [25], that all non perturbative effects are in operator matrix elements. According to this assumption the Wilson coefficients are determined by perturbation theory also at the non leading level in $1/Q^2$, where they are not protected by scale invariance against effects of order Λ^2/Q^2, with Λ some non perturbative scale, typically the confinement scale $\Lambda_{\overline{MS}}$. What is special about the τ-decay channel is that operators of dimension 2 are absent in the expansion of two electroweak currents. If not for this gap the above assumption would not play such a dramatic role, because in presence of an operator of dimension 2 what is non leading c-number coefficient and what is operator matrix element is largely conventional. If one accepts this assumption there are no $1/Q^2$ corrections in the massless quark limit neither in τ decay nor in e^+e^- annihilation (in the totally inclusive rates). To appreciate the impact of this assumption in the case of τ decay, note that $\alpha_s(m_\tau)/\pi \sim 0.10 \sim (500\ \text{MeV}/m_\tau)^2$ and that $\alpha_s(m_\tau) = 0.32$ corresponds to $\Lambda^{(3)} = 378$ MeV.

I think it is a fair statement that there is no theorem that guarantees the absence of $1/m_\tau^2$ terms in R_τ in the massless limit by proving that terms of order $(\Lambda/m_\tau)^2$ cannot arise. Some insight on this problem is obtained starting from the behaviour of the perturbative series at large orders, in order to obtain, by Borel transformation, properties of the resummed series which are intrinsically different than those at any fixed order. This approach has the obvious limit that it does not produce a theory of confinement. The classic papers based on the "rinormalon" method concluded [26], as recently emphasised by Mueller [27], that indeed no non perturbative $1/Q^2$ terms are induced by resummation . Recently there has been a revival of interest on the matter, following the interesting work by West [28] on large orders in QCD. The authors of Ref. [29] conclude that the issue cannot be decided. However, Zakharov, one of the fathers of the QCD sum rules, has argued [30] in favour of $1/Q^2$ terms being generated by the so called ultraviolet rinormalons, leading to an effective value of order Λ for the mass m in eq. 17. Thus the issue if far from being settled.

I also stress that the advantage from the absence of $1/m_\tau^2$ terms in R_τ could be an illusion when comparing $\alpha_s(m_\tau)$ measured from R_τ with $\alpha_s(Q)$ derived from some other process, because one needs control of $\alpha_s(Q)$ down to terms of order Λ^2/m_τ^2, while only the asymptotic form of $\alpha_s(Q)$ is known. For example, any freezing mechanism at $Q \sim \Lambda$ introduces typical corrections of order $\alpha_s(m_\tau)^2\Lambda^2/m_\tau^2$.

As for the contribution from operators of dimension 4, 6... , as estimated in Ref. [22] , I find that the quoted uncertainties are in some cases optimistic. For example, the four quark condensate of dimension 6 is estimated by vacuum saturation. The result is small due to a large cancellation between the vector and axial contributions. The vacuum saturation approximation is not much reliable, so that the reality of the cancellation is very questionable. In addition to that it seems to me that the error on the small result should not be taken smaller than the error on the vector and axial contributions separately.

For all the above reasons I would conclude that a $\sim$ 30% error on $\alpha_s(m_\tau)$ is a more realistic estimate, or $\alpha_s(m_\tau) = 0.32 \pm 0.10$, which also includes the uncertainties from possible corrections of order Λ^2/m_τ^2. This leads to:

$$\alpha_s(m_Z) = 0.117\pm^{0.010}_{0.016} \tag{18}$$

which is still a quite respectable result. The confidence on this method would be increased by the experimental study of the τ-decay mass distributions. By taking suitable moments, the different power behaved contributions can in principle be separated, so that the condensates as well as the $1/m_\tau^2$ terms could be measured. Some work in this direction has been already started [22] . Similarly, it is interesting to apply similar techniques to e^+e^- cross-sections at low energy to test the reliability of the method. A study of this type was performed in Ref. [31] with encouraging results.

I 5. Quarkonium Decay

Recently new precise measurements of quarkonium branching ratios done at SPEAR, CESR, DORIS and VEPP have been published. As the rate of quarkonium into 3 gluons, which is crucial for the determination of α_s, is obtained by subtracting all other channels from the total width , an improvement in any branching ratio reflects itself in a better determination of the 3-gluon width. All results are by now virtually finalised and little improvement is to be expected in the near future. A new comprehensive analysis has recently been presented in Ref. [32]. The results of this analysis are summarised in the following. We shall see that the extracted value of $\Lambda^{(4)}$ is increased with respect to the old compilations [2] , [33] of a few years ago, in better agreement with the other measurements of α_s.

The decays of the Υ are the most interesting for the determination of $\alpha_s(Q)$, while the mass of charmonium is so low that we prefer to use its decays merely as complementary information for a semiquantitative check of the method. In the nonrelativistic approximation, when taking the ratio of two suitable channels, the dependence on the quarkonium wave function drops. The most important ratios are [34]:

$$R_\mu = \frac{\Gamma(\Upsilon \to ggg)}{\Gamma(\Upsilon \to \mu\mu)} = \frac{10(\pi^2-9)}{9\pi}\frac{\alpha_s(\mu)^3}{\alpha^2}\left\{1 + \frac{\alpha_s(\mu)}{\pi}\left[3\pi b_4 \log\frac{\mu^2}{m_b^2} + 0.43\right] + \dots\right\} \tag{19}$$

$$R_\gamma = \frac{\Gamma(\Upsilon \to ggg)}{\Gamma(\Upsilon \to \gamma gg)} = \frac{5\alpha_s(\mu)}{4\alpha}\left\{1 + \frac{\alpha_s(\mu)}{\pi}\left[\pi b_4 \log\frac{\mu^2}{m_b^2} + 2.6\right] + \dots\right\} \tag{20}$$

where the terms proportional to b_4 (the beta function coefficient for f=4) take the scale dependence into account, when μ is varied around the natural scale m_b (recall eq.2).

For the Υ system, the present experimental average values, together with the corresponding old figures, are given by:

$$\begin{array}{lll} R_\mu : \Upsilon & 31.3 \pm 1.3 \text{ (old [2])} & 32.5 \pm 0.9 \text{ (new [32])} \\ \quad\Upsilon'' & 32.5 \pm 8.3 & 34.7 \pm 6.9 \\ \quad\Upsilon''' & 31.0 \pm 2.7 & 26.7 \pm 3.3 \\ R_\gamma : \Upsilon & 36.0 \pm 2.0 & 36.5 \pm 2.0 \end{array} \tag{21}$$

Quarkonium decays are nominally very precise (note the cubic dependence of R_μ on $\alpha_s(\mu)$). The problem, once more, has to do with the theoretical errors. Obvious sources of error are the limitations of the non relativistic approximation, the impact of higher order perturbative terms and possible non perturbative corrections.

The consistency of the non relativistic approximation :

$$\Gamma(\Upsilon \to F) = |\Psi(0)|^2 |M(b\bar{b} \to F)|^2 \tag{22}$$

is to some extent experimentally checked by observing reasonably compatible results for R_μ in Υ, Υ' and Υ'' decays, as seen from eq. 21. Yet sizeable corrections are well possible, given that $v^2/c^2 \sim 10\%$ for the Υ and $\sim 23\%$ for the J/Ψ. In Ref. [32], as in other similar analyses, a corrective factor $(1+Dv^2/c^2)$ is included. In order to fit D one must consider both J/Ψ and Υ decays. If we want to restrict ourselves to Υ decays, than the error from non relativistic corrections is estimated in Ref. [32] by taking $|D| < 3$ at 90% c.l.

As usual, one can estimate the possible effect of higher order terms by studying the scale dependence of the result on $\Lambda^{(4)}$, or by writing the correction to R_μ or R_γ in different ways which are perturbatively equivalent (the factor $\{1+\alpha_s/\pi[\ldots]\} = 1+\epsilon$ can be written as $1/1-\epsilon$ or $[1+(1+a)\epsilon]/[1+a\epsilon]$ depending on whether we expand the ratio or its inverse or each process separately in the numerator and the denominator). The scale dependence is obviously much stronger in R_μ than in R_γ because of the cubic versus linear $\alpha_s(\mu)$ dependence. Kobel uses the following, uncommon but reasonable, procedure to estimate the error [32]. In the curly brackets in eqs. 19,20, at order $[\alpha_s(\mu)/\pi]^2$ he adds those terms in b_4 and b_4' which are needed to cancel the μ dependence also at that order. The remaining part C of the second order coefficient, which is μ-independent, is varied within generous limits to estimate the error from its ignorance. He takes $|C| \leq 100$ for R_μ and $|C| \leq 50$ for R_γ, as the ambiguity corresponding to a 90% c.l.

Non perturbative corrections can be non negligible because m_b is not so large. There is no theory of such corrections. As far as I know they could be present even at the level of $1/m_b$ rather than $1/m_b^2$. As well known, the spectrum of the photon in $\Upsilon \to \gamma gg$ is much softer than its perturbative estimate [35]. This could indicate sizeable non perturbative corrections also in the totally inclusive width. No estimate of errors from non perturbative corrections was included in the analysis of Ref. [32].

Some particularly significant results quoted by Kobel, expressed in terms of $\alpha_s(m_Z)$, are:

$$\begin{aligned} &\alpha_s(m_Z) = 0.108 \pm 0.001 \pm^{0.008}_{0.005} && \text{from } R_\mu \text{for the} \Upsilon, \Upsilon', \Upsilon'' \\ &\alpha_s(m_Z) = 0.111 \pm 0.003 \pm^{0.019}_{0.009} && \text{from } R_\gamma \text{ for the } \Upsilon \\ &\alpha_s(m_Z) = 0.113 \pm 0.001 \pm^{0.007}_{0.005} && \text{from a combined fit of } \Upsilon \text{ and } J/\Psi \end{aligned} \tag{23}$$

Note that in all cases the central value corresponds to $C = 0$. For the first two results, that refer to the Υ system, the central value also corresponds to $D = 0$, while in the third case D was fitted in order to get the best simultaneous agreement for charmonium and bottomonium.

In conclusion, I think that the results quoted show a remarkable consistency which sounds encouraging. The estimate of the errors associated with the non relativistic approximation and with higher orders appears to be reasonable to me. I stress again that no error for non perturbative corrections was included. Also in view of that, I would summarise this chapter by stating that the value of $\alpha_s(m_Z)$ which emerges from quarkonium decays is something like:

$$\alpha_s(m_Z) = 0.11 \pm 0.01 \tag{24}$$

I 6. Jets at LEP

The measurements of jet rates, event shapes and energy correlations at LEP have marked a very sharp progress with respect to similar studies at lower energies. Contrary to all measurements discussed in the previous sections, jet studies are clearly not completely inclusive. This necessarily implies that some model must be used to correct for hadronisation effects. The advantage of LEP over previous experiments is that the energy is larger, the hadronisation effects are smaller and there is a more limited ambiguity on the corresponding corrections. Due to the existence of hadronisation effects which one can neither really compute nor disregard, but only model with some uncertainties, it is important to measure many different observables and check that consistent results are obtained. Thus at LEP α_s is extracted [3] , [36] from jet rates (jet multiplicity fractions, average jet multiplicity) with different jet finding algorithms, event shape variables (thrust, oblateness, jet masses,...), energy correlations (energy-energy correlations and relative asymmetry, planar triple energy correlation). It is true that there are strong correlations among the different observables but certainly a lot of useful redundancy is achieved.

For each observable O_i used to measure α_s a next to leading calculation exists at the parton level, of the form [36]:

$$O_i = \frac{\alpha_s(\mu)}{\pi}\left\{A_i + \frac{\alpha_s(\mu)}{\pi}\left[A_i\pi\ b_5 \log\frac{\mu^2}{m_Z^2} + B_i + \ldots\right]\right\} \tag{25}$$

with A_i and B_i computed. Recently, for some of the observables, an improvement has been implemented. Near the boundary of phase space logarithmic terms like $\alpha_s\log(1-T)$ and $\alpha_s^2\log(1-T)$ at leading or next to leading level become large (here T is thrust and T=1 corresponds to the 2-jet limit). Leading and next to leading logs of this type have been computed and resummed for energy-energy correlations [37], thrust [38] , heavy-jet mass [39], and the multiplicity of jets [40]. The advantage is to increase the fiducial domain where the perturbative calculations can be trusted. In some cases the apparent preference for surprisingly low fitted values of μ, the scale parameter, is eliminated when the resummation is included.

The problem of hadronisation corrections appears when the theoretical calculations in terms of partons have to be compared with the experimental observations in terms of hadrons. In general the effect of hadronisation is simulated by Monte Carlo generators. For each observable the fit range is restricted to an interval of values where the impact of hadronisation is relatively small (fig. 7). The magnitude at LEP of the hadronisation corrections is of order 5-15%. The error on the correction is estimated by comparing the 3 or 4 sufficiently sophisticated and updated generators which are available. In general the performance of these generators in reproducing the observed distributions is really good. In spite of this, since many of the ingredients of these codes are similar, I think that the errors on hadronisation corrections estimated by this comparison could be optimistic.

The theoretical error from our ignorance of higher order terms in the perturbative expansion is estimated, as usual, from the scale dependence of the fitted value of $\alpha_s(m_Z)$ (figs. 8, 9) or in other similar ways.The observed scale dependence is in most cases less pronounced when resummed formulae are used. Without resummation the best fit is usually obtained for very small values of μ, often so small that $\alpha_s \log\mu^2/m_Z^2$ becomes rather large. In fact the points that drive the fit toward small values of μ sit in a region where perturbation theory is of marginal validity. Since small values of μ in general lead to relatively small values of $\alpha_s(m_Z)$, there has been some tendency to argue that values of $\alpha_s(m_Z)$ on the low side are preferred by the data. With the advent of resummed formulae it has been convincingly shown that the preference for small values of μ disappears. In the region near the boundary of phase space, where the preference for small values of μ was originated, the corrective terms are non negligible and their effect removes the need for small μ. As a result the fits with resummed formulae indicate a larger central value for $\alpha_s(m_Z)$.

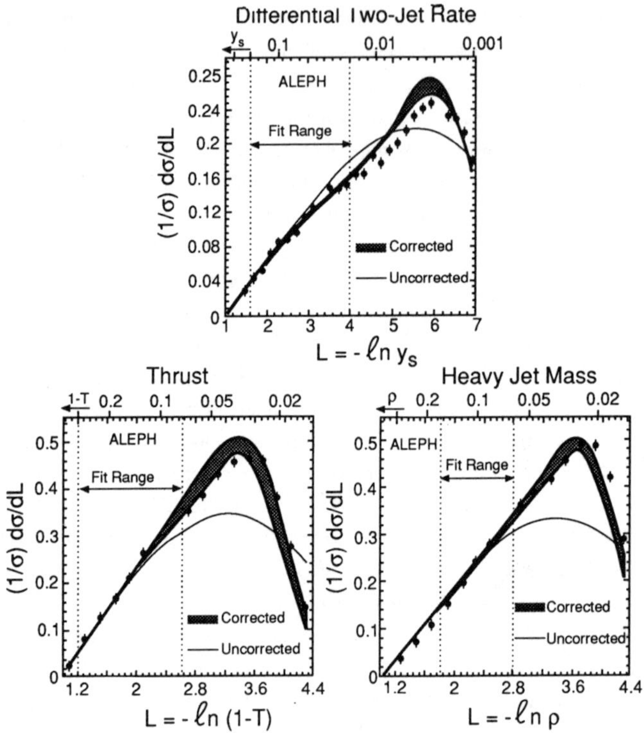

Fig. 7. Experimental distributions (statistical errors only) together with bands covering the predictions using three hadronisation models. The curves refer to the same values of α_s but without hadronisation corrections.

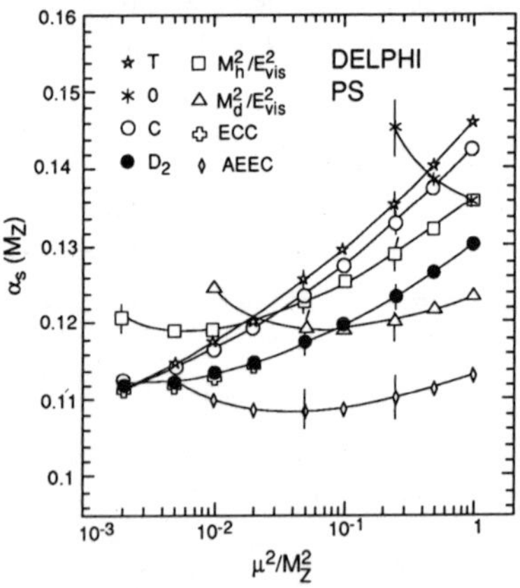

Fig. 8. Observed scale dependence on the results for $\alpha_s(m_Z)$ from different observable measured by DELPHI and fitted with second order QCD predictions, i.e. with no resummation.

The results obtained by the four LEP collaborations and by SLD on $\alpha_s(m_Z)$ from second order perturbative formulae are summarised in Table 3, while the results from expressions including the resummation of logs, which are only available for a subset of observables (see, for example, fig.10), are displayed in Table 4. The split up of the total error into experimental and theoretical, when possible, is also shown.

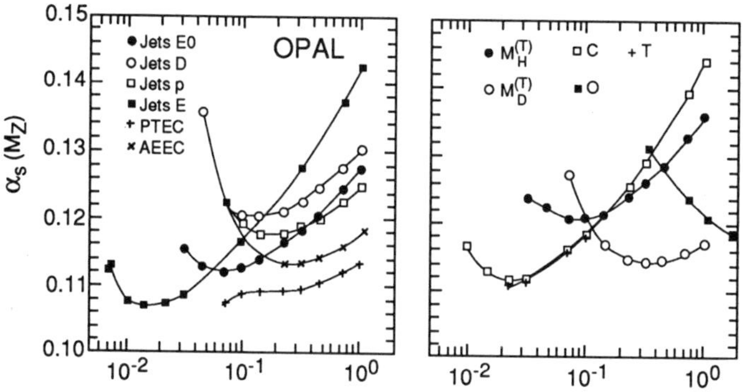

Fig. 9. Observed scale dependence on the results for $\alpha_s(m_Z)$ from different observables measured by OPAL and fitted with second order QCD predictions, i.e. with no resummation.

Table 3

Experiment	$\alpha_s(m_Z)$	Exp. Error	Theor.Error
ALEPH [41]	0.118 ± 0.009		
DELPHI [42]	0.113 ± 0.007	± 0.002	± 0.007
L3 [43]	0.118 ± 0.010		
OPAL [44]	$0.122 \pm^{0.006}_{0.005}$	± 0.001	$\pm^{0.006}_{0.005}$
SLD [45]	0.121 ± 0.014	± 0.009	$\pm^{0.012}_{0.009}$

Table 3. Values of $\alpha_s(m_Z)$ from second order perturbative formulae.

Table 4

Experiment	$\alpha_s(m_Z)$	Exp. Error	Theor.Error
ALEPH [41]	0.125 ± 0.005	± 0.002	± 0.004
DELPHI [42]	0.125 ± 0.006	± 0.002	± 0.006
L3 [43]	0.125 ± 0.009	± 0.003	± 0.008
OPAL[44]	$0.122 \pm^{0.003}_{0.006}$	± 0.001	$\pm^{0.003}_{0.006}$

Table 4. Values of $\alpha_s(m_Z)$ from resummed formulae.

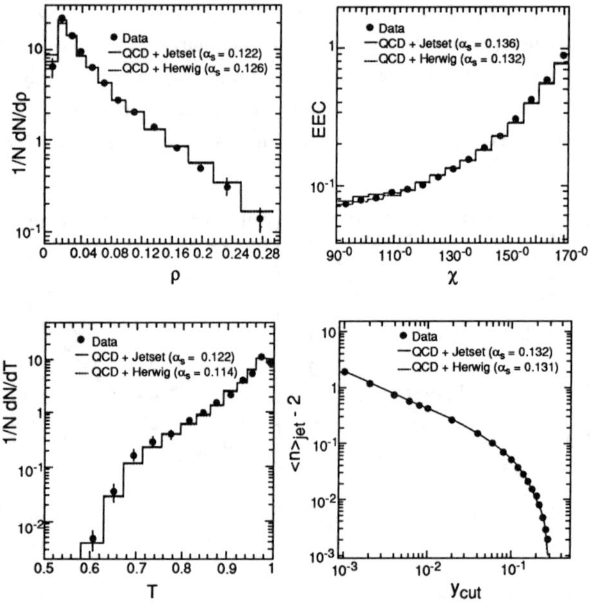

Fig. 10. Heavy jet mass (ρ), thrust (T), energy-energy correlation (EEC) and average jet multiplicity ($< \mathrm{n_{jet}} >$) distributions measured by L3 and compared with QCD predictions with resummation.

The "official" averages presented [46] at the winter conferences are:

$$\begin{aligned} \mathrm{o}(\alpha_s^2) : \alpha_s(\mathrm{m_Z}) &= 0.119 \pm 0.006 \\ \mathrm{o}(\alpha_s^2) + \text{ resumm.} : \alpha_s(\mathrm{m_Z}) &= 0.123 \pm 0.005 \end{aligned} \tag{26}$$

As seen from Tables 3 and 4 the theoretical error is by far the dominant uncertainty. I think that a realistic error is somewhat larger. First, the uncertainty on the hadronisation corrections, which was estimated from the differences among a few Monte Carlo generators could well be larger, due to common biases, in spite of the good overall fitting. Second, the error from higher order perturbative terms, as estimated from scale dependence and other similar methods, should be taken as a lower bound on the error (as already discussed, the error cannot be smaller than that, otherwise one would be confronted with a contradiction among apriori equally reliable determinations of $\alpha_s(\mathrm{m_Z})$). In conclusion I would replace eqs. 26 with something like:

$$\alpha_s(\mathrm{m_Z}) = 0.122 \pm 0.009 \tag{27}$$

I 7. Other Determinations of α_s

We now list a number of additional determinations of α_s which, for reasons either of precision or of reliability or simply because they are still in a rather exploratory phase, are not as important as the ones discussed in the previous sections.

$p\bar{p} \to W + \text{jets}$ [47]. The UA2 Collaboration has measured $r = \sigma(W+1\text{jet})/\sigma(W+0\text{jets}) = (3.91 \pm 0.40)\%$, with a specified jet-counting criterium. The value of α_s can be obtained from this quantity by taking advantage of some recent theoretical calculations, i.e. the 2-loop Drell-Yan QCD correction [48], the p_T distribution of W's at order α_s^2 [49] (see also [50]), which enters because there is a minimum p_T for the jet to be identified, plus a good deal of Monte Carlo simulation in order to properly correct for the underlying event, fragmentation effects and so on. The result, expressed in terms of $\alpha_s(m_Z)$, is given by:

$$\alpha_s(m_Z) = 0.121 \pm 0.024 \tag{28}$$

$p\bar{p} \to b\bar{b}X$ [51]. The value of α_s is extracted from the $b\bar{b}$ production cross-section with $p_T > p_{T\min}$, measured by UA1. The sample of $b\bar{b}$ events is subdivided into 2-body or 3-body, according to their azimuthal distribution. The 2-body are nearly back to back: $\Delta\phi > 150^0$, while for 3-body $\Delta\phi < 150^0$. The 2-body cross-section σ_2 starts at order α_s^2 and the theoretical calculations are available [52] up to order α_s^3. The 3-body cross-section σ_3 starts at order α_s^3 and the next to leading corrections are not known. Two main methods are described. One uses the ratio σ_3/σ_2, in order to reduce the dependence on the structure functions, the other is directly from σ_2. Clearly only the second method is acceptable, because σ_3 is not known beyond leading accuracy. The second method leads to:

$$\alpha_s(m_Z) = 0.109 \pm^{0.016}_{0.012} \tag{29}$$

σ_2 is sensitive to α_s because of its quadratic dependence. The error allegedly includes the ambiguities from the poor knowledge of the gluon structure function, the scale dependence etc. There are some words of caution to be spelled because it is well known that conventional gluon structure functions fit the UA1 data but fail to fit the CDF data [53] on b production at large p_T. To reconcile the theory with the CDF data, that probe the gluon density at smaller values of x, the gluon density has to be modified [54]. It would be interesting to check that the resulting change on the present determination of α_s is still within the quoted error.

Scaling Violations in Fragmentation Functions. In principle the scaling violations for fragmentation functions are predicted by QCD in close analogy with the case of structure functions. An interesting first attempt at the measurement of α_s from the fragmentation function into the sum of all charged tracks has been presented very recently [55]. The necessary lever arm in Q^2 is obtained by combining LEP data with the results of PEP, Petra and Tristan. The error quoted in the determination of $\alpha_s(m_Z)$, which is ±0.006, is in my opinion not plausible. One problem is that the value of α_s is not derived by measuring the logarithmic slopes and then comparing them with the QCD evolution equations, which relate the slopes (i.e. the derivatives of the fragmentation functions) with α_S. Rather the observed behaviour in Q^2 is directly compared with a Monte Carlo of the matrix element type, i.e. one that contain the exact QCD matrix element squared at the next to leading level. The Monte Carlo makes the analysis simpler because it also provides a built in model for the non perturbative component of the fragmentation functions, for the effect of the change of composition of the final state (for example, takes into account that one goes from photon exchange at low energies to Z dominance at LEP) and so on. But one has to trust the Monte Carlo. Among other things, the QCD resummation of logs is truncated at the second order, which is what is contained in the Monte Carlo. As for all cases where a Monte Carlo is crucial, it is difficult to understand what the real error can be. Thus I quote the result of Ref. [55] with a big question mark added to the error:

$$\alpha_s(m_Z) = 0.119 \pm 0.006 \pm ? \tag{30}$$

Photon Structure Function. There are no great news on this front. Apparently a small bug in the existing non leading corrections to the evolution equations was found and corrected [56]. The existing data sets and analyses [57] are not homogeneous and it is often difficult to understand what has been done and even what the precise result is (e.g. $\Lambda^{(3)}$ or $\Lambda^{(4)}$?). Also the uncertainties on the hadronic part remain [58]. It would be great if the data could be reanalysed with the present knowledge and experience. I will repeat here the tentative conclusion that I have proposed in Ref. [2], i.e. that $\Lambda^{(4)}$ ~50-300 GeV, or:

$$\alpha_s(m_Z) = 0.103 \pm 0.013 \tag{31}$$

Charmonium Splittings and Lattice Calculations. In a recent paper [59] it is claimed that lattice calculations are by now sufficiently accurate that a precise determination of α_s is possible from the observed 1P-1S splitting in charmonium. The experimental input is the value $m_{hc} - (3m_{J/\Psi} + m\eta_c)/4 = 458.6 \pm 0.4$ MeV, where h_c is the spin singlet 1P state. This splitting can be computed on the lattice in the quenched approximation. This is equivalent to determining a discrete version of Λ for zero flavours. Then one must apply corrections to go from the lattice to the continuum limit in the $\overline{\text{MS}}$ definition and from 0 to 3 or 4 flavours of quarks. The claim is that one obtains a result equivalent to $\alpha_s(m_Z) = 0.105 \pm 0.004$. I think that there are too many technical assumptions and it is difficult to assess the actual size of the error.Certainly I find the error by far too optimistic, but the result is interesting, as a promise for the future.

I 8. Summary and Conclusion on α_s

In the following I collect those which I consider the main methods of determining α_s with my subjective estimate of the theoretical errors (while, of course, the experimental errors were in all cases left intact!). For each process the label (Exp) or (Th) qualifies whether the dominant source of error is experimental or theoretical, respectively. We have:

$$
\begin{array}{ll}
R_\tau & \alpha_s(m_Z) = 0.117 \pm^{0.010}_{0.016} \text{ (Th)} \\
\text{Deep Inelastic scattering} & 0.112 \pm 0.07 \text{ (Th)} \\
\Upsilon \text{ Decays} & 0.11 \pm 0.01 \text{ (Th)} \\
Re^+e^-(\sqrt{s} < 62 \text{ GeV}) & 0.14 \pm 0.02 \text{ (Exp)} \\
p\bar{p} \to W + \text{jets} & 0.121 \pm 0.024 \text{ (Exp} \sim \text{Th)} \\
\Gamma(Z \to \text{hadrons})/\Gamma(Z \to \ell\ell) & 0.132 \pm 0.012 \text{ (Exp)} \\
\text{Jets at LEP} & 0.122 \pm 0.009 \text{ (Th)} \\
\textbf{Average}^{2} & \mathbf{0.118 \pm 0.007}
\end{array}
\tag{32}
$$

The error on the average was set close to the typical size of the theoretical error involved in each of the best experiments, because I do not like to average theoretical errors (a naive average would give $\pm$ 0.004)[2]. The corresponding range of $\Lambda^{(5)}$ is:

$$\Lambda^{(5)} = (240 \pm 90) \text{ MeV} \tag{33}$$

We see that in spite of the fact that I was extremely conservative at each step in this review, the resulting determination of as is very precise, with a 6% relative error on $\alpha_s(m_Z)$. While, on the one hand, I am supporting conservative estimates of theoretical errors, on the other hand I warn skeptical people that the value of $\alpha_s(m_Z)$ and the error given by eq. 32 appear by now quite solid, and the error cannot be much larger than that (for example larger than 0.01). The consistency of the results on α_s from all the available experiments provides

[2] It is interesting that exactly the same central value and the same error is also proposed as a final result by S. Bethke in his talk at the Dallas conference [104], following an independent discussion of the existing data and their errors.

a remarkable quantitative test of the theory. The list of results on $\alpha_s(m_Z)$ in eq. (32) are reproduced in fig. 11, while in fig. 12 the corresponding value of α_s at the scale of each experiment was plotted.

In conclusion, the overall picture looks very favourably for QCD. The determination of α_s from many different experiments coincide within errors. The value of α_s found at LEP energies is small enough to correspond to a very mild interaction. This is the best evidence for asymptotic freedom and the running of α_s, because such a weak force could not explain the observed binding of hadrons. To me this is a better argument than the consistency with the running of α_s predicted by QCD which is visible in fig. 12.

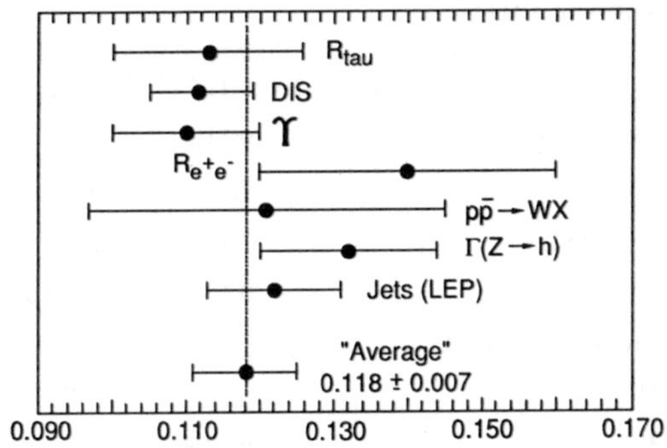

Fig. 11. Main measurements of $\alpha_s(m_Z)$ and their weighted average.

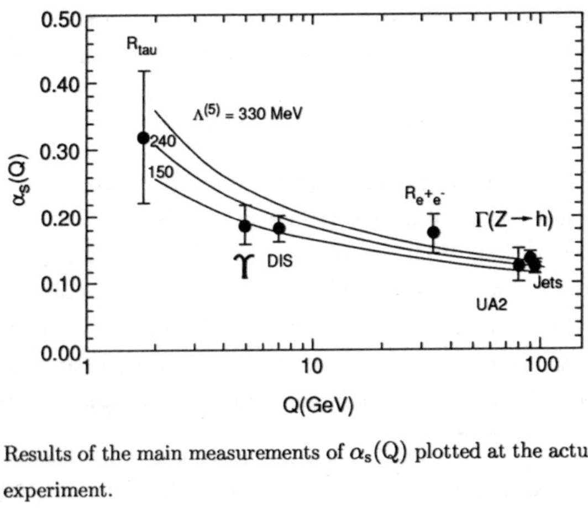

Fig. 12. Results of the main measurements of $\alpha_s(Q)$ plotted at the actual scale of the experiment.

PART II. HARD PROCESSES

The set of consistent measurements of $\alpha_s(m_Z)$ derived from many different processes, that has been reviewed in Part I, provides a very solid quantitative evidence in support of QCD. A lot more of convincing experimental successes for the theory are obtained from the phenomenology of hard processes. QCD provides a solid field theoretical basis for the parton approach. The beautiful "naïve" parton model of Bjorken, Feynman and others (see e.g. Ref. [60]) has evolved into the "QCD improved" parton model (see e.g. Refs. [1] , [61]). This powerful approach has become such a familiar and widely used tool for the everyday practice of high energy physics that one is led to take all its continuously increasing list of successes for granted and, in a way, to think of them as obvious.Actually the very definite and caracteristic pattern of testable predictions given by the parton model already makes quite a strong experimental evidence for QCD. But in addition of inheriting the beautiful experimental score of the parton approach, QCD is also tested in a growing set of phenomena that go beyond the naive parton model and therefore provide an additional and qualitatively different experimental basis for the specific dynamical structure of the theory.

An enormous amount of theoretical and experimental work on hard processes in QCD has been accumulated over the years. Here we do not even attempt to give a systematic review of the subject [1] , [2] , [3] . Rather we concentrate on some recent developments and on a few particularly significant examples.

II 1. Jets in e^+e^- Annihilation

Experiments on e^+e^- annihilation at high energy provide a wonderful laboratory for a systematic test of the distinct signatures predicted by QCD on the structure of the final state. Beyond the important application of measuring $\alpha_s(m_Z)$, one is able to quantitatively verify in detail the matrix elements squared obtained from QCD for the measurable distributions of final state variables. In the following we shall update this item (e.g. with respect to Ref. [2]) by reporting on the results recently obtained at LEP [3] .

One set of results is obtained from the experimental study of the final state with three jets (with a given jet counting definition) by comparing the observed distributions of suitable 3-jet variables with the QCD matrix elements squared and with alternative toy models (so called because purely ad hoc, not really theoretically tenable and extremely unrealistic) which are used just to have an indication of the discriminating power of the data. The variables which are used are listed in the following.

The energy fractions x_1, x_2 and x_3, defined as $x = 2E/\sqrt{s}$, are ordered in decreasing order $x_1 \geq x_2 \geq x_3$ (i.e. the first jet is the most energetic one and so on). The angles θ and χ are defined as the polar angle of the first jet with respect to the beam direction and the angle between the jet plane and the one spanned by the first jet and the beam, respectively (see fig. 13). The Ellis-Karliner angle λ [62] is defined as the angle between jet 1 and the common direction of jets 2 and 3 in their rest system (fig. 14a). Finally, related to θ and χ is $\overline{\theta}$, the angle between the normal to the 3-jet plane and the e- direction (fig. 14b).

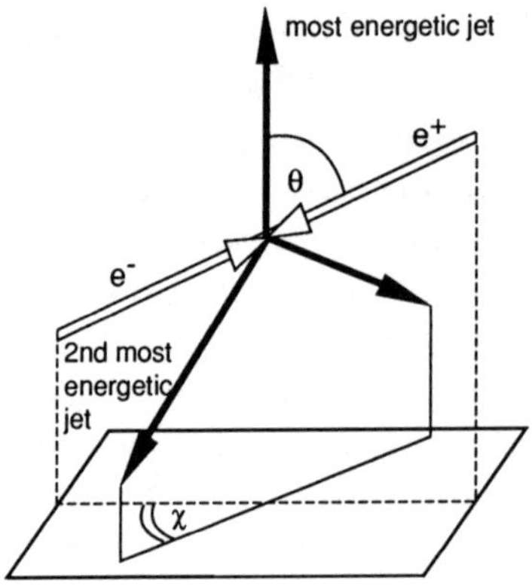

Fig. 13.Definition of θ and χ

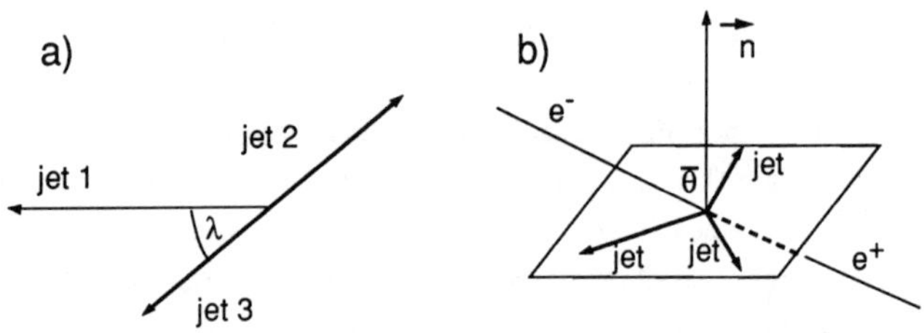

Fig. 14. a) Ellis-Karliner angle λ; b) polar angle of the normal to the 3-jet plane $\overline{\theta}$.

The 3-jet distributions are sensitive to the spin of the gluon. For orientation, in the case of order α_s, massless partons and pure photon exchange one has (see e.g. eq.9-39 of Ref. [1] and, for scalar gluons, Ref. [63]):

$$\begin{aligned}\frac{d\sigma}{dx_q dx_{\bar{q}}} &\sim \frac{x_q^2 + x_{\bar{q}}^2}{(1-x_q)(1-x_{\bar{q}})} \quad \text{(vector gluons)} \\ &\sim \frac{(2-x_q-x_{\bar{q}})^2}{(1-x_q)(1-x_{\bar{q}})} \quad \text{(scalar gluons)}\end{aligned} \tag{34}$$

Note the absence of the double pole at $x_q = x_{\bar{q}} = 1$ in the case of scalar gluons. Similarly, by taking into account that for massless partons the Ellis-Karliner angle λ is given by $|\cos\lambda| = (x_2 - x_3)/x_1$, one obtains [62]:

$$\begin{aligned}\frac{d\sigma}{dx_1 d\cos\lambda} &\sim \frac{x_1^2 + \left[1 + \frac{x_1}{2}\,(\cos\lambda - 1)\right]^2}{(1-x_1)(1-\cos\lambda)} \quad \text{(vector gluons)} \\ &\sim \frac{\left[1 - x_1 + \frac{x_1}{2}\,(1-\cos\lambda)\right]^2}{(1-x_1)(1-\cos\lambda)} \quad \text{(scalar gluons)}\end{aligned} \tag{35}$$

The above formulae can be modified for Z-exchange [64], mass corrections [65] and next to leading corrections [66].

The 3-jet distributions observed by the L3 [67] and OPAL [68] collaborations are shown in fig. 15 and 16 respectively. A related result obtained by DELPHI [69] is shown in fig. 17. The measured quantity is $\bar{\alpha}$ as a function of thrust T, defined from the $\cos\overline{\theta}$ distribution given by $1 + \bar{\alpha}\cos^2\overline{\theta}$. We see that the experimental distributions strongly favour vector gluons and contradict scalar gluons.

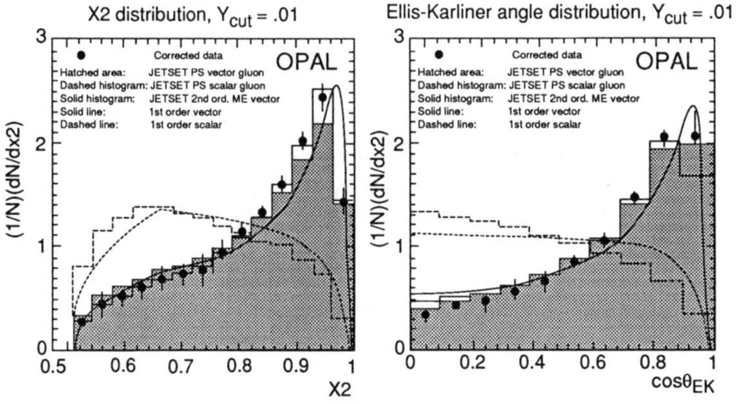

Fig. 15.

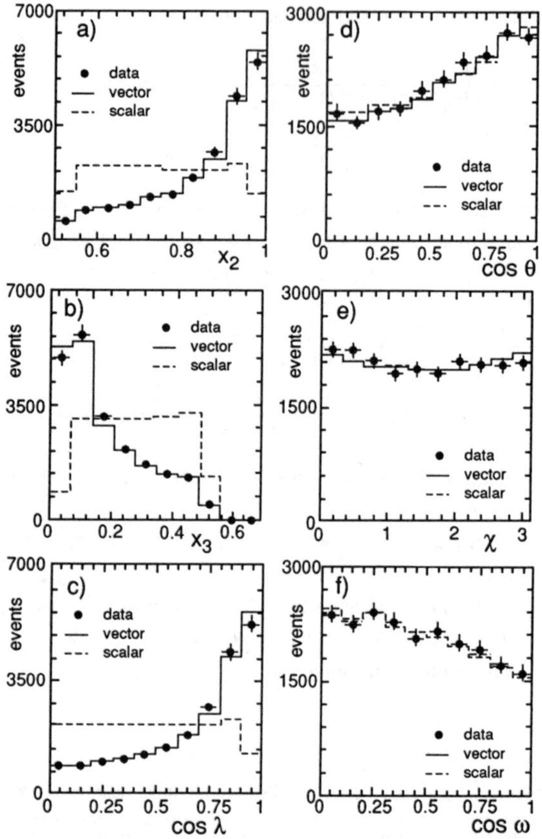

Fig. 16. Distibutions measured by L3

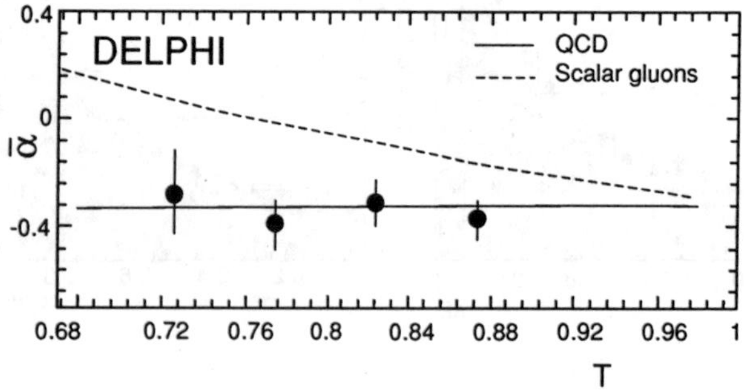

Fig. 17. Distribution of $\bar{\alpha}$ measured by DELPHI.

While 3-jet distributions are particularly interesting for the spin of the gluon, those for 4-jets are studied with the main goal of disentangling the effect of the 3-gluon vertex and of measuring the caracteristic colour factors of the SU(3) group. Clearly the 3-gluon vertex first appears at tree level in the 4-jets final state. The four jets can be either two $q\bar{q}$ pairs or one $q\bar{q}$ pair and two gluons. In the latter case a particular contribution is from diagrams where the two final gluons are generated via the 3-gluon vertex by a virtual gluon radiated from the quark line. Due to the 3-gluon vertex the fraction of all 4-jet events with two gluons in the final state is larger in QCD than in a Abelian model. In order to construct an Abelian model with the closest resemblance to QCD one replaces the colour factors $N_C = 3, C_F = 4/3$ and $T_F = 1/2$ which appear in the QCD rates by $N_C = 0$ (which makes the 3-gluon vertex to vanish), $C_F = 1$ and $T_F = 3$, while, at the same time α_s is rescaled into $4/3\alpha_s$ (the number of quark flavours is kept to N_F in both cases). This prescription makes the total hadronic cross-section to order α_s, the 3-jet rate in lowest order as well as the leading logs in the Q^2 evolution equations the same in the Abelian model and in QCD.

A couple of angles χ_{BZ} and θ_{NR} are commonly used because the corresponding distributions are especially sensitive to the effect of the 3-gluon vertex. With the jets ordered according to decreasing energies $E_1 \geq E_2 \geq E_3 \geq E_4$, χ_{BZ} is the angle between the planes spanned by the jets 1,2 and 3,4 respectively [70]. Similarly θ_{NR} is the angle between the vectors $p_1 - p_2$ and $p_3 - p_4$ [71]. The sensitivity to the 3-gluon vertex arises from the fact that the two gluons from gluon splitting tend to be the less energetic ones in a rather large fraction of events. The results by L3 [72] on the χ_{BZ} and $\cos\theta_{NR}$ distributions are shown in fig.18. Similarly in fig.19 the average values $< \chi_{BZ} >$ and $< \cos\theta_{NR}$ obtained by OPAL [73] are plotted versus y_{cut} (which is the cut appearing in the jet definition criterium: two particles i and j are in the same jet if $2M_{ij}^2/s < y_{cut}$.The data are compared with several QCD predictions obtained by different event generators or by perturbative calculations, which define an error band for the QCD prediction, and with the analogous results from the Abelian model. We see that in general the largest discriminating power between the QCD and the Abelian model predictions is provided by the variable χ_{BZ}. A clear indication in favour of QCD emerges from this test.

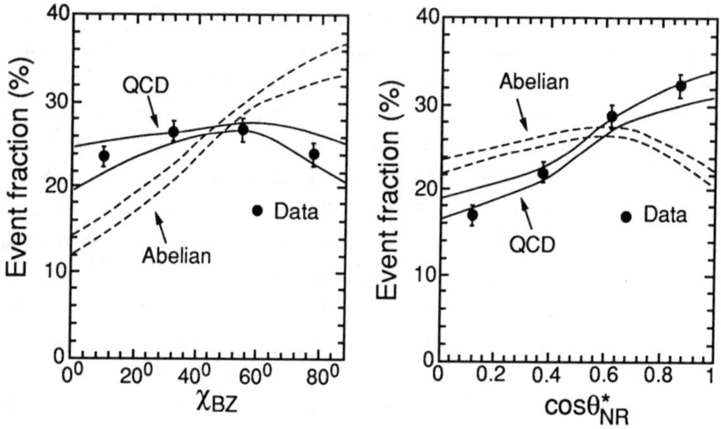

Fig. 18. Measured distribution of χ_{BZ} and θ^*_{NR} (L3). The predictions for QCD and the Abelian model are shown as bands indicating the theoretical uncertainties.

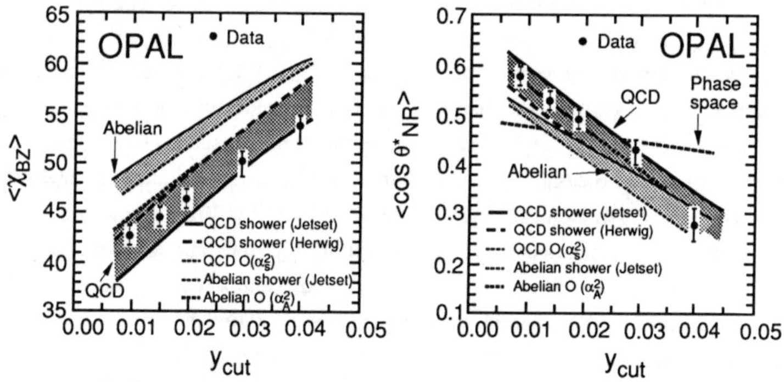

Fig. 19. Mean values of the χ_{BZ} and $\cos\theta^*_{NR}$ distributions as a function of the jet resolution parameter y_{cut}; the data are corrected for detector and hadronisation effects.

An alternative way of quantitatively testing the squared matrix element is to extract from the data the ratios of the colour factors C_A/C_F and T_F/C_F and compare them with the theoretical values. In QCD $C_A = N_C = 3, C_F = (N_C^2 - 1)/2N_C$ and $T_F = 1/2$. The QCD coulour factors were measured by ALEPH [74] and DELPHI [75] in the 4-jet distributions. The average values found are $C_A/C_F = 2.0 \pm 0.3$ and $T_F/C_F = 0.3 \pm 0.2$. The observed values are in good agreement with the QCD predictions 2.25 and 0.375, respectively, and many standard deviations away from those of the Abelian model, which are 0 and 3, respectively. In fig. 20 we see a plot of the observed values versus the predictions of several alternative gauge groups. This is a purely indicative plot, in the sense of showing the discriminating power of this measurement of the matrix element squared, but certainly we possess much better arguments to show that the correct gauge group is SU(3) [1].

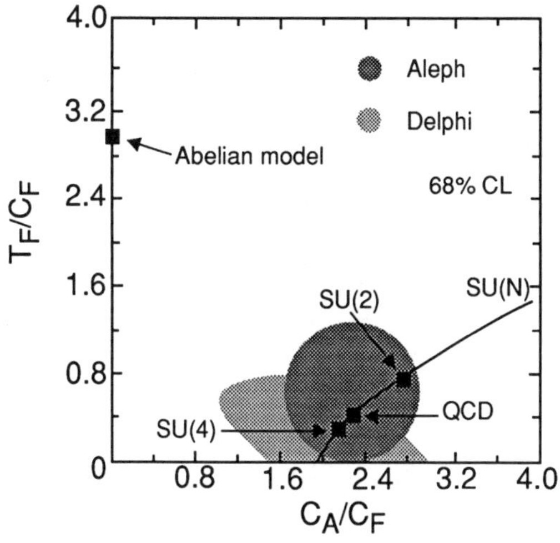

Fig. 20. QCD color factors as measured by ALEPH and DELPHI.

II 2. The QCD-Improved Parton Model of Hard Processes

In recent years hadron colliders have provided very much support for QCD by a number of important experiments that turned into spectacular successes for the QCD-improved parton model. Examples are the Drell-Yan process and its analogue, the W and Z production process (total cross-sections σ_W and σ_Z, x and p_T distributions), jet rates at large p_T, hadron- and photo-production of heavy quarks, large p_T photons and so on. The basic starting point is the factorisation theorem [76], [77] that relates the measured hadronic cross-section $d\sigma$ to its parton analogue $d\hat{\sigma}$ via the parton densities $p_{A,B}$ in the incoming hadrons A and B:

$$d\sigma(s\ldots) = \int dx_1 dx_2 p_A(x_1, Q^2) p_B(x_2, Q^2) d\hat{\sigma}(x_1 x_2 s, Q^2 \ldots) \tag{36}$$

The parton densities $p_{A,B}$ stand for the quark, antiquark and gluon densities in the hadron A or B, which in the most relevant cases is a proton or an antiproton. These densities are measured in deep inelastic scattering. In leading order the structure functions are given by [1]:

$$\begin{aligned} 2xF_1(x, Q^2) &= F_2(x, Q^2) \sim q_i(x, Q^2) + \bar{q}_j(x, Q^2) \\ F_3(x, Q^2) &\sim -q_i(x, Q^2) + \bar{q}_j(x, Q^2) \end{aligned} \tag{37}$$

where appropriate flavours appear on the right hand side in a sum weighted by the appropriate electroweak charges, depending on whether we consider charged lepton or neutrino or antineutrino beams and electromagnetic or weak charged- or neutral-current deep inelastic scattering. In leading order in $\alpha_s(Q)$ the gluon density only appears in the small longitudinal structure function:

$$F_L(x, Q^2) = \frac{\alpha_s(Q^2)}{2\pi} x^2 \int_x^1 \frac{dy}{y^3} \left[\frac{8}{3} F_2(y, Q^2) + 2 \sum_1^{2N_f} e_i^2 \left(1 - \frac{x}{y} \right) y \, g(y, Q^2) \right] \quad (38)$$

and in the QCD evolution equations [1] ,[77]:

$$\begin{aligned} \frac{d}{d\log Q^2} q_i(x, Q^2) &= \frac{\alpha_s(Q^2)}{2\pi} [P_{qq} \otimes q_i + P_{qg} \otimes g] \\ \frac{d}{d\log Q^2} g(x, Q^2) &= \frac{\alpha_s(Q^2)}{2\pi} [P_{gq} \otimes \sum_i q_i + P_{gg} \otimes g] \end{aligned} \quad (39)$$

where:

$$P \otimes f = \int_x^1 \frac{dy}{y} P\left(\frac{x}{y}\right) f(y)$$

Thus the gluon density is measured from deep inelastic scattering with less precision than quark and antiquark densities . The definition of parton densities and their Q^2 evolution can be made more precise by adding the computed next to the leading corrections [1]. Once the parton densities are measured in deep inelastic scattering the partonic cross-sections can be computed in a perturbative expansion in $\alpha_s(Q)$. Finally the hadronic cross-section can be obtained by the convolution integral in eq. 36. Only the knowledge of $\Lambda_{\overline{MS}}$ is in principle needed beyond the measured parton densities. In the following we discuss some examples of hard processes where some recent developments have taken place.

II 3. W and Z Production

The experimental study of W/Z production at the Sp$\bar{p}$S (CERN) and the Tevatron (Fermilab) colliders is an ideal testing ground for QCD. The relevant energy scale is quite large $Q \sim M_{W/Z}$ i.e. as large as at LEP. There is a good balance between Q and $\sqrt{s}$, so that the typical values where the parton densities are probed, $x_{typical} \sim M_{W/Z}/\sqrt{s} \sim 0.13$ at the Sp$\bar{p}$S and ~ 0.05 at the Tevatron, are not too small and well inside the measured region (the corresponding numbers become 0.005 and 0.002 at the LHC and the SSC respectively, and the situation is less favourable). Moreover, in $p\bar{p}$ collisions the dominant contribution is from valence-valence subprocesses and valence densities are the best measured. Thus the input are the quark and gluon densities and $\Lambda_{\overline{MS}}$ and the output are the predicted

σ_W, σ_Z with the corresponding rapidity and p_T distributions. The sources of error are the experimental uncertainties on the parton densities at $Q \sim M_W$ (which is the largest uncertainty) and on $\Lambda_{\overline{MS}}$ and the theoretical ambiguities especially from higher orders in perturbation theory (scale ambiguities, non leading terms in the definition and the evolution of parton densities).

On the total cross-sections σ_W and σ_Z an important progress has been recently accomplished by the calculation of the second order QCD corrections. The first order corrections were computed long ago [78]. They represented a surprise because they were large. The K-factor, defined as the ratio of the corrected versus uncorrected cross-sections,was so large (more than 2 in some cases) at the relatively low energies of the Drell-Yan processes which were measured at that time that the validity of the perturbative expansion was obviously under question. At the energy scale relevant for W and Z production the K factor is more reasonable, of order 30%. The calculation of the next order correction was long awaited for in order to check that the correction stabilises and that the dependence on the scale Q at which the parton densities and the coupling constant are all taken is also mitigated. This calculation [79] [80] is very complicated because many subprocesses are involved ($q+\bar{q}$ starting at order 0, $q+g$ and $g+g$ starting at order $\alpha_s, q+q, \bar{q}+\bar{q}$ and $g+g$ starting at order α_s^2) and the individual diagrams are complicated involving up to two loops. The results show that the additional correction of order α_s^2 is not large and has the positive effect of drastically reducing the scale dependence (fig.21 [81]). It was also checked [79] , [80] that the dependence on the definition of the parton densities beyond the leading approximation is small (that is, the dependence on the precise relation between deep inelastic scattering structure functions and parton densities when corrections of order α_s are included). Thus, the crossections can be predicted with much more accuracy and reliability than previously. The predicted cross-sections are shown in fig. 22 [82] and compared with the data from the Sp$\bar{p}$S and the Tevatron. The largest theoretical error is from the limited knowledge of parton densities. The error is estimated to be ± 8% at CERN and ± 12% at FNAL. Because of the smaller x-values the projections to LHC and SSC energies are affected by larger errors. The ratio $r = \sigma_W/\sigma_Z$ can be computed with much more accuracy, because the largest uncertainties drop in taking the ratio. The resulting theoretical error on r is of about ± 3%. Actually one can use the identity:

$$R \equiv r\,\frac{B(W \to \ell\nu)}{B(Z \to \ell\ell)} = r\,\frac{\Gamma(W \to \ell\nu)}{\Gamma(Z \to \ell\ell)}\,\frac{\Gamma_Z}{\Gamma_W} \tag{40}$$

and derive Γ_W from the data on R by taking Γ_Z and $\Gamma(Z \to \ell\ell)$ from LEP and using the computations of r in QCD and of $\Gamma(W \to \ell\nu)$ from the standard electroweak theory. The result $\Gamma_W = 2.15 \pm 0.11$ GeV is the best available value for Γ_W [82].

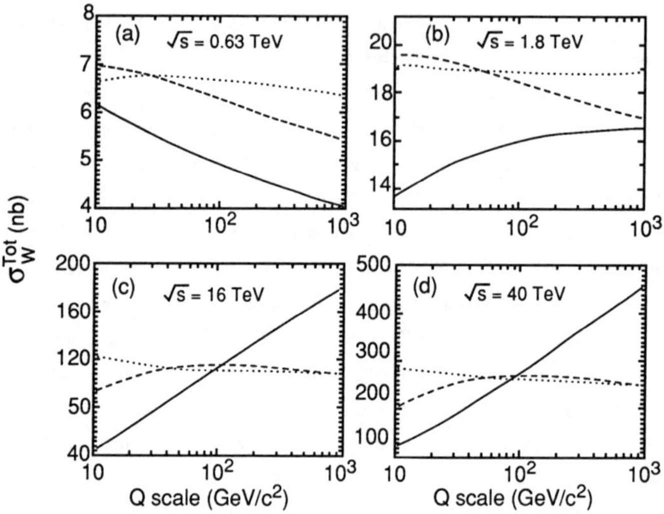

Fig. 21. Total W cross-section at Born level, 1st and 2nd order QCD corrections as a function of the scale Q at $\sqrt{s} = 0.630$ (a), 1.8 (b), 16 (c) and 40 TeV (d) [81].

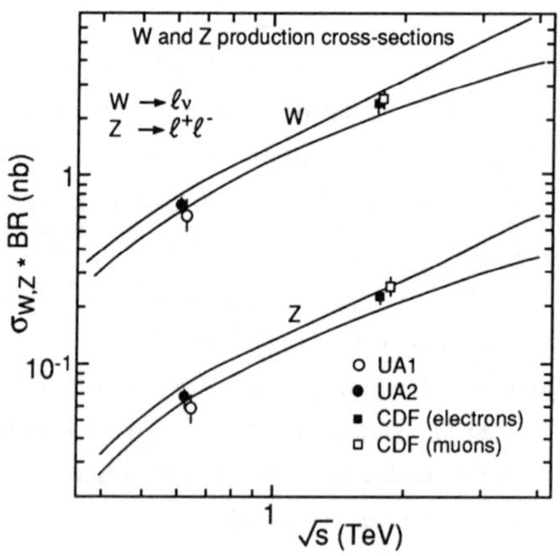

Fig. 22. W, Z production cross-sections as a function of $\sqrt{s}$ [82].

The distributions in rapidity or on Feynman $x_F = 2p_L/\sqrt{s}$ for the W and the Z are simply determined by the x shape of the parton densities. In fact, schematically:

$$\frac{d\sigma}{dQ^2 dx_F} \sim \frac{1}{x_1 + x_2} \sum_1 q_1(x_1, Q^2)\bar{q}_i(x_2, Q^2) + \dots \tag{41}$$

with $x_{1,2} = \frac{1}{2}\left[\pm x_F + \sqrt{x_F^2 + 4\tau}\right]$ and $\tau = Q^2/s$.. These distributions are in good agreement with the experimental results [83] but in a sense are trivial because the normalised distributions are essentially determined by the naive parton model.

Much subtler is the problem of determining the p_T distribution. In this case the naive parton model would only admit the presence of very small transverse momenta (of the order of a few hundred MeV) from the intrinsic p_T smearing of partons inside the proton. The purely perturbative calculations are only valid at $p_T \sim M_{W/Z}$. At smaller p_T values, in the region $\Lambda \ll p_T \ll M_{W/Z}$, where the bulk of the data on $d\sigma/dp_T$ is concentrated, the sequence of logarithmic terms of all orders arising from the soft gluon radiation from the parton legs must be resummed [84]. The corresponding Sudakov exponent [1] is typical of vector gluons. The prediction of the p_T distribution with the correct perturbative limit [85], the soft gluon resummation and the exact integral under the curve to reproduce the corrected total cross-section (with the $o(\alpha_s)$ K-factor included) was obtained at leading order in Ref. [86]. Recently the complete $o(\alpha_s^2)$ next to the leading corrections to the perturbative p_T distribution were obtained [87]. This improved large p_T tail can be matched with a similarly improved Sudakov factor. Indeed the first non leading corrections to the exponent are known [see the last few articles in Ref. [84]]. As a result a quite reliable prediction is now available for comparison with the detailed measurements performed by UA1 and UA2 at CERN [83] and by CDF at Fermilab [88]. In particular the CDF analysis has become available only recently. The CDF p_T distributions for both the W and the Z at $\sqrt{s} = 1.8$ TeV are presented in figs. 23a and b [88] . The bands correspond to the next to leading theoretical prediction obtained with HMRS(B) structure functions [89], $\Lambda^{(5)} = 190$ MeV and the scale $Q = p_T$. The agreement between theory and experiment is really impressive, given that both the parton densities and the value of Λ are fixed from other experiments. We stress again that the average value of p_T at $\sqrt{s} = 1.8$ TeV is 11.5 ± 1.0 GeV [88] , very much larger than any hadronic scale. Previous work [90] had shown that the p_T distribution of Drell-Yan pairs measured at the ISR at $\sqrt{s} = 62$ GeV, where the average p_T is of only 1-2 GeV, is also well described by the theory. Thus the QCD dynamics of gluon radiation is shown to be able to reproduce the buildup with energy of the average transverse momentum up to values which are clearly very far from the intrinsic p_T values of the naïve parton model.

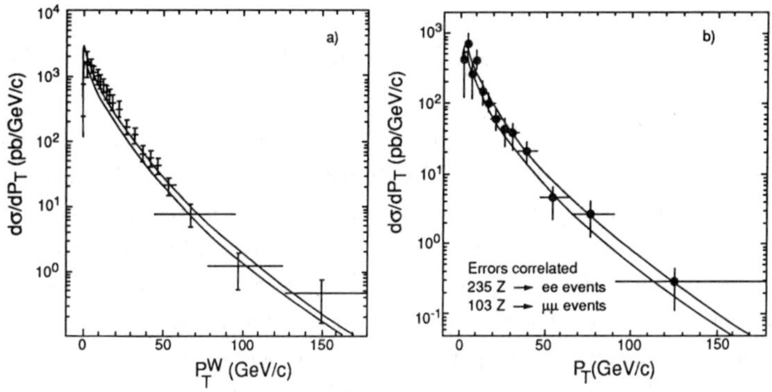

Fig. 23. The differential cross sections for $W(a)$ and for Z(b) production. The next-to-leading-order QCD calculation is shown as a band;the width of the band indicates the theoretical uncertainty.

II 4. Inclusive Jet Yield

The study of jet production at hadron colliders has opened a new territory for probing the validity of the QCD version of the parton model. In $p\bar{p}$ reactions the amplitudes for $2 \to 2$ and $2 \to 3$ quarks/gluons are made accessible to experimental study. In presence of 4 to 5 coloured legs the effects of gluon exchange or radiation are dynamically much more complex than when a single current is present (as in the case of W/Z production). The successful predictions by QCD of the jet rate at large p_T, of their angular distribution, of the ratio of 3 to 2 jet events and of the 3 jet distributions are certainly very relevant tests of the theory [83]. Note that, at $\sqrt{s} = 1.8$ TeV, the jet rate is measured up to $p_T \sim 400$ GeV, the largest scale sofar accessible to experiment. Until recently one could object that the prediction of the jet rate at large p_T had only been obtained at the leading approximation, with the related consequence of a relatively rough accuracy. By now the complex work needed for the complete calculation of the jet rate at large p_T has been completed. Starting from the whole set of partonic cross-sections at order $o(\alpha_s^3)$ computed by R.K.Ellis and Sexton [91], two groups [92] [93] started the jet dressing procedure, each of them with a different definition of jets and with results in agreement for quantities independent of the jet criterium. The scale dependence of the results is much reduced by the inclusion of the next to leading correction. In fig. 24 the comparison is shown of the theoretical curve of Ref. [93] with the data by

CDF on the jet rate at large transverse energy (data and theory obviously refer to the same jet definition). In this case also, the prediction is obtained from standard structure functions and normal values of Λ. To better appreciate the discriminating power of this test, one can look at fig.17 of Ref. [2] where the p_T distributions of jets at widely different energies, ranging from the ISR up to the Tevatron, are shown. From there we see that two very steep dependences, the $\sqrt{s}$ and the p_T dependences, are both correctly reproduced.

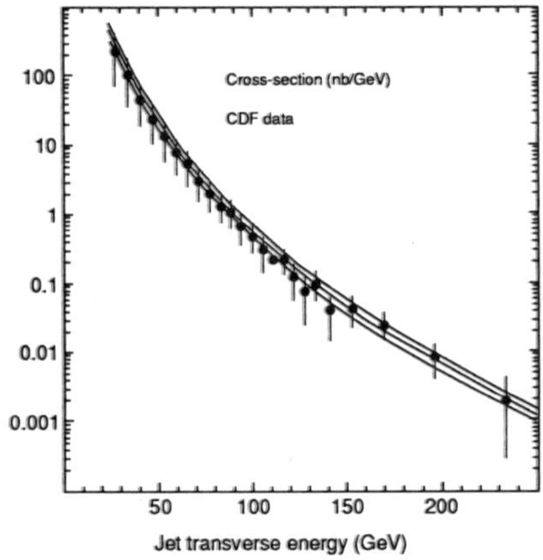

Fig. 24. Cross-section $d\sigma/d\eta dE_T$ averaged over CDF fiducial region $0.1 < |\eta| < 0.7$.

II 5. Prompt Photons

The production of prompt photons at large p_T ($p\bar{p} \to \gamma X$) is in some way similar to that of jets. In this case the prediction is especially sensitive to the gluon density due to the QCD Compton process $g + q(\text{or } \bar{q}) \to \gamma + q(\text{or } \bar{q})$. The calculation of the rate at the next to leading accuracy has also been completed [94]. The available data are from fixed target experiments [95], from the ISR [96] and from UA1 and UA2 [97]. The news here is the publication of the CDF data at $\sqrt{s} = 1.8$ TeV [98], which probe p_T values of the photon up to $p_T \sim 60$ GeV (fig. 25). At all explored energies the agreement between theory and experiment is quite good.

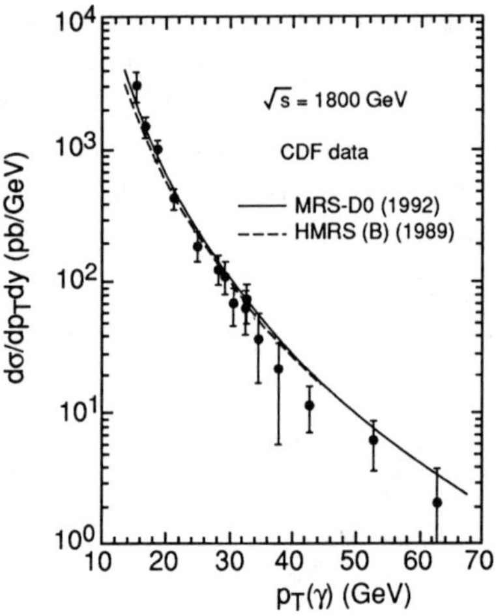

Fig. 25. The p_T distribution of photons measured by CDF and compared with the QCD prediction.

II 6. Heavy Quark Production

The next to leading calculations for heavy flavour production ($p\bar{p} \to Q\overline{Q}X$) [99] and photo-production ($p\bar{p} \to Q\overline{Q}X$) [100] are already available since a few years and their phenomenological implications have been discussed [101]. For a good accuracy of the theoretical predictions for the total cross-section of a heavy quark pair of mass m_Q at a centre of mass energy $\sqrt{s}$, it is necessary that m_Q is large and the ratio $m_Q/\sqrt{s}$ is not too small. If the ratio $m_Q/\sqrt{s}$ is small then one can study the heavy quark production at large p_T. In that case the relevant ratio becomes $p_T/\sqrt{s}$ or $\sqrt{m_Q^2 + p_T^2}/\sqrt{s}$. One such case is b-quark production at $\sqrt{s} = 1.8$ TeV. Some disagreement with existing calculations was recently reported by the CDF collaboration, as shown in fig. 26 [102] [103]. Note that the data and the calculations are in agreement at $\sqrt{s} = 0.63$ TeV, corresponding to the UA1 data. A solution to this problem has been indicated in Refs. [102] , [103] . Gluon fusion is the main contributing parton subprocess. As the CDF data do not extend beyond $p_T = 30$ GeV, the gluon structure function is probed at smaller values of x at CDF than at UA1. It turns out that a small modification of the gluon parton density at small x can bring data and calculations into much better agreement (fig. 27) and yet provide a satisfactory fit of all existing data which probe the gluon density at small x: prompt photons and deep inelastic scattering data, including the most recent NMC data [9] (at especially small values of x) and CCFR data [10].

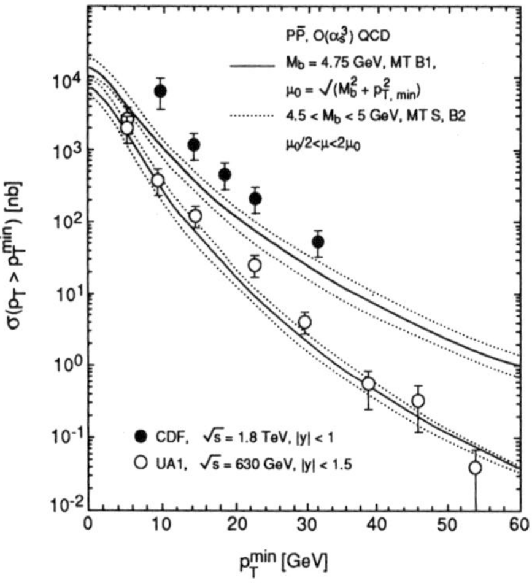

Fig. 26. b production at large p_T measured by UA1 and CDF versus QCD with conventional gluons.

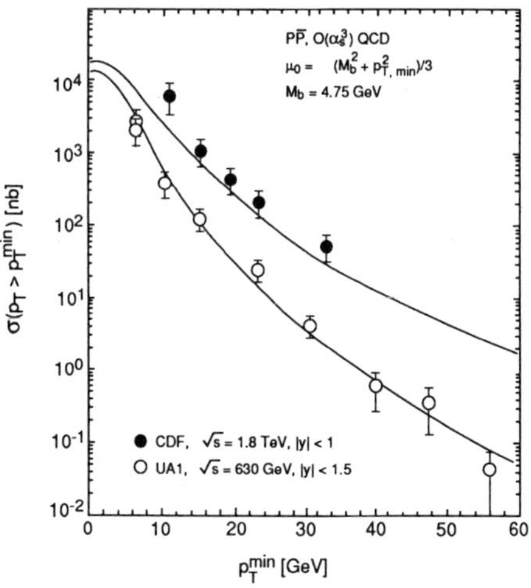

Fig. 27. The same as Fig.26 but with a modified gluon distribution at small x [102] [103]

II 7. Conclusion

In these lectures we have reviewed the measurements of α_s and discussed some recent developments on the QCD description of hard processes and its experimental tests. The overall picture is extremely favourable for QCD. By now the list of successful quantitative tests of the theory has become really long. The series of measurements of α_s from many different processes is imposing by itself. But we have seen in the second part of the lectures that a variety of highly non trivial predictions of perturbative QCD, which cover quite different aspects of hard processes, are indeed also supported by experiment. In all fairness no evident failures of the theory have ever been observed. Certainly the achieved precision is at best of the order of 5-10%, quite far from, say, the precision of electroweak or QED tests. But the fact remains that within the feasible accuracy the theory has indeed been verified by experiment.

References

[1] See, for example, G. Altarelli, *Phys. Reports* **81** (1982) 1.

[2] G. Altarelli, *Ann. Rev. Nucl. Part. Sci.* **39** (1989) 357.

[3] See, for example, S. Bethke and J.E. Pilcher, *Ann. Rev. Nucl. Parti. Sci.* **42** (1992), to be published;
T. Hebbeker, *Phys. Rep.*, to be published.

[4] BCDMS Collaboration, A.C. Benvenuti et al., *Phys. Lett.* **B195** (1987) 91; **B195** (1987) 97; **B223** (1989) 485; **B223** (1989) 490; **B237** (1990) 592; **B237** (1990) 599; preprints CERN/EP-89/06, CERN/EP-89/170.

[5] L.W. Whitlow, Ph. D. Thesis, SLAC Report 357 (1990);
A. Bodek, in Proc. Workshop on Hadron Structure Functions, FNAL (1990);
L.W. Whitlow et al., *Phys. Lett.* **B250** (1990) 193.

[6] M. Virchaux and A. Milsztajn, *Phys. Lett.* **B274** (1992) 221.

[7] EMC Collaboration, J.J. Aubert et al., *Nucl. Phys.* **B259** (1985) 189; **B293** (1987) 740; *Phys. Lett.* **B202** (1988) 603.

[8] S. Wimpenny, Proc. of the LP-HEP 91 Conference, Geneva (1991), **Vol. 1** p. 145.

[9] M.A.J. Botje, Proc. of the XXVIIth Rencontres de Moriond, Les Arcs, France (1992).

[10] CCFR Collaboration, J.R. Mishra et al., Nevis preprints 1459, 1460, 1461 (1992).

[11] CDHSW Collaboration, P. Berge et al., *Z Phys.* **C49** (1991) 187.

[12] CHARM Collaboration, F. Bergsma et al., *Phys. Lett.* **153B** (1985) 111.

[13] A.D. Martin, R.G. Roberts and W.J. Stirling, *Phys. Rev* **D43** (1991) 3658; *Phys. Lett.* **B266** (1991) 273; RAL preprint-9-021 (1992).

[14] S.G. Gorishny, A.L. Kataev and S.A. Larin, *Phys. Lett.* **B259** (1991) 144; see also L.R. Surguladze and M.A. Samuel, *Phys. Rev. Lett.* **66** (1991) 560.

[15] A. Djouadi, J.H. Kuhn and P.M. Zerwas, *Z. Phys.* **C46** (1990) 411.

[16] T. Hebbeker, Aachen preprint PITHA 91-08 (1991).

[17] B.A. Kniehl and J.H. Kuhn, *Phys. Lett.* **B224** (1989) 229; *Nucl. Phys.* **B329** (1990) 557.

[18] J. Nash, Proc. of the XVII Rencontres de Moriond, Les Arcs, France (1992);
D. Schaile, Proc. of the Rencontre de la Vallee d'Aoste, La Thuile, Italy (1992).

[19] E. Fernandez et al., *Phys. Rev.* **D31** (1985) 1537;
H.J. Behrend et al., *Phys. Lett.* **B183** (1987) 400;
G. D'Agostini, W. De Boer and G. Grindhammer, *Phys. Lett.* **B229** (1989) 160;
R. Marshall, *Z. Phys.* **C43** (189) 595.

[20] V. Branchina et al., *Phys. Rev.* **D46** (1992) 75.

[21] W. De. Boer, Proc. of the XXIV Int. Conf. on High Energy Physics, Munich, Germany (1988), p. 905.

[22] E. Braaten, *Phys. Rev. Lett.* **60** (1988) 1606; *Phys. Rev.* **D39** (1989) 1458;
S. Narison and A. Pich, *Phys. Lett.* **B211** (1988) 183;
A. Pich, "Hadronic Tau-Decays and QCD", in Proc. Workshop on Tau Lepton Physics, Orsay, 1990, eds. M. Davier and B. Jean-Marie (Ed. Frontieres, Gif-sur-Yvette, 1991), p. 321.

[23] A. Pich, Proc. of the XVII Rencontres de Moriond, Les Arcs, France (1992); see also M. Luo and W.J. Marciano, Brookhaven preprint BNL-47187 (1992).

[24] J. Pumplin, *Phys. Rev.* **D41** (1990) 900.

[25] M.A. Shifman, A.I. Vainshtein and V.I. Zakharov, *Nucl. Phys.* **B147** (1979) 385, 448, 519.

[26] F. David, *Nucl. Phys.* **B235** (1984) 237;
A.H. Mueller, *Nucl. Phys.* **B250** (1985) 327.

[27] A.H. Mueller, Proceedings of QCD-20 Years Later, Aachen (1992).

[28] G.B. West, *Phys. Rev. Lett.* **67** (1991) 1388, E67 (1991) 3732.

[29] L.S. Brown, L.G. Yaffe and C. Zhai, Univ. of Washington preprint UW/PT-92-07 (1992).

[30] V.I. Zakharov, Max-Planck Institut preprint MPI-Ph/92-33 (1992); see also M.Beneke,V.I.Zakharov, MPI-Ph/92-53; M.Beneke, MPI-Ph/92-64

[31] S.Narison,Heidelberg Preprint HD-THEP-92-34; see however T.N. Truong, Ecole Polytechnique preprint EP-CPTh.A172.0492 (1992).

[32] M. Kobel, Proc. of the XXVII Rencontres de Moriond, Les Arcs, France (1992) and Thesis Univ. of Erlangen, Nurnberg (1991).

[33] W. Kwong et al., *Phys. Rev.***D37** (1980) 3210.

[34] B.P. Mackenzie and G.P. Lepage, *Phys. Rev. Lett.* **47** (1981) 1244;
R. Barbieri, R. Gatto, R. Kogerler and Z. Kunszt, *Phys. Lett.* **57B** (1975) 455;
W. Celmaster, *Phys. Rev.* **D19** (1979) 1517;
R. Barbieri, M. Caffo, R. Gatto and E. Remiddi, *Phys. Lett.* **95B** (1980) 93.

[35] A. Bizzeti, Proceedings of the XXIV International Conference on High Energy Physics, Munich, Springer Verlag, p. 909 (1988).

[36] Z. Kunszt and P. Nason in "Z physics at LEP1" ed. by G. Altarelli, R. Kleiss and C.Verzegnassi, CERN report 89-08 (1989).

[37] For EEC: J. Kodaira and L. Trentadue, *Phys. Lett.* **B112** (1982) 66, B123 (1982) 335;
J. Collins and D. Soper, *Nucl. Phys.* **B197** (1982) 446;
G. Turnock, Cavendish preprint HEP-91/6 (1991).

[38] For thrust: S. Catani et al., *Phys. Lett.* **B263** (1991) 491.

[39] G. Turnock and B.R. Webber, *Phys. Lett.* **B272** (1991) 368.

[40] For jet multiplicities: S. Catani et al., *Phys. Lett.* **B269** (1991) 432;
N. Brown and W.J. Stirling, RAL preprints RAL-91-049 and DTP/91/30 (1991).

[41] ALEPH Collaboration: R. Miquel, Proc. of the XXVII Rencontres de Moriond, Les Arcs, France (1992);
D. Decamp et al., CERN-PPE/92-33 (1992).

[42] H. Fürstenau, Proc. of the XXVII Rencontres de Moriond, Les Arcs, France (1992);
P. Abreu et al., CERN PPE/91/181/rev.

[43] L3 Collaboration: D. Kirkby, Proc. of the XXVII Rencontres de Moriond, Les Arcs, France (1992);
B. Adeva et al., CERN-PPE/92/58 (1992).

[44] OPAL Collaboration: S. Sanghera, Proc. of the XXVII Rencontres de Moriond, Les Arcs, France (1992);
P.D. Acton et al., CERN-PPE 92-18 (1992).

[45] SLD Collaboration: P.N. Burrows, Proc. of the XXVII Rencontres de Moriond, Les Arcs, France (1992);
P.N. Burrows et al., SLAC-Pub-5802 (1992).

[46] S. Bethke and S. Catani, Proc. of the XXVII Rencontres de Moriond, Les Arcs, France (1992).

[47] UA2 Collaboration: J. Alitti et al., *Phys. Lett.* **B263** (1991) 563.

[48] T. Matsuura and W.L. van Neerven, *Z. Phys.* **C38** (1988) 623;
T. Matsuura, S.C. van der Marck and W.L. van Neerven, *Phys. Lett.* **B211** (1988) 171; *Nucl. Phys.* **B319** (1989) 570;
R. Hamberg, W.L. van Neerven and T. Matsuura, *Nucl. Phys.* **B349** (1991) 343.

[49] P.B. Arnold and M.H. Reno, *Nucl. Phys.* **B319** (1989) 37.

[50] S.D. Ellis, R. Kleiss and W.J. Stirling, *Phys. Lett.* **B154** (1985) 435;
F.A. Berends, W.T. Giele and H. Kiujf, *Nucl. Phys.* **B321** (1989) 39;
F.A. Berends, W.T. Giele, H. Kuijf, R. Kleiss and W.J. Stirling, *Phys. Lett.* **B224** (1989) 237.)

[51] A. Geiser, Proc. of the XXVII Rencontres de Moriond, Les Arcs, France, 1992 (Aachen preprint, PITHA 92/19).

[52] P. Nason, S. Dawson and R.K. Ellis, *Nucl. Phys.* **B303** (1988) 607, B327 (1989) 49;
G. Altarelli et al., *Nucl. Phys.* **B308** (1988) 724;
M.C. Mangano, P. Nason and G. Ridolfi, Univ. of Parma preprint IFUP-TH 32/91 (1991).

[53] CDF Collaboration: M.J. Shochet, Fermilab-Conf-91/341-E (1991).

[54] E.L. Berger and R. Meng, Argonne preprint ANL-HEP-PR-92-32 (1992).

[55] DELPHI Collaboration: W. De , Proc. of QCD 20 Years Later, Aachen, 1992.

[56] M. Fontannaz and E. Pilon, Orsay preprint LPTHE 91/21 (1991).

[57] C. Berger and W. Wagner, *Phys. Rep.* **146** (1987) 1;
H. Kolanoski and P. Zerwas, in Adv. Ser. on Directions in High Energy

Physics, eds. A. Ali and P,. Soding, World. Sci., Vol. 1 (1988).
[58] I. Antoniadis and G. Grunberg, *Nucl. Phys.* **B213** (1983) 445;
I. Antoniadis and L. Marleua, *Phys. Lett.* **161B** (1985) 163;
J.H. Field, F. Kapusta and L. Poggioli, *Phys. Lett.* **181B** (1986); *Z. Phys.* **C36** (1987) 121;
M. Gluck and E. Reya, Dortmund Univ. preprint (1988).
[59] A.X. El-Khadra et al., Fermilab-pub-91/354-T (1991); see also G.S.Bali,K.Schilling, Wuppertal Preprint WUB-92-29(1992).
[60] R.P. Feynman, "Photon Hadron Interactions", New York, Benjamin (1972).
[61] S. Bethke, Proceedings of "QCD 20 Years Later", Aachen, Germany (1992);
W.J. Stirling, Proceedings of the E. Majorana Summer School, Erice 1992
[62] J. Ellis and I. Karliner, *Nucl. Phys.* **B148** (1979) 141.
[63] P. Hoyer et al., *Nucl. Phys.* **B161** (1979) 349.
[64] E. Laerman, K.H. Streng and P.M. Zerwas, *Z. Phys.* **C3** (1980) 289;
E. Laerman, K.H. Streng and P.M. Zerwas, to be published in *Z. Phys.* C (Erratum);
K. Koller et al., *Z. Phys.* **C6** (1980)131;
J.G. Körner and D.H. Schiller, "Helicity Description of $e^+e^- \to q\bar{q}g$ and $e^+e^- \to Q\overline{Q}(1^{--}) \to qqq$ on and off the Z^0: Quark, Gluon and Beam Polarization Effects", DESY Report 81-043 (1981).
[65] G. Kramer, G. Schierholz and J. Willrodt, *Z. Phys.* **C4** (1980) 149.
[66] J.G. Körner and G.A. Schuler, *Z. Phys.* **C26** (1985) 559;
J.G. Körner, G.A. Schuler and F. Barreiro, *Phys. Lett.* **B188** (1987) 272;
G.A. Schuler and J.G. Körner, *Nucl. Phys.* **B325** (1989) 557.
[67] L3 Collaboration, Adeva et al., *Phys. Lett* **263** (1991) 551.
[68] OPAL Collaboration, G. Alexander et al., *Z. Phys.* **C52** (1991) 543.
[69] P. Abreu et al., *Phys. Lett.* **B274** (1992) 498.
[70] M. Bengtsson and P.M. Zerwas, *Phys. Lett.* **B208** (1988) 306.
[71] O. Nachtmann and A. Reiter, *Z. Phys.* **C16** (1982) 45;
G. Rudolph in "Physics at LEP", CERN Report CERN-86-02, Vol. II, p. 150.
[72] L3 Collaboration, B. Adeva et al., *Phys. Lett.* **B248** (1990) 227.
[73] OPAL Collaboration, M.Z. Akrawy et al., *Z. Phys.* **C49** (1991) 49.
[74] ALEPH Collaboration, D. Decamp et al., "Testing the Non-Abelian Structure of QCD Based on Four-Jet Events - Evidence for the Triple Gluon Vertex", ALEPH internal note 91-84, June 1991.
[75] DELPHI Collaboration, P. Abreu et al., *Phys. Lett.* **B255** (1991) 466;
DELPHI Collaboration, P. Abreu et al., "Evidence for the Triple-Gluon Vertex by Measuring the Color Factor", DELPHI note 91-52 PHYS 107, paper contributed to the Joint International Lepton-Photon Symposium and Europhysics Conference on High Energy Physics, Geneva, 1991.
[76] A.V. Radyushkin, *Phys. Lett.* **69B**: 461 (1978);

H.D. Politzer, *Nucl. Phys.* **B129**: 301 (1977);
D. Amati, R. Petronzio and G. Veneziano, *Nucl. Phys.* **B140** (1978) 54; **B146** (1978) 29;
R.K. Ellis, H. Georgi, M. Machacek, H. D. Politzer, G.G. Ross, *Nucl. Phys.* **B152** (1979) 285;
S. Libby, G. Sterman, *Phys. Rev.* **D18** (1978) 3252, 4737.

[77] G. Altarelli and G. Parisi, *Nucl. Phys.* **B126** (1977) 298.

[78] G. Altarelli, R.K. Ellis and G. Martinelli, *Nucl. Phys.* **B143** (1978) 521; **EB146** (1978) 544; **B157** (1979) 461;
J. Kubar-André and F. Paige, *Phys. Rev.* **D19** (1979) 221;
J. Kubar-André, M. Le Bellac, J.L. Meunier and G. Plaut, *Nucl. Phys.* **B157** (1980) 251.

[79] R. Hamberg, W.L. van Neerven and T. Matsuura, *Nucl. Phys.* **B359** (1991) 343;
W.L. van Neerven and E.B. Zijestra, INLO-PUB14/91.

[80] W.L. van Neerven, INLO-PUB-9/92, Proceedings of the Rencontres de Moriond, Les Arcs, France, 1992.

[81] T. Matsuura and H. Plothow-Besch, in Proceedings of the LHC Workshop, Aachen, Germany, 1990 CERN 90-10, Vol II, page 94.

[82] H. Plotow-Besch, CERN-PPE/91-206.

[83] See e.g., G. Altarelli and L. Di Lella, "Adv. Ser. on Directions in High Energy Physics", Singapore, World Scientific 4 (1989) 177.

[84] Yu. L. Dokshitzer, D.I. Dyakonov and S.I. Troyan, *Phys. Lett.* **78B** (1978) 290; *Phys. Rep.* **58** (1980) 58;
G. Parisi and R. Petronzio, *Nucl. Phys.* **B154** (1979) 427;
G. Curci, M. Greco and Y. Srivastava, *Phys. Rev. Lett.* **43** (1979) 434;
J. Collins and D.E. Soper, *Nucl. Phys.* **B139** (1981) 381 **B194** (1982) 445; **B197** (1982) 446;
J. Collins, D.E. Soper and G. Sterman, CERN preprint TH.3923/84 (1984);
J. Kodaira, L. Trentadue, *Phys. Lett.* **112b** (1982) 66; **123B** (1983) 335;;
C.T. Davies and W.J. Stirling, *Nucl. Phys.* **B244** (1984) 337.

[85] R.K. Ellis, G. Martinelli and R. Petronzio, *Nucl. Phys.* **B211** (1983) 116.

[86] G. Altarelli et al., *Nucl. Phys.* **B246** (1984) 12;
G. Altarelli, R.K. Ellis and G. Martinelli *Z. Phys.* **C27** (1985) 617.

[87] P. B. Arnold and M.H. Reno, *Nucl. Phys.* **B319** (1989) 37; **B330** (1990) 284;
P.B. Arnold and R.P. Kauffman, *Phys. Lett.* **B349** (1991) 381;
R.J. Gonsalves, J. Powlowski and C. F. Wai, *Phys. Rev.* **D40** (1989) 2245.

[88] CDF Collaboration, V.F.Abe et al., *Phys. Rev. Lett.* **66** (1991) 2932, 2952.

[89] N. Harriman et al., *Phys. Rev.* **D42** (1990) 798.

[90] G. Altarelli, R.K. Ellis and G. Martinelli, *Phys. Lett.* **151B** (1985) 457.

[91] R.K. Ellis and J.C. Sexton, *Nucl. Phys.* **B269** (1986) 445.

[92] P. Aversa et al., *Nucl. Phys.* **B327** (1989) 105; *Z. Phys.* **C46** (1990) 253; *Phys. Rev. Lett.* **65** (1990) 401; *Z. Phys.* **C49** (1991) 459.

[93] S. D. Ellis, Z. Kunszt and D. E. Soper, *Phys. Rev. Lett.* **64** (1990) 2121; *Phys. Rev.* **D40** (1989) 2188; *Phys. Rev. Lett.* **62** (1989) 726.

[94] P. Aurenche, R. Baier and M. Fontannaz, *Phys. Rev.* **D42** (1990) 1440.

[95] WA70 Collaboration: M. Bonesini et al., *Z. Phys.* **C38** (1988) 371;
UA6 Collaboration: A. Bernasconi et al., *Phys. Lett.* **206B** (1988) 163;
NA3 Collaboration: J. Badier et al., *Z. Phys.* **C31** (1986) 341;
NA4 Collaboration: C. De Mazzo et al., *Phys. Rev.* **D36** (1987) 8.

[96] AFS Collaboration: T. Akesson et al., *Phys. Lett.* **158B** (1985) 282.

[97] UA1 Collaboration: C. Albajar et al., *Phys. Lett.* **209B** (1988) 385;
UA2 Collaboration: J.A. Appel et al., *Phys. Lett.* **176B** (1986) 239.

[98] CDF Collaboration: F. Abe et al., *Phys. Rev. Lett.* **68** (1992) 2734.

[99] P. Nason, S. Dawson and R.K. Ellis, *Nucl. Phys.* **B303** (1988) 607;
P. Nason, S. Dawson and R.K. Ellis, *Nucl. Phys.* **B237** (1988) 49;
W. Beenakker, H. Kuijf, W.L. van Neerven and J. Smith, *Nucl. Phys.* **B351** (1991) 507;
W. Beenakker et al., *Phys. Rev.* **D40** (1989) 54;
M. Mangano, P. Nason and G. Ridolfi, *Nucl. Phys.* **B373** (1992) 295.

[100] R.K. Ellis and P. Nason, *Nucl. Phys.* **B312** (1989) 551;
J. Smith and W.L. van Neerven, ITP-SB-91-40.

[101] G. Altarelli, M. Diemoz, G. Martinelli and P. Nason, *Nucl. Phys.* **B308** (1988) 607.

[102] E.L. Berger, R. Meng and W.K. Tung, *Phys. Rev.* **D46** (1992) 1895.

[103] E.L. Berger, R. Meng and J. Qiu, ANL-HEP-CP92-79 (1992).

[104] S. Bethke, DH-PY 92/13 and Proceedings of the XXXVI Int. Conf. on High Energy Physics, Dallas, Texas, 1992.

CHAIRMAN: G. Altarelli

Scientific Secretaries: N. Misiak, S. Shotkin

DISCUSSION

– *Ball:*

What is the meaning of singlet and non-singlet structure function?

– *Altarelli:*

Recall that the structure of the QCD evolution equations in leading order in α_s, for a quark or antiquark density $q_i(x, Q^2)$, where the index i runs over quarks and antiquarks of different flavours, is given by:

$$Q^2 \frac{d}{dQ^2} q_i(x, Q^2) = \frac{\alpha_s(Q^2)}{2\pi}\left[P_{qq} \otimes q_i + P_{qg} \otimes g\right] + 0(\alpha_s^2)$$

$$Q^2 \frac{d}{dQ^2} g(x, Q^2) = \frac{\alpha_s(Q^2)}{2\pi}\left[P_{gq} \otimes \sum_i q_i + P_{gg} \otimes g\right] + 0(\alpha_s^2)$$

where the symbol $\otimes$ indicates the convolution:

$$P \otimes q = \int_x^1 \frac{dy}{y}\, q\left(\frac{x}{y}, Q^2\right) P(y)$$

$g(x, Q^2)$ is the gluon density and the kernels $P_{qq}, P_{qg} \dots$ are known distributions. The crucial point is that the kernels are independent of the index i. This follows from the fact that gluons are coupled in the same way to all flavours. Thus, for example, if one takes any difference $q_i - q_j$ the equations simplify very much:

$$Q^2 \frac{d}{dQ^2} (q_i - q_j) = \frac{\alpha_s(Q^2)}{2\pi} P_{qq} \otimes (q_i - q_j) + 0(\alpha_s^2)$$

This homogeneous equation describes the "non-singlet" evolution. Any structure function given by:

$$F(x, Q^2) = \sum_i C_i q_i(x, Q^2)$$

can be decomposed into a non-singlet and a singlet part according to:

$$F = F^{NS} + F^S = \sum_i^{2f} (C_i - < C >)\, q_i + < C > \Sigma$$

where $< C >= \sum_{i=1}^{2f} C_i/2f$ and $\sum = \sum_i q_i$. The density $\sum$ obeys the singlet evolution equation, obtained from the equation for q_i, by summing over all values of i:

$$Q^2 \frac{d}{dQ^2}\Sigma = \frac{\alpha_s(Q^2)}{2\pi}\left[P_{qq} \otimes \Sigma + P_{qg} \otimes g\right] + 0(\alpha_s^2)$$

$$Q^2 \frac{d}{dQ^2}g = \frac{\alpha_s(Q^2)}{2\pi}\left[P_{gq} \otimes \Sigma + P_{gg} \otimes g\right] + 0(\alpha_s^2)$$

– *Ball:*

Are there some calculations of higher twist effects and how do they compare with experiment?

– *Altarelli:*

Not very much is known about higher twist corrections, which are suppressed by powers of $1/Q^2$. They are called higher twist because in the light-cone expansion they correspond to operators of larger "twist", i.e. dimension minus spin, so by dimensional analysis they lead to contributions which are suppressed by powers of $1/Q^2$. The difficulty of getting hold of these terms is that the relative weight of this part in comparison with the leading part depends on the hadronic matrix elements of these higher twist operators in the light-cone expansion. So the problem comes down to the calculation of non-perturbative quantities like the matrix element between proton states of these operators constructed from the fields of the theory. We are not very good at computing hadronic matrix elements of operators which depend on soft physics. I do not know many interesting experimental results, or many reliable calculations of higher twist terms, so that I agree that we do not know much about higher twist effects.

– *Word:*

You showed results for $\Lambda^{(4)}$ and $\alpha_s(M_Z)$ derived from various Υ and J/ψ decays, yet the results of the "combined fit" were larger than any of the individual values. Why is that?

– *Altarelli:*

First of all there are data on J/ψ decays and corresponding analyses within QCD, but I find that they are less reliable than Υ decays because when a J/ψ of 3 GeV goes into 3 gluons (3-jet final state), there is really little space for perturbative QCD to be reliably applied. In addition there are non-relativistic corrections which are huge. Thus I would give more emphasis to Υ decays, as far as measuring α_s is concerned.

When it comes to your question, when only Υ decays are fitted, the D parameter which gives the non-relativistic corrections is fixed to its central value D=0. For the combined J/ψ and Υ fit, D is treated as a free parameter. So the explanation of what you say is due to this different way of fitting. These numbers were taken from the paper by Kobel.

– *Word:*

Does this mean that you can only do this kind of a fit if you have multiple results or can you also let your values float for a single result?

– *Altarelli:*

If you have a single species, like the upsilon, $(1 + Dv^2/c^2)$ is a constant, so you cannot fit this term. It is only when you have both the J/ψ and the upsilon that you can make a fit. However, I do not think that the combined fit is in any way superior, because myself I would not include the J/ψ. I would prefer to stay within the upsilon, and use the J/ψ as a check that nothing absurd pops out about the values that you find from these relativistic corrections.

– *Vuaridel:*

The error of α_s determined by the Deep Inelastic Scattering data is dominated by the theoretical uncertainties. Do you think this error can be further reduced? Would a new generation of more precise DIS experiments make sense to improve the α_s results?

– *Altarelli:*

That's a very important question because what you are asking is whether a new generation of DIS experiments would be important or not. Well, with a new generation of DIS experiments we can check these results in a different region of x and Q^2. That already would be a very important achievement. In other words, for example, part of the theoretical error is connected to the fact that Q^2, although being somewhat large, is not that large. So, an experiment that collects large statistics at larger average values of Q^2 would be extremely important. With measurements at higher values of Q^2, you will also reduce the theoretical error from the problem of being in the middle of passing the threshold, for example, for bottom production, or, at low energy, for charm production. So if you go to different Q^2, you can eliminate this type of error. Also, it would be important to test the behavior of the QCD-scaling violations for smaller values of x. An important problem is to find the limit of validity of the QCD evolution equations at small x and their modifications. An important part of the experimental program at HERA is devoted to these problems. Also, with time, there can be progress in the theoretical calculations. For example, at least the non-singlet parts of the evolution kernels could perhaps be computed to the next order of accuracy thereby cutting down drastically the main source of error, i.e. the higher orders. So, I think it would be worthwhile to continue the experimental study of Deep Inelastic Scattering structure functions. Also, this study is being kept interesting by the fact that knowing better the parton densities (especially the sea and the gluon parton densities) at small x gives really new interesting phenomena. For example, recently, the NMC collaboration went down to smaller values of x, and they've got values for the Gottfried sum rule which seem to imply that the $\bar{u}$ content of the proton is different than the $\bar{d}$ content of the proton, so that the sea appears not to be an isospin-neutral, which is not a big problem in principle, but is an interesting thing. So, what I mean is that we keep discovering interesting phenomena in Deep Inelastic Scattering. In my opinion it would be important to continue the study of DIS structure functions, especially at small x.

– *Vuaridel:*

Can you explain how the current quark masses are computed?

– *Altarelli:*

You know that the proton has a mass of about 1GeV. If the quark masses were switched off from the QCD lagrangian, the mass of the proton would not change much. The proton is made up of three constituent quarks. It weighs 1 GeV, so the mass of each constituent quark is about 300 MeV. This mass is large due to the interaction with the gluonic field, and not due to the mass of the quark that you put into the lagrangian. The mass of the quark that you put into the lagrangian is much smaller. This we know starting from the matrix element of the divergence of the axial current between the vacuum and one pion. One obtains a relation among f_π, the quark masses and a pseudoscalar density. Estimating by different methods this density, for example, by QCD sum rules, leads to consistent values of m_u and m_d. Ratios of quark masses can be obtained from the observed hadron masses. The current masses of the quarks obtained in this way are 5 to 10 MeV for the up- and down-quarks, some 190 MeV for the s-quark, 1.2 GeV for the charm quark, 4.7 GeV for the b-quark.

– *Malik:*

How reliable is the value of $\alpha_s(m_Z)$ measured by Γ_h/Γ_ℓ, if the experimental error is reduced to be very small?

– *Altarelli:*

The ratio of the Z-width into hadrons divided by the width into leptons is, from the point of view of QCD, conceptually equivalent to $R(e^+e^-)$. Namely, the correction factors (apart from some details coming from the presence of an axial current and some minor correction from the b-quark mass, which are different in two cases), from the pure QCD point of view are the same things. Of course, physically they are strictly not the same. For example, Γ_h/Γ_ℓ is a measurement done on top of the Z, while measurements done at low energies could be, for example, more affected if a heavier Z′ were present at some larger mass, because on the peak of the Z you would not be so sensitive as in the continuum region. For example, in the TRISTAN region, which is a region of minimum cross-section, they would have perhaps the most sensitive probe of some heavier Z′. I repeat, if we take the electroweak part of the Standard Model as granted to this level of precision, and we look into QCD only, then these quantities should be equivalent. And since the LEP experiments have prepared themselves for the much more difficult task of measuring the electroweak parameters of the theory, the LEP experiments are really much more accurate than the previous QCD experiments at lower energies. Thus I think that the LEP results are more reliable than collecting the old results in the lower energy domain.

– *Peccei:*

I also have a question: why is there 1.05 in the correction proportional to α_s/π?

– *Altarelli:*

It comes from the b-quark mass corrections which is particularly large in the axial part of the width.

– *Eskut:*

You have given α_s measured by PEP, PETRA, TRISTAN and LEP. They have used $\Lambda = \Lambda_5$, while Υ decay, deep-inelastic scattering uses $\Lambda = \Lambda_4$. The relationship $\Lambda_5 = 0.65\ \Lambda_4$ is obtained by using the threshold at $1 \div 2m_b$. How is it that the values of α_s for 4- and 5-flavours are equal at the threshold?

– *Altarelli:*

$\alpha_s(Q^2)$ is the effective value for the QCD coupling as a function of the transferred momentum. This is a physical quantity and must be continuous. For Q below one or two m_b, the b quarks do not affect the running and do not appear in the beta function. This is why the perturbative expression in terms of Λ_4 is appropriate when Q^2 is much larger than the masses of the four flavors of quark which are excited. In the region below Q=1 or 2 m_b which I call Q-threshold, the derivative of α_s is ruled by only the light quark loops. Then, when you are far above this threshold, again you are in an asymptotic region because Q is much larger than m_b and here you have five flavours. The precise evolution in between is affected by some non-perturbative terms. If you apply a matching condition, by taking some point where you extrapolate these two asymptotic branches, and impose that they are continuous, you then obtain the stated relationship between Λ_4 and Λ_5 .

– *Wadhwa:*

What happens to the Λ value after the discovery of the top quark?

– *Altarelli:*

It depends. Suppose that the top quark mass is 130 GeV which is the most popular value at present. And if you do, for example, consider e^+e^- into hadrons at the Q value of 500 GeV, then the evolution of α_s (Q) in that region of energy which is above the threshold for m_{top} will also be influenced by m_{top}. So when I go to a Q much larger than m_{top} I will have a formula in terms of Λ_6, with b_6, b_6' known, and I will have another matching to be done, and Λ_6 will be again smaller than Λ_5. The essential point is that in QCD there is the decoupling theorem that works. States of higher mass, in particular coloured states of higher mass, do not contribute to the running of α_s at energies that are less than the threshold energy for production of these new states. As soon as you cross the threshold, you must also take them into account.

– *Buttery:*

I am working with photon fragmentation functions. How low in Q^2 can you take the evolution equations (e.g. LO, NLO)?

– *Altarelli:*

Well, you can judge for yourself. The usual evolution equations hold true apart from terms that are of order $1/Q^2$. So if you have $Q^2 = 4$ GeV^2, $1/Q^2$ is of the order of 25%. I think that Q^2 is a little bit too small in this case. It depends on the precision of your data. If you are working at the 30% level, you can feel safe. If you want to go beyond, then you cannot neglect these terms suppressed by $1/Q^2$ because order $1/Q^2$ is presumably 25% but it can well be 50%, it can even be 75%. When deep inelastic scattering started to be studied, everybody was happy with Q^2 of (1 GeV)^2. Then with time,we have seen that in order to really see the so-called asymptotic behavior, it is better to have the terms of order $1/Q^2$ really down. Remember that in e^+e^- annihilation, no one was able to see two jets before a centre of mass energy of about 7 GeV was reached at SPEAR.

– *Buttery:*

So can I start the evolution at 1 or 4 GeV^2 and evolve to high Q^2 (e.g. 70 GeV^2)?

– *Altarelli:*

I would advise you to start from 4 rather than 1 GeV^2.

– *Liu:*

Could you explain a little bit about the reason for adding the experimental error and the theoretical error together?

– *Altarelli:*

This is what is normally done. If you read the experimental papers, it is described how the error is computed. But what is finally quoted is often a combination of the experimental and theoretical errors done in quadrature. The experimental and theoretical errors are completely different things and they are assumed to be uncorrelated. Of course the experimentalists should always quote the experimental and theoretical errors separately, describe the way they got the estimated theoretical errors, and leave the theorists to give their estimate of theoretical errors if they want. Also the message to experimentalists is that an experiment will never become better by pushing down the theoretical error. This is not the way to become a better experimentalist by taking smaller and smaller theoretical errors.

– *Shabelski:*

In LEP experiments we measure the α_s value not at m_z but at some argument which is 3-4 times smaller (for example it can be the quark virtuality, transverse momentum of hard gluons, etc.) But in your transparencies the values of $\alpha_s(m_z)$ are presented. The question: Was such extrapolation from Q^2 of order

$[m_z/(3 \text{ to } 4)]^2$ accounted for in the results on Γ_h/Γ_ℓ? And if yes, how can you estimate the accuracy?

– *Altarelli:*

One writes the expansion for Γ_h/Γ_ℓ with coefficients appropriate to the choice of scale of α_s which corresponds to m_z. If you want to choose another scale you can do that, for example, $m_z/2$. But then the coefficients change because there is a precise perturbative relationship between $\alpha_s(m_z/2)$, and $\alpha_s(m_z)$. So once you have computed the order α^2 and the order α^3, the ambiguity of the choice of scale is an ambiguity which is of order α_s^4. So, you are only risking a term of order α_s^4, when you decide which scale to use. We have expressed the results in terms of $\alpha_s(m_z)$ but, the result could be obtained by writing down the corresponding expansion in terms of $\alpha_s(m_z/2)$. Then the expansion coefficients would be slightly different. You could redo the fit and obtain $\alpha_s(m_z/2)$. You can make the evolution up to m_z and compare the final value for $\alpha_s(m_z)$ that you obtain by this path with the value obtained through the direct expansion in terms of $\alpha_s(m_z)$. The difference between these two values, which is of order α_s^4, is what is called the error from the scale dependence. In case of Γ_h/Γ_ℓ, these scale dependence ambiguities are particularly small because we have two things. First the experimental error is very large, because α/π is a small correction to one. Secondly the calculation up to three loops makes the ambiguity particularly small. As a result, there is not a big assumption under this choice.

– *Misiak:*

The measurements of α_s from Γ_h/Γ_ℓ and $R(e^+e^-)$ have dominant experimental errors. Are these errors mainly statistical or mainly systematic? Are there any quantitative perspectives for improvement?

– *Altarelli:*

At present on Γ_h/Γ_ℓ the statistical error is still a big part of the total. The final error could be cut down by a factor of about 2.

– *Hanany:*

How do you determine the passage in the beta function from 4 to 5 flavours?

– *Altarelli:*

You work with the β function. The evolution of α is given by $d\alpha(Q^2)/d\log(Q^2) = \beta(\alpha(Q^2))$. One computes $\beta(\alpha)$ in perturbation theory. β is given by $-b \times \alpha^2 \times$

$(1+b'\alpha)$. This sort of perturbative expansion is valid when Q is much larger than m_f. In the region where Q approaches one mass one cannot rely on perturbation theory. When I go from Q much larger than some light mass fermion to a Q comparable to a new fermion being excited, I have an interval of energies where the beta function cannot be given by this type of perturbative expression. And so, I have to make a matching of two different asymptotic regions.

– *Sopczak:*

You emphasized that the simulation of hadronization corrections is crucial for α_s measurements at LEP from jets. Iwould like to point out that also the simulation of hadronization leads to the main uncertainty in the search for new particles which decay multihadronically. What are the possibilities that new theoretical input improves existing hadronization programs?

– *Altarelli:*

The hadronization programs that we have are constantly improved, because they are confronted with more and more data. Before LEP the existing event generators were optimized on the PETRA data, then when LEP started these programs were taken as they were and confronted with the LEP data. First one looked at how the measured distributions were reproduced by the event generators, and the general agreement was surprisingly good. Then the event generators were re-tuned, in such a way as to bring the best combined fit of the PETRA and the LEP distributions together. Also, by making more experiments you learn better values for the parameters or for the assumptions that were made when constructing the generator. For example, you measure a new process that before was only computed or was only guessed. So, the event generators are continually updated, and the more experiments you do, the more experience you get and more information is actually stored in these event generators so that they become more and more reliable. But they are always very similar one to the other. It is true that one has LUND string fragmentation, the other has more care with the coherent effects, a third one has exact matrix elements up to non-leading order. They have some different characteristics. But they also have very many assumptions that are common, very many tricks of the game which are the same in all of them, so there is a common bias in these particular programs. Even if you don't see the biases when you observe bulk properties, these biases can be important when you go to a precise measurement of α_s.

– *Sopczak:*

If you rely on the experimental input to tune your generators, then clearly, you will miss any unexpected signals.

– *Altarelli:*

Well, it depends of what kind of effect you see, because you know what can be predicted and what cannot be really predicted. So if you find a surprise in the area where predictions are not reliable, you are ready to change the parameters to accomodate that. If you find a surprise in something that is not negotiable (because it is a prediction that comes from the principles of the theory) then you say that it cannot be accepted. For example, if you find a value of R that is wrong by a factor of 2, there is no question - you cannot readjust the parameters. The theory is seriously wrong.

CHAIRMAN: G. Altarelli

Scientific Secretaries: E. Eskut, R. Malik, R. Sehgal

DISCUSSION

– *Hanany:*

Is the Abelian QCD model in fact QED-like with 3 flavours?

– *Altarelli:*

Yes, it is similar to QED-like with 3 flavours. More precisely it is defined by the group constants C_F and T_F (with $C_A = 0$) as given in the text.

– *Misiak:*

I have a question concerning the "measurement" of SU(3) group constants. Does the figure you have presented mean that QCD based on some representation of SU(4) is not yet excluded by experiment?

– *Altarelli:*

It is not excluded by this measurement but many other measurements exclude SU(4). For example the observed spectrum of hadrons.

– *Zichichi, comment:*

It should be noted that the figure shows measurements with 68% confidence level, ie. just one σ. Such results are not very significant.

New Phenomena Below the Mega TeV Scale*

R .D. Peccei
Department of Physics, University of California, Los Angeles, CA 90024

Abstract

After some musings about a Mega TeV scale, I contrast the difference expected between theories where the $SU(2) \times U(1)$ group is broken dynamically and theories where this breakdown is induced by the vacuum expectation value of an elementary Higgs field. Although high precision electroweak data, whose impact I review, do not favor the dynamical option they also do not rule it out, making it imperative to explore the Higgs-dynamical symmetry breaking dichotomy at higher energies. Real differences between these options emerge in $W_L W_L$ scattering, with this process showing both resonance formation and multiparticle production if $SU(2) \times (1)$ is broken dynamically. Expectations for multi W_L production in the Eloisatron energy range are discussed, as well as some of the difficulties and challenges one faces in trying to generate fermion masses dynamically.

*Lecture given at the XXX Erice Summer School, Erice, Italy, July 1992. To appear in the Proceedings of the Summer School.

1. Musings on a Mega TeV scale

Thinking about physics below a Mega TeV (10^9 GeV) is a nice, liberating experience. For one thing, viewed from this scale, electroweak symmetry breaking is truly a "low energy" phenomenon, whose manifestations should by then all be apparent. Of course, it is possible that a scale around a Mega TeV has physical significance per se. For instance, $M = 10^9$ GeV is numerically the geometric average of the Planck Scale and the dynamical scale of QCD,

$$M = \sqrt{M_P \Lambda_{QCD}} \tag{1}$$

As such it can serve as an ideal intermediate scale.[1] It is also a "see saw" scale for neutrino masses, since it gives a mass for ν_τ which renders this particle a plausible dark matter candidate[2]:

$$m_{\nu_\tau} \simeq \frac{m_\tau^2}{M} \sim 0(eV) \tag{2}$$

For the purposes of this lecture, however, the principal virtue of M is really that it is a scale which is sufficiently far above the electroweak symmetry breaking scale $v \simeq 250$ GeV, so that the statement $M >> v$ truly applies. As we shall see, being able to consider electroweak symmetry breaking effects far above threshold has its rewards. Furthermore, M is sufficiently large that it is possible (likely ?) that, if the electroweak symmetry breaking is dynamical, the scale Λ_m associated with the dynamics of fermion mass generation is also below M. Unfortunately, the fermion mass issue in models of dynamical symmetry breaking is still poorly understood,[3] so I will concentrate for the most part in this lecture on effects connected purely with the weak boson sector. Nevertheless, even with this restriction, we will encounter quite a rich and exciting physics.

2. What do we know: electroweak symmetry breaking and the equivalence theorem

The very successful comparison of data from LEP with the standard model predictions,[4] provides strong evidence that electroweak interactions are described by a spontaneously broken $SU(2) \times U(1)$ gauge theory.[5] In addition, the fact that the measured ρ parameter[4]

$$\rho = 0.9968 \pm 0.0048 \tag{3}$$

is so close to unity argues that the theory possesses some (approximate) custodial global symmetry to guarantee this property.[6] These results notwithstanding, however, one still has no knowledge of what the detailed nature of the spontaneous symmetry breaking is. Thus one can still contemplate the alternatives that the order parameter for the $SU(2) \times U(1) \to U(1)_{em}$ breakdown is either:

i) the vacuum expectation value of an elementary scalar field: $< \Phi > \neq 0$

or

ii) the condensate of fermion-antifermion pairs of some underlying theory: $< \bar{T}_L T_R > \neq 0$

The physics associated with these two options is quite distinct. The first option, the elementary Higgs option,[7] is perfectly compatible with weak coupling. That is, one can imagine the formation of a vacuum expectation value for the Higgs field Φ, even though the interaction strength λ in the Higgs potential

$$V = \lambda(\Phi^\dagger \Phi - \frac{v^2}{2})^2 \tag{4}$$

is small, $\lambda << 1$. In the second option, called here Technicolor[8] for short, however, one cannot avoid strong coupling in the underlying theory, since only in this regime will the $< \bar{T}_L T_R >$ condensate form. The strength of this condensate is measured by the dynamical scale Λ_T at which the coupling constant $\alpha_T(q^2)$ becomes of order unity:

$$< \bar{T}_L T_R > \simeq \Lambda_T^3 \tag{5}$$

If this condensate is to be associated with electroweak symmetry breaking, then Λ_T must be of order of the electroweak symmetry breaking parameter v:

$$\Lambda_T \sim v = (\sqrt{2} G_F)^{-1/2} \simeq 250 \ GeV \tag{6}$$

So, in this case, we expect that effects associated with the strong coupling underlying theory be visible in this energy range.

In both the Higgs case and in Technicolor, the physics of the spontaneous breakdown will be reflected in the amplitudes for longitudinal gauge boson scattering - $W_L W_L$ scattering. The amplitudes for processes involving longitudinal gauge bosons, including $W_L W_L$ scattering itself, are governed by an important "equivalence theorem", proven long ago by Cornwall, Levin and Tiktopoulos.[9] This theorem states that, in the limit in which the energies of all longitudinal gauge bosons are much bigger than their masses ($E_i >> M_W$), one can obtain the longitudinal gauge boson amplitude directly from an amplitude in which each W_L is replaced by its corresponding Goldstone boson, ensuing from the spontaneous breakdown. That is, schematically, if one denotes the Goldstone boson corresponding to W_L by w one has, in the high energy limit,

$$A(W_{L_1}, W_{L_2}, \cdots, W_{L_n}) = A(w_1, w_2, \cdots, w_n) \tag{7}$$

For the elementary Higgs case, one can write the Higgs field Φ as

$$\Phi = \frac{1}{\sqrt{2}} \begin{pmatrix} v + H + i w_3 \\ w_1 + i w_2 \end{pmatrix} \tag{8}$$

where H is the physical Higgs field. Then the potential (4) becomes simply

$$V = \frac{\lambda}{4}(\vec{w}^2 + H^2)^2 + \lambda v H(\vec{w}^2 + H^2) + \lambda v^2 H^2 \tag{9}$$

One sees that, if λ is small, one will have weak interactions among the Goldstone fields and with the Higgs field H. Hence, from the equivalence theorem one expects also weak interactions among the longitudinal gauge bosons W_L. If, on the other hand, the spontaneous breakdown of the electroweak theory occurs dynamically, through the formation of condensates in some strongly interacting underlying theory, then one would expect that the interactions among the Goldstone bosons in the underlying theory will also be strong. In turn this will lead to strong interactions among the W_L. It is this phenomena which we want to explore in some detail here.

Before doing so, however, it is important to note a feature of the Higgs potential which needs to be incorporated in all underlying dynamically symmetry breaking theories. The potential (4), with Φ a complex $SU(2)$ doublet, has a larger symmetry that $SU(2) \times U(1)$, namely $0(4)$. When Φ acquires a vacuum expectation value $< \Phi >$, $0(4) \rightarrow 0(3) \sim SU(2)$. It is this custodial $SU(2)$ symmetry[6] which, neglecting fermion mass effects, guarantees that $\rho = 1$. For the dynamical symmetry breaking theory one wants to retain this desirable feature. This is most easily done by requiring that the Technicolor theory possess a global $SU(2)_L \times SU(2)_R$ symmetry, under which T_L and T_R transform as $(2,1)$ and $(1,2)$, respectively. The condensate $< \bar{T}_L T_R > \neq 0$ then breaks this symmetry to the diagonal subgroup, $SU(2)_L \times SU(2)_R \rightarrow SU(2)_{L+R}$, and one has again a custodial $SU(2)$ symmetry to guarantee that $\rho = 1$.

3. Signals of dynamical symmetry breaking near threshold

As I mentioned above, the tests of the electroweak theory at $\sqrt{s} \simeq 100\ GeV$ are not terrifically sensitive to the symmetry breaking sector. One hopes to learn indirectly about this sector through its effects on radiative corrections. However, for the Higgs case one knows that these effects are small as a result of the, so called, Veltman's screening theorem.[10] This theorem basically states that, at $0(\alpha)$, radiative corrections depend on the Higgs mass only logarithmically. Although there are also quadratic dependences on the Higgs mass, these only enter at $0(\alpha^2)$, so that

$$\text{Rad. Corr} \sim \frac{\alpha}{\pi}\, \ell\text{n}\ \mathrm{M_H/M_Z} + (\frac{\alpha}{\pi})^2 \frac{\mathrm{M_H^2}}{\mathrm{M_Z^2}} \tag{10}$$

Because of Eq. (10), one does not expect significant problems with the usual perturbative treatment of the Higgs sector until the Higgs mass is sufficiently big, so that the first term above is of order of the second:

$$M_H^2 \sim \frac{\pi}{\alpha}\, M_Z^2 \tag{11}$$

This corresponds to the regime where the strength λ of the Higgs potential is much greater than unity ($\lambda >> 1$), corresponding to strong coupling for the elementary Higgs theory. If λ is small, however, the radiative corrections are largely insensitive to the value of the Higgs mass.

There is a roughly analogous situation in the case of dynamical symmetry breaking. It is rather easy to show that, at low energies, there is weak scattering among the Goldstone bosons, arising from the fact that because they are Goldstone bosons they couple to themselves and to other fields derivatively. So, for instance, the amplitude for $w^+w^- \rightarrow zz$ scattering is given by

$$\mathcal{A}(w^+w^- \rightarrow zz) = \frac{s}{v^2} \tag{12}$$

where $\sqrt{s}$ is the w^+w^- CM energy. This amplitude is weak as long as $\sqrt{s} < v$. Furthermore, even if it grows with energy, in fact it does not hit the elastic unitarity limit until $\sqrt{s} = 4\sqrt{\pi}\, v \simeq 1.7\ TeV$.[11] Nevertheless, in general there are differences in the contributions to the radiative corrections to the electroweak theory depending on whether the symmetry breakdown is caused by an elementary Higgs field acquiring a vacuum expectation value or whether this breakdown is more dynamical.

To appreciate these differences, it is useful to focus on the gauge propagators of the electroweak theory. Let us denote the self energy functions for these propagators as $\Pi_{AB}(q^2)$, where the indices AB span the 4 possibilities of the $SU(2) \times U(1)$ theory: ($AB = WW; ZZ; \gamma\gamma; \gamma Z$). Since one is interested in the effects of the physics of the symmetry breaking sector for the relatively low q^2 values represented by the W and Z masses, it suffices to retain just the first two terms in a q^2 expansion of $\Pi_{AB}(q^2)$. Thus, let us approximate[12]

$$\Pi_{AB}(q^2) \simeq \Pi_{AB}(0) + q^2\Pi'_{AB}(0) \ . \tag{13}$$

Because of photonic gauge invariance, one can check that $\Pi_{\gamma\gamma}(0) = \Pi_{\gamma Z}(0) = 0$, so Eq. (13) contain 6 independent parameters. Three of these parameters can be expressed in terms of the 3 independent parameters of the $SU(2)\times U(1)$ theory, the coupling constants g_2 and g_1, and the Fermi scale v. In fact, it is more convenient instead of using the set $\{g_1, g_2, v\}$ to replace these parameters by three equivalent physical constants which are well measured experimentally: α, M_Z and G_F.[13] The remaining 3 constants in Eq. (13) have been chosen variously in the literature, and I find it useful here to follow the conventions of Peskin and Takeuchi[14] who have denoted these additional parameters simply as S, T and U.

The parameter S, which is defined as[15]

$$S = 16\pi\{\Pi'_{33}(0) - \Pi'_{3\gamma}(0)\} \tag{14}$$

is particularly interesting because it is rather small for a large range of Higgs masses, but is quite sensitive to possible chirality breaking contributions in a theory where the symmetry breaking is done dynamically. It is useful to pick a reference Higgs mass (and a reference top mass, although that will not be relevant for our discussion here) and try to predict where S will be relative to this standard. For the case of pure Higgs breaking, of course, this is trivial. Keeping leading contributions only, one has

$$\Delta S \simeq \frac{1}{12\pi} \ell n\, M_H^2/(M_H^2)_{\rm ref} \tag{15}$$

Picking $(M_H)_{\rm ref} = 100\ GeV$, one sees that $\Delta S \leq 0.1$ even for $M_H = 1000\ GeV$.

For any dynamical symmetry breaking model one can, in principle, try to compute the expected S value relative to the $(M_H)_{\rm ref}$ standard. This is not straightforward because it requires making various assumptions on the underlying strong dynamics. Nevertheless, Peskin and Takeuchi[14] made some progress in this direction for a class of theories in which a vectorial $SU(2)$ is conserved by the underlying theory. In view of the dynamical need for such an $SU(2)$ to obtain $\rho = 1$, these theories are certainly interesting candidates. In these theories only the vector-vector (VV) and axial-axial (AA) components enter in the vacuum polarization Π_{33}, while $\Pi_{3\gamma}$ only contains a VV piece. Thus S can be rewritten entirely in terms of the vacuum polarization functions which enter in the current correlation functions for the $SU(2)_V$ and $SU(2)_A$ currents in the underlying theory. A simple calculation then gives

$$S = -4\pi\{\Pi'_{VV}(0) - \Pi'_{AA}(0)\} \tag{16}$$

This nice formula for S allows one to attempt estimating this quantity in theories of dynamical symmetry breakdown by "scaling up" what happens in QCD. Two flavor QCD with $m_u = m_d = 0$ has a global $SU(2)_V \times SU(2)_A$ symmetry which is broken spontaneously to $SU(2)_V$ by the presence of condensates $< \bar{u}u > = < \bar{d}d > \neq 0$. In QCD, the vacuum polarization difference entering in Eq. (16) can be rewritten in terms of an integral difference between the vector and axial vector spectral functions of the theory

$$\Pi'(0)_{VV} - \Pi'_{AA}(0) = -\frac{1}{4\pi}\int_0^\infty \frac{ds}{s}[v(s) - a(s)] \tag{17}$$

This integral can be estimated by making use of the two Weinberg spectral function sum rules[16] obeyed by $v(s)$ and $a(s)$

$$\int_0^\infty ds[v(s) - a(s)] = 4\pi f_\pi^2 \tag{18a}$$

$$\int_0^\infty s\,ds[v(s) - a(s)] = 0 \quad , \tag{18b}$$

where f_π is the pion decay constant, along with the assumption that $v(s)$ and $a(s)$ are saturated by the ρ and A_1 poles. Saturating Eqs. (18) with the ρ and A_1, poles, respectively,

$$v(s) = g_\rho^2\delta(s - m_\rho^2) \;\; ; \;\; a(s) = g_{A_1}^2\delta(s - m_{A_1}^2) \tag{19}$$

allows one to determine the strengths g_ρ^2 and $g_{A_1}^2$ in terms of m_ρ^2, $m_{A_1}^2$ and f_π^2, and gives for the integral that enters in Eq. (17) the formula

$$\int_0^\infty \frac{ds}{s}[v(s)-a(s)] = \frac{4\pi f_\pi^2}{m_\rho^2}[1+\frac{m_\rho^2}{m_{A_1}^2}] \tag{20}$$

The original technicolor theory of Susskind and Weinberg[8] was precisely just a "scaled" up version of 2 flavor QCD where f_π is replaced by v, the Fermi scale, and the Goldstone pions are replaced by technipions which, effectively, get absorbed to become the longitudinal components of the weak gauged bosons. Many other dynamical symmetry breaking theories have the same structure, except that perhaps the technicolor group is $SU(N)_{TC}$ not simply $SU(3)$ and there may be a variety of species of technicolor doublets, not just simply T_u and T_d to mimic the u and d quarks of QCD. For all of these theories, one should be able to estimates S by just "scaling up" the formula (20) with some simple modifications. Large N arguments suggest that[17]

$$\frac{f_\pi^2}{m_\rho^2} \to \frac{3}{N_T}\frac{v^2}{m_{\rho_T}^2} \quad ; \quad \frac{m_\rho^2}{m_{A_1}^2} \to \frac{m_{\rho_T}^2}{m_{A_T}^2} \quad , \tag{21}$$

with m_{ρ_T} and m_{A_T} the analogons of m_ρ and m_{A_1}. Furthermore, if there are n technicolor doublets, each of these doublets should give an equivalent contribution. Hence, one arrives at the simple formula

$$S \simeq \frac{N_T}{3}\, n \int_0^\infty \frac{ds}{s}[v(s)-a(s)]\Big|_{QCD} \simeq 0.25[\frac{N_T}{3}n] \tag{22}$$

The above rough estimate of S agrees qualitatively with the more careful evaluation of the QCD integral of Eq. (17) done by Peskin and Takeuchi.[14] It is also in general accord with the work of Cahn and Suzuki,[18] although these authors argue that the errors in evaluating the QCD integral could actually be consistent with a much smaller value for S than that given in Eq. (22). At any rate, two things regarding S are apparent from this exercise of "scaling up" QCD:

i) $S > 0$

ii) In technicolor models containing a full complement of techniquarks, to account for both color and lepton number, $n = 4$ and then $S \sim 0(1)$.

It is possible to extract values for S (and T, U) allowed by experiment by comparing observable parameters, like the mass of the W boson or the leptonic width of the Z boson, to the theoretical predictions of this generalized $SU(2) \times U(1)$ analysis. The value of any experimental observable $0_{\rm exp}$ can be written as

$$0_{\rm exp} = 0_{SM}(M_H^{\rm ref}\ , m_t^{\rm ref}) + \alpha_0 S + \beta_0\, T + \gamma_0 U \ , \tag{23}$$

where $0_{SM}(M_H^{\text{ref}}\ ,\ m_t^{\text{ref}})$ is the standard model prediction, assuming a single Higgs doublet does the breaking, with some reference choice for the values of the Higgs and top masses. The coefficients α_0, β_0 and γ_0 depend on the observable in question, but are fixed once M_H^{ref} and m_t^{ref} are chosen. Table 1, taken from Peskin and Takeuchi,[14] gives the dependence of various observable on S, T and U for the choice of $M_H^{\text{ref}} = 1\ TeV$ and $m_t^{\text{ref}} = 150\ GeV$.

Table 1: Dependence of Observables on S, T and U

$$\begin{aligned}
M_W/M_Z &= 0.8787 - 3.15 \times 10^{-3} S + 4.86 \times 10^{-3} T + 3.70 \times 10^{-3} U \\
\Gamma_Z(GeV) &= 2.484 - 9.58 \times 10^{-3} S + 2.615 \times 10^{-2} T \\
\Gamma_\ell(MeV) &= 83.5 - 1.91 \times 10^{-1} S + 7.83 \times 10^{-1} T \\
\Gamma_u(MeV) &= 296.2 - 1.92 S + 3.67 T \\
\Gamma_d(MeV) &= 382.3 - 1.72 S + 4.20 T \\
\Gamma_b(MeV) &= 377.9 - 1.72 S + 4.20 T \\
\Gamma_{\text{had}}(GeV) &= 1.7348 - 9.00 \times 10^{-3} S + 1.993 \times 10^{-2} T \\
R_Z = \frac{\Gamma_{\text{had}}}{\Gamma_\ell} &= 20.78 - 5.99 \times 10^{-2} S + 4.24 \times 10^{-2} T \\
\sin^2 \theta_w^*(M_Z) &= 0.2337 + 3.59 \times 10^{-3} S - 2.54 \times 10^{-3} T \\
A_{LR} &= 0.1297 - 2.82 \times 10^{-2} S + 2.00 \times 10^{-2} T \\
A_{FB}^b &= 0.0848 - 1.97 \times 10^{-2} S + 1.40 \times 10^{-2} T \\
A_{FB}^\ell &= 0.0126 - 6.72 \times 10^{-3} + 4.76 \times 10^{-3} T \\
g_L^2 &= 0.3001 - 2.67 \times 10^{-3} S + 6.53 \times 10^{-3} T \\
g_R^2 &= 0.0302 + 9.17 \times 10^{-4} S - 1.94 \times 10^{-4} T \\
R_\nu &= 0.3126 - 2.32 \times 10^{-2} S + 6.46 \times 10^{-2} T \\
R_{\bar\nu} &= 0.3824 - 2.77 \times 10^{-3} S + 6.03 \times 10^{-3} T \\
Q_W(Cs) &= -73.31 - 0.79 S - 0.01 T
\end{aligned}$$

In Fig. 1, taken again from Peskin and Takeuchi,[14] I show the allowed range (both at 68% C.L. and 90 % C.L.) for S and T from a combined analysis of all extant data. This analysis clearly favors **negative** S values, with $S \simeq -0(1)$. Similar results to these have been obtained also by Altarelli, Barbieri and Jadach[19] and by Langacker and Luo.[20] Also shown in Fig. 1 are the expectations for S and T in the standard model, as one lets M_H and m_t vary from their reference values of 1000 GeV and 150 GeV, respectively. Note that T is very sensitive to changes in m_t, but quite insensitive to M_H changes. S is largely insensitive to changes in both m_t and M_H. Note also that the expected values for S for a full generation ($n = 4$) of techniquarks of an $SU(4)_{TC}$ theory is about 3σ away from the best fit.

Although the above result is disappointing if one believes in dynamical symmetry breaking, it is not totally disastrous. The result for S, one is comparing

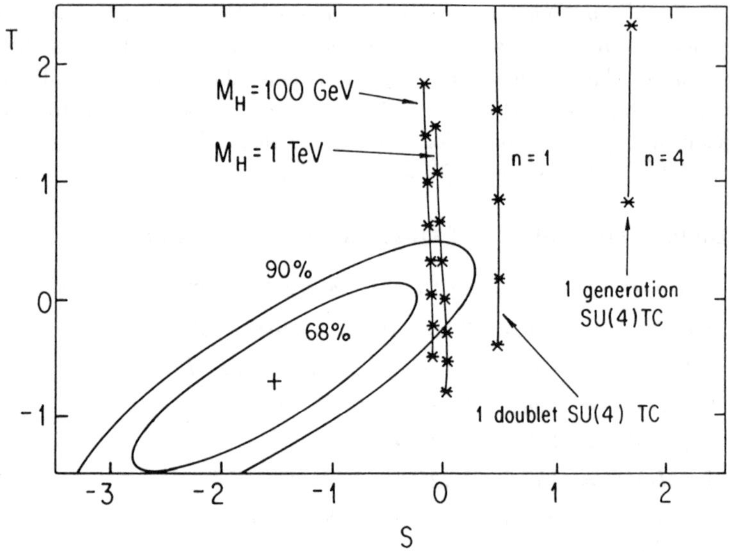

Figure 1: 68% C.L. and 90% C.L. curves in the S and T plane determined by the global fit to all data by Peskin and Takeuchi.[14] Also shown are the predictions for the standard model and two dynamical symmetry breaking models, as a function of m_t and M_H. The values of m_t are shown by the crosses which, starting from the bottom, increase by 20 GeV from $m_t = 90\ GeV$.

data to, was obtained by "scaling up" QCD and there is no guarantee that this is necessarily applicable for the "correct" theory of dynamical symmetry breaking. Furthermore, even in QCD, one knows that the Weinberg sum rules are rather slowly convergent,[21] so a saturation of the integral in Eq. (17) solely by low energy states may not be totally accurate. This lack of convergence may, in fact, be even more of a problem in the more sophisticated versions of Technicolor, so called Walking Technicolor, which are needed to address the issue of fermion masses. I will briefly discuss some aspects of Walking Technicolor theories at the end of this lecture, but I note here only that precisely because of the slow running of $\alpha_{WTC}(q^2)$ with q^2 one can argue[22] that it is likely that $S_{WTC} < S_{TC}$. In fact, in the model analysis of Hsu and Sundrum,[22] values of $S_{WTC} \sim (\frac{1}{4} - \frac{1}{10})S_{TC}$ were obtained

4. Signals for dynamical symmetry breaking far above threshold

As the above discussion demonstrates, it is difficult to learn much about the symmetry breaking sector even from the very precise data obtained at LEP, mostly because LEP is really a low energy machine! Things will be considerably amelio-

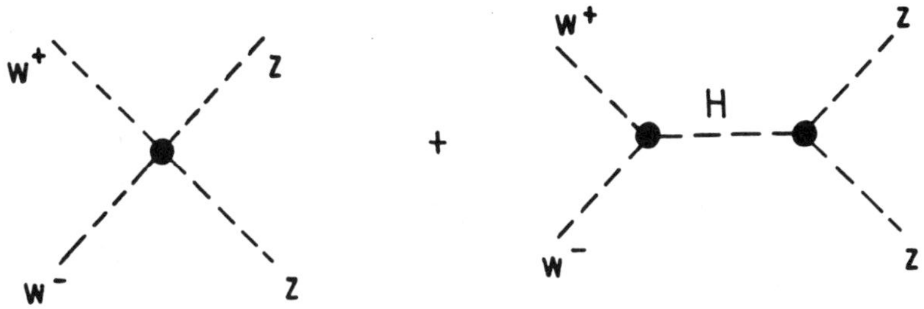

Figure 2: Feynman graphs contributing to the process $w^+w^- \rightarrow zz$ in the purely Higgs theory

rated at the LHC and the SSC, because one can begin to probe TeV subenergies in the $q\bar{q}$ subsystem. However, even here, despite the stupendous CM energies of the LHC and the SSC one is not very far above threshold for probing W_LW_L scattering, which is so crucial to ascertain whether the symmetry breaking sector is that of a weakly coupled or a strongly coupled theory. Only at even higher energies, those of the Eloisatron and a fantomatic Mega TeV machine, will the distinction become really manifest.

In this section I want to examine W_LW_L scattering far above threshold, to learn what are the characteristics which distinguish between having a weakly coupled or a strongly coupled symmetry breaking dynamics. By the equivalence theorem,[9] for these purposes it suffices to study the dynamics of the associated Goldstone bosons. As we discussed earlier, at threshold the behaviour of the scattering amplitudes for Goldstone bosons is solely fixed by the coset space of the breakdown (e.g. $SU(2)_L \times SU_R(2)/SU(2)_{L+R}$) and not by the specific dynamics of the theory. For instance for the process $w^+w^- \rightarrow zz$, in the pure Higgs theory, the evaluation of the Feynman graphs of Fig. 2 gives the scattering amplitude

$$\mathcal{A}(w^+w^- \rightarrow zz) = -2\lambda\left\{1 + \frac{2\lambda v^2}{s - 2\lambda v^2}\right\} \tag{24}$$

This amplitude agrees at threshold, $s << 2\lambda v^2$, with that of Eq. (12) obtained by considering only the dynamics of the nonlinearly coupled Goldstone bosons

$$\mathcal{A}(w^+w^- \rightarrow zz) \rightarrow \frac{s}{v^2} \tag{25}$$

So, obviously, nothing can be learned about the nature of the mechanism for symmetry breakdown precisely at this point.

Above the threshold region, however, there are real distinctions between the scattering amplitude for $w^+w^- \rightarrow zz$ predicted by the Higgs theory and that precisely the same pattern to emerge in a dynamically broken $SU(2) \times U(1)$ theory. The underlying theory which governs the interactions of the technipions, which eventually become the longitudinal gauge bosons, should manifest both **resonance formation** and **multiparticle production**, with the former process being more prevalent at lower CM energies in the scattering of technipions, and the latter processes dominating at higher CM energies.

Hadronic machines like the SSC and LHC are primarily suited to probe the hard scattering of quarks and gluons. One can, of course, also probe the scattering of virtual weak gauge bosons in these machines, but the relevant accessible CM subenergy is much degraded since the $W's$ only carry a fraction of the energy of the quarks, which themselves only carry a fraction of the energy of the initial protons. As a result the invariant subenergy effectively available for $W_L W_L$ scattering at the SSC and LHC is small, and even searching for resonance formation in the $W_L W_L$ scattering process is challenging. This is particularly so because resonance detection in the $W - W$ channel is hampered by a copious background of non resonant W pairs produced by quark annihilation, $q\bar{q} \to WW$. Furthermore, experimentally one must isolate the W signal through their decay into leptons, which effectively reduces much further the number of events one can study. Nevertheless, there is broad agreement that, although difficult, a resonant signal in the $W_L W_L$ channel with resonance mass in the range 2 – 4 TeV is observable.[23] For instance, a $\rho^{\pm}_{TC}$ with mass of 1.8 TeV, decaying into $W^{\pm}Z^0$, for a total integrated luminosity of $10 fb^{-1}$ gives a signal of 30 trilepton events $\ell^{\pm}\ell^{+}\ell^{-}$, from the further decays of the $W^{\pm}Z^0$, compared to a background of about 9 events, so reduced by appropriate cuts. For a $\rho^{\pm}_{TC}$ of mass of 4 TeV, the signal is down to 13 events, but the background stays the same, so this effect is just marginally detectable.[23]

At the LHC and SSC the available subenergy for $W_L W_L$ scattering is too small to hope to see the opening up of inelastic channels, to be expected if there is strong dynamics in the symmetry breaking sector. However, if one had a machine of sufficient energy, like the Eloisatron, it may be that the signal provided by this second aspect of strong dynamics could be simpler to detect than the presence of resonances in the W_L W_L channel. In collaboration with Duncan Morris, and Rogerio Rosenfeld, I have recently investigated this issue at UCLA.[24] What we imagined is that, above a certain threshold energy $\sqrt{s_0}$, instead of having purely resonant dominated, elastic W_L W_L scattering, W_L W_L scattering would give rise predominantly to multiparticle production: $W_L \, W_L \to nW_L \; (n > 2)$.

If our experience with hadronic scattering is any guide, $\sqrt{s_0}$, for $W_L W_L$ scattering will not be, comparatively, a terrifically high energy. Typically in strong interaction physics, multiparticle production begins to dominate at around $\sqrt{s_0} \simeq 2\, GeV$, which for a Technicolor theory, with the scaling implied by Eq. (21), gives a threshold $\sqrt{s_0} \simeq 4-5$ TeV. Furthermore, at $\sqrt{s_0} \simeq 2\, GeV$ in hadron physics the average multiplicity of produced states, which are dominantly pions, is already quite high. Typically $< n >_{\rm charged} \simeq 4$ and $< n >_{\rm total} \simeq 6$. So, analogously, we are

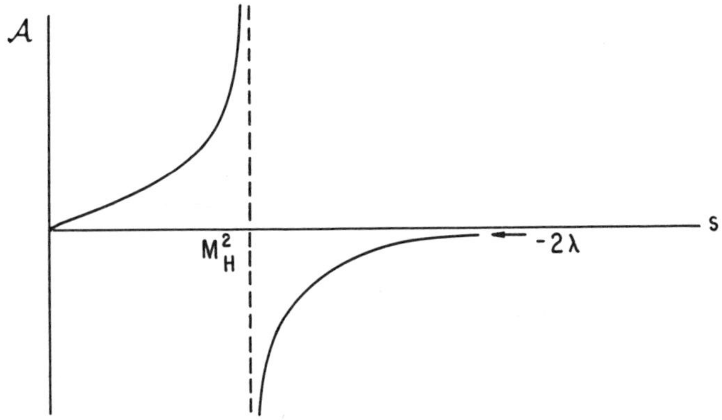

Figure 3: Behaviour of the amplitude $\mathcal{A}(w^+w^- \to zz)$ as a function of s for the Higgs model. The amplitude displays resonance behaviour as $s \to M_H^2$ and goes to an asymptotic value $\mathcal{A} \to -2\lambda$.

predicted by some model of dynamical symmetry breakdown. The amplitude (24) for the Higgs case, when plotted as a function of s, displays only one remarkable feature, a resonance pole at the Higgs mass $M_H^2 = 2\lambda v^2$, as shown in Fig. 3. Of course, even in this case, there is not a real pole in $\mathcal{A}$ because as $s \to M_H^2$ the result (24) ceases to be valid since one must include the effect of higher order in λ graphs which build up the Higgs width. For the case of dynamical symmetry breakdown, on the other hand, in complete analogy to what happens in the known strongly interacting theory which describes $\pi - \pi$ interactions (QCD), one expects that $\mathcal{A}$ should contain resonances in other partial waves (like the f_2 in $\pi^+\pi^- \to \pi^0\pi^0$), besides the $J = 0$ Higgs boson:

$$\mathcal{A} = 32\pi \sum_J (2J+1) P_J(\cos\Theta) a_J(s) \quad . \tag{26}$$

Secondly, again in analogy to $\pi - \pi$ scattering, one expects the opening up of inelastic channels, so that the partial waves $a_J(s)$ should not have unit strength

$$a_J(s) \;=\; \frac{1}{2i}\{\eta_J(s) e^{2i\delta_J(s)} - 1\} \;\; ; \;\; \eta_J(s) \;<\; 1 \tag{27}$$

In $\pi - \pi$ scattering, as in other strongly interacting scattering processes, resonance formation is the dominant process at lower energies. Eventually, however, when the CM energy is high enough the principal feature of the scattering process is the copious opening up of inelastic channels. Indeed, at high energy, the dominant feature of hadronic interactions is multiparticle production, which is mostly that of pions, since these particles are the lightest hadrons. One should expect

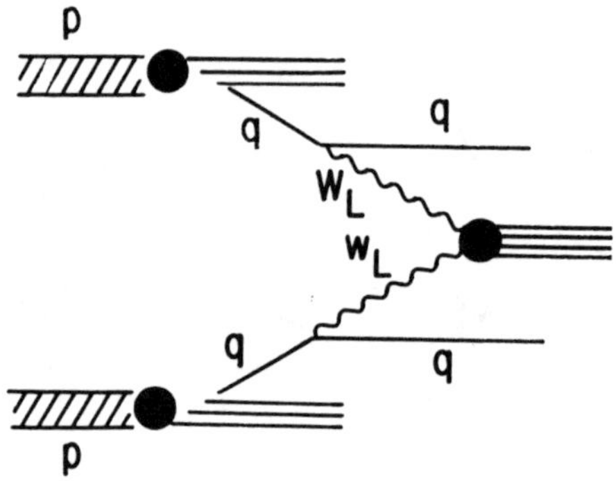

Figure 4: W_L W_L scattering arising from quark bremsstrahlung

led to expect that if we can access energies of the order of $4-5$ TeV in the W_LW_L CM system, the average multiplicity of produced $W's$ should also be given by similar numbers. The existence of copious multi W production at these energies is a clear signal that one has a strongly interacting symmetry breaking sector, for otherwise a process like $W_LW_L \to 6W_L$ would be hopelessly suppressed by the weakness of the electroweak coupling constant.

Morris, Rosenfeld and I[24] computed the cross section for W_LW_L scattering due to the quark bremsstrahlung process of Fig. 4 in the effective W approximation.[25] In this approximation, the contribution of W_LW_L scattering processes in a pp scattering process, of total center of mass energy $\sqrt{s}$, is given by

$$\sigma_{pp}^{W_L}(s) = \int_{\tau_{\min}} \mathcal{L}_{W_LW_L}(\tau)\, \sigma_{W_LW_L}(\hat{s} = \tau s) \tag{28}$$

Here $\sqrt{\hat{s}}$ is the CM subenergy of the W_LW_L system and $\mathcal{L}_{W_LW_L}(\tau)$ is the effective W_L W_L luminosity, for producing a W_LW_L system of a given invariant mass $\hat{s} = \tau s$.

The W_LW_L luminosity function is readily calculable as the square of the convolution of the probability of finding a W_L in a given quark P_{W_Lq}, times the quark distribution function in the proton

$$\mathcal{L}_{W_LW_L} = (f_q \bigotimes P_{W_Lq})^2 \tag{29}$$

The result of this calculation is shown in Fig. 5, plotted as a function of τ. As expected, this is a sharply falling function of τ, as τ increases towards unity. For a fixed value of $\sqrt{s}$, it is always most favorable to fuse together two W_L 's, with the smallest possible invariant mass. Wanting to access high values of $\sqrt{\hat{s}}$, to see the opening up of multi W_L production, forces τ to larger values with a consequent decrease in the effective $\mathcal{L}_{W_LW_L}$ luminosity. For a putative threshold of $\sqrt{s_0} = 5\ TeV$, the minimum value of τ reached at the LHC ($\sqrt{s} = 16\ TeV$)

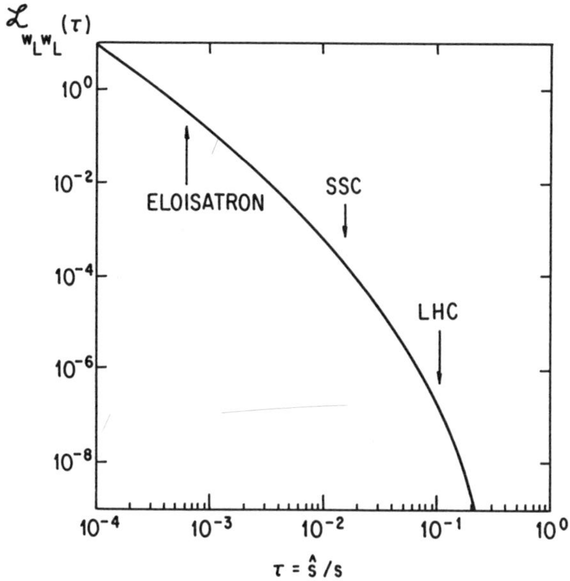

Figure 5: Effective W_LW_L luminosity, plotted as a function of $\tau = \hat{s}/s$, from.[24] Also indicated are the values of $\tau_{\min}$ (for $\sqrt{\hat{s}_0} = 5\ TeV$) reachable at the LHC, SSC and Eloisatron.

SSC ($\sqrt{s} = 40\ TeV$) and Eloisatron ($\sqrt{s} = 200\ TeV$) are, respectively, $\tau_{\min} = 9.77 \times 10^{-2}$, 1.56×10^{-2}, 6.25×10^{-4}. These minimum values, which correspond to the maximal W_LW_L luminosity available in each machine for producing W_LW_L interactions with at least $\sqrt{\hat{s}_0} = 5\ TeV$, are also indicated in Fig. 5.

Because we are only interested in the signal for multiparticle production in $W_L\ W_L$ scattering, it is perfectly sensible to entirely neglect any $W_L\ W_L$ scattering **below** $\sqrt{\hat{s}} = \sqrt{\hat{s}_0}$. Above this threshold, on the other hand, we assume that the $W_L\ W_L$ scattering is dominated by multiparticle production with a constant total cross section. That is, the simple model adopted[24] is that

$$[\sigma_{W_LW_L}(\hat{s})]^{\text{Multi}} = \theta(\hat{s} - \hat{s}_0)\sigma_0 \tag{30}$$

Scaling from $\pi - \pi$ scattering, using Eq. (21), suggest a value $\sigma_0 \simeq 1\ nb$ for the $W_L\ W_L$ multiparticle cross section. Using the model of Eq. (30) in Eq. (28) gives for the component of the pp scattering cross section which involves multi W_L production the formula

$$[\sigma_{pp}^{W_L}(s)]^{\text{Multi}} = \sigma_0 \int_{\tau_{\min}}^{1} d\tau \mathcal{L}_{W_L\ W_L}(\tau) \simeq \frac{2\tau_{\min}}{3}\sigma_0 \mathcal{L}_{W_L\ W_L}(\tau_{\min}), \tag{31}$$

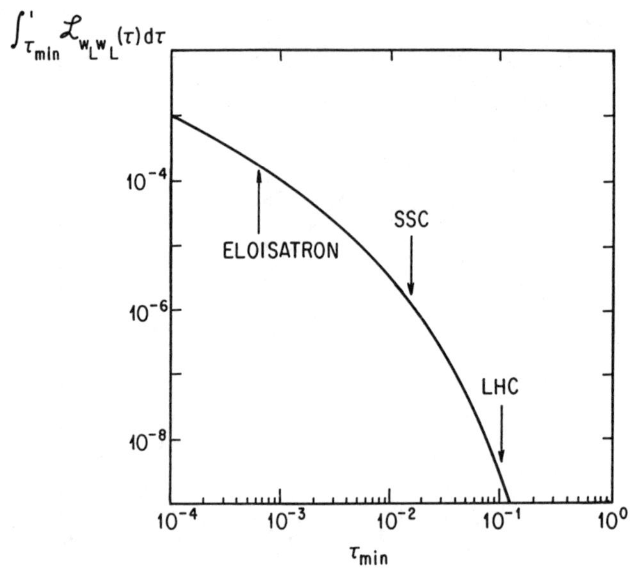

Figure 6: Plot of $\int_{\tau_{min}}^{1} \mathcal{L}_{W_L\ W_L}(\tau)d\tau$ as a function of $\tau_{\rm min}$. The arrows indicate the value of $\tau_{\rm min}$ for the Eloisatron, SSC and LIIC, assuming $\sqrt{\hat{s}_0} = 5\ TeV$.

where the last line uses the fact that the luminosity function decreases approximately as $\tau^{-5/2}$ for small τ. For the same amount of integrated luminosity, it is clear from Eq. (31) that to see this phenomena one really needs the energy of the Eloisatron, with the number of events scaling roughly as the CM energy of the machine cubed. This is demonstrated explicitly in Fig. 6 where the integral of $\int_{\tau_{\rm min}}^{1} \mathcal{L}_{W_L W_L}(\tau)$ is given as a function of $\tau_{\rm min}$. Assuming a given threshold value for multi W_L production, the value of this integral for the Eloisatron, SSC and LHC can be read off from the corresponding $\tau_{\rm min} = \hat{s}_0/s$ in the figure. The arrows in Fig. 6 for these accelerators assume as a threshold $\sqrt{\hat{s}_0} = 5\ TeV$.

The assumptions made concerning multi W_L production, Eq. (30), are rather conservative, in that both σ_0 could be larger and the threshold energy $\sqrt{\hat{s}_0}$ could be smaller. Nevertheless, even in this case, assuming an equally conservative yearly integrated luminosity of $\int \mathcal{L}dt = 10^{40} cm^{-2} = 10 fb^{-1}$ one gets about 2000 multi W events per year.[24] These events will mostly contain $< n >_{\rm total} \simeq\ 6\ W_L$, if the W_L multiparticle distributions are rather narrowly peaked around $< n >$ as are the pion distributions. Fig. 7, taken from the paper of Morris, Rosenfeld and I,[24] shows a guesstimate of what the distribution of W_L events as a function of n might look like. Although the tails are not large, there are of 0(100) events expected with 8 W_L and a handful of events containing 12 W_L *'s*. Just for fun, one can also try

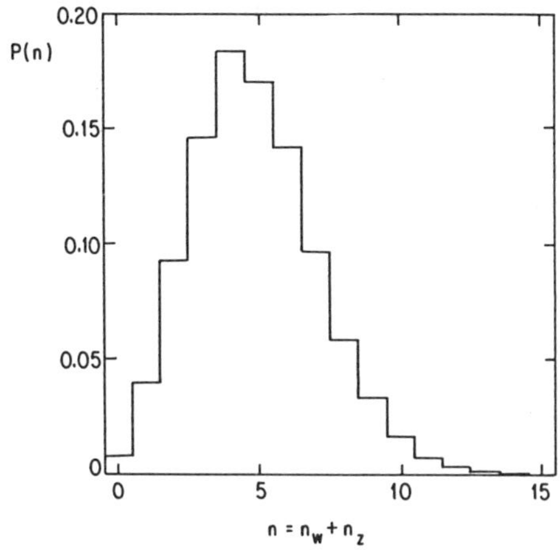

Figure 7: Guesstimates for the distribution of multi W_L events at the Eloisatron. From.[24]

to estimate the corresponding number of events at a Mega TeV machine (a good name for it might be Megatron, or even better, Megaton!). This number should be in the neighborhood of 10^{12} multi W_L events, still mostly containing 6 or so W_L - the threshold value of $< n >_{\text{total}}$ - but now events with large numbers of W_L (20 or above) in the tails are in the thousands!

Although multi W signals are characteristic of a theory having strong $W_L\, W_L$ interactions, their observation (based solely on the number of events expected) only appears to begin to be possible at the Eloisatron. Even then, it is not at all clear that one can really dig out this signal from the background. Because the number of events is small it appears to be necessary to use the multi jet nature of the events, coming from hadronic decays of the $W's$, as a signature combined with the presence of multileptons. Multiple top production is then a natural background, but fortunately, one does not expect an overwhelming number of events containing 3 or more $t\bar{t}$ pairs.[24] Clearly before one can make any more concrete statements about the observability of this signal, a rather thorough phenomenological analysis will be necessary. Some aspects of this analysis are contained in ref. (24). There we show that events containing 2 high p_T leptons coming from W decay and ten or more jets, or one leptonically reconstructed Z with one or more high p_T lepton and eight or more jets, have a combined cross section of a few fb. These events, furthermore, are a factor of 3-5 above background.

5. The Messy problem of fermion mass

Fermion masses are the unresolved crux in models of spontaneous symmetry breakdown. Because this issue is so central, and so difficult to resolve in a satisfactory fashion, it is important to contrast the "solutions" which have been put forward in these models with the way fermion masses are generated in the standard model.[5] Because of the presence of non trivial Yukawa couplings between the Higgs doublet field and the quark and lepton fields, in the standard model the Higgs doublet vacuum expectation value $<\Phi>=\frac{1}{\sqrt{2}}v$ serves to give mass to all fermions. Schematically, the fermion mass matrix is given by

$$M_f = \Gamma_f < \Phi > \quad , \tag{32}$$

where the Γ_f are the Yukawa couplings. Because the Yukawa couplings are arbitrary, although the masses are proportional to the Fermi scale v, the masses themselves are not predictable in the model.

The above negative feature of the standard model is offset by two very nice properties. Because the Yukawa couplings Γ_f have no symmetry reason to be diagonal, the fermion mass matrix M_f will, in general, be non diagonal. The diagonalization of M_f, necessary to go to a physical basis, then introduces naturally into the theory a mixing matrix for the quarks - the Cabibbo Kobayashi Maskawa[26] matrix V_{CKM}. Furthermore, if only one Higgs doublet contributes to M_f, as in Eq. (32), then diagonalizing M_f necessarily also diagonalizes the Yukawa couplings Γ_f. Thus, in the standard model, Higgs exchange never generates any flavor changing neutral currents (FCNC), in complete accord with experiment where one knows that these processes are very suppressed.[27]

Fermion masses cannot be generated by Technicolor models[8] without introducing further structure. Although technifermion condensates $<\bar{T}T>$ can serve to break down $SU(2)\times U(1)$ - thus playing an analogous role to the Higgs vacuum expectation value $<\Phi>$ - to generate mass for the fermions it is necessary to be able to connect the technifermions to the quarks and leptons. That is, one must be able to introduce something analogous to the Yukawa couplings Γ_f of the standard model.

The connection between ordinary fermions and technifermions can be achieved in a variety of ways. The most popular approach, called Extended Technicolor (ETC), is to introduce further gauge interactions under which ordinary fermions and technifermions transform irreducibly.[28] Another idea is that there are some residual interactions between these states because both the quarks and leptons and the technifermions are bound states of some still more elementary fermions.[29] Here, for concreteness, I will concentrate only on the ETC suggestion describing briefly both the idea, some of its problems and some of their putative solutions in terms of Walking Technicolor models.[30]

In ETC models[28] one introduces a further non Abelian gauge theory (ETC) besides Technicolor. Both quarks and leptons and the technifermions are presumed to transform according to some irreducible ETC representation. Furthermore, the ETC gauge theory is assumed to be broken spontaneously at a scale Λ_m. If we let F denote generically the fermions ETC multiplets, then the exchange of an ETC

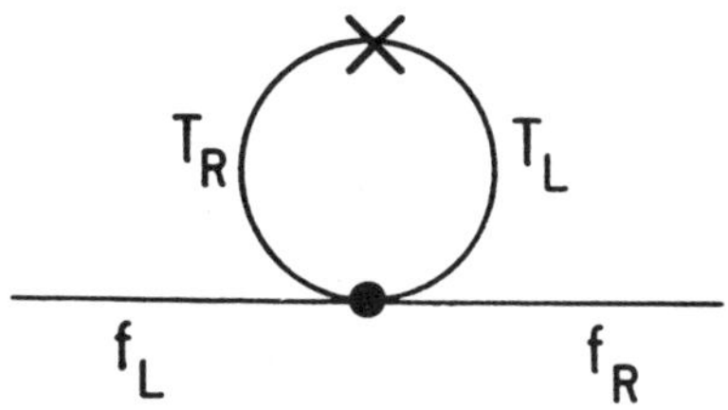

Figure 8: Mass generation in ETC models

gauge boson will give rise to effective interactions among these fermions of the form

$$\mathcal{L}_{\text{ETC}} \sim \frac{1}{\Lambda_m^2} \bar{F}_L \lambda_a \gamma_\mu F_L \bar{F}_R \lambda_a \gamma^\mu F_R \tag{33}$$

As a result of the formation of technicolor condensates $< \bar{T}_L T_R > \sim \Lambda_{TC}^3 \simeq v^3$ the interaction in Eq. (33) will give rise to mass terms for the ordinary fermions. This process is shown schematically in Fig. 8. and the magnitude of the dynamically generated mass is clearly of the order

$$m_f \sim \frac{< \bar{T}_L T_R >}{\Lambda_m^2} \sim \frac{v^3}{\Lambda_M^2} \tag{34}$$

Unfortunately, the interaction (33) will generate not only masses for the fermions f, but once fermion mixing is taken into account, also FCNC among these fermions. To avoid too large FCNC one needs to make the ETC dynamical scale rather large,[31] typically $\Lambda_m \sim 10^2 - 10^3 \ TeV$. However, if Λ_m is so large then it is very difficult to generate via Eq. (34) large masses for the fermions, since this equation gives masses in the MeV not GeV range!

The apparent incompatibility of having sizable fermions masses and small FCNC interactions is not the only problem of ETC models. In general, it is extremely difficult to construct any realistic models. Furthermore, even semirealistic models - like the Farhi Susskind model[32] - are plagued by the presence of very light pseudo Goldstone bosons arising from the formation of the Technicolor condensates.[33] All of these problems have, rather naturally, discouraged people from thinking about dynamical symmetry breaking theories. However, as Holdom and others have pointed out,[30] many of the above problems may in fact not arise in more realistic theories. They are only problems for theories which, in their dynamics, resemble rather closely QCD. As we saw already, when we considered the S parameter, such theories run into other experimental difficulties purely related to the weak interactions.

In QCD the expectation value of u - quarks in the vacuum and the pion decay constant, f_π are closely related, since they both are determined by the dynamical scale of QCD, Λ_{QCD}:

$$< \bar{u}u > \sim \Lambda_{QCD}^3 \; ; \; f_\pi \sim \Lambda_{QCD} \tag{35}$$

If this also obtains in the underlying Technicolor theory then, most likely, not only will one run into trouble experimentally with the S parameter, but also any attempts to build realistic ETC models will fail - for the reasons discussed above. It is therefore important to try to understand under what conditions the QCD relation (35) may be obviated by the dynamics of the underlying theory one is using to trigger the spontaneous breakdown of the electroweak $SU(2) \times U(1)$ theory.

For these purposes it is important to note that f_π and $< \bar{u}u >$ are, in principle, quite distinct parameters. The pion decay constant typifies the strength of the coupling of the Goldstone pions to the broken axial current J_A^μ:

$$i f_\pi k^\mu = < \pi(k) | J_A^\mu | 0 > \quad . \tag{36}$$

$< \bar{u}u >$, on the other hand, is the order parameter for the breaking of the chiral symmetry associated with the currents J_A^μ in QCD. Both f_π and $< \bar{u}u >$, however, can be related to the quark self energy $\Sigma(p)$, defined as usual as the part of the full inverse quark propagator, $S^{-1}(p)$, which is not proportional to $\gamma \cdot p$

$$S^{-1}(p) = \gamma \cdot p A(p^2) + \Sigma(p) \tag{37}$$

Graphically, the relation of $< \bar{u}u >$ and f_π^2 to $\Sigma(p)$ can be easily inferred from Fig. 9. Retaining only the leading momentum behaviour one sees that

$$< \bar{u}u > \sim \int d^4p \, \frac{\Sigma(p)}{p^2} \sim \int dp^2 \Sigma(p) \tag{38a}$$

$$f_\pi^2 \sim \int d^4p \, \frac{\Sigma^2(p)}{(p^2)^2} \sim \int \frac{dp^2}{p^2} \Sigma^2(p) \tag{38b}$$

The presence of a nontrivial quark self energy $\Sigma(p)$ is a signal of the breaking of the chiral symmetry present in the absence of quark masses. Thus $\Sigma(0)$ serves as a chiral breaking parameter and naturally one expects in QCD that $\Sigma(0) \sim \Lambda_{QCD}$. One can show in QCD[34] further that, up to logarithmic factors, for large momentum

$$\Sigma(p) \sim \Sigma^3(0)/p^2 \quad . \tag{39}$$

Because of the different weighting of the p^2 integrals associated with $< \bar{u}u >$ and f_π^2, one sees that f_π^2 probes the region of the quark self energy near $p^2 = 0$, while

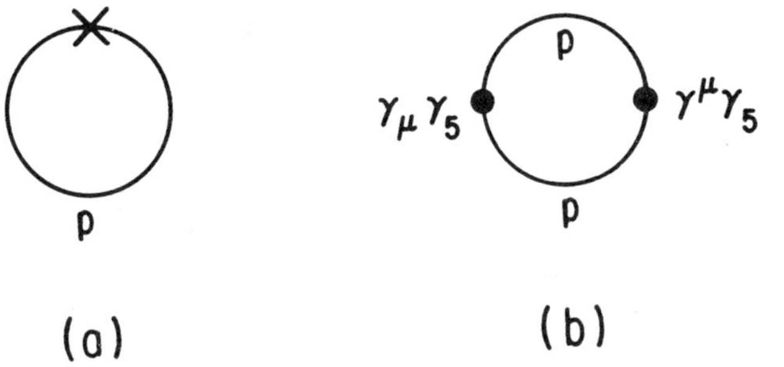

Figure 9: Graphs relating $<\bar{u}u>$ (a) and f_π^2 (b) to the quark self energy $\sum(p)$.

$<\bar{u}u>$ feels more the large momentum structure of $\Sigma(p)$. Using Eq. (37), and (38) one deduces immediately that, up to logarithmic factors,

$$<\bar{u}u> \quad \sim \int dp^2 \sum(p) \quad \sim \Sigma^3(0) \tag{40a}$$

$$f_\pi^2 \quad \sim \int \frac{dp^2}{p^2}(\Sigma(p))^2 \quad \sim \Sigma^2(0) \tag{40b}$$

The above natural relationship between $<\bar{u}u>$ and f_π need not apply in theories where the self energy of the underlying fermions in the important region for mass generation falls much slower with p^2 than does the quark self energy in QCD., This is the basic idea behind Walking Technicolor theories.[30] Although purely on dimensional grounds $\Sigma(p) \sim \Sigma^3(0)/p^2$ at large p^2, the region of p^2 important for fermion mass generation due to the ETC interactions (33) is $p^2 \sim \Lambda_m^2$. It may well be that for a realistic Technicolor theory at these large, but not yet asymptotic, values of p^2, the techniquark self energy is much larger than its asymptotic value. Although the ordinary fermion masses are still given by Eq. (34), the second approximate equality in this formula no longer necessarily applies. Rather, more properly, one has

$$m_f \sim \frac{<\bar{T}T>}{\Lambda_m^2} \sim \frac{1}{\Lambda_m^2}\int^{\Lambda_m^2} dp^2\Sigma(p) \sim \Sigma(\Lambda_m) \tag{41}$$

$\Sigma(\Lambda_m)$ is not known a priori without a dynamical calculation. Of particular interest is its relation with the technipion decay constant v, which is the scale that is related directly to the gauge boson masses. This scale, as is f_π in QCD, is connected [cf Eq. (40)] to the technifermion self energy at zero momentum transfer

$$v \simeq \Sigma(0) \tag{42}$$

Using (42), it is clear that, if at scales of $0(\Lambda_m)$ the technifermion self energy has already approached its asymptotic value (39), then

$$\Sigma(\Lambda_m) \sim \frac{\Sigma^3(0)}{\Lambda_m^2} \sim \frac{v^3}{\Lambda_m^2} \tag{43}$$

and one recovers the estimate made previously in Eq. (34). The assumption one makes in Walking Technicolor theories,[30] however, is that the dynamics is such that $\Sigma(\Lambda_m)$ is still much above this value. That is

$$[\Sigma(\Lambda_m)]_{WTC} >> \frac{v^3}{\Lambda_m^2} \tag{44}$$

To obtain a behaviour like (44) it is necessary that all physical quantities evolve slowly with momentum - hence the moniker walking for these theories. This can be achieved if the underlying theory is very nearly not asymptotically free, or, perhaps, a fixed point theory.[30] I do not have the time, or the relevant expertise, to enter into the details of the dynamics of these theories here. However, I note that (44) neatly resolves one of the difficulties of ETC models. Namely, one can have Λ_m be sufficiently large to avoid trouble with FCNC and yet get fermion masses which are sufficiently large. Furthermore, since the fermion masses are given by $\Sigma(\Lambda_m)$ and this self energy can be different for different technifermions[35] one can begin building in this way some of the observed hierarchies in the masses of the quarks and leptons.

6. Concluding Remarks

Although Walking Technicolor models resolve some of the dynamical conundrums of ETC models, they are not a panacea and many issues are still unclear. For instance, despite intense activity in the last few years no real full models are yet in sight.[36] Furthermore, even in semirealistic models the dynamics and group structure introduced are in general quite complex.[37] In addition, many problems continue to be difficult to tackle. For example, although it is possible to build a hierarchy between m_t and m_b, it is difficult to do so without distabilizing by the dynamics the result that $\rho \simeq 1$.

These comments not withstanding, it is important not to let the difficult issue of fermion masses discourage one from contemplating some of the possible physical consequences of theories where the symmetry breaking of $SU(2) \times U(1)$ is done more dynamically. After all, it could well be that fermion mass generation really requires a new idea. Much in the same way in which quark models of hadrons were dynamically untenable before the discovery of asymptotic freedom, it could well be that to understand fermion masses we will need to break away from our present conventional mode of thinking.

Because fermion masses most probably arise due to new physics at a scale $\Lambda_m >> v$, it is important to try to explore the consequences of dynamical symmetry breaking in the Eloisatron energy range, which lies between Λ_m and v. Because of the energy degradation that occurs when one focuses on the W_L constituents of the colliding protons [cf Fig. 4], for the telltale $W_L W_L$ scattering processes high energy machines like the LHC and SSC are only threshold machines for probing dynamical symmetry breaking. However, it is quite likely that new and interesting phenomena may begin to open up at Eloisatron energies (and would be spectacular at $M = 10^6$ TeV!) As we discussed, beside the presence of characteristic resonances in the $W_L W_L$ channel, one should expect to see the first signs that the $W_L W_L$ total cross section remains approximately constant through the opening of inelastic channels. The discovery of multi W_L events ($n_{W_L} > 6 - 8$) at the Eloisatron would be a spectacular confirmation of the existance of a new strongly interacting sector in the physics of fundamental particles.

7. Acknowledgement

I am grateful to Duncan Morris and Rogerio Rosenfeld for their collaboration on some of the material presented in this lecture. This work was supported in part by the Department of Energy under grant AT03-88ER 40384 Task C.

References

1. H. P. Nilles, Phys. Rept 110 (1984) 1
2. For a recent review, see for example B. Sadoulet, Proceedings of the XI Moriond Workshop, Les Arcs, France, 1991, (Ed. Frontiers, Gif sur Yvette, France)
3. The fermion mass problem in dynamical symmetry breaking theories is nicely reviewed by T. Applequist, Proceedings of the IV Mexican School of Particles and Fields, Oaxtepec, Mexico, 1990 (World Scientific, Singapore 1991)
4. L. Rolandi in Proceedings of the XXVI International Conference in High Energy Physics, Dallas, Texas, Aug. 1992
5. S. L. Glashow, Nucl Phys. 22 (1961) 579; A. Salam in **Elementary Particle Theory** ed. M. Svartholm (Almquist and Wiksell, Stockholm, 1968): S. Weinberg, Phys. Rev. Lett. 19 (1967) 1264
6. P. Sikivie, L. Susskind, M. Voloshin and Z. Zakharov Nucl. Phys. B173 (1980) 189
7. P. W. Higgs, Phys. Lett. 13 (1964) 508
8. L. Susskind, Phys. Rev D20 (1979) 2019; S. Weinberg, Phys. Rev. D16 (1976) 974, D19 (1979) 1277
9. J. M. Cornwall, D. M. Levin and G. Tiktopolous, Phys. Rev D10 (1974) 1145; D11 (1975) 972(E)
10. M. Veltman, Nucl. Phys. B123 (1977) 89; Acta Phys. Pol B12 (1981) 437
11. B. W. Lee, C. Quigg and H. Thacker, Phys. Rev. D16 (1977) 1519
12. D. Kennedy and P. Langacker, Phys. Rev. D44 (1991) 1591
13. Here α is the fine structure constant measured at zero momentum transfer; M_Z is the Z mass precisely measured at LEP; and G_F is the Fermi constant one infers from μ-decay after some purely electromagnetic corrections are

removed. For details see, Review of Particle Properties, Phys. Rev. D45 (1992) 1

14. M. Peskin and T. Takeuchi, Phys. Rev. Lett. 65 (1990) 964; Phys. Rev. D46 (1992) 381
15. In Eq (14) the index 3 refers to the 3rd component of the weak $SU(2)$ group
16. S. Weinberg, Phys. Rev. Lett. 18 (1967) 507
17. G.'t Hooft in **Recent Developments in Gauge Theories** ed. G.'t Hooft et al (Plenium Press, New York 1980); E. Witten, Nucl. Phys. B160 (1979) 15
18. R. N. Cahn and M. Suzuki, Phys. Rev. Lett. 67 (1991) 169
19. G. Altarelli, R. Barbieri and S. Jadach, Nucl. Phys. B369 (1992) 3
20. P. Langacker and M. Luo, Phys. Rev. D44 (1991) 817; D45 (1992) 278
21. R. D. Peccei and J. Sola', Nucl. Phys. B281 (1987) 1
22. T. Applequist and G. Triantaphyllou, Phys. Rev. Lett 278B (1992) 345; S. Hsu and R. Sundrum, LBL preprint 1992
23. M. S. Chanowitz in **Perspections in Higgs Physics** ed G. Kane (World Scientific, Singapore) in press
24. D. A. Morris, R. D. Peccei and R. Rosenfeld, UCLA preprint, UCLA 92/TEP/45
25. G. L. Kane. W. W. Repko and W. B. Rolnick, Phys. Lett 148B (1984) 367; S. Dawson, Nucl. Phys. B249 (1985) 42
26. N. Cabibbo, Phys. Rev. Lett. 10 (1963) 531; M. Kobayashi and T. Maskawa, Prog. Theor. Phys. 49 (1973) 65
27. For a compilation of limits on FCNC processes see, for example W. Buchmuller and D. Wyler, Nucl. Phys. B268 (1986) 621
28. S. Dimopulous and L. Susskind, Nucl. Phys. B155 (1979) 237; E. Eichten and K. Lane Phys. Lett 90B (1980) 237
29. For a discussion, see for example, R. D. Peccei in the Proceedings of the 1987 Lake Louise Winter Institute, ed. J. M. Cameron et al, (World Scientific, Singapore 1988)
30. B. Holdom, Phys. Rev. D24 (1981) 1441; Y. Yamawaki, M. Bando, and K. Matumoto, Phys. Rev. Lett. 56 (1986) 1335; T. Akiba and T. Yanagida, Phys. Lett 169B (1986) 432; T. Applequist, D. Karabali and L. Wijewardhana, Phys. Rev. Lett. 57 (1986) 952
31. S. Dimopoulos and J. Ellis, Nucl. Phys. B182 (1981) 505
32. E. Farhi and L. Susskind, Phys. Rept. 74C (1981) 277
33. S. Dimopoulos, S. Raby and G. Kane, Nucl. Phys. B182 (1981) 77
34. K. D. Lane, Phys. D10(1974) 2605; H. D. Politzer, Nucl. Phys. B 117 (1976) 397
35. B. Holdom, Phy. Lett B246 (1990) 169
36. M. Einhorn in the Proceedings of the 1991 Meeting of the Division of Particles and Fields of the American Physical Society, Vancouver, B.C., August 1991, Ed. D. Axen et al., (World Scientific, Singapore 1992)
37. For a recent example, see for instance, R. Sundrum, LBL preprint 1992

CHAIRMAN: R. Peccei

Scientific Secretaries: P. Cho, G. Word

DISCUSSION

– *Benakli:*

What is the philosophy of Technicolour and how does it solve any problems?

– *Peccei:*

It is useful to recall the history of superconductivity. The original Landau-Ginzberg theory of superconductivity has an order parameter which looks like an elementary Higgs expectation value. That phenomenological model was later replaced by the BCS theory in which the symmetry of the ground state is dynamically broken via Cooper pair formation. The idea behind Technicolour is to replace the scalar excitation in the Standard Model by something that is more dynamical.

The Technicolour idea is also motivated by the problem of naturalness. Scalar theories are not natural since they depend quadratically upon a cutoff scale. On the other hand, theories with dynamically determined scales are natural. Chiral condensates form at the scale where a coupling constant in the underlying theory grows large. So condensate formation is explained in the same way as chiral symmetry breaking in QCD.

– *Benakli:*

Why have Technifermions not been seen yet if they are light?

– *Peccei:*

Technifermions cannot be observed since they are confined at the 1 TeV scale. One never sees Techniquarks just as one never sees quarks.

– *Malik:*

You showed a table which indicated the dependence of observables on S, T and U. How should these parameters be treated when comparing with the experimental results of LEP?

Table 1. Dependence of observable on S, T, U
(numbers by Peskin-Tacheushi) $m_t = 150$ GeV, $m_H = 1$ TeV

$$\frac{m_W}{m_Z} = 0.8787 - [3.15\times10^{-3}]\,S + [4.86\times10^{-3}]\,T + [3.70\times10^{-3}]U$$
$$\Gamma_Z = 2.484 - [9.58\times10^{-3}]\,S + [2.615\times10^{-2}]\,T\ \text{(GeV)}$$
$$\Gamma_{\ell^+\ell^-} = 0.0835 - [1.91\times10^{-4}]\,S + [7.83\times10^{-4}]\,T\ \text{(GeV)}$$
$$\Gamma_{u\bar{u}} = 0.2962 - [1.91\times10^{-3}]\,S + [3.67\times10^{-3}]\,T\ \text{(GeV)}$$
$$\Gamma_{d\bar{d}} = 0.3823 - [1.72\times10^{-3}]\,S + [4.20\times10^{-3}]\,T\ \text{(GeV)}$$
$$\Gamma_{b\bar{b}} = 0.3779 - [1.72\times10^{-3}]\,S + [4.20\times10^{-3}]\,T\ \text{(GeV)}$$
$$\Gamma_{\rm had} = 1.7348 - [9.00\times10^{-3}]\,S + [1.993\times10^{-2}]\,T\ \text{(GeV)}$$
$$R_Z = \Gamma_{\rm had}/\Gamma_{\ell^+\ell^-} = 20.78 - [5.99\times10^{-2}]\,S + [4.24\times10^{-2}]\,T$$
$$s_*^2(m_Z^2) = 0.2337 + [3.59\times10^{-3}]\,S - [2.54\times10^{-3}]\,T$$
$$A_{LR} = -P_\tau = 0.1297 - [2.82\times10^{-2}]\,S + [2.00\times10^{-2}]\,T$$
$$A_{FB}^b = 0.0848 - [1.97\times10^{-2}]\,S + [1.40\times10^{-2}]\,T$$
$$A_{FB}^\ell = 0.0126 - [6.72\times10^{-3}]\,S + [4.76\times10^{-3}]\,T$$
$$g_L^2 = 0.3001 - [2.67\times10^{-3}]\,S + [6.53\times10^{-3}]\,T$$
$$g_R^2 = 0.0302 + [9.17\times10^{-4}]\,S - [1.94\times10^{-4}]\,T$$
$$R_\nu = 0.3126 - [2.32\times10^{-2}]\,S + [6.46\times10^{-2}]\,T \qquad (r = 0.383)$$
$$R_{\bar{\nu}} = 0.3824 - [2.77\times10^{-3}]\,S + [6.03\times10^{-3}]\,T \qquad (r = 0.371)$$
$$Q_W({}^{133}_{55}\mathrm{Cs}) = -73.31 - 0.790\,S - 0.011\,T$$

– *Peccei:*

Using the electroweak observables shown in the table in the text, you can determine a best fit for S and T along with error ellipses that result from combining the errors on all of the experimental measurements. Two such error ellipses are shown in the S–T plane figure (Fig.1). The predictions of the Standard Model are plotted in this figure for various top and Higgs mass values. Results from a Technicolour model with one doublet and a model with one Technicolour generation are also illustrated for various values of the top mass. In order to assess the validity of any particular model, one should plot the model's predictions in the S–T plane and then see how well they agree with the data.

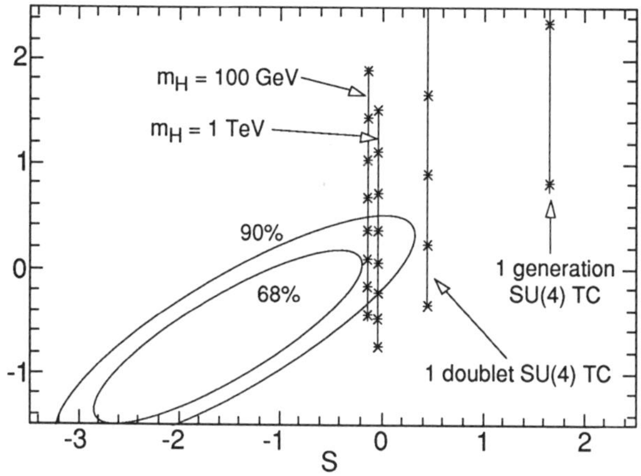

Figure 1

–*do Couto e Silva:*

Why would one want to build a machine such as the Eloisatron?

– *Peccei:*

We would like to determine the physics responsible for breaking $SU(2) \times U(1)$. Weakly coupled theories like normal Higgs models with or without SUSY and theories where the breakdown is dynamical in origin are clearly distinct. One can distinguish between these two types of theories by measuring longitudinal W_L scattering. Such scattering is strong if there is an underlying theory and weak in a Higgs model. In order to study W_L scattering, one needs to probe at very high energies. W_L bosons can be produced out of p–p scattering, but high luminosity is essential. Luminosity drops very rapidly with $\hat{s}/s$. However, if we can produce large values for s, then we will be able to see greater number of events. This is why we would like to build a machine like the Eloisatron.

– *Wadhwa:*

You mentioned that the Eloisatron would be a good machine for multi W_L production. How optimistic are you about the number of W_L bosons that can be produced at 200 TeV?

– *Peccei:*

If there is an underlying strongly interacting theory, then a total cross section of 1 nb is a very reasonable value to scale up. One knows the approximate size of the order parameter to be scaled up from QCD. The appearance of a threshold for multiparticle production after the resonance region is also an extrapolation from strong interaction scattering. I do not see why it should not go in this way also here. Strong interactions tend to produce final multiparticle states rather than remain in one channel. So once you believe that the underlying dynamics is strongly interacting, then the number of events that I quoted for W_L production is quite reasonable.

– *Leike:*

You reported that about 300 to 400 four–W boson events are to be expected at the Eloisatron. I suppose that you will try to reconstruct them from eight jet events. How many of your 400 produced events do you hope to reconstruct?

– *Peccei:*

Obviously it is not easy to extract the four-W processes from the large number of events required to produce them. One needs to explore such characteristics as increases in multi-jet production. Clearly more work needs to be done to identify other handles. Nevertheless, signals of strong interactions among W_L bosons are certainly quite different from what one would expect in a weakly coupled theory where it is difficult to produce many W bosons. More work is required to determine whether such signals can be identified.

– *Tao:*

In Walking Technicolour, how do you make the self energy $\Sigma(p)$ fall very slowly with p?

– *Peccei:*

Imagine QCD had 16 flavours. In this case, the QCD coupling constant is almost not asymptotically free, and the running of the QCD coupling is very slow. In Walking Technicolour models, the coupling also runs slowly due to the inclusion of many Technifermions.

– *Baillie:*

Do you believe that Technicolour theories will be able to reproduce the fermion spectrum in a natural way, or will some higher, perhaps string-like theory be necessary?

– *Peccei:*

There are many difficulties associated with fermion mass generation in Technicolour models. Remember that in QCD we have an enormous amount of data at our disposal. Yet we still cannot compute the mass of the proton! I think that it's unlikely that if Technicolour is discovered in our lifetimes that we will be able to compute some realistic fermion mass spectrum. In my lifetime, I hope to find out whether the dynamical approach to breaking $SU(2) \times U(1)$ is true or whether SUSY is the better theory. I would be very happy to just know the answer to this basic question. Working out the details in either case is extraordinarily difficult.

– *Hanany:*

Why are the masses of quarks and leptons so small?

– *Peccei:*

It's easier to generate low rather than high masses in Technicolour models. Large differences between quarks and leptons are not difficult to produce. For instance, the splitting between the top quark and the tau lepton can be fairly easily understood. But it is much more difficult to generate a top–bottom mass difference since the t and b quarks are only distinguished by their $U(1)_Y$ quantum numbers. It is very difficult to generate a realistic fermion mass spectrum as a result of a complex dynamical calculation.

– *Sopczak:*

What is the underlying reason for the similarities between theories with dynamical symmetry breaking and QCD?

– *Peccei:*

The only real similarity is a protective SU(2). It is simplest to assume that a dynamical theory with a chiral $SU(2)_L \times SU(2)_R$ symmetry breaks down to $SU(2)_{L+R}$. We know of precisely one such theory: QCD. So it makes sense to try to scale it up because is has precisely the same global symmetry structure and because it is strongly interacting. Such a scaled up theory should be adequate in the threshold region. Of course, we are interested in looking above threshold where it's not obvious that a scaled up version of QCD is valid.

– *Sopczak:*

In the figure showing the S–T plane, the expected values for S and T are confronted with data. The data favour small S values whereas the Minimal Standard

Model predictions are larger. Even with a very light Higgs, the one sigma contour cannot be reached. Where would the S prediction for a two doublet Higgs model lie?

– *Peccei:*

The Higgs dependence on S and T is very mild. There are no significant differences in the S and T predictions from models which contain one or two Higgs doublets. Any sensible multi-Higgs model cannot be excluded based upon the current S and T information.

– *Liu:*

How big would an accelerator have to be to probe the mega-TeV range?

– *Peccei:*

Well, the Eloisatron would have a 300 km tunnel and could reach approximately 200 TeV. It represents a possible next generation accelerator.

– *Liu:*

Is it possible for a *very large area* underground detector to see this new phenomena?

– *Peccei:*

I don't think so. Maybe one could see a few W–W scattering events coming from very energetic neutrinos. Of course, we know nothing about the flux of such neutrinos. One might be able to see some dimuon events coming from underground.

– *Block:*

When talking about the scattering amplitude for $W^+W^- \to ZZ$, you made the analogy to the scattering amplitude of $\pi^+\pi^- \to \pi^0\pi^0$. Why could you do that?

– *Peccei:*

Consider again the analogy with QCD. The pions of QCD are the Goldstone bosons associated with the $SU(2)_L \times SU(2)_R$ breakdown to $SU(2)_{L+R}$. So if you believe that the dynamically broken theory is analogous to QCD, you would expect to be able to relate $A(W^+W^- \to ZZ)$ to $A(\pi^+\pi^- \to \pi^0\pi^0)$. Many studies of LHC and SSC physics have been based upon this analogy with QCD.

- *Cho:*

Are SU(3) and $SU(N)_{TC}$ embedded inside $G_{\text{Extended Technicolour}}$?

- *Peccei:*

The SU(3) factor in the gauge group $G = G_{ETC} \times \{G_{UT} \times G_{UT'}\} \times G_{TG} \times SU(3) \times SU(2) \times U(1)$ considered recently by Sundrum does not necessarily have to be colour. In many theories, $SU(3)_{colour}$ is a diagonal subgroup of an SU(3) factor in G_{ETC} and an SU(3) factor in G. On the other hand, $SU(3)_{colour}$ could sit completely inside of G_{ETC}. It depends upon the details of the specific model.

- *Cho:*

Has anyone tried to partially unify $SU(N)_{TC}$ or G_{ETC} with any of the factors in $SU(3) \times SU(2) \times U(1)$?

- *Peccei:*

Unifying Technicolour and colour is generally problematic. Technicolour models do not mesh particularly well with GUT theories. I think some models have been developed which have somewhat simpler group structure than that of the group written above.

- *Cho:*

Is Technibaryon number conserved and is the lightest Technibaryon stable? What are the cosmological implications of stable Technibaryons?

- *Peccei:*

I think there will be some stable Technibaryons. While I do not remember the exact details, I think that they are cosmologically safe.

- *Petrache:*

Why have you taken $\langle \overline{U}U \rangle = \langle \overline{D}D \rangle$?

- *Peccei:*

The breakdown of chiral $SU(2)_L \times SU(2)_R$ in QCD preserves a custodial $SU(2)_{L+R}$ which says that there is no distinction between up and down quarks. In theories like QCD, you can show that global chiral symmetries break down to vector subgroups which remain unbroken. This is known as the Witten-Vafa theorem.

CHAIRMAN: R. Peccei

Scientific Secretaries: E. do Couto e Silva, Y. Zamora Garcia

DISCUSSION

– *A. Hanany:*

Could you explain what is done with Einstein-Prodovsky-Rosen paradox?

– *Peccei:*

There is no EPR paradox. According to QM, your knowledge on the initial states is that you have a combination of K^0 and $\overline{K}^0$ in both decays.

– *Baillie:*

On the first part of your notes, the assumptions of the CPT theorem include Lorentz invariance. Is a positive norm on the set of states also required?

– *Peccei:*

Yes.

– *Baillie:*

So could gauge theories escape the CPT theorem?

– *Peccei:*

You must be careful. Take a physical subset of gauge theories. You can apply Gauss Law and you will get states with positive norm in your physical gauge, but you are only looking at the polarization and not really properly quantizing the gauge. The CPT theorem certainly applies to gauge theories too.

– *Cho:*

1) Can one directly test CPT by measuring masses of particles and their antiparticles to high precision (e.g. p and $\bar{p}$)?

2) The most basic ingredients of Quantum Field Theory are unitarity, locality, causality, Lorentz invariance. Which, if any of these, could be abandoned to arrive at a more general framework?

– *Peccei:*

1) One can measure the mass difference between a positron and an antipositron and take the ratio to the mass of the positron and one gets a number of the order of 10^{-4}. The same comparison can be done with pions, and numbers of the order of 10^{-5} can be found. But what one really wants to do is to measure the difference in the mass of K^0 and $\overline{K}^0$ relative to the $K_S - K_L$ mass difference (of the order of $10^{-12} - 10^{-14}$ GeV), which is essentially zero. Therefore measuring effects to the accuracy of 10^{-4} actually gives an enormous enhancement and that is why one is measuring things at the order of 10^{-18}. This is the reason for looking at the kaon system.

2) If things are based on Quantum Field Theory one should expect CPT to hold but they might as well be based on strings. I do not think locality is that important, but people have proved that CPT is conserved in string theory at very short distances but they basically go to perturbative limits, in other words, they go back to the field theory limit. What really interests me is what happens when you compactify CPT from a higher number of dimensions where everything is fine to a small scale. As an example one can see what happens with chirality: an asymmetry appears when one reduces the number of dimensions. Chirality is much simpler to understand than CPT. It is important to do experiments of the order of 10^{-18} and maybe effects that go like 1 GeV/M_{Pl} can be seen.

– *Sopczak:*

If there is experimental evidence for CPT violation, is it possible to incorporate it in the Standard Model?

– *Peccei:*

Not in the Standard Model. If CPT is violated people will first attack the experimentalists. They will question how well the measurements were done. Next they will attack the theorists to understand why this is CPT violation. There could also be an environmental CPT violation: some long range force which is slightly asymmetric between K^0 and $\overline{K}^0$. Given the fact that it looks like that, then people will be wondering about assumptions that I did not tell you having to do with a simple Schrödinger problem. But they will eventually worry about hermeticity. There is for sure physics beyond the Standard Model, but in the Standard Model CPT is really conserved.

– *do Couto e Silva:*

Can you compare CP violation in the K^0 and B^0 systems?

– *Peccei:*

The advantage of looking for CP violation in the K^0 system is the large number of kaons, but on the other hand this system is made up of just the first two generations. Because of that, you will have to incorporate loops and virtual effects in order to get CP violation. The reason why a parameter ϵ is so small is not because the CP phaser is small (it could also be equal to one), but because the mixings needed to connect to the third family lead to a small effect.

In the B^0 system there are decays to CP conjugate states. One can therefore measure directly the CP phase that enters in the CKM matrix by looking at the asymmetries in the B^0 system. The difficulty is that you do not have many B^0's, but in principle you expect to see much clear asymmetries.

– *Tao:*

1) Is the analysis of CPT violation model independent?

2) If CPT is violated, is there any significant implication on cosmology?

– *Peccei:*

1) Yes, the formalism is perfectly general, but of course it depends on the assumptions of the Standard Model.

2) Soon after Andrei Sakharov's famous paper on baryogenesis, there was a paper which pointed out that one of the conditions which Sakharov used, namely that you have to be out of equilibrium, is not necessary if you have CPT violation. Yes, there are interesting cosmological consequences but I cannot tell you what they really are.

– *Azimov:*

I have no question, but two comments:

1) The value given in Particle Data Table, indeed deviates from the superweak value. This deviation is rather large and could be, by itself, considered as some evidence for CPT violation. But the experimental value is linearly dependent on $\Delta m = m_{K_L} - m_{K_s}$. A quite admissible shift of Δm eliminates the deviation of phases between ϕ_{+-} and ϕ_{SW}. Therefore, an important task for the ϕ-factory is to lower the errors of Δm.

2) CPT-theorem gives clear and simple relations only for decays into isolated channels, i.e. diagonalizing the S-matrix of the final state interactions. Usually one assumes that this is the case for the channels $\pi^0\pi^\mp\ell^\pm\nu$. As a result, due to CP-violation (even with CPT conservation) a difference between the absolute values of amplitudes $K^0(\overline{K}^0) \to \pi^\mp\ell^\pm\nu$ appears. This difference should be kept in mind when comparing semileptonic decays with very high precision. An exact

estimate of possible relative effects is not straightforward, but it could be roughly of the order of $10^{-3} - 10^{-4}$.

– *Peccei:*

1) The kind of error you can expect in a ϕ factory with integrated luminosity of 10^{40} is shown in Table 1. For ϕ_{+-} itself it is $\pm$ 1.2 degrees. That is not very good, but you can do much better in Δm.

2) Usually they worry about the fact that you can have some $\Delta S = -\Delta Q$ transitions which can be estimated using the Standard Model, and they are indeed too small to give corrections of the order of 10^{-7}. The range you mentioned is very interesting and it is worth making a more careful measurement.

– *Liu:*

Looking at fig. 1 there are three quantities that must be measured. We just know one of these quantities, namely ϵ_{CPT}. What is the difficulty in measuring the other two?

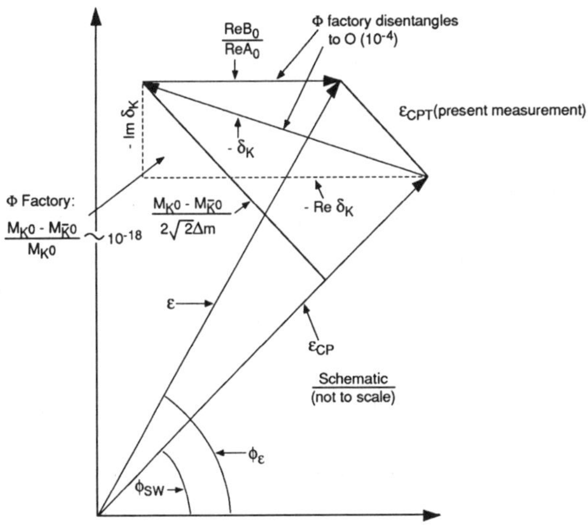

Figure 1

– *Peccei:*

What you measure is the kinematical parameter ϕ_{SW} which depends on the mass difference and the width difference of the $K_L - K_S$; there you know very well: up to less than a tenth of a degree. The other parameter is the phase of ϵ, or $\eta_{\pm}$. Theoretically $(M_{K^0} - M_{\overline{K}^0})/(2\sqrt{2}\Delta m)$ is made up of two pieces: δ_K and $\text{Re}(B_0)/\text{Re}(A_0)$. You would like to get an idea whether in fact you do not have a large cancellation by these two pieces. For that you need to measure $-\delta_K$. This can be done in the ϕ-factory measuring separately the K_s semileptonic asymmetry. Knowing this vector you can project it out. Therefore you estimate, as the particle data group does, $(M_{K^0}) - M_{\overline{K}^0}/(2\sqrt{2}\Delta m)$ neglecting the other part $\text{Re}(B_0)/\text{Re}(A_0)$. But in principle you might like to measure both pieces.

– *Dreiner:*

1) There has been a proposal for measuring the time-of-flight of kaons at CERN in order to test for T-violation. If we see T-invariance this would be an observation of the CPT violation. Is this experiment being done and can you comment on its prospects for CPT violation tests?

2) Could you comment on your statement: "I have shown that it is not possible to test Bell's inequality at a collider, unless you explicitly measure non-commuting variables (more than one spin component). Since we usually just measure momenta at colliders this is not possible."

– *Peccei:*

1) No, I do not know in detail but this is an experiment you can do in the ϕ factory. It is a marvellous time reversal experiment. The ϕ decays into a $K^0\overline{K}^0$. The K^0 can oscillate into a $\overline{K}^0$ and vice-versa. Therefore you can have as a final state two K^0 or two $\overline{K}^0$, one is precisely the time reversal process of the other. If these states decay semileptonically the signature for two K^0 will always be two positive leptons and for the two $\overline{K}_0$ you will have two negative leptons. So if you measure the difference in the number of positive leptons minus the number of negative leptons this will be the most accurate test of time reversal, because it is directly T violating. This is expected to be roughly twice the semi-leptonic asymmetry, that is to say 10^{-3} (almost 1%). This lovely experiment can be easily done in the ϕ factory.

2) You cannot really do it. Maybe a strangeness polarizer would do it. In measuring K_L or a K_S you will find that there is a given contribution and you cannot really change it. It is just fixed. Maybe you could change it using a regenerator or something that can affect the proportion of K^0 and $\overline{K}^0$. As far as I know even in a non-realistic experiment you always get into some problem and you cannot actually do it.

– *Misiak:*

Could you compare the values of CPT parameters that can be measured at the high luminosity ϕ-factory to the values that one could expect from CPT violation at the M_{Pl} scale?

– *Peccei:*

Each of these vectors (fig. 1) can be measured at the level of 10^{-4} which means that $(M_{K^0} - M_{\overline{K}^0})/2\sqrt{2}\Delta m$ is of the order of 10^{-4}. When you include the value of the mass difference $\Delta m(\sim 10^{-14}$ GeV), you will get $M_{K^0} - M_{\overline{K}^0}$ of the order of 10^{-18}.

– *Sannino:*

Can you show a CPT violating operator in four dimension and not in the two dimensional string theory?

– *Peccei:*

I do not know. If I write an operator in terms of the Field Theory that piece of the Lagrangian is CPT conserved. But I can nevertheless tell you experimentally what to measure: you just measure the mass of the particle and the mass of the antiparticle and look at their difference. I cannot write down an operator but I do not care, I can do measurements that will be a measurement of CPT violation.

– *Paz:*

When we compare a process with an antiprocess you assume you changed the signs of all internal charges, so if there is a CPT violation, maybe it is due to a new unknown internal quantum number that has not been changed in the experiment.

– *Peccei:*

That could be correct. This is one way to easily explain some apparent violations of CPT. In fact that will be one of the first questions that will be asked if such an effect is seen. It is very similar to having $\ell = 1$ ground state and then you have to rely on spin statistics. In fact there is a close connection between thinking about CPT violation and thinking about violating the Pauli principle. If you put an extra quantum number you can always undo violation on the Pauli principle.

TESTS OF SU(5) SUPERGRAVITY GRAND UNIFICATION

Pran Nath*† and R. Arnowitt**

*Department of Physics, Northeastern University
Boston, MA 02115, USA

**Center for Theoretical Physics, Department of Physics,
Texas A & M University, College Station, TX 77843

ABSTRACT

A pedagogical review is given of supergravity grand unification. Specific tests of SU(5) supergravity grand unification, beyond the unification of the electro-weak and the strong coupling constants implied by the high precision LEP data, are proposed. The analysis is carried out within the framework of radiative breaking of the electroweak symmetry and the contraints of CDF and LEP data on the SUSY spectrum and the contraint of proton lifetime limit. The following two theoretical assumptions are also made: (i) The higgs triplet mass has an upper bound of few times the unification mass M_G and (ii) excessive finetuning is avoided by requiring that the squark ($\tilde{q}$) and gluino ($\tilde{g}$) masses $\leq$ 1 TeV. It is found that the analysis restricts severely the allowed parameter space and the theory becomes very predictive. Over a significant domain of the parameter space, a number of mass relations arise between the gaugino masses. Thus one finds that $m_{\tilde{Z}_1} = \frac{1}{2} m_{\tilde{Z}_2}$, $m_{\tilde{Z}_2} = m_{\tilde{W}_1}$ and $m_{\tilde{Z}_3} = m_{\tilde{Z}_4} = m_{\tilde{W}_2}$. Further it is found that $m_{\tilde{W}_1} = \frac{1}{3} m_{\tilde{g}}$ for $\mu < o$ and $m_{\tilde{W}_1} = \frac{1}{4} m_{\tilde{g}}$ for $\mu > o$, where μ is the Higgs mixing parameter. The analysis also gives an upper bound on the top quark mass of 180 GeV. The light Higgs has an upper bound of 110 GeV. Another important consequence of the analysis is the correlation between the light Higgs mass and the light Wino mass. One finds that a light Higgs or a light Wino or both should be observable at LEP2 for the case when $m_t \leq 140$ GeV.

†Speaker at the Erice School

The above restriction on m_t is removed if $\tau(p \to \bar{\nu}K)$ exceeds 1.3×10^{32} yr. Thus the prospect for observing partially or fully the light sector of supergravity SU(5) at LEP2 appears very strong. Further, essentially the full spectrum i.e. 32 new particles should be accessible at SSC. The prospect for observing proton decay in the next round of proton decay experiments also appears very strong.

I. INTRODUCTION

While the Standard Model of electro-weak[1] and strong interactions is a phenomenologically successful model of particle interactions up to the current acclerator energies, one expects the Standard Model to arise from a more unified structure. One of the fundamental ingredients for such a formalism is the idea of grand unification[2,3] where one attempts to merge the three forces of microscopic physics into a single theoretical structure. However, it is well known that the standard grand unification suffers severly from the so-called gauge heirarchy problem involving the Higgs boson mass[4]. In the Standard Model quantum corrections to the Higgs mass possess a quadratic divergence, and in a grand unified theory with $M_G \approx 10^{15}$ GeV one needs to fine tune parameters to more than 24 decimal places to control the Higgs boson mass. The above problem of quadratic divergence is eliminated in a natural way in supersymmetric theories.[5] Thus it appears attractive that one combine the ideas of grand unifcation and sypersymmetry.

The first attempts to build realistic models using supersymmetry were made as early as 1977[6] and were later extended to include grand unification[7,8]. Models of Refs (6)-(8) are all based on global supersymmetry. However, it is well known that a phenomenologically satisfactory breaking of supersymmetry does not exist within the framework of global supersymmetry at the semi-classical (tree) level. Further, it can be shown that if supersymmetry does not break at the semi-classically (tree) level, it also remains unbroken for a wide class of models at the quantum level[9]. In order to overcome the difficulty discussed above one needs to extend global supersymmetry to a local supersyymetry[10]. The theory of Ref (10) is labelled as N = 1 supergravity, where N refers to the number of Majorana gravitino fields.

Extended supergravities with values of N up to a maximum of 8 also exist. Unfortunately, even the largest (i.e. N = 8) supergravity does not have the particle content large enough to accomodate all the physical degrees of freedom of the standard $SU(3)_C \times SU(2)_L \times U(1)$

model. The above difficulty leads one to the idea that one must couple supergravity with matter to achieve realistic models. Only N = 1 supergravity in relevant since this is the only framework which naturally accomodates chiral fermions of electro-weak physics. This means that to build realistic supergravity models one must couple N = 1 supergravity with N = 1 matter containing quarks, leptons and Higgs bosons as well as N = 1 gauge boson multiplets belonging to the adjoint representation of a gauge group. The above is achieved by using techniques of applied supergravity[11-14]. Specifically in the work of Refs (11)-(12) one used the techniques of tensor calculus[15] to couple the minimal N = 1 supergravity[16] to N = 1 Yang-Mills and N = 1 matter.

Using the techniques of applied supergravity an SU(5) supergravity model was proposed in Ref 12. One of the remarkable aspects of supergravity unification is that spontaneous supersymmetry breaking can arise via the hidden sector[11,13,17,18]. Thus in supergravity unification the arbitrariness of having to add soft supersymmetry breaking terms by hand to the Lagrangian is removed, and a set of well defined soft SUSY breaking terms arise naturally in supergravity unification after spontaneous breaking. A second remarkable aspect of supergravity unification first demostrated in Ref 11, is that the spontaneous breaking of supersymmetry triggers the breaking of the electro-weak symmetry. This phenomenon implies that the scale of supersymmetry breaking and the scale of electro-weak symmetry breaking are comparable and thus one expects the masses of the superparticles to lie in the O(100 GeV) - O(1 TeV) mass range. Currently the most attractive mechanism of achieving the breaking of electro-weak symmetry is to have it arise radiatively via renormalization group effects[19-22].

The recent very precise LEP data gives indirect support to the ideas of grand unification and supersymmetry and specifically to supergravity Gut[11]. The LEP data gives the following values for the three coupling constants α_1, α_2 and α_3 of $SU(3)_C \times SU(2)_L \times U(1)$ at the Z boson mass:[23,24]

$$\begin{aligned} \alpha_1(M_Z) &= 0.016887 \pm 0.000040, \\ \alpha_2(M_Z) &= 0.03322 \pm 0.00025 \\ \alpha_3(M_Z) &= 0.118 \pm 0.007 \end{aligned} \tag{1.1}$$

Here $\alpha_1 \equiv (5/3)\ \alpha_y$ where α_y is the hypercharge coupling constant. Analysis of the above data indicates that using the renormalization group equations, the three coupling constants become equal to within one standard deviation for the supergravity SU(5) model[25-29]. Unification of the couplings can also be achieved within the non-supersymmetric framework[25,30]. However, the non-supersymmetric grand unification still suffers from the severe gauge hierarchy problem. For this reason we shall follow here only the supersymmetric option.

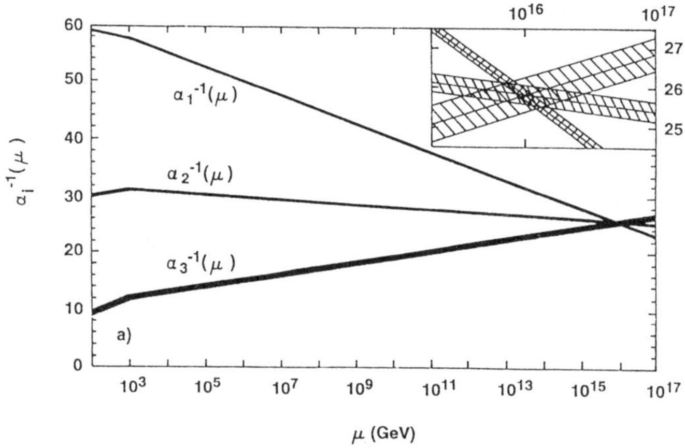

Fig. 1. Unification of the electro-weak and strong coupling constants α_1, α_2 α_3 in supergravity SU(5) theory using the very precise LEP data

As is well known proton stability puts stringent constraints on grand unified models. For example the non-supersymmetric SU(5) model[3] predicts that proton decays into $e^+\pi^o$ at a much more rapid rate than the current experimental bounds. Thus the model of Ref 3 predicts $\tau(p \to e^+\pi^o) \cong 4 \times 10^{29\pm2}$ yr. This lifetime is significally lower than the experimental limit from IMB[31] of 5.5×10^{32} yr (90%CL). For the supergravity SU(5) model a preliminary analysis yields[27]

$$\tau(p \to e^+\pi^o) = 2 \times 10^{35\,\pm\,2}\ \text{yr} \tag{1.2}$$

This result is consistent with the current experimental bounds and may become accessible in the next generation of proton decay experiments which aspire for a sensitivity of 10^{34}yr for this mode[33]. However, in supersymmetric SU(5) model the dominant proton decay proceeds via dimension five operators[34,35] (see Fig 2) which arise via exchange of superheavy Higgs triplets of SU(5) model. The major decay mode here is $p \to \bar{\nu}K^+$. The experimental lower bound for this decay mode is[32]

$$\tau(p \to \bar{\nu}K^+) > 9 \times 10^{31}\ \text{yr} \qquad (90\%\ \text{CL}) \tag{1.3}$$

The lower limit of Eq (1.3) acts as a stringent constraint on supergravity grand unified models. In fact the constraint of Eq (1.3) is found stringent enough that the (unflipped) SU(5) No-Scale[36-38] model is essentially found to be not viable[39]. For the standard SU(5) model[11] the proton decay constraints are found to be strong enough to significantly restrict the gut parameter space of the theory[40-42,64].

The outline of the remainder of this paper is as follows: In Sec. II we discuss the minimal supersymmetric standard mode (MSSM) and point out the ad hoc nature of its assumptions. In Sec III we review the basic formalism of supergravity grand unification and give the relevant equations needed for further discussion. In Sec IV we review radiative breaking of the electro-weak $SU(2)_L \times U(1)$ symmetry. In Sec V we discuss loop corrections to radiative breaking. Proton decay constraints are discussed in Sec VI. The analysis of SUSY mass spectra computed from radiative breaking under the combined constraints of proton lifetime limits and the experimental lower limit constraints from CDF and LEP1 is given in Sec VII. In Sec VIII we give an analysis of top quark and Higgs mass limits. Discussion of results and conclusions are given in Sec IX. This section also contains an addendum on the constraints of cosmology on the SU(5) supergravity model.

II. THE MINIMAL SUPERSYMMETRIC STANDARD MODEL

In this section we discuss how one builds models in global supersymmetry. To build globally supersymmetric models one needs two types of massless SUSY multiplets i.e. the chiral multiplets with spins (0, 1/2) corresponding to a complex scalar field $z(x)$ and a Weyl left-handed spinor $\chi(x)$, and the vector multiplet with spins (1/2, 1) corresponding to a Majorana gaugino field $\lambda(x)$ and a vector boson field $V_\mu(x)$. In building models of particle interactions one accomodates the quarks, leptons and the Higgs in chiral multiplets and the gauge vector bosons in vector multiplets. Thus for the supersymmetric extension of the Standard Model with the gauge group $SU(3)_C \times SU(2)_L \times U(1)_Y$ the particle content is as given in Tables 1 and 2.

Table 1: Components of $SU(3)_C \times SU(2)_L \times U(1)$ Vector Multiplets

Vector Multiplet	$J = \frac{1}{2}$	$j = 1$
$SU(3)_C$	λ^a, $a = 1..8$	V^a_μ, $a = 1...8$
$SU(2)_L$	λ^α, $a = 1, 2, 3$	B^a_μ, $a = 1, 2, 3$
$U(1)_y$	λ^y	B^y_μ

Table 2: Chiral Multiplets

Chiral Multiplets	$J = 0$	$J = \frac{1}{2}$
lepton multiplets	$\begin{pmatrix} \tilde{\nu}_{iL} \\ \tilde{e}_{iL} \end{pmatrix}$, $\tilde{e}_{iR}$	$\begin{pmatrix} \nu_{iL} \\ e_{iL} \end{pmatrix}$, e_{iR}
quark multiplets	$\begin{pmatrix} \tilde{u}_{iL} \\ \tilde{d}_{iL} \end{pmatrix}$, $\tilde{u}_{iR}$, $\tilde{d}_{iR}$	$\begin{pmatrix} u_{iL} \\ d_{iL} \end{pmatrix}$, u_{iR}, d_{iR}
Higgs multiplets	$H_1 = \begin{pmatrix} H_1{}^0 \\ H_1{}^- \end{pmatrix}$, $H_2 = \begin{pmatrix} H_2{}^+ \\ H_2{}^0 \end{pmatrix}$	$\tilde{H}_1 = \begin{pmatrix} \tilde{H}_1{}^0 \\ \tilde{H}_1{}^- \end{pmatrix}$, $\tilde{H}_2 = \begin{pmatrix} \tilde{H}_2{}^+ \\ \tilde{H}_2{}^0 \end{pmatrix}$

Dynamics in a supersymmetric theory is determined by the gauge interactions and by the superpotential function W which is a function of the scalar fields of the theory i.e. $W=W(z_a)$. The gauge interactions of the supersymmetric theory consist of the usual gauge interactions of the vector bosons and in addition the following set of interactions:

(a) Interactions of gauginos which are given by

$$L_\lambda = -i\sqrt{2}\, g_A \bar{\lambda}^A z_b{}^\dagger (T^A)_{ba} \chi^a + \text{h.c.} \tag{2.1}$$

where T^A are the group generations in the representation of the chiral multiplet $\{z_a, \chi^a\}$ and g_A is the corresponding coupling constant.

(b) D terms in the effective potential given by

$$V_D = \frac{1}{2} g_A^2 \; D_A D_A^\dagger \quad ; \quad D_A = z_a^\dagger (T^A)_{ab} z_b \tag{2.2}$$

In addition the superpotential contributes the following set of terms:

(c) A bosonic contribution to the tree potential of the form

$$V_W = \sum_a \left| \frac{\partial W}{\partial z_a} \right|^2 \tag{2.3}$$

(d) A Yukawa interaction among the chiral fermions given by:

$$L_y = -\frac{1}{2} \sum_{a,b} \left(\bar{\chi}^{ac} \frac{\partial^2 W}{\partial z_a \partial z_b} \chi^b + \text{h.c.} \right) \tag{2.4}$$

The total tree effective potential of the theory is then given by the form:

$$V = V_W + V_D + V_{SB} \tag{2.5}$$

where V_{SB} denotes the supersymmetry breaking part of the potential. For the standard SUSY Model the general form of W is

$$W = \mu H_1 H_2 + \lambda_{ij}^{(e)} \ell_i H_1 e_j^c + \lambda_{ij}^{(u)} q_i H_2 u^c + \lambda_{ij}^{(d)} q_i H_1 d_j^c \tag{2.6}$$

With the determination of W given by Eq. (2.6), V is completely known except for V_{SB}. As pointed out in Sec. I, there is no fundamental way to deduce a phenomenologically acceptable form of V_{SB} in global supersymmetry. If one uses the principle that the addition of V_{SB} not spoil the cancellation of the quardratic divergence in the scalar boson masses, then V_{SB} is restricted to soft SUSY breaking terms of the following form:

$$V_{SB} = m_{ab}^2 z_a^* z_b + [A_{ij}^{(e)} \lambda_{ij}^{(e)} \ell_i H_1 e^c{}_j + A_{ij}^{(u)} \lambda_{ij}^{(u)} q_i H_2 u_j^c$$

$$+ A_{ij}^{(d)} \lambda_{ij}^{(d)} q_i H_1 d_j^c + B\mu H_1 H_2 + \text{h.c.}] + \tilde{m}_A \bar{\lambda}^A \lambda^A \tag{2.7}$$

where m^2_{ab} is the scalar boson $(\text{mass})^2$ matrix, $A^{(e)}_{ij}$, $A^{(d)}_{ij}$, $A^{(u)}_{ij}$ and B are Polonyi constants and $\tilde{m}_A$ are Majorana masses. These constitute a huge number of arbitrary parameters. For an $SU(3)_C \times SU(2)_L \times U(1)_Y$ invariant theory the number of parameters in V_{SB} are 137 if we allow for complex parameters and 87 if we restrict the parameters to be real. In the so-called Minimal Supersymmetric Standard Model (MSSM) one places the following restrictions on the arbitrary parameters: (a) The scalar masses are assumed degenerate (except possibly for the third generation squarks), (b) the Polonyi matrices $A^{(e)}_{ij}$ etc are all assumed diagonal, and (c) the gaugino masses are assumed to be scaled by the coupling constants i.e.

$$\tilde{m}_1 : \tilde{m}_2 : \tilde{m}_3 = \alpha_1 : \alpha_2 : \alpha_3 \tag{2.8}$$

These assumptions reduce drastically the arbitrary number of parameters. We see from the preceeding discussion that the MSSM is arrived at in a rather adhoc fashion. <u>However, in our analysis later we shall find that the MSSM arises as a low energy approximation from the supergravity grand unified model</u>.

III. SUPERGRAVITY GRAND UNIFICATION

As discussed in Sec. I and II there is no phenomenologically acceptable way of breaking supersymmetry spontaneously in a globally supersymmetric framework. To achieve a spontaneous breaking of supersymmetry one must promote global supersymmetry to a local supersymmetry. For the purpose of model building one needs to couple N=1 supergravity multiplet, with N=1 matter and N=1 Yang-Mills.[12-14] The full form of the resulting Lagrangian is rather complicated. We highlight here the relevant features. The Lagrangian depends on three functions. These are the (i) the superpotential $W(z_i)$ which is a function only of the complex scalar fields, (ii) the Kahler potential $d(z_i, z_i^*)$ which is a function of the complex scalar fields and their conjugates and (iii) the gauge kinetic function $f_{\alpha\beta}(z_i, z_i^*)$ which is also a function of the fields and their conjugates. The functions W and D enter in a specific combination

$$G = -\ell n[\kappa^6 WW^*] - \kappa^2 d \tag{3.1}$$

where $\kappa = (8\pi G_N)^{1/2} = 0.41 \times 10^{-18}$ Gev and G_N is the Newtonian constant. Some of the relevant parts of the Lagrangian are as follows:

(i) The Gauge Kinetic Energy

$$-\frac{1}{4}\,\mathrm{Ref}_{\alpha\beta}\,F_{\mu\nu}{}^{\alpha}\,F_{\mu\nu}{}^{\beta} \tag{3.2}$$

(ii) The Effective Potential

$$V=\frac{-1}{\chi^4}\,e^{-G}\left[(G^{-1})^a{}_b G,_a G,^b+3\right]+\frac{g^2}{2}\,(\mathrm{Ref}^{-1}{}_{\alpha\beta})D_\alpha D_\beta \tag{3.3}$$

where D_α is given by

$$D_\alpha=-\frac{1}{\chi^4}\,G,_a\,(T^\alpha)_{ab}\,z_b \tag{3.4}$$

(iii) The Gluino Mass term

$$\frac{1}{4}\,e^{-G/2}(G^{-1})^a{}_b\,G,^a f^*{}_{\alpha\beta,b}\,\bar{\lambda}^{\alpha}\lambda^{\beta} \tag{3.5}$$

<u>Breaking of supersymmetry</u>: Unlike global supersymmetry the formalism of applied supergravity allows for a spontaneous breaking of supersymmetry via a superhiggs effect.[18] It is found convenient to decompose the scalar fields into matter fields z_i and superhiggs field z i.e. $\{z_a\}=\{z_i, z\}$. In order to protect the low energy physics from mass growths $O(M_{P\ell})$, one postulates the following separable form for the superpotential[11]

$$W(z_a)=W(z_i)+W_{\mathrm{Polonyi}}(z) \tag{3.6}$$

where $W_{\mathrm{Polonyi}}(z)$ corresponds to the "hidden" sector and $W(z_i)$ to the "visible" sector. Because of the separable form of Eq. (3.6) the interaction between the hidden and the visible sectors are gravitational in nature, and thus although the Polonyi field z develops a VEV of $O(M_{P\ell})$, the visible sector is protected from superheavy mass growths. The simplest choice of W_{Poloyni} is $m^2(z+c)$, where m and c are constants.

After spontaneous breaking a mass term is generated for the gravitino field with mass

$$m_{3/2}=\frac{1}{2}\,\chi^2 W(z_o)\exp\left(\frac{\chi^2}{4}\,d\right) \tag{3.7}$$

where $z_0=\langle z\rangle$. Additionally the gauginos acquire a mass term via Eq. (3.5) since in general $f_{\alpha\beta}$ may have a non-trivial z-dependence i.e.

$$f_{\alpha\beta} = \delta_{\alpha\beta}(1 + xz + ..) \tag{3.8}$$

Effective theory below M_G: We assume that we have a grand unified supergravity theory with the gauge invariance of a semi-simple GUT group G. After the spontaneous breaking of supersymmetry at the Planck scale $M_{P\ell}$ as described above, one proceeds to the energy scale M_G where the grand unification group G breaks spontaneously to the Standard Model gauge group i.e. $G \to SU(3)_C x SU(2)_L x U(1)_Y$ at M_G. After the Goldstone and the Higgs phenomena one finds at M_C that the vector bosons corresponding to the coset space $G/SU(3)_C x SU(2)_L x U(1)_Y$ and some of the scalar fields develop superheavy masses $O(M_G)$. To achieve a low energy effective theory one must integrate on the Polonyi fields[11,17,43,44] as well as on the superheavy fields[11,43,44]. At energy just below the gut scale the effective Lagrangian is given by[43,44]

$$V_{eff} = [\Sigma \, |\frac{\partial W_{eff}}{\partial z_i}|^2 + V_D]$$

$$+ \{m_o{}^2 z_i z_i{}^\dagger + (A_o W^{(3)} + B_o W^{(2)} + h.c.)\} \tag{3.9}$$

where W_{eff} is the low energy effective superpotential with the decomposition

$$W_{eff} = W^{(2)} + W^{(3)} \tag{3.10}$$

and where $W^{(2)}$ and $W^{(3)}$ are the quadratic and the cubic parts of W_{eff}. In addition to Eq. (3.9) one has a gaugino mass term of the form

$$L^{mass}_{gaugino} = -m_{1/2} \, \bar{\lambda}^\alpha \lambda^\alpha \tag{3.11}$$

The effective theory given by Eqs. (3.9)-(3.11) depends on four arbitrary soft SUSY breaking parameters:

$$m_o, \; m_{1/2} \; , \; A_o, \; B_o \tag{3.12}$$

We see then that there is a drastic reduction in the number of soft SUSY breaking parameters that one could have in the global SUSY theory (i.e. there are 137 arbitrary parameters for the CP violating case and 87 arbitrary parameters for the CP preserving case for the globally supersymmetric theory).

A Specific SU(5) GUT Model: We consider now a specific form for an

SU(5) gut theory which contains the particle spectrum discussed in Tables 1 and 2. We take for W the form

$$W = W_Y + W_G \tag{3.13}$$

where W_Y is the Yukawa part and W_G is the GUT part. W_y is given by

$$W_Y = \epsilon_{uvwxy} H_1^u M_i^{vw} f_{1ij} M_j^{xy} + H_{2x} \bar{M}_{iy} f_{2ij} M_j^{xy} \tag{3.14}$$

and where $M_i^{xy}=\underline{10}$ and $\bar{M}_{iy}=\underline{\bar{5}}$ are matter fields and $H_1^x=5$ and $H_{2x}=\bar{5}$ are Higgs fields. A simple model for W_G is[8]

$$W_G = \lambda_1[\frac{1}{3}\,\mathrm{Tr}\,\Sigma^3 + \frac{1}{2}\,M\mathrm{Tr}\Sigma^2] + \lambda_2 H_{2x}(\Sigma^x{}_y + 3M'\delta^x_y)H_1^y \tag{3.15}$$

and where $\Sigma^x{}_y=\underline{24}$ of Higgs. W_{GUT} generates spontaneous breaking of SU(5) and $SU(5)\to SU(3)_C \times SU(2)_L \times U(1)_Y$ corresponding to

$$\mathrm{diag}\langle\Sigma^x{}_y\rangle = M(2, 2, 2, -3, -3) \tag{3.16}$$

For this model W_2 and W_3 of the low energy effective theory are given by:

$$W_2 = \mu_o H_1 H_2 \tag{3.17}$$

$$W_3 = \lambda_{ij}^{(e)} \ell_i H_1 e_j^c + \lambda_{ij}^{(u)} q_i H_2 u_j^c + \lambda_{ij}^{(d)} q_i H_1 d_j^c \tag{3.18}$$

where H_1 and H_2 are the Higgs doublet components of H_1^x and H_{2x}. We note that this theory now depends on five GUT parameters M_0, $m_{1/2}$, A_0, B_0 and μ_0 in addition to the usual yukawa couplings.

We discuss briefly the issue of fine tuning needed to acheive a light Higgs doublet. In order to acheive a light Higgs doublet one fine tunes M′ so that $M'=M+\mu_0/3\lambda_2$. Since the fine tuning is affected in the superpotential it is technically natural due to the no-renormalization theorems.[45] There are also other possibilities where light Higgs arise without the necessity of fine tuning. An example of this is the so called missing partner mechanism[46–48] where one uses 50, $\overline{50}$ and 75 dimensional representations with the 75 breaking the SU(5) symmetry. Finally the finetuning of the type discussed above might naturally arise in the framework of a higher symmetry, e.g. if the GUT sector possesed an SU(6) symmetry,[49] or an SO(10) symmetry.[49,50] The Higgs doublet fields in the models appear as pseudo Goldstone bosons and are thus automatically light.

IV. RADIATIVE BREAKING OF THE ELECTROWEAK SYMMETRY

As indicated in Sec. I an interesting feature of supergravity models is that spontaneous breaking of electro-weak symmetry can arise from soft SUSY breaking terms. The most attractive mechanism for realizing this is via radiative breaking. If we confine ourselves to vacua that do not break either the color or charge invariance then the vacuum state is determinded by minimizing the effective potential V_1 which depends only on the Higgs VeVs:

$$V_1 = V_0 + \Delta V_1 \tag{4.1}$$

where $V_0(Q)$ is the renormalization group improved tree potential at scale Q and is given by

$$V_0(Q) = m_1^{\,2}(Q)|H_1|^2 + m_2^{\,2}\,(Q)\;|H_2|^2 - m_3^{\,2}\,(Q)\;(H_1H_2 + \text{h.c.}) + \frac{1}{8}\,(g_2^2 + g_Y^2)(|H_1|^2-|H_2|^2)^2 \tag{4.2}$$

and $\Delta V_1(Q)$ is the one loop correction[51]

$$\Delta V_1 = \frac{1}{64\pi^2}\,\text{STr}\,[M^4(\log\frac{M^2}{Q^2} - \frac{3}{2})] \tag{4.3}$$

In the radiative breaking analysis one runs the renormalization group equations on the gauge couplings, Yukawa couplings and the soft SUSY breaking parameters downwards starting at M_G. Spontaneous symmetry breaking occurs when below a scale Q_0 the symmetric point of the potential is no longer the absolute minimum of the potential. Thus for scales $Q<Q_0$ where $D(Q_0)=0$ symmetry breaking occurs when

$$D(Q) = m_1^{\,2}(Q)m_2^2(Q) - m_3^{\,4}(Q) < 0 \tag{4.4}$$

The analysis of radiative breaking also assumes that the potential is bounded from below both at the GUT scale as well as at the scale Q where one is minimizing the potential. The latter condition requires that

$$m_1^2(Q) + m_2^2(Q) - 2m_3^2\ (Q) \geq 0 \tag{4.5}$$

Electro-weak symmetry breaking is then characterized by the following relations:

$$\frac{1}{2} M_Z^2 = \frac{\mu_1^2 - \mu_2^2 \tan^2\beta}{\tan^2\beta - 1}\ , \ \sin 2\beta = \frac{2m_3^3}{\mu_1^2 + \mu_2^2} \tag{4.6}$$

where

$$\mu_i^2 = m_i^2 + \partial\Delta V_1/\partial v_i^2\ , \ \tan\beta = \langle H_2\rangle/\langle H_1\rangle \tag{4.7}$$

and $v_i = \langle H_i\rangle$. The m_i^2 may be viewed as functions of the running parameter $t = \ell n\ (M_G^2/Q^2)$ and are related to m_{H_i} and μ as follows:

$$m_i^2\ (t) = m_{H_i}^2\ (t) + \mu^2(t);\ i=1,2 \tag{4.8}$$

$$m_3^2\ (t) = -\ B(t)\mu(t) \tag{4.9}$$

Thus m_i^2 obey the following GUT boundary conditions:

$$m_i^2\ (M_G) = m_0^2 + \mu_0^2\ ;\ m_3^2 = -\ B_0\mu_0\ ;\ i=1,2 \tag{4.10}$$

Returning to the criterion stated at the beginning of this section that spontaneous symmetry breaking not break charge or color invariance, it should be noted that a sufficient set of conditions to avoid this undesirable possibility are not known. However, some of the necessary conditions to avoid color and charge breaking are known. For example for the third generation the squarks masses must satisfy the inequalities[52,53]

$$A_t^2 < 3(m_{\tilde{t}_L}^2 + m_{\tilde{t}_R}^2 + \mu_2^2) \tag{4.11}$$

$$A_b^2 < 3(m_{\tilde{b}_L}^2 + m_{\tilde{b}_R}^2 + \mu_1^2) \tag{4.12}$$

and similar relations hold for the first two generation of squarks, while the slepton masses must satisfy the inequalities:

$$A_{\ell}^{2} < 3\ (m_{\tilde{\ell}_L}^{2} + m_{\tilde{\ell}_R}^{2} + \mu_1^2) \tag{4.13}$$

$$m_{H_1}^{2} + m_{\tilde{\ell}_L}^{2} > 0 \tag{4.14}$$

The relevant renormalization group equations governing the evolution of the gauge couplings, Yukawa couplings and soft SUSY breaking parameters are given in papers of Ref. 21. We discuss below the solutions to the renormalization group equations for soft SUSY breaking masses[54].

Gauginos: Integration of the one-loop RGE for the gaugino masses gives

$$\tilde{m}_i(t) = \frac{\alpha_i(t)}{\alpha_G}\, m_{\frac{1}{2}} \quad ; \; i=1,2,3 \tag{4.15}$$

Eq. (4.15) implies that the $SU(3)_C$, $SU(2)_L$ and $U(1)$ gauginos satisfy the relation[55,56]

$$\tilde{m}_1 : \tilde{m}_2 : \tilde{m}_3 = \alpha_1 : \alpha_2 : \alpha_3 \tag{4.16}$$

This relation is postulated in MSSM but is deduced in supergravity unification[55,56].

Charginos: Chargino masses are given by the relation

$$m_{\tilde{W}_{1,2}} = \frac{1}{2}\ |[4\nu_+^{2} + (\mu - \tilde{m}_2)^2]^{1/2} \mp [4\nu_-^2 + (\mu + \tilde{m}_2)^2]^{1/2} \tag{4.17}$$

where

$$\sqrt{2}\ \nu_\pm = M_W\ (\sin\beta \pm \cos\beta) \tag{4.18}$$

Neutralinos: There are four neutralino mass eigen-states and they are given by solutions of the quartic equation $f(\lambda) = 0$ where $f(\lambda)$ is given by

$$f(\lambda) = \lambda^4 - (m_{\tilde{\gamma}} + m_{\tilde{Z}})\lambda^3 - (M_Z^2 + \mu^2 + m^2_{\tilde{\gamma}\tilde{Z}} - m_{\tilde{\gamma}} m_{\tilde{Z}})\lambda^2$$
$$+ [(m_{\tilde{\gamma}} - \mu \sin 2\beta) M_Z^2 + (m_{\tilde{\gamma}} + m_{\tilde{Z}}) \mu^2]\lambda$$
$$+ (\mu^2 (m^2_{\tilde{\gamma}\tilde{Z}} - m_{\tilde{\gamma}} m_{\tilde{Z}}) + \mu m_{\tilde{\gamma}} M_Z^2 \sin 2\beta) \quad (4.19)$$

where

$$m_{\tilde{\gamma}} = \tilde{m}_2 \sin^2\theta_W + \tilde{m}_1 \cos\theta_W, \ \tilde{m}_Z = \tilde{m}_2 \cos^2\theta_W + \tilde{m}_1 \sin^2\theta_W$$
$$\tilde{m}_{\gamma Z} = \sin\theta_W \cos\theta_W (\tilde{m}_1 - \tilde{m}_2) \quad (4.20)$$

Sleptons: Solutions to the RGE for slepton masses give the following:

$$m^2_{\tilde{e}_{L_i}} = m_0^2 + m_{e_i}^2 + (\tfrac{3}{2}\tilde{\alpha}_2 f_2 + \tfrac{1}{2}\tilde{\alpha}_1 f_1) m_{1/2}^2$$
$$+ (-\tfrac{1}{2} + \sin^2\theta_W) \cos 2\beta M_Z^2 \quad (4.21)$$

$$m^2_{\tilde{e}_{R_i}} = m_0^2 + m_{e_i}^2 + 2\tilde{\alpha}_1 f_1 \, m_{1/2}^2 - \sin^2\theta_W \cos 2\beta \, M_Z^2 \quad (4.22)$$

$$m^2_{\tilde{\nu}_{L_i}} = m_0^2 + (\tfrac{3}{2}\tilde{\alpha}_2 \, f_2 + \tfrac{1}{2}\tilde{\alpha}_1 \, f_1) m_{1/2}^2 + \tfrac{1}{2} \cos 2\beta \, M_Z^2 \quad (4.23)$$

where $\beta_i = b_i \tilde{\alpha}_i$, $\tilde{\alpha}_i = \alpha_i/4\pi$ and b_i are determined by

$$b_i = \begin{pmatrix} 0 \\ -6 \\ 9 \end{pmatrix} + N_{fam} \begin{pmatrix} 2 \\ 2 \\ 2 \end{pmatrix} + N_{Higgs} \begin{pmatrix} 3/10 \\ 1/2 \\ 0 \end{pmatrix} \quad (4.24)$$

and f_i are defined by

$$f_i = t \frac{2 - \beta_i t}{(1 + \beta_i t)^2} \quad (4.25)$$

Generations i=1,2 squarks: The soft SUSY breaking masses for generation i=1,2 squarks are:[12,20,21]

$$m_{\tilde{u}_{L_i}}^2 = m_0^2 + m_{u_i}^2 + (\tfrac{8}{3}\,\tilde{\alpha}_3 f_3 + \tfrac{3}{2}\,\tilde{\alpha}_2 f_2 + \tfrac{1}{18}\,\tilde{\alpha}_1 f_1)\, m_{1/2}^2$$

$$+ (\tfrac{1}{2} - \tfrac{2}{3}\sin^2\theta_W)\cos 2\beta\, M_Z^2 \qquad (4.26)$$

$$m_{\tilde{d}_{L_i}}^2 = m_0^2 + m_{d_i}^2 + (\tfrac{8}{3}\,\tilde{\alpha}_3 f_3 + \tfrac{3}{2}\,\tilde{\alpha}_2 f_2 + \tfrac{1}{18}\,\tilde{\alpha}_1 f_1) m_{1/2}^2$$

$$+ (-\tfrac{1}{2} + \tfrac{1}{3}\sin^2\theta_W)\cos 2\beta\, M_Z^2 \qquad (4.27)$$

$$m_{\tilde{u}_{R_i}}^2 = m_0^2 + m_{u_i}^2 + (\tfrac{8}{3}\,\tilde{\alpha}_3\, f_3 + \tfrac{8}{9}\,\tilde{\alpha}_1 f_1) m_{1/2}^2$$

$$+ (\tfrac{2}{3}\sin^2\theta_W)\cos 2\beta\, M_Z^2 \qquad (4.28)$$

$$m_{\tilde{d}_{R_i}}^2 = m_0^2 + m_{d_i}^2 + (\tfrac{8}{3}\,\tilde{\alpha}_3 f_3 + \tfrac{2}{9}\,\tilde{\alpha}_1 f_1) m_{1/2}^2$$

$$+ (-\tfrac{1}{3}\sin^2\theta_W)\cos 2\beta\, M_Z^2 \qquad (4.29)$$

We can make two observations from the above relations. First me notice that for the situation when m_0^2, $m_{1/2}^2 \gg M_Z^2$, the masses of the squarks are nearly degenerate. Thus in this limit one recovers the result which is assumed in the MSSM. Second, we notice from above that the difference between the first and second generation squark masses is

$$m_{\tilde{c}}^2 - m_{\tilde{u}}^2 = m_c - m_u^2 \qquad (4.30)$$

Thus one has a super-GIM mechanism contained in this theory for suppression of the flavor changing neutral currents (FCNC).

3rd Generation Squarks: For the b-squarks one has the relations

$$m_{\tilde{b}_R}^2 = m_0^2 + m_b^2 + (\tfrac{8}{3}\,\alpha_3 f_3 + \tfrac{2}{9}\,\tilde{\alpha}_1 f_1) m_{1/2}^2 - \tfrac{1}{3}\sin^2\theta_W \cos 2\beta\, M_Z^2 \qquad (4.31)$$

$$m_{\tilde{b}_L}^2 = \tfrac{1}{2}\, m_0^2 + m_b^2 + \tfrac{1}{2}\, m_U^2 + (\tfrac{4}{3}\,\tilde{\alpha}_3 f_3 + \tfrac{1}{9}\,\tilde{\alpha}_1 f_1) m_{1/2}^2$$

$$+ (-\tfrac{1}{2} + \tfrac{1}{3}\sin^2\theta_W)\cos 2\beta\, M_Z^2 \qquad (4.32)$$

where m_U is as defined in Ref. 21. For the third generation up-squarks (stops) one must take account of mixing between $\tilde{t}_L$, $\tilde{t}_R$ due to t-quark yukawa couplings given by the matrix:

$$\begin{pmatrix} m_{\tilde{t}_L}{}^2 & m_t(A_t + \mu\cot\beta) \\ m_t(A_t + \mu\cot\beta) & m^2_{\tilde{t}_R} \end{pmatrix} \tag{4.33}$$

where

$$m_{\tilde{t}_L} = m_Q{}^2 + m_t{}^2 \left(-\frac{1}{2} + \frac{2}{3}\sin^2\theta_W\right)M_Z{}^2\cos\beta \tag{4.34}$$

$$m_{\tilde{t}_R}{}^2 = m_U{}^2 + m_t{}^2 + \left(-\frac{2}{3}\sin^2\theta_W\right) M_Z{}^2\cos 2\beta \tag{4.35}$$

where $m_U{}^2$ and $m_Q{}^2$ are as defined in Ref. 21.

Higgs Bosons: After electro-weak symmetry breaking one is left with one charged Higgs, one neutral CP odd Higgs and two neutral CP even Higgs. The tree masses for the charged Higgs and for the neutral CP even Higgs bosons are

$$m_{H^\pm}{}^2 = m_A{}^2 + M_W{}^2 \tag{4.34}$$

$$m_A{}^2 = \frac{m_3{}^2}{\sin 2\beta} \qquad \text{CP odd Higgs} \tag{4.35}$$

For computation of neutral CP even Higgs one must include, at least for the lighter Higgs, one loop corrections to the Higgs mass. To one loop order one has[57]

$$m^2_{h,H} = \frac{1}{2}\left[M_Z{}^2 + m + \varepsilon \mp \{(M_Z{}^2 + M_A{}^2 + \varepsilon)^2 - 4m_A{}^2M_Z{}^2\cos 2\beta + \varepsilon_1\}^{\frac{1}{2}}\right] \tag{4.36}$$

where

$$\varepsilon = \mathrm{Tr}\,\Delta\ ;\ \varepsilon_1 = -4(\mathrm{Tr}\nu\Delta + \det\Delta)$$

$$\nu_{11} = s^2M_Z{}^2 + c^2m_A{}^2;\ \nu_{22} = c^2M_Z{}^2 + s^2m_A{}^2$$

$$\nu_{12} = \nu_{21} = sc(M_Z{}^2 + m_A{}^2)$$

$$\Delta_{11} = x\mu^2y^2z\ ,\ \Delta_{12} = \Delta_{21} = x\mu y(w + A_t yz)$$

$$\Delta_{22} = x\,(v + 2\,A_t yw + A_t{}^2y^2z) \tag{4.37}$$

and

$$x = \frac{3\alpha_2}{4\pi} \frac{m_t^4}{M_W^2 s^2} \; , \; y = \frac{A_t + \mu\cot\beta}{m_{\tilde{t}_1}^2 - m_{\tilde{t}_2}^2} \; , \; z = 2 - \omega \frac{m_{\tilde{t}_1}^2 + m_{\tilde{t}_2}^2}{m_{\tilde{t}_1}^2 - m_{\tilde{t}_2}^2}$$

$$\omega = \ell n \frac{m_{\tilde{t}_1}^2}{m_{\tilde{t}_2}^2} \; ; \; v = \ell n \left(\frac{m_{\tilde{t}_1}^2 m_{\tilde{t}_2}^2}{m_t^4} \right) \tag{4.38}$$

where $(s,c)=(\sin\beta, \cos\beta)$. One notes that terms containing x have m_t^4 dependence and thus can produce large effects when m_t is large and the Higgs is light. There are also loop corrections for m_A and $m_{H^\pm}$ but these loop effects are generally small except when A and $H^\pm$ are light. However, this is generally not the case for the solutions satisfying laboratory and proton stability constraints.

V. LOOP CORRECTIONS TO RADIATIVE BREAKING

One loop corrections to symmetry breaking enter Eq. (4.6) via the term Σ^a in μ_a^2 i.e. $\mu_a^2 = m_a^2 + \Sigma^a$, a=1,2, where $\Sigma^a = \partial V_1/\partial v_a$ and is given by

$$\Sigma^a = \frac{1}{32\pi^2} \sum_i (-1)^{2j_i} n_i M_i^2 \ell n \left(\frac{M_i^2}{e^{\frac{1}{2}} Q^2} \right) \frac{\partial M_i^2}{\partial v_a^2} \tag{5.1}$$

Here j_i is the spin and n_i the multiplicity of particle of type i entering the sum of Eq. (5.1). A measure of the loop effects is given by R_μ where

$$R_\mu = \Delta\mu/\mu_t \tag{5.2}$$

and where μ_t is the tree value and $\Delta\mu = \mu_\ell - \mu_t$ is the one loop correction to the tree. $\Delta\mu$ is determined from Eq. (4.6) since

$$\Delta\mu^2 = \mu_\ell^2 - \mu_t^2 = (\Sigma^1 - \Sigma^2 \tan^2\beta)/(\tan^2\beta - 1) \tag{5.3}$$

Because of the large number of states in the sum in Eq. (5.1), there is the possibilty that the loop correction may indeed be large[58]. Thus if a single state contributed ≈ 1% to the loop, one may end up with ≈ 30% correction to the tree results. However, one finds that for almost all the parameter space there is a significant cancellation among various loop corrections, reducing the net loop contribution to only a few percent of the tree[59]. By and large these cancellations are not of the usual supersymmetric type i.e. between bose and fermi contributions. Rather major cancellations occur in the bose sector alone. For clarity, let us consider the limit $m_0^2 >> M_Z^2$, $m_{1/2}^2$ and the first two generations of squarks and sleptons where one may neglect the mixings between the left and right fields. In this limit one finds for the first two generation of squarks[59].

$$\Sigma^a_{\tilde{q}_i} = -\frac{s_a}{8\pi\cos^2\theta_W}(\sin^2\theta_W Y_i - \cos^2\theta_W T_{3i})M^2_{\tilde{q}_i}\ell n\left(\frac{M^2_{\tilde{q}i}}{e^{\frac{1}{2}}Q^2}\right) \tag{5.4}$$

where $s_a=(1,-1)$ for $a=1,2$, $i=u_L$, d_L, u_R, d_R and (T_{3i},Y_i) are the $(SU(2)_L, U(1)_Y)$ quantum numbers. For $M_{\tilde{q}_i}$ the RG analysis gives

$$M^2_{\tilde{q}_i} = m_0^2 + (\sin^2\theta\ Y_i - \cos^2\theta\ T_{3i})M_Z^2\cos 2\beta + \gamma_i m^2_{1/2} \tag{5.5}$$

In Eq. (5.5) γ_i is defined by $\gamma_{3i}=\gamma_{3i}+\gamma_{2i}+\gamma_{1i}$ where $\gamma_{3,2,1}$ are the SU(3), SU(2) and U(1) gaugino contributions and $\gamma_{3i}=\gamma_3$ is independent of the squark index i. We note that $M_{\tilde{q}}^2$ is of the form $Mq_i^2=m_0^2+\Delta_i^2$. Expanding Eq. (5.4) in the limit $m_0>>\Delta_i^2$ one gets

$$\Sigma^a_{\tilde{q}_i} = -\frac{s_a}{8\pi\cos^2\theta_W}(\sin^2\theta_W Y_i - \cos^2\theta_W T_{3i})[m_0^2\ell n\left(\frac{m_0^2}{e^{\frac{1}{2}}Q^2}\right)$$

$$+ \Delta_i^2\left\{1 + \ell n\left(\frac{m_o^2}{e^{\frac{1}{2}}Q^2}\right)\right\} + O\left(\frac{\Delta^4}{m_0^2}\right)] \tag{5.6}$$

Now we note that the trace over all the squarks of a given generation cancels the largest term in Eq. (5.6) proportional to $m_0^2\ \ell n\ m_o^2$. A similar analysis holds for all three generations of sleptons and thus one has left over in the trace of Σ^a for the first two generations of squarks and three generation of sleptons, only small terms:

$$\Sigma^a \text{ (2 generations } \tilde{q}, \text{ 3 generations } \tilde{\ell}) \cong$$

$$- \frac{s_a \alpha_2}{8\pi} (2.32 \cos 2\beta M_Z^2 - 0.26 m_{1/2}^2) \left(1 + \ell n \frac{m_0^2}{e^{\frac{1}{2}} Q^2}\right) + O\left(\frac{\Delta^4}{m_0^2}\right) \tag{5.7}$$

A similar but only partial cancellation due to the large L-R mixing from the top mass occurs in the case of the third generation of squarks. Further there are also the usual SUSY type cancellations of the charged Higgs and the chargino contributions and between the neutral Higgs and the neutralino contributions.

A numerical illustration of the large cancellations that occur in Eq. (5.7) is given in Table 3 where contributions to Σ^1 from the first two generation of squarks and from three generation of sleptons are shown (contributions to Σ^2 are just opposite in sign to Σ^1). One finds that while each of the contributions are substantial their overall sum is negligible due to cancellations.

Table 3: Contributions to one loop corrections Σ^1 for $m_{\tilde{g}} = 248$ GeV, $m_0 = 600$ GeV, $A_t = 0.0$, $\tan\beta = 1.732$ and $m_t = 130$ GeV.

Particles	Mass (GeV)	Σ^1 (GeV^2)
$\tilde{u}_L$ (gen. 1, 2)	630.1	1587.2
$\tilde{u}_R$ (gen. 1, 2)	628.4	711.7
$\tilde{d}_L$ (gen. 1, 2)	632.6	-1965.8
$\tilde{d}_R$ (gen. 1, 2, 3)	628.9	-356.6
Σ_q (gen. 1 + 2)		-23.5
$\tilde{e}_L$ (gen. 1, 2, 3)	603.9	-1650.9
$\tilde{e}_R$ (gen. 1, 2, 3)	601.7	-1430.3
ν (gen. 1, 2, 3)	601.3	3059.8
$\Sigma_{\tilde{\ell}}$ (gen. 1 + 2 + 3)		-21.4

VI. PROTON STABILITY

We begin with a discussion of the unification of couplings, and of the SUSY scale M_S and the GUT scale M_G. The current fits to the LEP data without threshold corrections [using Ref. (30)] give

$$\log_{10}M_S = 2.368 + 17.4062\left(1 - \frac{\alpha_s}{0.118}\right) \tag{6.1}$$

$$\log_{10}M_G = 16.186 + 5.719\left(1 - \frac{\alpha_s}{0.118}\right) \tag{6.2}$$

where M_S and M_G are given in GeV units. For 1σ=0.007 variation one has for M_S and M_G

$$M_S = 10^{2.37\pm1.03}\text{GeV}; \qquad M_G = 10^{16.185\pm0.34}\text{GeV} \tag{6.3}$$

The analysis above shows an anti-correlation between M_S and M_G as exhibited in Table 4.

Table 4: Anti-correlation of M_S and M_G

α_s	M_S (GeV)	M_G (GeV)
0.107	9.79×10^3	4.49×10^{15}
0.111	2.52×10^3	7.03×10^{15}
0.113	1.28×10^3	8.77×10^{15}
0.118	233.51	1.53×10^{16}
0.125	21.66	3.35×10^{16}

From Table 4 we note that the anti-correlation of M_S and M_G can be expressed by the following relation:

$$M_G = 10^{16.185}\left(\frac{10^{2.37}}{M_S}\right)^{0.33} \tag{6.4}$$

We turn now to the analysis of proton stability. As discussed in Sec. I, the most dominant proton decay in SUSY theory is from dimension five operators when these operators are dressed by chargino, neutralino and gluino exchange diagrams. (See Fig. 2). The most dominant decay mode in this case is the $p \to \bar{\nu} K^+$ mode whose decay width is given by

$$\Gamma(p \to \bar{\nu} K^+) = \sum_{i=e,\mu,\tau} \Gamma(p \to \nu_i K^+) \tag{6.5}$$

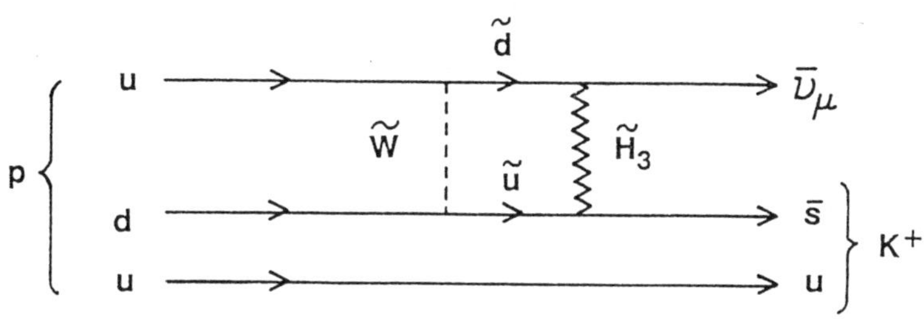

Fig. 2. Example of proton decay $p \to \bar{\nu} K^+$ via dimension five operators dressed by wino exchange in supergravity SU(5) theory.

In Eq. (6.5), the $\bar{\nu}_e K^+$ decay mode contribution is negligible. The remaining modes $\bar{\nu}_{\mu,\tau} K^+$ can be determined using the chiral Lagrangian methods[60,35]. One has for $i=\mu,\tau$

$$\Gamma(p \to \bar{\nu}_i K^+) = \frac{\beta_p^2}{M_{H_3}^2} \frac{m_N}{32\pi f_\pi^2} \left(1 - \frac{m_K^2}{m_N^2}\right) \left|1 + \frac{M_N(D+3F)}{3M_B} - \frac{2}{3}\frac{M_N}{M_B} D\right|^2$$

$$\left|\alpha_2^2 (2M_W^2)^{-1} m_{d_i} m_c V_{i1}^{\dagger} V_{21} V_{22}\right|^2 A_L^2 (A_S^L)^2 \left(\sum_{A=1,2,3} B_{Ai}\right)^2 \tag{6.6}$$

The notation of Eq. (6.6) is as follows: β_p is the three-quark matrix element of the proton defined by

$$\epsilon_{abc}\, \epsilon_{\alpha\beta} \langle o | d^{\alpha}_{aL} u^{\beta}_{bL} u^{\gamma}_{cL} | p \rangle = \beta_p U^{\gamma}_L \tag{6.7}$$

where U_L^{γ} is the proton spinor. The most recent lattice gauge calculations of β_p in the quenched approximation give[61]

$$\beta_p = (5.6 \pm 0.8) \times 10^{-3}\mathrm{GeV}^3 \qquad \left[\begin{array}{l}\text{lattice gauge}\\ \text{quenched approx}\end{array}\right] \tag{6.8}$$

In Eq. (6.6) D,F, f_π etc. are the chiral Lagrangian factors,[60] V_{ij} are the K-M matrix elements, $A_L(A_S)$ are the long (short) distance suppression factors and B_{Ai} are the dressing loop functions to be defined below. The numerical values of the chiral Lagrangian factors used are

$$\begin{array}{l} D = 0.76,\ F = 0.48,\ \ f_\pi = 139\ \mathrm{MeV} \\ M_N = 938\ \mathrm{MeV},\ M_B = 1154\ \mathrm{MeV} \end{array} \qquad \left[\begin{array}{c}\text{Chiral Lagrangian}\\ \text{Parameters}\end{array}\right] \tag{6.9}$$

For the $A_L(A_S)$ suppression factors [see Ellis, Nanopoulos and Rudaz in Ref. 34] we find the following values from a two loop analysis:[62]

$$A_S^L = 0.883 \qquad A_L = 0.283 \qquad \text{(2-loop)} \tag{6.10}$$

For the K-M matrix we use the central value of elements.[63] For the dressing loop functions the first generation contributions are essentially negligible i.e.

$$B_{1i} \ll B_{2i},\ B_{3i} \tag{6.11}$$

The second generation contributions B_{2i}, where one may neglect the L-R mixing, are straightforwardly given by

$$B_{2i} = (\sin 2\beta)^{-1}[F_2(\tilde{c},\ \tilde{d}_i,\ \tilde{W}) + \tilde{d}_i \to \tilde{\ell}_i] \tag{6.12}$$

where

$$F_2(\tilde{c},\ \tilde{d}_i,\ \tilde{W}) = \sin\gamma_+\cos\gamma_-\, f(\tilde{c},\ \tilde{d}_i,\ \tilde{W}_1) + \cos\gamma_+\sin\gamma_-\, f(\tilde{c},\ \tilde{d}_i,\ \tilde{W}_2) \tag{6.13}$$

and where $f(a,b,c)$ is a one loop integral defined by

$$f(a,\ b,\ c) = m_c/(m_b^2 - m_c^2)[[m_b^2/(m_a^2 - m_b^2)]\ \ell n(m_a^2/m_b^2) - (b \leftrightarrow c)] \tag{6.14}$$

In Eq. (6.13) $\gamma_\pm = \beta_+ \pm \beta_-$ and $\beta_\pm$ are defined by

$$\sin(2\beta_\pm) = (\mu \pm \tilde{m}_2)/[4\nu_\pm^2 + (\mu \mp \tilde{m}_2)^2]^{\frac{1}{2}} \tag{6.15}$$

where $\nu_\pm$ are given by Eq (4.18). The third generation dressing loop contributions are more complex because of the L-R mixing. Here one has

$$\frac{B_{3i}}{B_{2i}} = \frac{P_3}{P_2}\ \frac{m_t V_{31} V_{32}}{m_c V_{21} V_{22}}\ \frac{(F_3(\tilde{t},\tilde{d}_i,\tilde{W}) + \tilde{d}_i \to \tilde{\ell}_i)}{(F_2(\tilde{c},\tilde{d}_i,\tilde{W}) + \tilde{d}_i \to \tilde{\ell}_i)} \quad (i = 2,3) \tag{6.16}$$

In Eq. (6.16) P_i=i=1,2,3 are elements of a diagonal phase matrix that eneters in the baryon-number violating dimension five operators and F_3 is given by

$$F_3(\tilde{t},\tilde{d}_i,\tilde{W}) = [E\cos\gamma_-\sin\gamma_+\tilde{f}(\tilde{t},\tilde{d}_i,\tilde{W}) + \tilde{f}(\tilde{t},\tilde{d}_i,\tilde{W})$$

$$\cos\gamma_+\sin\gamma_-] - \frac{1}{2}\sin2\delta_t\left(\frac{m_t}{\sqrt{2}\,\sin\beta M_W}\right)[\{f(\tilde{t}_1,\tilde{d}_i,\tilde{W}_1)$$

$$E\sin\gamma_-\sin\gamma_+ - f(\tilde{t}_1,\tilde{d}_i,\tilde{W}_2)\cos\gamma_-\cos\gamma_+\} - \{\tilde{t}_1 \leftrightarrow \tilde{t}_2\}] \tag{6.17}$$

where $\tilde{t}_{1,2}$ are the squark mass eigen-states, and

$$\tilde{f}(\tilde{t},\tilde{d}_i,\tilde{W}) = \sin^2\delta_t f(\tilde{t}_1,\tilde{d}_{iL},\tilde{W}) + \cos^2\delta_t f(\tilde{t}_2,\tilde{d}_{iL},\tilde{W}) \tag{6.18}$$

and δ_t is given by

$$\sin2\delta_t = -2m_t(A_t + \mu\cot\beta)\,(m_{\tilde{t}_1}{}^2 - m_{\tilde{t}_2}{}^2)^{-1} \tag{6.19}$$

The factor E that enters in Eq. (6.17) is defined by

$$\begin{aligned} E &= (-1)^\Theta \\ \Theta &= 0,\ \sin2\beta > \mu\tilde{m}_2/M_W{}^2 \\ &= 1,\ \sin2\beta < \mu\tilde{m}_2/M_W{}^2 \end{aligned} \tag{6.20}$$

It is useful to define an effective B as follows:

$$B = \left((B_{2\mu} + B_{3\mu})^2 + \left(\frac{m_b^2 V_{31}^{\dagger 2}}{m_s^2 V_{21}^{\dagger 2}} \right) (B_{2\tau} + B_{3\tau})^2 \right)^{\frac{1}{2}}$$

$$M_S(\text{GeV})/10^a)^b \times 10^6 \text{ GeV}^{-1} \tag{6.21}$$

where the parameterization of Eq. (6.4) gives a=2.368 and b=0.329. Using the current experimental lower bound of Kamiokande, one can deduce the following limit on B:

$$B < \frac{381}{(M_s)^{0.0286}} \left(\frac{M_{H_3}}{3M_G} \right) \text{GeV}^{-1} \tag{6.22}$$

The value of $B_0 = 381(M_S)^{-0.0285}$ for various M_S values is exhibited in Table 5.

Table 5: Values of B_o for various M_S

M_S(Gev)	B_o
200	327
600	317
1200	311

From Table 5 we find that the effect of the variation of M_S is not very significant. We note however, that B is rather sensitive to the choice of the Higgs triplet mass. We also note that in SU(5) supergravity model with the GUT sector of Eq. (3.15), one obtains the following relations between the massive vector boson mass and the Higgs triplet mass

$$M_{H_3}/M_V = (\alpha_\lambda/\alpha_G)^{\frac{1}{2}} \tag{6.23}$$

where $\alpha_\lambda = \lambda_2^2/4\pi$. The condition that Yukawa couplings be perturbative at the GUT scale puts a limit on α_λ. The criterion $\alpha_\lambda < 1/3$ gives $M_{H3} < 3M_G$ and $\alpha_\lambda < 1$ gives $M_{H3} \leq 5M_G$, although larger values of M_{H_3} have also been explored in the literature.[64,65]

VII. ANALYSIS OF RADIATIVE BREAKING WITH CDF, LEP AND PROTON STABILITY CONSTRAINTS

In this section we discuss radiative breaking solutions to the electro-weak symmetry breaking under the constraints of proton stability discussed in Sec. VI and under the CDF and LEP constraints on the SUSY mass spectrum. The LEP[66] and CDF[67] constraints imposed on the SUSY spectrum are indicated below:

Sleptons:	$m_{\tilde{e}}, m_{\tilde{\mu}}, m_{\tilde{\tau}} > 45$ GeV	[LEP]
	$m_{\tilde{\nu}} > 32$ GeV	[LEP]
Charginos:	$m_{\tilde{W}_1} > 45$ GeV	[LEP]
Neutralinos:	$m_{\tilde{Z}_1} > 20$ GeV; $\tan\beta>3$	[LEP]
	$m_{\tilde{Z}_2} > 51$ GeV; $\tan\beta > 3$	
Higgs (SUSY):	$m_h > 43$ GeV	[LEP]
	$m_A > 20\text{–}44$ GeV	
	$m_{H^\pm} > 42$ GeV	
Gluino:	$m_{\tilde{g}} \geq 130$ GeV	[CDF; BTW[68]]
Squark:	$m_{\tilde{q}} \geq 150$ GeV	[CDF; BTW[68]]
	$m_{\tilde{t}_i} > 45$ GeV	[LEP]
Top:	$m_t \geq 90$ GeV	[CDF] (7.1)

Regarding the top quark mass a detailed fit to all the data assuming SM gives[69]

$$m_t = 130\,^{+25}_{-28}\ \text{GeV}\quad [m_H = M_Z] \tag{7.2}$$

and for the Higgs one has the limit[69]

$$1.3\text{GeV} < m_H < 160\ \text{GeV},\ 68\%\ \text{CL}\ [m_t = 130\ \text{GeV}] \tag{7.3}$$

We shall see below that while the CDF bound on m_t in Eq (7.1) is imposed as the lower limit on the SUSY spectrum, an upper limit on m_t consistent with the upper limit of Eq. (7.2) within 1 std arises from radiative breaking while for m_h a much more stringent upper bound arises.

SUSY Mass Limits Attainable in Future: An important question concerns the highest SUSY masses that can be explored in the near future in collider experiments. We give below the limits for the Tevatron and LEP2.

(A) Tevatron: With a data sample of 100 pb^{-1} one can attain the following limits on squark masses:

$$m_{\tilde{q}} \approx 200\text{GeV},\ m_{\tilde{t}_1} \approx 100\text{GeV} \tag{7.4}$$

The analysis of the trileptonic signal in $W \to \tilde{Z}\tilde{W}$ can allow one to explore Wino masses up to[70,71]

$$m_{\tilde{W}_1} \approx 70\text{GeV} \tag{7.5}$$

(B) LEP2: Using maximum energy, optimum luminosity and efficient triggers LEP2 may be able to explore Higgs and chargino masses up to limits shown below

$$m_h \approx 95\ \text{GeV}^{72},\ m_{\tilde{W}_1} \approx 100\text{GeV} \tag{7.6}$$

Parameter space: For the standard SU(5) supergravity GUT theory the parameter space consists of the following:

$$m_o,\ m_{\frac{1}{2}},\ A_o,\ B_o,\ \mu,\ m_t \tag{7.7}$$

and of course one has in addition as usual the Yukawa couplings of the lighter quarks and leptons. Using the electro-weak symmetry breaking Eqs. (4.6) one can trade the parameters μ and B_0 in favor of M_Z and $\tan\beta$ and thus one is left with only a five dimensional parameter space

$$m_o,\ m_{\frac{1}{2}},\ A_o,\ \tan\beta,\ m_t \tag{7.8}$$

Hopefully in the near future the value of m_t will be determined by CDF and D0. And thus one will be left with only a four dimensional parameter space consisting of m_0, $m_{1/2}$, A_0 and $\tan\beta$. Masses of 32 new particles (12 squarks, 9 sleptons, 2 Winos, 4 Zinos, 1 gluino and 4 Higgs) are then determined in terms of the above. The case m_0=0=A_0

belongs to a class of the so called No-Scale models. This is just a restricted subspace of the more general parameter space and here one predicts the mass spectra of the particles in terms of only the parameters $m_{1/2}$ and $\tan\beta$.

Computations of radiative breaking within SU(5) supergravity unification has been carried out recently by several authors.[73-77] However, in all of these works the constraints of proton stability were not included. The first analyses of radiative breaking which also include constraints of proton stability were given in Refs. (39-42). [For recent reviews see Refs. (78) and (79).] Inclusion of the proton lifetime constraints show that the parameter space of the SU(5) supergravity models is significantly reduced and the theory becomes much more predictive. In the following we give a detailed discussion of results within SU(5) supergravity theory including proton decay constraints.

$m_0=0=A_0$ Models: This case is a restricted version of the general supergravity model[11] and is viewed as belonging to the class of the No-Scale models.[38] As discussed already the model is essentially defined by the parameters $m_{1/2}$ and $\tan\beta$. In Ref(39) a detailed analysis of this model was carried out. It was found that under the restriction that $M_{H3}\leq 3M_G$ and the squark and gluino masses $\leq$1TeV, that proton decay in this model is in conflict with the current experimental lower limits of Eq. (1.3). The difficulty the $m_0=0=A_0$ model has in satisfying proton decay bounds can be understood as follows: The dressing loop contribution is controlled by a single parameter $m_{1/2}$ and falls off as $(m_{1/2})^{-1}$. However, $m_{1/2}$ cannot get too large so as not to violate the fine tuning condition $m_{\tilde{q},\tilde{g}}\leq$1TeV. Under this restriction the dressing loop function is too large to satisfy the experimental proton lifetime constraint if one restricts $M_{H3}\leq 3M_G$. It also turns out[39,75] that for $m_{1/2}$ greater than ~500 GeV, the LSP is no longer the lightest neutralino but rather it is the right selection $\tilde{e}_R$ which is the LSP, which violates the cosmological requirement that LSP be electrically neutral. Thus when this constraint is included, one need not impose a fine tuning condition.

Standard SU(5) Supergravity Model: In the analysis of the standard supergravity SU(5) model[11] we again carry out radiative breaking of the electro-weak symmetry under the constraints discussed above i.e. the CDF and LEP constraints and the proton stability constraints. Additionally we impose the finetuning restriction as before i.e. that $m_{\tilde{q},\tilde{g}}\leq$1TeV.

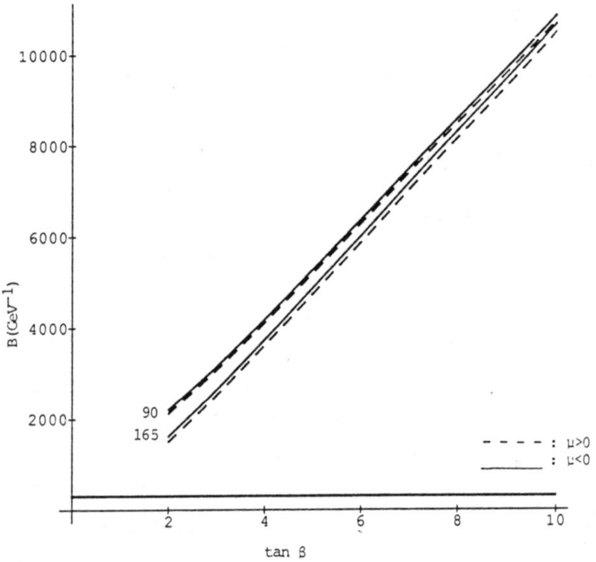

Figure 3. B for the m_0=0=A_0 model as a function of tanβ for values of m_t of 90 GeV and 160GeV for μ>0 and μ<0 and P_3/P_2=-1. Solutions must be below the solid horizontal line to satisfy Kamiokande lifetime limits.

A typical analysis of constraints on m_0 as a function of $m_{\tilde{g}}$ are shown in Fig. 4. One finds that a significant domain of the parameter space is eliminated by proton decay constraints.

Mass Relations for Charginos and Neutralinos: In the standard SU(5) supergravity grand unification constrained by proton decay one observes a pattern of chargino and neutralino masses and mass relations. We discuss next these relations. We note that the history of chargino and neutralino mass relations co-incides with the development of supergravity models In 1982 a mass inequality on the lightest chargino was derived by Weinberg[80] (in the massless photino limit) and by Chamseddine and the authors[80] (without the $m_{\tilde{\gamma}}$=0 limit) and mass relations for neutralinos were also found.[80] The mass relations we discuss here are dynamical in that they arise from radiative breaking satisfying all of the current experimental constraints as discussed already.

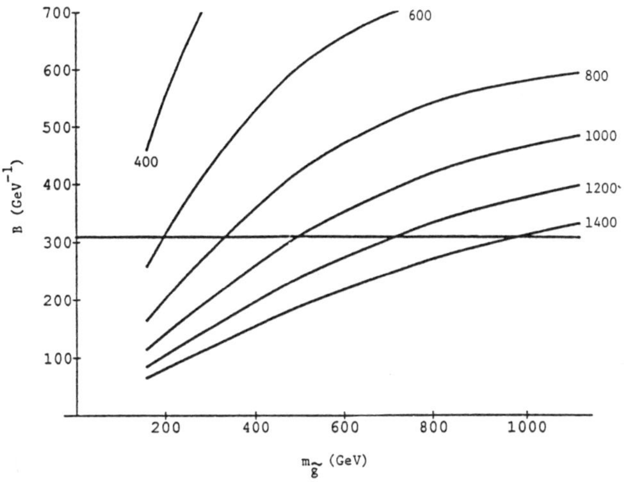

Fig. 4: B is plotted as a function of $m_{\tilde{g}}$ for various values of m_0 when A_t=-0.6, $\tan\beta$=1.73, m_t=125 GeV, P_3/P_2=-1 and μ<0. The allowed solutions lie below the horizontal solid line.

One of the consequences of the proton stability constraint is that the μ parameter turns out to be generally large i.e. $\mu^2 \gg M_Z^2$ over a major domain of the allowed parameter space. Expanding for the case $\mu^2 \gg M_Z^2$, $\tilde{m}_2^2$ one finds that $m_{\tilde{W}_i}$ are given by[40,41]

$$m_{\tilde{W}_1} \cong \tilde{m}_2 - M_W^2 \sin 2\beta/\mu; \quad m_{\tilde{W}_2} \cong \mu + M_W^2/\mu \tag{7.9}$$

where

$$\tilde{m}_2 = (\alpha_2/\alpha_3)\, m_{\tilde{g}} \cong 0.294\, m_{\tilde{g}} \tag{7.10}$$

Thus for μ<0, $m_{\tilde{W}1}$ increases to roughly $\sim\frac{1}{3}\, m_{\tilde{g}}$ while for μ>0, $m_{\tilde{W}1}$ decreases to roughly $\sim\frac{1}{4} m_{\tilde{g}}$. Of course if there are surviving cases with μ=O(M_Z) where proton stability is satisfied, one may get a large deviation from the above estimate.

A similar analysis can be carried out for the neutralino masses. Here the four roots of the neutralino eigen-value equation decompose into to small roots and two large roots. Thus when $\mu^2 \gg M_Z^2$, $\tilde{m}_2^2$ one has[40,41]

$$m_{\tilde{Z}_1} \cong \tilde{m}_1 - (M_Z^2/\mu) \sin 2\beta \sin^2\theta_W$$

$$m_{\tilde{Z}_2} \cong \tilde{m}_2 - (M_W^2/\mu) \sin 2\beta$$

$$m_{\tilde{Z}_{3,4}} \cong |\mu + \frac{1}{2} (M_Z^2/\mu)(1 \pm \sin 2\beta)| \qquad (7.11)$$

From Eq. (7.9) and (7.11) one finds that $m_{\tilde{W}1}$ and $m_{\tilde{Z}_2}$ eigen-masses are identical up to $0(1/\mu)$ in the μ-expansion:

$$m_{\tilde{W}_1} \approx m_{\tilde{Z}_2} \qquad (7.12)$$

To zeroth order in $(1/\mu)$ expansion one also finds that

$$m_{\tilde{Z}_1}/m_{\tilde{Z}_2} \cong \tilde{m}_1/\tilde{m}_2 = 0.508 \qquad (7.13)$$

and for the heavier Winos and Zinos one has

$$m_{\tilde{W}_2} \cong m_{\tilde{Z}_3} \cong m_{\tilde{Z}_4} \qquad (7.14)$$

VIII. TOP AND HIGGS MASS

To get an insight into constraint that the proton life time limit imposes on the top and Higgs mass, we consider the form of the dressing loop function B for the case when m_0 is large:

$$B = -2(\alpha_2/\alpha_3 \sin 2\beta)(m_{\tilde{g}}/m_{\tilde{q}}^2) \times 10^6 \qquad (8.1)$$

where one may also approximate the squark $(\text{mass})^2$ as follows:

$$m_{\tilde{q}}^2 = m_0^2 + a\, m_{\tilde{g}}^2,\ a \approx 2/3 \qquad (8.2)$$

The proton decay constraints limit $\tan\beta$ and one finds $\tan\beta \leq 4.7$ for $M_{H3} < 3M_G$. Now in the stop mass matrix the L-R mixing is controlled by the off diagonal term $m_t(A_t m_0 + \mu\cot\beta)$. Thus as m_t increases the off diagonal term increases and the mass of the lighter stop decreases which eventually falls below the LEP bound of 45GeV and becomes tachyonic. Thus the proton decay constraint along with the LEP-bound on $\tilde{t}_1$ places an upper bound on m_t which is found to be $m_t \leq 180$ GeV. We note that this result is consistent with the value of m_t extracted from radiative corrections Eq. (7.2). It is also consistent with the requirement that the Landau pole for the Yukawa couplings not be generated which gives a limit of $m_t \leq 200$ GeV[21,81].

Next we discuss the limits on the Higgs mass. An interesing prediction of the model is that there is an upper bound on the Higgs mass m_h as a function of m_t. The results are exhibited in Fig. 5. As discussed in Sec. VII, Higgs mass up to 95 GeV and a Wino mass of up to 100 GeV may be accessible at LEP2. Now the analysis shows that for m_t=125 GeV, one has m_h<95GeV for $\alpha_H \geq 16^0$ and that m_h exceeds 95GeV only for a small region of the parameter space. Exploration of the full parameter space shows that the Higgs mass always lies below 110 GeV.

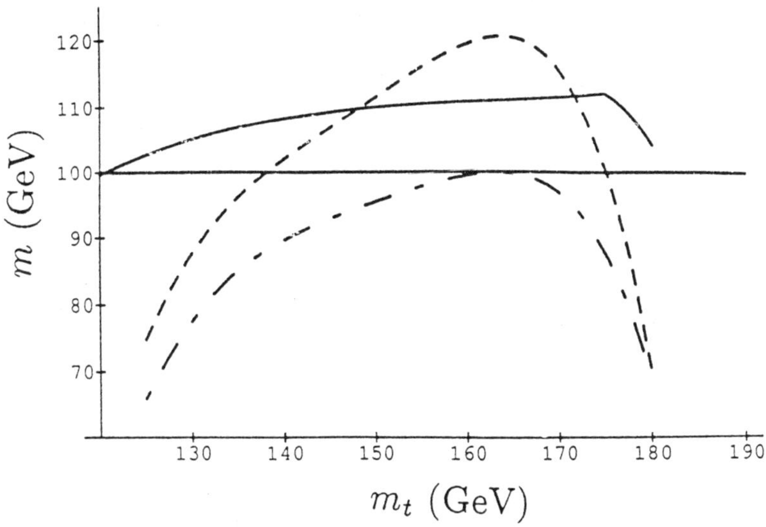

Fig. 5: The maximum higgs mass m_h as a functions of m_t (sold curve) and the maximum value of chargino mass $m_{\tilde{W}_1}$ when m_h>95GeV and B<293 GeV^{-1} (dashed curve) and the maximum of chargino mass $m_{\tilde{W}1}$ when m_h>95GeV and B<256 GeV^{-1} (dash-dot curve).

The results are exhibited in Eq. 5 by the solid curve as a function of m_t. Also plotted in Fig. 5 is the maximum value of the Wino mass for two cases. For the first case one has B<293 GeV^{-1} (corresponding to the current proton decay bound) and here one finds that $m_{\tilde{W}1}$<100GeV when m_h > 95 GeV for top masses less than 140 GeV. For the second case one has B<256GeV^{-1} (which corresponds to $\tau(p\rightarrow\bar{\nu}K)>1.3\times10^{32}$ yr or 1σ below the current value of B) and here we find that $m_{\tilde{W}1}$<100GeV when m_h > 95 GeV for all top masses. The above analysis indicates that if the Kamiokande experimental lower bound on proton lifetime exceeds 1.3×10^{32}yr, then we expect on the basis of the analysis presented above that either the light Higgs or the light Wino and possibly both should be observable at LEP2.

Finally we give in Table 6 an illustration of a typical computation of the mass spectrum of the 32 particles which shows a wide dispersion in the SUSY masses.

Table 6: Mass spectra of 32 new particles (12 squarks, q sleptons, 2 winos, 4 zinos, 1 gluino and 4 Higgs) as an illustrative example with values of parameters taken as $m_{\tilde{g}}$=248 GeV, m_o=600GeV, A_t=0.0, $\tan\beta$=1.732 and m_t=130GeV.

Particles	Masses(GeV)	Particles	Masses(GeV)
$\tilde{W}_1$	57.95	$\tilde{U}_L$ gens=1,2	630.08
$\tilde{W}_2$	452.74	$\tilde{U}_L$ gens=1,2	628.39
$\tilde{Z}_1$	30.84	$\tilde{d}_L$ gens=1,2	632.60
$\tilde{Z}_2$	60.38	$\tilde{b}_L$	525.54
$\tilde{Z}_3$	455.97	$\tilde{d}_R$ gens=1,2,3	628.87
$\tilde{Z}_4$	437.23	$\tilde{t}_1$	394.71
h	63.61	$\tilde{t}_2$	545.79
H	861.56	$\tilde{e}_L$ gens=1,2,3	603.93
A	857.70	$\tilde{e}_R$ gens=1,2,3	601.70
$H^{\pm}$	861.43	$\tilde{\nu}_L$ gens=1,2,3	601.29

IX. DISCUSSION OF RESULTS AND CONCLUSIONS

Currently the Standard SU(5) Supergravity Model[11] is the leading candidate for the unification of the electro-weak and the strong interactions. Using the very precise LEP data on the couplings α_3, α_2 and α_1 one finds that unification does occur within supergravity SU(5) model when one extrapolates using the supersymmetry renormalization group equations with one pair of Higgs bosons. Although threshold effects have not been taken into account fully especially at low energy, one expects these effects to be typically $O(1\sigma)$ on the current calculations and thus not modify in any drastic fashion the overall unification picture within SU(5) supergravity model.

One of the remarkable results of the recent analyses of the model constrained by proton decay and the CDF and LEP data constraints are some very definitive predictions. The first of these is the existence of a light sector of the SUSY and Higgs mass spectrum which should be accessible at LEP2 if one achieves the maximal energy and liminosity and efficient detection for Higgs and charginos are used. Under these conditions we expect that either a light Higgs or a light chargino should be accessible at LEP2[41]. A second prediction of the model is that the $\bar{\nu}K^+$ mode of the proton should be accessible in improved proton decay experiments. Thus the next generation of proton decay experiments should observe this decay mode.

The SU(5) supergravity model also has specific predictions for the observation of new phenomena at the SSC and LHC. As can be seen from Table 6, the SU(5) supergravity model has a majority of the spectrum which is not accessible at either the Tevatron or LEP2 i.e. all particle except for $\tilde{Z}_{1,2}, \tilde{W}_1$, h and possibly $\tilde{t}_1$ belong to this category. Thus one expects a great deal of SUSY phenomena at SSC and LHC energies. Also there are obvious implications of the SU(5) supergravity mass spectra at the next generation of high energy e^+e^- colliders. Finally our analysis has been specifically confined to SU(5). For a recent analysis for SO(10) the reader is referred to Ref. (82)

ADDENDUM: Issues related to cosmology were discussed only in passing at the lecture given at the Erice School. A very brief discussion is given here for completeness. In the analysis given thus far we have relied on constraints from laboratory experiments and not included the cosmological constraints which are of a more speculative nature. Essentially the cosmological constraints arise from the condition that the neutralino relic density not exceed the critical mass density needed to close the universe i.e. one requires $\Omega_{\tilde{Z}1}h^2 \leq 1$ where $\Omega_{\tilde{Z}1}$ is the ratio of the relic density due to the lightest neutralino and the critical mass density and h is the Hubble constant in units of 100km/sec Mpc. This subject has been investigated by several authors recently[42,83,84]. We discuss here the result of our investigation of SU(5) supergravity including both the proton decay and the relic density constraints[85,86]. We find that there exists a wide region of the parameter space where both the proton decay and the relic density constraints are satisfied.

The analysis of Ref. (42) also finds a significant domain of the parameter space where both the relic density and proton decay constraints are satisfied.

We give below an illustration of our results taken from Ref. (85). In Fig. 6, the domains of $m_{\tilde{g}}$ consistent with CDF, LEP, proton decay and relic density constraints are exhibited as a function of A_t/m_0 for α_H=30^0 and three different values of the top mass. One finds in each case a significant domain where all the constraints are satisfied. Typically the relic density constraints require that $m_{h/2}$ is close to the lightest Zino mass.

Fig. 6. Domains in gluino mass $m_{\tilde{g}}$ as a function of Polonyi paramter A_t (at the electro-weak scale) consistent with proton lifetime limits and relic density constraints. The solid curve is for m_t=125 GEV, the dashed curved for m_t=110 GeV and the dot-dash curve for 140 Gev.

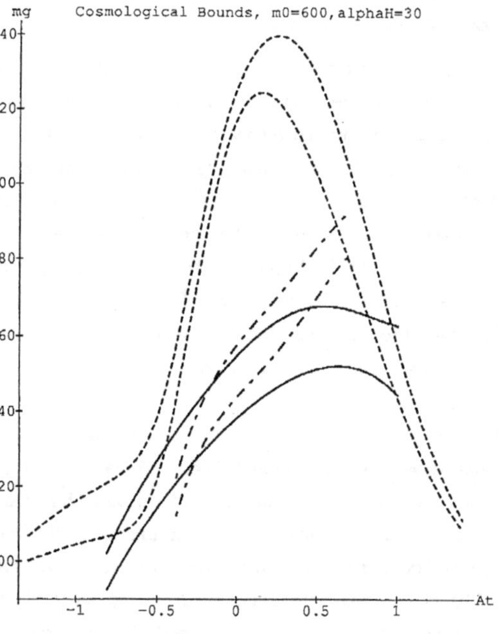

In general one finds that the allowed domain is a shell in $m_{\tilde{g}}$ 5-20 GeV wide in the five dimensional parameter space defined by $m_0, m_{\tilde{g}}, A_0, m_t$ and $\tan\beta$. It should also be noted that solutions satisfying relic density constraints and proton stability also exist with large squark and gluino mass, i.e. in excess of 5 TeV, due to the large number of open channels as recently pointed out in Ref. (84).

There also exists a second possibility by which the relic density constraints can be satisfied. This possibility concerns the fact that the lightest Zino may be unstable due to the existence of "Planck slop". It is fairly easy to generate Planck slop terms which lead to d-D and d^c-D^c mixing (where D and D^c are the Higgs triplet particles of the SU(5) 5 and $\overline{5}$ representations) and thus produce an R-parity violating interaction of the type

$$L_{int} = g(\tilde{Z}_1\mu_L)(\bar{s}^a c_L^a); \; g = \frac{2e}{m_{\tilde{\mu}}^2} f_{22} U s_1 \qquad (9.1)$$

where f_{22} is an element of the SU(5) coupling in Eq. (3.14) and U is the projection of the Zino $\tilde{Z}_1$ state onto the photino state and s_1 is a small mixing $0(10^{-4})$. The above leads to a Zino-lifetime of

$$\tau(\tilde{Z}_1 \to \bar{c}s\mu^+) \approx (1 \times 10^{-19}\text{sec})(f_{22}s_1U)^{-2}\left(\frac{m_{\tilde{\mu}}}{m_{\tilde{Z}_1}}\right)^4\left(\frac{1\text{GeV}}{m_{\tilde{Z}_1}}\right) \qquad (9.2)$$

For some typical values e.g. $m_{\tilde{\mu}}$=500 GeV, $m_{\tilde{Z}}$=50GeV, one finds $\tau=10^{-4}$sec so that all of the neutralinos decay without leaving any significant cosmological trace. Further for this case the typical supersymmetric signals with missing E_T will still operate since the neutralino will decay well outside the detector.

ACKNOWLEDGEMENTS: This research was supported in part by NSF Grant Nos. PHY-917809 and PHY-916593.

REFERENCES

1. S.L. Glashow, Nucl. Phys. B22: 579 (1986); S. Weinberg, Phys. Rev. Lett. 19: 1264 (1967); A. Salam, Proc. 8th Nobel Symposium Stockholm 1968, ed. N. Svartholm (Almquist and Wiksells, Stockholm, 1968).
2. J.C. Pati and A. Salam, Phys. Rev. D8: 1240 (1973).
3. H. Georgi and S.L. Glashow, Phys. Rev. Lett. 32: 448 (1974).
4. S. Weinberg, in "Gauge Theories and Modern Field Theory", edited by R. Arnowitt and P. Nath, p.1.
5. Yu, a. Gol'fand and E. P. Likhtman, JETP Lett. 13: (1971); P. Ramond, Phys. Rev. D3: 2415 (1971); A. Neveu and J. Schwarz, Nucl. Phys. B31: 86 (1971); D. V. Volkov and V.P. Akulov, Phys. Lett. 46B: 109 (1973); J. Wess and B. Zumino, Nucl. Phys. B70: 39 (1974).
6. P. Fayet, Phys. Lett. 69B: 489 (1977); 70B: 461 (1977); M. Sohnius, Nucl. Phys. B122: 291 (1977).
7. E. Witten, Nucl. Phys. B177: 477 (1981); B185: 513 (1981).
8. S. Dimopoulos and H. Georgi, Nucl. Phys. B193: 150 (1981); N. Sakai, Zeit. F. Phys. C11: (1981).
9. S. Weinberg, Phys. Lett. 62B: 111 (1976); E Witten, Nucl. Phys.

B202: 253 (1982).
10. D.Z. Freedman, P. van Nieuwenhuizen and S. Ferrara, Phys. Rev. D16: 3214 (1976); S. Deser and B. Zumino, Phys. Lett. 62B: 335 (1976).
11. A. H. Chamseddine, R. Arnowitt and P. Nath, Phys. Rev. Lett. 49: 970 (1982).
12. P. Nath, R. Arnowitt and A.H. Chamseddine, APPLIED N=1 SUPERGRAVITY, ICTP Lecture Series Vol. I 1983 (World Scientific, Singapore, 1984).
13. E. Cremmer, S. Ferrara, L. Girardello and A. van Proeyen, Phys. Lett. 116B: 231 (1982); Nucl. Phys. B212: 413 (1983).
14. E. Witten and J. Bagger, Nucl. Phys. B222: 125(1983)
15. K.S. Stelle and P.C. West, Nucl. Phys. B145: 175 (1978).
16. P. van Nieuwenhuizen and S. Ferrara, Phys. Lett. 76: 404(1978).
17. R. Barbieri, S. Ferrara and C.A. Savoy, Phys. Lett. 119B: 343 (1982).
18. J. Polonyi, University of Budapest Report No. KFKI-1977-93, 1977 (unpublished).
19. L.E. Ibanez, Phys. Lett. 118B: 73 (1982); J. Ellis, D.V. Nanopoulos and K. Tamvakis, Phys. Lett. 121B: 123 (1983).
20. L. Avarez-Gaume, J. Polchinski and M. Wise, Nucl. Phys. B221: 495(1983); J. Ellis, J. Hagelin, D. V. Nanopoulos and K. Tamvakis, Phys. Lett. 125B: 275 (1983).
21. L.E. Ibanez and C. Lopez, Phys. Lett. B126: 54(183); L. Ibañez, C. Lopez and C. Muñoz, Nucl. Phys. B256: 218 (1985).
22. J. Ellis, A.B. Lahanas, D.V. Nanopoulos and K. Tamvakis, Phys. Lett. 134B: 429(1984).
23. XXVI International Conference on High Energy Physics, Dallas, August, 1992.
24. The value of α_3 is taken from analysis of G. Altarelli at this school.
25. P. Langacker, Proc. PASCOS 90-Symposium, P. 231: Eds. P. Nath and S. Reucroft, (World Scientific, Singapore 1990); P. Langacker and M. Luo, Phys. rev. D44: 817 (1991).
26. J. Ellis, S. Kelley and D. V. Nanopoulos, Phys. Lett. 249B: 441(1990); B260: 131(1991).
27. U. Amaldi, W. de Boer and H. Furstenau, Phys. Lett. 260B: 447(1991).
28. F. Anselmo, L. Cifarelli, A. Peterman and A. Zichichi, N.C. 115A: N.4: 581(1992); N.C. 104A: N.12;1817(1991).
29. A. Zichichi, "Where We Stand with The Real Superworld", in the Erice School from Superstrings to The Real Superworld, Erice-Sicily 14-22 July 1992.
30. U. Amaldi, W. de Boer, P. Frampton, H. Furstenau and Liu, CERN-PPE/91-233/rev.
31. R. Becker-Szendy et. al., Phys. Rev. D42: 2974(1990).
32. Particle Data Group, Phys. Rev. D45: (1992).
33. Y. Totsuka, Proc. of XXIV Conf. on High Energy Physics, Munich, 1988 Eds. R. Kotthaus and J. H. Kuhn (Spring Verlag, Berlin, Heidelberg, 1989).
34. S. Weinberg, Phys. Rev. D.26: 287 (1982); N. Sakai and T. Yanagida, Nucl. Phys. B197: 533(1982); S. Dimopoulos, S. Raby and F. Wilczek, Phys. Lett. 112B: 133 (1982); J. Ellis, D.V. Nanopoulos and S. Rudaz, Nucl. Phys. B202: 43(1982). B.A Campbell, J. Ellis and D.V. Nanopoulos, Phys. Lett. 141B: 229 (1984); S. Chadha, G.D. Coughlan, M. Daniel and G.G. Ross, Phys. Lett. 149B: 47, (1984).
35. R. Arnowitt, A.H. Chamseddine and P. Nath, Phys. Lett 156B: 215(1985); P. Nath, R. Arnowitt and A.H. Chamseddine, Phys. Rev. 32D: 2348(1985).
36. E. Cremmer, S. Ferrara, C. Kounnas and D.V. Nanopoulos, Phys. Lett. B133: 61(1983); N.P. Chang, S. Ouvry and X. Wu, Phys. Lett. 51:

327(1983).
37. J. Ellis, C. Kounnas and D.V. Nanopoulos, Nucl. Phys. B241: 405(1984); Nucl. Phys. B247: 373 (1984).
38. For a review of No-scale models see A.B. Lahanas and D.V. Nanopoulos, Phys. Rev. 145: 1(1987).
39. P. Nath and R. Arnowitt, Phys. Lett. B287; 3282(1992).
40. R. Arnowitt and P. Nath, Phys. Rev. Letts. 69: 725(1992).
41. P. Nath and R. Arnowitt, Phys. Lett. B289: 308 (1992).
42. J. Lopez, D. V. Nanopoulos and H. Pois, CERN-TH-6628/929.
43. L. Hall, J. Lykken and S. Weinberg, Phys. Rev. D. 27: 2359(1983).
44. P. Nath, R. Arnowitt and A.H. Chamseddine, Nucl. Phys. B227: 121(1983).
45. M.T. Grisaru, W. Siegel and M.Rocek, Nucl. Phys. B159: 420 (1979).
46. G. Grinstein, Nucl. Phys. B206: 387 (1982).
47. H. Georgi, Phys. Lett. 108B: 283(1982).
48. A. Masiero et. al., Phys. Lett. 115B: 380(1982).
49. K. Inoue, A. Kakuto and T. Tankano, Prog. Theor. Phys. 75: 664 (1986); A. Anselm and A. Johansen, Phys. Lett. 200B: 331(1988); A. Anselm, Sov. Phys. JEPT 67: 663 (1988).
50. R. Barbieri, G. Dvali and A. Strumia, CERN-TH-6526/92.
51. S. Coleman and E. Weinberg, Phys. Rev. D7: 1888 (1973); S. Weinberg, Phys. Rev. D7: 2887(1973).
52. J.M. Frere, D. Jones and S. Raby, Nucl. Phys. B222: 1(1983); M. Claudson, L. Hall and I. Hinchliffe, Nucl. Phys. B228: 501 (1983).
53. M. Drees, M. Gluck and K. Brassie, Phys. Lett. B157: 164 (1985).
54. For review see Ref. 12 and H. Nilles, Phys. Reports 110: 1(1984); H. Haber and G. Kane, Phys. Rep. 117: 75(1985).
55. P. Nath, R. Arnowitt and A.H. Chamseddine, HUTP-83/A077-NUB#2588.
56. Alvarez-Gaume et. al. in Ref. 20
57. Y. Ikada M. Yamaguchi and T. Yanagida, Prog. Theor. Phys. 85: 1(1991); J. Ellis, G. Ridolfi and F. Zwirner, Phys. Lett. 257B: 83(1991); H.E. Haber and R. Hampling, Phys. Rev. Lett. 66: 1815(1991).
58. G. Gamberini, G. Ridolfi and F. Zwirner, Nucl. Phys. B331: 331(1990).
59. R. Arnowitt and P. Nath, Phys. Rev. D. (to be published).
60. S. Chadha and M. Daniels, Nucl. Phys. B229; 105(1983).
61. M.B. Gavela, et. al., Nucl. Phys. B312: 269 (1989).
62. The result of Eq. (6.10) is on the basis of an updated analysis and may be compared to the previous evaluations of Ellis et. al. in Ref. 34 which gives A_S=0.91, A_L=0.20.
63. Particle Data Group, Phys. Rev. D45(1992). Specifically for V_{13}, V_{31} and V_{32} we use V_{13}=-0.002, V_{31}=0.011, V_{32}=-0.042.
64. H. Hisano, H. Murayama, and T. Yanagida, Tohuku University Preprint TU-400-July, 1992.
65. A.E. Faraggi, B. Grinstein and S. Meshkov, SSCL-Preprint-126 (1992).
66. Lepton-Photon High Energy Physics Conference (LP-HEP91) Geneva, 1991.
67. See M. Davier in Ref (66).
68. H.Baer, X. Tata and J. Woodside Phys. Rev. Lett. 63, 352 (1989) and FSU-HEP-910211/UH-511-721-91/IS-J4185.
69. J.Ellis, G.L. Fogli and E. Lisi, Phys. Lett. B274: 456(1992); ibid B285: 238(1992); ibid B288: 85(1992), BARI-TH-111-92.
70. P. Nath and R. Arnowitt, Mod. Phys. Lett. A, Vol. 2, No. 5: 331 (1987); R. Barbieri, F. Caravaglio, M. Frigeni and M. Mangano, Nucl. Phys. B367: 28 (1991).
71. H. Baer and X. Tata, FSU-HEP-920907/UH-511-749-92(1992).
72. J. Hilgart et. al. Z. Phys. C35 (1987); K.J.F. Gaemers and N. de Groot, Z. Phys. C49(1991) 267; H. Haber, SCIPP 92/10(1992).

73. M. Drees and M.M. Nojiri, Nucl. Phys. B369; 54(1992); Phys. Rev. D45: 2482 (1992).
74. G.G. Ross and R.G. Roberts, Nucl. Phys. B377; 571(1992).
75. S. Kelley, J.L. Lopez, D.V. Nanopoulos H. Pois and K. Yuan, Phys. Lett. B273: 423(1991).
76. K. Inoue, M. Kawasaki, M. Yamaguchi, and T. Yanagida, Phys. Rev. D45: 328 (1992).
77. H. Arason, D. Castano, B. Keszthelyi, S. Mikaelian, E. Picard, P. Ramond, and E. Wright, Phys. Rev. Lett. 67: 2933(1991).
78. R. Arnowitt and P. Nath, "Testing Supergravity Grand Unification", in Proc. of the Seventh Lake Louise Winter Institute, Ed. B.A. Campbell et. al. World Scientific 1992.
79. P. Nath and R. Arnowitt, "Radiative Breaking of the Electro-Weak Symmetry and Proton Stability in SUSY Unification", in Proc. of Les Recontres de Physique de la Vallee D'Aosta, Lathuile March 8-14, 1992, edited by M. Greco and R. Bertolini, Editions Frontieres.
80. S. Weinberg, Phys. Rev. Letts, 50: 387(1983), R. Arnowitt, A.H. Chamseddine and P. Nath, Phys. Rev. Lett. 50, 232(1983).
81. V. Barger, M.S. Berger and P. Ohmann, MD/PH/711 (1992).
82. B. Ananthanarayan, G. Lazarides and Q. Shafi, B-92-29-PRL-TH-92/16; Q. Shafi and B. Ananthanarayan, BA-91-76.
83. J. Lopez, D. V. Nanopoulos and A. Zichichi, Phys. Lett. B291: 255 (1992).
84. M. Drees and M. M. Nojiri, DESY-92-101/SLAC PUB-5860 (1992).
85. R. Arnowitt and P. Nath, CTP-TAMU-65/92-NUB-TH-3055/92.
86. P. Nath and R. Arnowitt, NUB-TH-3056/92-CTP-TAMU-66/92

CHAIRMAN: P. Nath

Scientific Secretaries: A. Leike, N. Yunus

DISCUSSION

– *do Couto e Silva:*

You mentioned that in Los Alamos lepton flavour violation has been looked at. What is the major constraint in this experiment, the number of events or the detection itself?

– *Nath:*

First, it is the observation of any event at all. If such a process is seen, then one can try to collect more events at higher luminosity. Then the angular distribution becomes important. Looking for whether one has forward or backward peaking, one can determine the handedness of the final lepton.

– *Benakli:*

1) In the so-called string test of your talk you talked about lepton number violation due to discrete symmetry of strings. Is there anything in this phenomena which would be a signal of string?

2) What is the missing partner mechanism?

3) Can you explain more about the Q^2 dependence of the vacuum expectation value ?

– *Nath:*

1) At the moment we do not have a string model which has this violation. However, string theory has so many vacua that there is a possibility that the above phenomenon can be realized in a concrete model.

2) In a missing partner mechanism one couples the 5-plet of Higgs with a higher representation of SU(5) which contains Higgs triplet but no Higgs doublet. An example is $\underline{50}\ \underline{24}\ \underline{5}$ coupling which produces a mass for the Higgs triplet but no mass for the Higgs doublet.

3) One expects the one loop corrected potential to be a better approximation to the tree potential than just the tree potential which may have a rapid Q^2 dependence. The one-loop contribution cancels the rapid Q^2 dependence of the tree potential and consequently the vacuum expectation value equations possess only a dilute Q^2 dependence as expected.

– *Leike:*

Are there some theoretical constraints on neutralino and chargino mixing?

– *Nath:*

In the supergravity SU(5) theory one has well defined neutralino and chargino mass matrices which are determined completely in terms of SU(2) $\times$ U(1) gaugino masses $\tilde{m}_i = (\alpha_i/\alpha_G)m_{1/2}$ $(i = 1, 2)$, the Higgs mixing parameter μ, M_Z and $\sin^2\theta_W$. Since there are only two arbitrary quantities μ and $m_{1/2}$, there are relations among the neutralino and chargino eigen-masses. These relations become dramatically simple in the limit when $\mu^2 \gg M_Z^2, m_{1/2}^2$. Then one predicts, for example, that the lightest neutralino is about half the mass of the lightest chargino. Such relations should help in experimental searches for SUSY particles.

– *Cho:*

How stringent are constraints imposed on SUSY GUT couplings of dimension 5 operators which mediate proton decay? Are SUSY models other than flipped SU(5) really viable?

– *Nath:*

The couplings of the Higgs triplets to matter are completely fixed in terms of quark masses and mixings. The only question is the mass of the Higgs triplet and the masses of the SUSY particles that circulate in the loops. One finds that there is a domain of the parameter space, which I exhibited in my transparency, where proton stability μ achieved is consistent with current experiments. The flipped SU(5) model is not really unified in Supergravity due to the extra U(1) factor.

– *Hanany:*

How do you treat the dimension 5 operators? Are they irrelevant?

– *Nath:*

Dimension five operators are very relevant. They arise from regular dimension four operators when one eliminates the heavy Higgs triplet at low energy. The effective interaction one obtains after the elimination is dimension five and enters into the computation of proton decay.

– *Hanany:*

What is the Coleman-Weinberg term?

– *Nath:*

The Coleman-Weinberg term is a one-loop correction to the tree level potential. As you know the vacuum state is defined by the minimization of the potential. Usually one minimizes just the potential at the tree level. However, sometimes this is not a good approximation to reality and addition of the one-loop correction significantly improves the situation.

– *Sopczak:*

What is the heaviest h^0 which can be seen at LEPII?

– *Nath:*

The optimistic limit of 95 GeV for the heaviest Higgs mass that can be seen at LEPII was based on the following papers: Hilgart et al. Z. Phys. **C35**, 347 (1987); K.J.F. Gaemers et al. Z. Phys. **C49**, 267 (1991), see also H. Haber, SCIPP92/10 Preprint. To realize the optimistic assessment of 95 GeV one must achieve the maximum energy of 200 GeV at LEPII, and use the integrated luminosity in excess of 500 pb^{-1}. Also an efficient detection of b quarks in Higgs decay is needed.

– *Misiak:*

1) What soft SUSY breaking terms do you use?

2) Do you include thresholds in your renormalization group equation evolution?

– *Nath:*

1) We use the full set of soft SUSY breaking terms characterized by $m_0, m_{1/2}$, A_0 and B_0 where B_0 can be eliminated in favour of $\tan\beta$. Including m_t, the parameter space of the theory is then characterized by $m_0, m_{1/2}, A_0, \tan\beta$ and m_t.

2) The values of the unification parameters α_G and M_G are taken from analyses which include threshold effects. These are then used to compute the superparticle spectrum.

– *Rind:*

What are the possible limits on proton lifetime from "Superkamiokande"?

– *Nath:*

The Superkamiokande is larger in size and will give better proton lifetime limits. For the $p \rightarrow e^+\pi^0$ mode the expectation is that the Superkamiokande can achieve a lifetime limit of 1×10^{34} yr. Since the experimental limit on $p \rightarrow e^+\pi^0$

decay mode is rougly 5×10^{32} yr, Superkamiokande limit if achieved will represent a gain of approximately 20 in improving the lifetime. For $p \to \bar{\nu}K^+$ the gain is roughly a factor of three smaller. Beyond Superkamiokande improvement in lifetime limits are possible but more difficult since the cosmic ray background effects become more significant. One needs very efficient simulations of the background effects to go beyond the Superkamiokande lifetime limits.

The Superworlds of SU(5) and SU(5) x U(1): A Critical Assessment and Overview

Jorge L. Lopez
Department of Physics, Texas A&M University, USA
D. V. Nanopoulos
Department of Physics, Texas A&M University, USA
A. Zichichi
CERN, Geneva, Switzerland

1 Introduction

The Standard Model of electroweak and strong interactions is well established by now. In fact, the effects of the top quark in one-loop electroweak processes predict its mass (within $\approx 20\%$) centered around $\approx 145\,\mathrm{GeV}$ [1]. Therefore, its expected direct experimental detection at the Tevatron in the near future will complete the set of Standard Model predictions for the vector and fermion sectors. The scalar sector is another story. The simplest electroweak symmetry breaking scenario with a single Higgs boson is only mildly constrained experimentally, with a lower bound of $m_H \gtrsim 60\,\mathrm{GeV}$ [2] and no firm indirect experimental upper bound, although this situation will change once the top quark mass is measured [3]. On the other hand, interesting upper bounds on m_H follow from various theoretical assumptions, such as perturbative unitarity at tree- ($m_H \lesssim 700\,\mathrm{GeV}$) [4] and one-loop ($m_H \lesssim 400\,\mathrm{GeV}$) [5] levels, and the stability of the Higgs potential ($m_H \lesssim 500\,\mathrm{GeV}$) [6]. In practice, with the advent of the SSC and LHC, experimental information about the TeV scale is likely to clarify the composition of the Higgs sector. Nevertheless, despite all these efforts the structure of the Standard Model and its corresponding Higgs sector will remain basically unexplained.

It has therefore become customary to turn to the physics at very high energies to search for answers to these theoretical questions. The most promising theories of this kind contain two new ingredients: supersymmetry and unification. Together these can explain the origin of the weak scale (*i.e.*, the gauge hierarchy problem) relative to the very high energy unification (M_U) or Planck (M_{Pl}) scales [7]. Furthermore, this class of theories predict a new set of relatively light ($\lesssim \mathcal{O}(1\,\mathrm{TeV})$) particles consisting of partners for the Standard Model particles but with spin offset by 1/2 unit. In fact, the new set of particles appears ever more likely to overlap little with the mass scales of the standard ones, thus their present unobserved status. Moreover, the Standard Model Higgs boson will then appear as one of the new particles but with mass close to M_Z, thus avoiding naturally the theoretical problems mentioned above.

Unfortunately, the introduction of supersymmetry also increases significantly the number of unknown parameters in the theory, mainly because this symmetry must be softly broken at low energies. Indeed, to describe a generic low-energy supersymmetric model (the so-called minimal supersymmetric standard model (MSSM)) neglecting the first- and second-generation Yukawa couplings, the CKM angles, and possible CP violating phases, we need the following set of parameters (the values of $\sin^2\theta_w, \alpha_3, \alpha_e, M_Z$ are taken as measured parameters):

(a) The Yukawa $(\lambda_t, \lambda_b, \lambda_\tau)$ and Higgs mixing (μ) superpotential couplings. (We can trade the Yukawa couplings for $m_t, \tan\beta$; m_b, m_τ, with $\tan\beta = v_2/v_1$ the ratio of Higgs vacuum expectation values, and m_b, m_τ given.)

(b) The soft-supersymmetry breaking trilinear (A_t, A_b, A_τ) and bilinear (B) scalar couplings (corresponding to the superpotential couplings in (a)).

(c) The soft-supersymmetry breaking left-left and right-right entries in the squark and slepton mass matrices for the first and second $(m_{Q,U^c,D^c},\, m_{L,E^c})$, and third $(m_{Q_3,U_3^c,D_3^c},\, m_{L_3,E_3^c})$ generations.

(d) The soft-supersymmetry breaking gaugino masses $m_{\tilde{g}}, m_{\widetilde{W}}, m_{\widetilde{B}}$.

(e) The Higgs sector parameter (at tree-level), *e.g.*, the pseudoscalar Higgs boson mass m_A.

The above 21 unknown parameters make any thorough analysis of this class of models rather impractical, and have allowed in the past only limited explorations of this parameter space. If we now add the gauge unification constraint $(\alpha_i(M_U) = \alpha_U,\ i = 1,2,3)$, the assumption of universal soft-supersymmetry breaking at a scale $\Lambda_{susy} = M_U$, and high-energy dynamics (in the form of renormalization group equations (RGEs) for all the parameters involved), the set of parameters in (b) reduces to $A = A_t = A_b = A_\tau$ and B, those in (c) to $m_0 = m_{Q,U^c,D^c} = m_{Q_3,U_3^c,D_3^c} = m_{L,E^c} = m_{L_3,E_3^c}$, and those in (d) to $m_{1/2} = m_{\tilde{g}} = m_{\widetilde{W}} = m_{\widetilde{B}}$; these relations are valid *only* at the scale Λ_{susy}. The number of parameters has been dramatically reduced down to eight.

Let us now add low-energy dynamics by demanding radiative breaking of the electroweak symmetry. The tree-level Higgs potential is given by

$$\begin{aligned} V_0 &= (m_{H_1}^2 + \mu^2)|H_1|^2 + (m_{H_2}^2 + \mu^2)|H_2|^2 + B\mu(H_1H_2 + \text{h.c.}) \\ &+ \tfrac{1}{8}g_2^2(H_2^\dagger\sigma H_2 + H_1^\dagger\sigma H_1)^2 + \tfrac{1}{8}g'^2\left(|H_2|^2 - |H_1|^2\right)^2, \end{aligned} \tag{1}$$

where $H_1 \equiv \binom{H_1^0}{H_1^-}$ and $H_2 \equiv \binom{H_2^+}{H_2^0}$ are the two complex Higgs doublet fields, $g' = \sqrt{5/3}\,g_1$ and g_2 are the $U(1)_Y$ and $SU(2)_L$ gauge couplings, and $B\mu$ is taken to be real and negative. This potential has a minimum[1] if $\partial V_0/\partial\phi_i = 0$, with ϕ_i denoting the eight real degrees of freedom of H_1 and H_2. In particular, for $\phi_i = \mathrm{Re}\, H_i^0$ one obtains two constraints which allow the determination of μ and B,

$$\mu^2 = \frac{m_{H_1}^2 - m_{H_2}^2 \tan^2\beta}{\tan^2\beta - 1} - \tfrac{1}{2}M_Z^2, \tag{2}$$

$$B\mu = -\tfrac{1}{2}\sin 2\beta (m_{H_1}^2 + m_{H_2}^2 + 2\mu^2) < 0, \tag{3}$$

up to the sign of μ. In these expressions, $m_{H_1}^2, m_{H_2}^2$ are soft-supersymmetry breaking masses equal to m_0^2 at Λ_{susy}. Since the whole set of Higgs masses and couplings (at tree-level) follows from m_A^2 (and $\tan\beta$), and one can easily show that $m_A^2 = -2B\mu/\sin 2\beta$, the parameter in (e) is also determined. (This result also holds at one-loop although the expression for m_A^2 is more complicated in this case.)

The final parameter count in the class of models we consider is then just five: $m_t, \tan\beta, m_{1/2}, m_0, A$ (plus the sign of μ). Note also that $\sin^2\theta_w$ (as well as M_U and α_U) gets determined (from α_3 and α_e) by the gauge unification condition. What are the *a priori* expected values of $m_{1/2}, m_0, A$? In principle choosing a suitable supergravity model (*i.e.*, a suitable hidden sector) one could have arbitrary values for these parameters. In fact, such generic scenaria are worthwhile studying. However, further well motivated constraints on this sector of the theories turn out to yield quite predictive models. Below we consider two string-inspired supersymmetry breaking scenaria where m_0 and A are fixed functions of $m_{1/2}$. We also consider a generic scenario which is constrained by the proton lifetime and which results in strict restrictions on the allowed choices for these three parameters.

In this paper we study and contrast two well motivated examples belonging to the class of unified supersymmetric models described above, namely the class of flipped $SU(5)$ supergravity models and the minimal $SU(5)$ supergravity model. Our purpose is to describe their features and compare their predictions. However, we emphasize throughout that the underlying motivation for these models is different: flipped $SU(5)$ is an archetypal string model, whereas minimal $SU(5)$ is an archetypal traditional grand unified model. The conceptual differences between these two frameworks are significant, *e.g.*:

1. In string models the gauge couplings of the various gauge groups (which are typical in these constructions [9]) all unify at a calculable scale near the Planck scale.[2] In fact, the decoupling effect of the infinite tower of string massive states can be incorporated exactly [11, 12, 13]. In contrast, in traditional GUT models the unification scale is not predicted by the theory, although it can be easily determined by running the low-energy gauge couplings until they unify.

[1]The parameters in V_0 must satisfy further consistency constraints to insure that this is a true minimum of the tree-level Higgs potential. As discussed in Ref. [8], the one-loop effective potential satisfies most of these constraints automatically.

[2]This result holds as long as all gauge groups are represented by level-one Kac-Moody algebras, otherwise various factors appear in the unification relations [10].

2. Grand unified gauge groups describing the observable sector of string models are rare [14], if not disfavored. One can easily obtain the Standard Model ($SU(3)\times SU(2)$) gauge groups [15, 16], the Pati-Salam ($SU(4)\times SU(2)\times SU(2)$) gauge group [17], and the flipped $SU(5)$ [18, 19] gauge group [20]. Of these, only flipped $SU(5)$ actually unifies the non-abelian gauge couplings of the Standard Model. A proliferation of $U(1)$ factors is the norm. It is amusing to note that flipped $SU(5)$ can be argued to be the simplest unified gauge group [19, 21], in that $SU(5)$ does not provide for neutrino masses, and $SO(10)$ includes flipped $SU(5)$.

3. The spectra of string models are correlated with the corresponding gauge groups [22] and the fermion matter representations are automatically anomaly-free. In contrast, in traditional GUT models anomalies are actual constraints on the possible models and the representations one can choose from are only limited by gauge invariance. For instance, the missing partner mechanism [23] to effect the doublet-triplet splitting of the Higgs pentaplets in $SU(5)$ requires the introduction of large representations which in a string model can only occur when complicated higher-level Kac–Moody algebra realizations of the gauge group are invoked [22].

4. All interactions in string models can be calculated once the model is specified. This full specification of the effective supergravity and the superpotential has no analog in GUT models, where all these are free parameters.

This paper is organized as follows. In Sec. 2 we discuss the typical characteristics of string models. In Sec. 3 we present in detail the flipped $SU(5)$ models which we consider here. In Sec. 4 we discuss the string-inspired supersymmetry breaking scenaria that we explore later. In Sec. 5 we consider the experimental predictions for all the sparticle and one-loop corrected Higgs boson masses in the flipped $SU(5)$ models, and deduce several simple relations among the various sparticle masses. In Sec. 6 we repeat this analysis for special more constrained cases of the chosen supersymmetry breaking scenaria. In Sec. 7 we discuss the minimal $SU(5)$ supergravity model, including the constraints from proton decay and the neutralino relic density. In Sec. 8 we discuss the prospects for direct experimental detection of these particles at Fermilab, LEPI,II, and HERA, while in Sec. 9 we consider the corresponding indirect detection signatures. Finally, in Sec. 10 we summarize our conclusions.

2 Typical Characteristics of String Models

The ultimate unification of all particles and interactions has string theory as the best candidate. If this theory were completely understood, we would be able to show that string theory is either inconsistent with the low-energy world or supported by experimental data. Since our present knowledge of string theory is at best fragmented and certainly incomplete, it is important to consider models which incorporate as many stringy ingredients as possible. The number of such models is expected to be large, however, the basic ingredients that such "string models" should incorporate fall into few categories: (i) gauge group and matter representations which unify at a calculable model-dependent string unification scale; (ii) a hidden sector which becomes strongly interacting at an intermediate scale and triggers supersymmetry breaking with vanishing vacuum energy and hierarchically small soft superpersymmetry breaking parameters; (iii) acceptable high-energy phenomenology, *e.g.*, gauge symmetry breaking to the Standard Model (if needed), not-too-rapid proton decay, decoupling of intermediate-mass-scale unobserved matter states, etc.; (iv) radiative electroweak symmetry breaking; (v) acceptable low-energy phenomenology, *e.g.*, reproduce the observed spectrum of quark and lepton masses and the quark mixing angles, sparticle and Higgs masses not in conflict with present experimental bounds, and not-too-large neutralino cosmological relic density.

All the above are to be understood as constraints on potentially realistic string models. Since some of the above constraints can be independently satisfied in specific models, the real power of a string model rests in the successful satisfaction of all these constraints within a single model. In what follows, "string-derived" models refer to models which can be derived rigorously from string, even if not all their interactions have been determined explicitly; "string-inspired" models are those field-theoretical models which are believed to be in principle derivable from string, although most likely not exactly reproducible; finally "string" models refer generically to both kinds, although perhaps describe more accurately the conceptual framework these models are examples of.

String model-building is at a state of development where large numbers of models can be constructed using various techniques (so-called formulations) [9]. Such models provide a gauge group and associated set of matter representations, as well as all interactions in the superpotential, the Kähler potential, and the gauge kinetic function. The effective string supergravity can then be worked out and thus all the above constraints can in principle be enforced. In practice this approach has never been followed in its entirety: sophisticated model-building techniques exist which can produce models satisfying constraints (i), (iii), (iv) and part of (v); detailed studies of supersymmetry breaking triggered by gaugino condensation have been performed for generic hidden sectors; and extensive explorations of the soft-supersymmetry breaking parameter space satisfying constraints (iii), (iv), and (v) have been conducted.

In searching for good string model candidates, we are faced with two kinds of choices to be made: the choice of the gauge and matter content of the model, and the choice of the supersymmetry breaking mechanism. Fortunately, a string theory theorem provides significant enlightenment regarding the first choice: models whose gauge groups are constructed from level-one Kac-Moody algebras do not allow adjoint or higher representations in their spectra [22]. This implies that the traditional GUT groups ($SU(5), SO(10), E_6$) are excluded since the GUT symmetry would remain unbroken. Exceptions to this theorem exist if one uses the technically complicated higher-level Kac-Moody algebras [14], but these models are beset with constraints [22]. If one imposes the aesthetic constraint of unification of the Standard

Model non-abelian gauge couplings, then flipped $SU(5)$ [18, 19, 20, 21] emerges as the prime candidate, as we shortly discuss. String models without non-abelian unification, such as the standard-like models of Refs. [15, 16] and the Pati-Salam–like model of Ref. [17] possess nonetheless gauge coupling unification at the string scale, even though no larger structure is revealed past this scale. However, the degree of phenomenological success which some of these models enjoy, usually rests on some fortuitous set of vanishing couplings which are best understood in terms of remnants of higher symmetries.

Besides the very economic GUT symmetry breaking mechanism in flipped $SU(5)$ [18, 19] – which allows it to be in principle derivable from superstring theory [20] – perhaps one of the more interesting motivations for considering such a unified gauge group is the natural avoidance of potentially dangerous dimension-five proton decay operators [21]. In Ref. [24] we constructed a supergravity model based on this gauge group, which has the additional property of unifying at a scale $M_U = \mathcal{O}(10^{18})$ GeV, as expected to occur in string-derived versions of this model [12]. As such, this model constitutes a blueprint for string model builders. In fact, in Ref. [25] one such model was derived from string and served as inspiration for the field theory model in Ref. [24]. The string unification scale should be contrasted with the naive unification scale, $M_U = \mathcal{O}(10^{16}\,\text{GeV})$, obtained by running the Standard Model particles and their superpartners to very high energies. This apparent discrepancy of two orders of magnitude [26, 27] creates a *gap* which needs to be bridged somehow in string models. It has been shown [28] that the simplest solution to this problem is the introduction in the spectrum of heavy vector-like particles with Standard Model quantum numbers. The minimal such choice [29, 30], a quark doublet pair $Q, \bar{Q}$ and a 1/3–charge quark singlet pair $D, \bar{D}$, fit snugly inside a $\mathbf{10},\overline{\mathbf{10}}$ pair of flipped $SU(5)$ representations, beyond the usual $3\cdot(\mathbf{10}+\overline{\mathbf{5}}+\mathbf{1})$ of matter and $\mathbf{10},\overline{\mathbf{10}}$ of Higgs.

In this model, gauge symmetry breaking occurs due to vacuum expectation values (vevs) of the neutral components of the $\mathbf{10},\overline{\mathbf{10}}$ Higgs representations, which develop along flat directions of the scalar potential. There are two known ways in which these vevs (and thus the symmetry breaking scale) could be determined:

(i) In the conventional way, radiative corrections to the scalar potential in the presence of soft supersymmetry breaking generate a global minimum of the potential for values of the vevs slightly below the scale where supersymmetry breaking effects are first felt in the observable sector [21]. If the latter scale is the Planck scale (in a suitable normalization) then $M_U \sim M_{Pl}/\sqrt{8\pi} \sim 10^{18}\,\text{GeV}$.

(ii) In string-derived models a pseudo $U_A(1)$ anomaly arises as a consequence of truncating the theory to just the massless degrees of freedom, and adds a contribution to its D-term, $D_A = \sum q_i^A |\langle\phi_i\rangle|^2 + \epsilon$, with $\epsilon = g^2 \mathrm{Tr}\, U_A(1)/192\pi^2 \sim (10^{18}\,\text{GeV})^2$ [31]. To avoid a huge breaking of supersymmetry we need to demand $D_A = 0$ and therefore the fields charged under $U_A(1)$ need to get suitable vevs. Among these one generally finds the symmetry breaking Higgs fields, and thus $M_U \sim 10^{18}\,\text{GeV}$ follows.

In general, both these mechanisms could produce somewhat lower values of M_U. However, $M_U \gtrsim 10^{16}\,\text{GeV}$ is necessary to avoid too rapid proton decay due to dimension-six operators [32]. In these more general cases the $SU(5)$ and $U(1)$

gauge couplings would not unify at M_U (only α_2 and α_3 would), although they would eventually "superunify" at the string scale $M_{SU} \sim 10^{18}\,\text{GeV}$. To simplify matters, below we consider the simplest possible case of $M_U = M_{SU} \sim 10^{18}\,\text{GeV}$. We also draw inspiration from string model-building and regard the Higgs mixing term $\mu h\bar{h}$ as a result of an effective higher-order coupling [33, 34, 35], instead of as a result of a light singlet field getting a small vev (*i.e.*, $\lambda h\bar{h}\phi \to \lambda \langle\phi\rangle h\bar{h}$) as originally considered [19, 21]. An additional contribution to μ is also generically present in supergravity models [36, 35, 37].

The choice of supersymmetry breaking scenario is less clear. Below we show that the phenomenologically acceptable choices basically fall in two categories:

1. The no-scale ansatz [38], which ensures the vanishing of the (tree-level) cosmological constant even after supersymmetry breaking. This framework also arises in the low-energy limit of superstring theory [39]. In a theory which contains heavy fields, the minimal no-scale structure $SU(1,1)$ [40] is generalized to $SU(N,1)$ [41] which implies that the scalar fields do not feel the supersymmetry breaking effects. In practice this means that the universal scalar mass (m_0) and the universal cubic scalar coupling (A) are set to zero. The sole source of supersymmetry breaking is the universal gaugino mass ($m_{1/2}$), *i.e.*,
$$m_0 = 0, \qquad A = 0. \tag{4}$$

2. The dilaton F-term scenario, which also leads to universal soft supersymmetry breaking parameters [37]
$$m_0 = \tfrac{1}{\sqrt{3}} m_{1/2}, \qquad A = -m_{1/2}. \tag{5}$$

In either case, after enforcement of the above constraints, the low-energy theory can be described in terms of just three parameters: the top-quark mass (m_t), the ratio of Higgs vacuum expectation values ($\tan\beta$), and the gluino mass ($m_{\tilde{g}} \propto m_{1/2}$). Therefore, measurement of only two sparticle or Higgs masses would determine the remaining thirty. Moreover, if the hidden sector responsible for these patterns of soft supersymmetry breaking is specified, the gravitino mass (m_0) will also be determined and the supersymmetry breaking sector of the theory will be completely fixed.

In sum, we see basically two *unified* string supergravity models emerging as good candidates for phenomenologically acceptable string models, both of which include a flipped $SU(5)$ observable gauge group supplemented by matter representations in order to unify at the string scale $M_U \sim 10^{18}\,\text{GeV}$ [28, 29, 30], and supersymmetry breaking is parametrized by either of the scenaria in Eqs. (4,5).

We should remark that a real string model will include a hidden sector in addition to the observable sector discussed in what follows. The model presented here tacitly assumes that such hidden sector is present and that it has suitable properties. For example, the superpotential in Eq. (9) below, in a string model will receive contributions from cubic and higher-order terms, with the latter generating effective observable sector couplings once hidden sector matter condensates develop [33]. The hidden sector is also assumed to play a fundamental role in triggering supersymmetry breaking via *e.g.*, gaugino condensation. This in turn would make possible the mechanism for gauge symmetry breaking discussed above. Probably the most important constraint on this sector of the theory is that it should yield one of the two supersymmetry breaking scenaria outlined above.

3 The SU(5) x U(1) Models

The model we consider is a generalization of that presented in Ref. [19], and contains the following flipped $SU(5)$ fields:

1. three generations of quark and lepton fields $F_i, \bar{f}_i, l_i^c$, $i = 1,2,3$;
2. two pairs of Higgs **10**,$\overline{\mathbf{10}}$ representations $H_i, \bar{H}_i$, $i = 1,2$;
3. one pair of "electroweak" Higgs **5**,$\overline{\mathbf{5}}$ representations $h, \bar{h}$;
4. three singlet fields $\phi_{1,2,3}$.

Under $SU(3) \times SU(2)$ the various flipped $SU(5)$ fields decompose as follows:

$$\begin{aligned} F_i &= \{Q_i, d_i^c, \nu_i^c\}, \quad \bar{f}_i = \{L_i, u_i^c\}, \quad l_i^c = e_i^c, & (6)\\ H_i &= \{Q_{H_i}, d_{H_i}^c, \nu_{H_i}^c\}, \quad \bar{H}_i = \{Q_{\bar{H}_i}, d_{\bar{H}_i}^c, \nu_{\bar{H}_i}^c\}, & (7)\\ h &= \{H, D\}, \quad \bar{h} = \{\bar{H}, \bar{D}\}. & (8)\end{aligned}$$

The most general effective[3] superpotential consistent with $SU(5) \times U(1)$ symmetry is given by

$$\begin{aligned} W &= \lambda_1^{ij} F_i F_j h + \lambda_2^{ij} F_i \bar{f}_j \bar{h} + \lambda_3^{ij} \bar{f}_i l_j^c h + \mu h\bar{h} + \lambda_4^{ij} H_i H_j h + \lambda_5^{ij} \bar{H}_i \bar{H}_j \bar{h} \\ &+ \lambda_{1'}^{ij} H_i F_j h + \lambda_{2'}^{ij} H_i \bar{f}_j \bar{h} + \lambda_6^{ijk} F_i \bar{H}_j \phi_k + w^{ij} H_i \bar{H}_j + \mu^{ij} \phi_i \phi_j. \qquad (9)\end{aligned}$$

Symmetry breaking is effected by non-zero vevs $\langle \nu_{H_i}^c \rangle = V_i$, $\langle \nu_{\bar{H}_i}^c \rangle = \bar{V}_i$, such that $V_1^2 + V_2^2 = \bar{V}_1^2 + \bar{V}_2^2$.

3.1 Higgs doublet and triplet mass matrices

The Higgs doublet mass matrix receives contributions from $\mu h\bar{h} \to \mu H\bar{H}$ and $\lambda_{2'}^{ij} H_i \bar{f}_j \bar{h}$ - $\lambda_{2'}^{ij} V_i L_j \bar{H}$. The resulting matrix is

$$\mathcal{M}_2 = \begin{matrix} \\ H \\ L_1 \\ L_2 \\ L_3 \end{matrix} \begin{matrix} \bar{H} \\ \begin{pmatrix} \mu \\ \lambda_{2'}^{i1} V_i \\ \lambda_{2'}^{i2} V_i \\ \lambda_{2'}^{i3} V_i \end{pmatrix} \end{matrix}. \qquad (10)$$

[3]To be understood in the string context as arising from cubic and higher order terms [42, 33].

To avoid fine-tunings of the $\lambda_{2'}^{ij}$ couplings we must demand $\lambda_{2'}^{ij} \equiv 0$, so that $\bar{H}$ remains light.

The Higgs triplet matrix receives several contributions: $\mu h\bar{h} \to \mu D\bar{D}$; $\lambda_{1'}^{ij} H_i F_j h \to \lambda_{1'}^{ij} V_i d_j^c D$; $\lambda_4^{ij} H_i H_j h \to \lambda_4^{ij} V_i d_{H_j}^c D$; $\lambda_5^{ij} \bar{H}_i \bar{H}_j \bar{h} \to \lambda_5^{ij} \bar{V}_i d_{\bar{H}_j}^c \bar{D}$; $w^{ij} d_{H_i}^c d_{\bar{H}_j}^c$. The resulting matrix is[4]

$$\mathcal{M}_3 = \begin{array}{c} \\ D \\ d^c_{\bar{H}_1} \\ d^c_{\bar{H}_2} \end{array} \begin{array}{c} \begin{array}{cccccc} \bar{D} & d^c_{H_1} & d^c_{H_2} & d^c_1 & d^c_2 & d^c_3 \end{array} \\ \left(\begin{array}{cccccc} \mu & \lambda_4^{i1} V_i & \lambda_4^{i2} V_i & \lambda_{1'}^{i1} V_i & \lambda_{1'}^{i2} V_i & \lambda_{1'}^{i3} V_i \\ \lambda_5^{i1} \bar{V}_i & w_{11} & w_{12} & 0 & 0 & 0 \\ \lambda_5^{i2} \bar{V}_i & w_{21} & w_{22} & 0 & 0 & 0 \end{array} \right) \end{array} . \tag{11}$$

Clearly three linear combinations of $\{\bar{D}, d^c_{H_{1,2}}, d^c_{1,2,3}\}$ will remain light. In fact, such a general situation will induce a mixing in the down-type Yukawa matrix $\lambda_1^{ij} F_i F_j h \to \lambda_1^{ij} Q_i d_j^c H$, since the d_j^c will need to be re-expressed in terms of these mixed light eigenstates.[5] This low-energy quark-mixing mechanism is an explicit realization of the general extra-vector-abeyance (EVA) mechanism of Ref. [43]. As a first approximation though, in what follows we will set $\lambda_{1'}^{ij} = 0$, so that the light eigenstates are $d^c_{1,2,3}$.

3.2 Neutrino see-saw matrix

The see-saw neutrino matrix receives contributions from: $\lambda_2^{ij} F_i \bar{f}_j \bar{h} \to m_u^{ij} \nu_i^c \nu_j$; $\lambda_6^{ijk} F_i \bar{H}_j \phi_k \to \lambda_6^{ijk} \bar{V}_j \nu_i^c \phi_k$; $\mu^{ij} \phi_i \phi_j$. The resulting matrix is[6]

$$\mathcal{M}_\nu = \begin{array}{c} \\ \nu_i \\ \nu_i^c \\ \phi_i \end{array} \begin{array}{c} \begin{array}{ccc} \nu_j & \nu_j^c & \phi_j \end{array} \\ \left(\begin{array}{ccc} 0 & m_u^{ji} & 0 \\ m_u^{ij} & 0 & \lambda_6^{ikj} \bar{V}_k \\ 0 & \lambda_6^{jki} \bar{V}_k & \mu^{ij} \end{array} \right) \end{array} . \tag{12}$$

3.3 Numerical scenario

To simplify the discussion we will assume, besides[7] $\lambda_{1'}^{ij} = \lambda_{2'}^{ij} \equiv 0$, that

$$\lambda_4^{ij} = \delta^{ij} \lambda_4^{(i)}, \quad \lambda_5^{ij} = \delta^{ij} \lambda_5^{(i)}, \quad \lambda_6^{ijk} = \delta^{ij} \delta^{ik} \lambda_6^{(i)}, \tag{13}$$

$$\mu^{ij} = \delta^{ij} \mu_i, \quad w^{ij} = \delta^{ij} w_i. \tag{14}$$

[4]The zero entries in $\mathcal{M}_3$ result from the assumption $\langle \phi_k \rangle = 0$ in $\lambda_6^{ijk} F_i \bar{H}_j \phi_k$.

[5]Note that this mixing is on top of any structure that λ_1^{ij} may have, and is the only source of mixing in the typical string model-building case of a diagonal λ_2 matrix.

[6]We neglect a possible higher-order contribution which could produce a non-vanishing $\nu_i^c \nu_j^c$ entry [44].

[7]In Ref. [19] the discrete symmetry $H_1 \to -H_1$ was imposed so that these couplings automatically vanish when $H_2, \bar{H}_2$ are not present. This symmetry (generalized to $H_i \to -H_i$) is not needed here since it would imply $w^{ij} \equiv 0$, which is shown below to be disastrous for gauge coupling unification.

These choices are likely to be realized in string versions of this model and will not alter our conclusions below. In this case the Higgs triplet mass matrix reduces to

$$\mathcal{M}_3 = \begin{array}{c} \\ D \\ d^c_{\bar{H}_1} \\ d^c_{\bar{H}_2} \end{array} \begin{array}{c} \begin{array}{ccc} \bar{D} & d^c_{H_1} & d^c_{H_2} \end{array} \\ \begin{pmatrix} \mu & \lambda_4^{(1)} V_1 & \lambda_4^{(2)} V_2 \\ \lambda_5^{(1)} \bar{V}_1 & w_1 & 0 \\ \lambda_5^{(2)} \bar{V}_2 & 0 & w_2 \end{pmatrix} \end{array} . \tag{15}$$

Regarding the $(3,2)$ states, the scalars get either eaten by the X, Y $SU(5)$ heavy gauge bosons or become heavy Higgs bosons, whereas the fermions interact with the $\widetilde{X}, \widetilde{Y}$ gauginos through the following mass matrix [25]

$$\mathcal{M}_{(3,2)} = \begin{array}{c} \\ Q_{H_1} \\ Q_{H_2} \\ \widetilde{X} \end{array} \begin{array}{c} \begin{array}{ccc} Q_{\bar{H}_1} & Q_{\bar{H}_2} & \widetilde{Y} \end{array} \\ \begin{pmatrix} w_1 & 0 & g_5 V_1 \\ 0 & w_2 & g_5 V_2 \\ g_5 \bar{V}_1 & g_5 \bar{V}_2 & 0 \end{pmatrix} \end{array} . \tag{16}$$

The lightest eigenvalues of these two matrices (denoted generally by d^c_H and Q_H respectively) constitute the new relatively light particles in the spectrum, which are hereafter referred to as the "*gap*" particles since with suitable masses they bridge the gap between unification masses at 10^{16} GeV and 10^{18} GeV.

Guided by the phenomenological requirement on the gap particle masses, *i.e.*, $M_{Q_H} \gg M_{d^c_H}$ [29], we consider the following explicit numerical scenario

$$\lambda_4^{(2)} = \lambda_5^{(2)} = 0, \quad V_1, \bar{V}_1, V_2, \bar{V}_2 \sim V \gg w_1 \gg w_2 \gg \mu, \tag{17}$$

which would need to be reproduced in a viable string-derived model. From Eq. (15) we then get $M_{d^c_{H_2}} = M_{d^c_{\bar{H}_2}} = w_2$, and all other mass eigenstates $\sim V$. Furthermore, $\mathcal{M}_{(3,2)}$ has a characteristic polynomial $\lambda^3 - \lambda^2(w_1 + w_2) - \lambda(2V^2 - w_1 w_2) + (w_1 + w_2)V^2 = 0$, which has two roots of $\mathcal{O}(V)$ and one root of $\mathcal{O}(w_1)$. The latter corresponds to $\sim (Q_{H_1} - Q_{H_2})$ and $\sim (Q_{\bar{H}_1} - Q_{\bar{H}_2})$. In sum then, the gap particles have masses $M_{Q_H} \sim w_1$ and $M_{d^c_H} \sim w_2$, whereas all other heavy particles have masses $\sim V$.

The see-saw matrix reduces to

$$\mathcal{M}_\nu = \begin{array}{c} \\ \nu_i \\ \nu^c_i \\ \phi_i \end{array} \begin{array}{c} \begin{array}{ccc} \nu_i & \nu^c_i & \phi_i \end{array} \\ \begin{pmatrix} 0 & m_u^i & 0 \\ m_u^i & 0 & \lambda^{(i)} \bar{V}_i \\ 0 & \lambda^{(i)} \bar{V}_i & \mu^i \end{pmatrix} \end{array} , \tag{18}$$

for each generation. The physics of this see-saw matrix has been discussed in Ref. [44] and more generally in Ref. [45], where it was shown to lead to an interesting amount of hot dark matter (ν_τ) and an MSW-effect (ν_e, ν_μ) compatible with all solar neutrino data. Moreover, the out-of-equilibrium decays of the ν^c "flipped neutrino" fields in the early Universe induce a lepton number asymmetry which is later processed into a baryon number asymmetry by non-perturbative electroweak processes [46, 45]. All these phenomena can occur in the same region of parameter space.

3.4 Proton decay

The dimension-six operators mediating proton decay in this model are highly suppressed due to the large mass of the X, Y gauge bosons ($\sim M_U = 10^{18}\,\mathrm{GeV}$). Higgsino mediated dimension-five operators exist and are naturally suppressed in the minimal model of Ref. [19]. The reason for this is that the Higgs triplet mixing term $\mu h\bar{h} \to \mu D\bar{D}$ is small ($\mu \sim M_Z$), whereas the Higgs triplet mass eigenstates obtained from Eq. (11) by just keeping the 2×2 submatrix in the upper left-hand corner, are always very heavy ($\sim V$). The dimension-five mediated operators are then proportional to μ/V^2 and thus the rate is suppressed by a factor or $(\mu/V)^2 \ll 1$ relative to the unsuppressed case found in the standard $SU(5)$ model.

In the generalized model presented here, the Higgs triplet mixing term is still $\mu D\bar{D}$. However, the exchanged mass eigenstates are not necessarily all very heavy. In fact, above we have demanded the existence of a relatively light ($\sim w_1$) Higgs triplet state (d^c_H). In this case the operators are proportional to $\mu\alpha_i\bar{\alpha}_i/\mathcal{M}_i^2$, where $\mathcal{M}_i$ is the mass of the i-th exchanged eigenstate and $\alpha_i, \bar{\alpha}_i$ are its $D, \bar{D}$ admixtures. In the scenario described above, the relatively light eigenstates ($d^c_{H_2}, d^c_{\bar{H}_2}$) contain no $D, \bar{D}$ admixtures, and the operator will again be $\propto \mu/V^2$.

Note however that if conditions (17) (or some analogous suitability requirement) are not satisfied, then diagonalization of $\mathcal{M}_3$ in Eq. (15) may re-introduce a sizeable dimension-five mediated proton decay rate, depending on the value of the $\alpha_i, \bar{\alpha}_i$ coefficients. To be safe one should demand [47, 48, 49]

$$\frac{\mu\alpha_i\bar{\alpha}_i}{\mathcal{M}_i^2} \lesssim \frac{1}{10^{17}\,\mathrm{GeV}}. \tag{19}$$

For the higher values of $M_{d^c_H}$ in Table 1 (see below), this constraint can be satisfied for not necessarily small values of $\alpha_i, \bar{\alpha}_i$.

3.5 Gauge coupling unification

Since we have chosen $V \sim M_U = M_{SU} = 10^{18}\,\mathrm{GeV}$, this means that the Standard Model gauge couplings should unify at the scale M_U. However, their running will be modified due to the presence of the gap particles. Note that the underlying flipped $SU(5)$ symmetry, even though not evident in this respect, is nevertheless essential in the above discussion. The masses M_Q and $M_{d^c_H}$ can then be determined, as follows [29]

$$\ln\frac{M_{Q_H}}{m_Z} = \pi\left(\frac{1}{2\alpha_e} - \frac{1}{3\alpha_3} - \frac{\sin^2\theta_w - 0.0029}{\alpha_e}\right) - 2\ln\frac{M_U}{m_Z} - 0.63, \tag{20}$$

$$\ln\frac{M_{d^c_H}}{m_Z} = \pi\left(\frac{1}{2\alpha_e} - \frac{7}{3\alpha_3} + \frac{\sin^2\theta_w - 0.0029}{\alpha_e}\right) - 6\ln\frac{M_U}{m_Z} - 1.47, \tag{21}$$

where α_e, α_3 and $\sin^2\theta_w$ are all measured at M_Z. This is a one-loop determination (the constants account for the dominant two-loop corrections) which neglects all low-

Table 1: The value of the gap particle masses and the unified coupling for $\alpha_3(M_Z) = 0.118 \pm 0.008$. We have taken $M_U = 10^{18}\,\mathrm{GeV}$, $\sin^2\theta_w = 0.233$, and $\alpha_e^{-1} = 127.9$.

$\alpha_3(M_Z)$	$M_{d^c_H}$ (GeV)	M_{Q_H} (GeV)	$\alpha(M_U)$
0.110	4.9×10^4 GeV	2.2×10^{12} GeV	0.0565
0.118	4.5×10^6 GeV	4.1×10^{12} GeV	0.0555
0.126	2.3×10^8 GeV	7.3×10^{12} GeV	0.0547

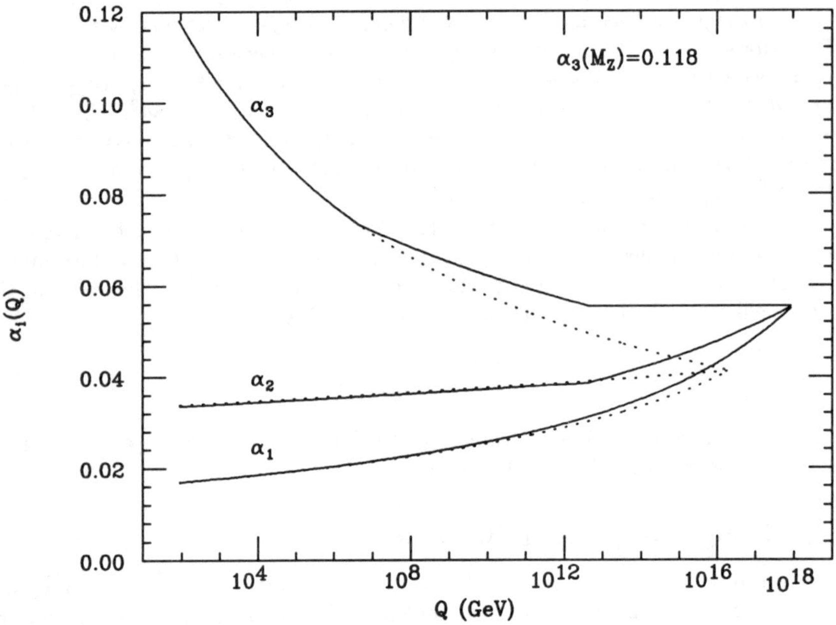

Figure 1: The running of the gauge couplings in the flipped $SU(5)$ model for $\alpha_3(M_Z) = 0.118$ (solid lines). The gap particle masses have been derived using the gauge coupling RGEs to achieve unification at $M_U = 10^{18}\,\mathrm{GeV}$. The case with no gap particles (dotted lines) is also shown; here $M_U \approx 10^{16}\,\mathrm{GeV}$.

and high-energy threshold effects,[8] but is quite adequate for our present purposes. As shown in Table 1 (and Eq. (21)) the d^c_H mass depends most sensitively on $\alpha_3(M_Z) = 0.118\pm0.008$ [50], whereas the Q_H mass and the unified coupling are rather insensitive to it. The unification of the gauge couplings is shown in Fig. 1 (solid lines) for the central value of $\alpha_3(M_Z)$. This figure also shows the case of no gap particles (dotted lines), for which $M_U \approx 10^{16}\,\mathrm{GeV}$.

4 String-inspired Supersymmetry Breaking Scenaria

Supersymmetry breaking in string models can generally be triggered in a phenomenologically acceptable way by non-zero F-terms for: (a) any of the moduli fields of the string model ($\langle F_M\rangle$) [51], (b) the dilaton field ($\langle F_D\rangle$) [37], or (c) the hidden matter fields ($\langle F_H\rangle$) [52]. It has been recently noted [37] that much model-independent information can be obtained about the structure of the soft supersymmetry breaking parameters in generic string supergravity models if one neglects the third possibility ($\langle F_H\rangle = 0$) and assumes that either: (i) $\langle F_M\rangle \gg \langle F_D\rangle$, or (ii) $\langle F_D\rangle \gg \langle F_M\rangle$.

In case (i) the scalar masses are generally not universal, *i.e.*, $m_i = f_i m_0$ where m_0 is the gravitino mass and f_i are calculable constants, and therefore large flavor-changing-neutral-currents (FCNCs) [53] are potentially dangerous [54]. The gaugino masses arise from the one-loop contribution to the gauge kinetic function and are thus suppressed ($m_{1/2} \sim (\alpha/4\pi)m_0$) [54, 55, 37]. The experimental constraints on the gaugino masses then force the squark and slepton masses into the TeV range [55]. It is interesting to note that this supersymmetry breaking scenario is not unlike that required for the minimal $SU(5)$ supergravity model in order to have the dimension-five proton decay operators under control [47, 56, 49], which requires $m_{1/2}/m_0 \lesssim \frac{1}{3}$, as discussed in Sec. 7.2. This constraint entails potential cosmological troubles: the neutralino relic density is large and one needs to tune the parameters to have the neutralino mass be very near the Higgs and Z resonances [57, 56, 49, 58, 59] (see Sec. 7.3). Clearly, such cosmological constraints are going to be exacerbated in the case (i) scenario ($m_{1/2}/m_0 \ll 1$) and will likely require real fine-tuning of the model parameters.

An important exception to case (i) occurs if $f_i \equiv 0$ and all scalar masses at the unification scale vanish ($\langle F_M\rangle_{m_0=0}$), as is the case in unified no-scale supergravity models [38], where the minimal no-scale structure $SU(1,1)$ [40] is generalized to $SU(N,1)$ [41] in the presence of the heavy GUT fields. This special case automatically restores the much needed universality of scalar masses, and in the context of no-scale models also entails $A = 0$, see Eq. (4). A special case of this scenario occurs when the bilinear soft-supersymmetry breaking mass parameter $B(M_U)$ is also required to vanish. With the additional ingredient of a flipped $SU(5)$ gauge group, all the above

[8]Here we assume a common supersymmetric threshold at M_Z. In fact, the supersymmetric threshold and the d^c_H mass are anticorrelated. See Ref. [29] for a discussion.

problems are naturally avoided [24], and interesting predictions for direct [60, 61, 62, 63] and indirect [64, 65, 66] experimental detection follow.

If supersymmetry breaking is triggered by $\langle F_D \rangle$ (case (ii)), one obtains *universal* soft-supersymmetry gaugino and scalar masses and trilinear interactions [37] and the soft-supersymmetry breaking parameters in Eq. (5) result. As well, there is a special more constrained case where $B(M_U) = 2m_0 = \frac{2}{\sqrt{3}} m_{1/2}$ is also required, if one demands that the μ parameter receive contributions solely from supergravity [37]. With the complement of a flipped $SU(5)$ structure, this model has also been seen to avoid all the difficulties of the generic $\langle F_M \rangle$ scenario [67]. This supersymmetry breaking scenario has been studied recently also in the context of the minimal supersymmetric Standard Model (MSSM) in Ref. [68].

More generally, presumably it should be possible to find suitable hidden sectors where an arbitrary choice of $m_0, m_{1/2}, A$ is realized, keeping the vacuum energy at zero. This is the attitude taken below when studying the minimal $SU(5)$ supergravity model, where the proton decay constraint imposes severe restrictions on the allowed choices of these three parameters. In the context of string-inspired models (*i.e.*, the flipped $SU(5)$ case), in what follows we restrict ourselves to the two supersymmetry breaking scenaria in Eqs. (4,5) and their special cases ($B(M_U) = 0$ and $B(M_U) = 2m_0$, respectively).

5 Flipped Phenomenology: General Case

The procedure to extract the low-energy predictions of the models outlined above is rather standard by now (see *e.g.*, Ref. [69]): (a) the bottom-quark and tau-lepton masses, together with the input values of m_t and $\tan\beta$ are used to determine the respective Yukawa couplings at the electroweak scale; (b) the gauge and Yukawa couplings are then run up to the unification scale $M_U = 10^{18}\,\mathrm{GeV}$ taking into account the extra vector-like quark doublet ($\sim 10^{12}\,\mathrm{GeV}$) and singlet ($\sim 10^{6}\,\mathrm{GeV}$) introduced above [29, 24]; (c) at the unification scale the soft-supersymmetry breaking parameters are introduced (according to Eqs. (4,5)) and the scalar masses are then run down to the electroweak scale; (d) radiative electroweak symmetry breaking is enforced by minimizing the one-loop effective potential which depends on the whole mass spectrum, and the values of the Higgs mixing term $|\mu|$ and the bilinear soft-supersymmetry breaking parameter B are determined from the minimization conditions; (e) all known phenomenological constraints on the sparticle and Higgs masses are applied (most importantly the LEP lower bounds on the chargino and Higgs masses), including the cosmological requirement of not-too-large neutralino relic density.

5.1 Mass ranges

We have scanned the parameter space for $m_t = 130, 150, 170\,\text{GeV}$, $\tan\beta = 2 \to 50$ and $m_{1/2} = 50 \to 500\,\text{GeV}$. Imposing the constraint $m_{\tilde{g},\tilde{q}} < 1\,\text{TeV}$ we find

$$\langle F_M\rangle_{m_0=0}: \qquad m_{1/2} < 475\,\text{GeV}, \quad \tan\beta \lesssim 32, \tag{22}$$

$$\langle F_D\rangle: \qquad m_{1/2} < 465\,\text{GeV}, \quad \tan\beta \lesssim 46. \tag{23}$$

These restrictions on $m_{1/2}$ cut off the growth of most of the sparticle and Higgs masses at $\approx 1\,\text{TeV}$. However, the sleptons, the lightest Higgs, the two lightest neutralinos, and the lightest chargino are cut off at a much lower mass, as follows[9]

$$\langle F_M\rangle_{m_0=0}: \left\{ \begin{array}{lll} m_{\tilde{e}_R} < 190\,\text{GeV}, & m_{\tilde{e}_L} < 305\,\text{GeV}, & m_{\tilde{\nu}} < 295\,\text{GeV} \\ m_{\tilde{\tau}_1} < 185\,\text{GeV}, & m_{\tilde{\tau}_2} < 315\,\text{GeV} & \\ m_h < 125\,\text{GeV} & & \\ m_{\chi^0_1} < 145\,\text{GeV}, & m_{\chi^0_2} < 290\,\text{GeV}, & m_{\chi^\pm_1} < 290\,\text{GeV} \end{array} \right. \tag{24}$$

$$\langle F_D\rangle: \left\{ \begin{array}{lll} m_{\tilde{e}_R} < 325\,\text{GeV}, & m_{\tilde{e}_L} < 400\,\text{GeV}, & m_{\tilde{\nu}} < 400\,\text{GeV} \\ m_{\tilde{\tau}_1} < 325\,\text{GeV}, & m_{\tilde{\tau}_2} < 400\,\text{GeV} & \\ m_h < 125\,\text{GeV} & & \\ m_{\chi^0_1} < 145\,\text{GeV}, & m_{\chi^0_2} < 285\,\text{GeV}, & m_{\chi^\pm_1} < 285\,\text{GeV} \end{array} \right. \tag{25}$$

It is interesting to note that due to the various constraints on the model, the gluino and (average) squark masses are bounded from below,

$$\langle F_M\rangle_{m_0=0}: \left\{ \begin{array}{l} m_{\tilde{g}} \gtrsim 245\,(260)\,\text{GeV} \\ m_{\tilde{q}} \gtrsim 240\,(250)\,\text{GeV} \end{array} \right. \qquad \langle F_D\rangle: \left\{ \begin{array}{l} m_{\tilde{g}} \gtrsim 195\,(235)\,\text{GeV} \\ m_{\tilde{q}} \gtrsim 195\,(235)\,\text{GeV} \end{array} \right. \tag{26}$$

for $\mu > 0 (\mu < 0)$. Relaxing the above conditions on $m_{1/2}$ simply allows all sparticle masses to grow further proportional to $m_{\tilde{g}}$.

5.2 Mass relations

The neutralino and chargino masses show a correlation observed before in this class of models [70, 24], namely

$$m_{\chi^0_1} \approx \tfrac{1}{2} m_{\chi^0_2}, \qquad m_{\chi^0_2} \approx m_{\chi^\pm_1} \approx M_2 = (\alpha_2/\alpha_3) m_{\tilde{g}} \approx 0.28 m_{\tilde{g}}. \tag{27}$$

This is because throughout the parameter space $|\mu|$ is generally much larger than M_W and $|\mu| > M_2$. In practice we find $m_{\chi^0_2} \approx m_{\chi^\pm_1}$ to be satisfied quite accurately, whereas $m_{\chi^0_1} \approx \frac{1}{2} m_{\chi^0_2}$ is only qualitatively satisfied, although the agreement is better in the $\langle F_D\rangle$ case. In fact, these two mass relations are much more reliable than the one that links them to $m_{\tilde{g}}$. The heavier neutralino ($\chi^0_{3,4}$) and chargino ($\chi^\pm_2$) masses

[9] In this class of supergravity models the three sneutrinos ($\tilde{\nu}$) are degenerate in mass. Also, $m_{\tilde{\mu}_L} = m_{\tilde{e}_L}$ and $m_{\tilde{\mu}_R} = m_{\tilde{e}_R}$.

Table 2: The value of the c_i coefficients appearing in Eq. (28), the ratio $c_{\tilde{g}} = m_{\tilde{g}}/m_{1/2}$, and the average squark coefficient $\bar{c}_{\tilde{q}}$, for $\alpha_3(M_Z) = 0.118 \pm 0.008$. Also shown are the a_i, b_i coefficients for the central value of $\alpha_3(M_Z)$ and both supersymmetry breaking scenaria ($M : \langle F_M \rangle_{m_0=0}$, $D : \langle F_D \rangle$). The results apply as well to the second-generation squark and slepton masses.

i	c_i (0.110)	c_i (0.118)	c_i (0.126)
$\tilde{\nu}, \tilde{e}_L$	0.406	0.409	0.413
$\tilde{e}_R$	0.153	0.153	0.153
$\tilde{u}_L, \tilde{d}_L$	3.98	4.41	4.97
$\tilde{u}_R$	3.68	4.11	4.66
$\tilde{d}_R$	3.63	4.06	4.61
$c_{\tilde{g}}$	1.95	2.12	2.30
$\bar{c}_{\tilde{q}}$	3.82	4.07	4.80

i	$a_i(M)$	$b_i(M)$	$a_i(D)$	$b_i(D)$
$\tilde{e}_L$	0.302	+1.115	0.406	+0.616
$\tilde{e}_R$	0.185	+2.602	0.329	+0.818
$\tilde{\nu}$	0.302	−2.089	0.406	−1.153
$\tilde{u}_L$	0.991	−0.118	1.027	−0.110
$\tilde{u}_R$	0.956	−0.016	0.994	−0.015
$\tilde{d}_L$	0.991	+0.164	1.027	+0.152
$\tilde{d}_R$	0.950	−0.033	0.989	−0.030

are determined by the value of $|\mu|$; they all approach this limit for large enough $|\mu|$. More precisely, $m_{\chi^0_3}$ approaches $|\mu|$ sooner than $m_{\chi^0_4}$ does. On the other hand, $m_{\chi^0_4}$ approaches $m_{\chi^\pm_2}$ rather quickly.

The first- and second-generation squark and slepton masses can be determined analytically

$$\widetilde{m}_i = \left[m^2_{1/2}(c_i + \xi^2_0) - d_i \frac{\tan^2\beta - 1}{\tan^2\beta + 1} M^2_W \right]^{1/2} = a_i m_{\tilde{g}} \left[1 + b_i \left(\frac{150}{m_{\tilde{g}}} \right)^2 \frac{\tan^2\beta - 1}{\tan^2\beta + 1} \right]^{1/2}, \tag{28}$$

where $d_i = (T_{3i} - Q)\tan^2\theta_w + T_{3i}$ (*e.g.*, $d_{\tilde{u}_L} = \frac{1}{2} - \frac{1}{6}\tan^2\theta_w$, $d_{\tilde{e}_R} = -\tan^2\theta_w$), and $\xi_0 = m_0/m_{1/2} = 0, \frac{1}{\sqrt{3}}$. The coefficients c_i can be calculated numerically in terms of the low-energy gauge couplings, and are given in Table 2[10] for $\alpha_3(M_Z) = 0.118 \pm 0.008$. In the table we also give $c_{\tilde{g}} = m_{\tilde{g}}/m_{1/2}$. Note that these values are smaller than what is obtained in the minimal $SU(5)$ supergravity model (where $c_{\tilde{g}} = 2.90$ for $\alpha_3(M_Z) = 0.118$) and therefore the numerical relations between the gluino mass and the neutralino masses are different in that model. In the table we also show the resulting values for a_i, b_i for the central value of $\alpha_3(M_Z)$. Note that the apparently larger $\tan\beta$ dependence in the $\langle F_M \rangle_{m_0=0}$ case (*i.e.*, $|b_i(M)| > |b_i(D)|$) is actually compensated by a larger minimum value of $m_{\tilde{g}}$ in this case (see Eq. (26)).

The "average" squark mass, $m_{\tilde{q}} \equiv \frac{1}{8}(m_{\tilde{u}_L} + m_{\tilde{u}_R} + m_{\tilde{d}_L} + m_{\tilde{d}_R} + m_{\tilde{c}_L} + m_{\tilde{c}_R} +$

[10] These are renormalized at the scale M_Z. In a more accurate treatment, the c_i would be renormalized at the physical sparticle mass scale, leading to second order shifts on the sparticle masses.

$m_{\tilde{s}_L} + m_{\tilde{s}_R}) = (m_{\tilde{g}}/c_{\tilde{q}})\sqrt{\bar{c}_{\tilde{q}} + \xi_0^2}$, with $\bar{c}_{\tilde{q}}$ given in Table 2, is determined to be

$$m_{\tilde{q}} = \begin{cases} (1.00, 0.95, 0.95)m_{\tilde{g}}, & \langle F_M \rangle_{m_0=0} \\ (1.05, 0.99, 0.98)m_{\tilde{g}}, & \langle F_D \rangle \end{cases} \tag{29}$$

for $\alpha_3(M_Z) = 0.110, 0.118, 0.126$ (the dependence on $\tan\beta$ is small). The squark splitting around the average is $\approx 2\%$.

These masses are plotted in Fig. 2. The thickness and straightness of the lines shows the small $\tan\beta$ dependence, except for $\tilde{\nu}$. The results do not depend on the sign of μ, except to the extent that some points in parameter space are not allowed for both signs of μ: the $\mu < 0$ lines start-off at larger mass values. Note that

$$\langle F_M \rangle_{m_0=0} : \begin{cases} m_{\tilde{e}_R} \approx 0.18 m_{\tilde{g}} \\ m_{\tilde{e}_L} \approx 0.30 m_{\tilde{g}} \\ m_{\tilde{e}_R}/m_{\tilde{e}_L} \approx 0.61 \end{cases} \qquad \langle F_D \rangle : \begin{cases} m_{\tilde{e}_R} \approx 0.33 m_{\tilde{g}} \\ m_{\tilde{e}_L} \approx 0.41 m_{\tilde{g}} \\ m_{\tilde{e}_R}/m_{\tilde{e}_L} \approx 0.81 \end{cases} \tag{30}$$

The third generation squark and slepton masses cannot be determined analytically. In Fig. 3 we show $\tilde{\tau}_{1,2}, \tilde{b}_{1,2}, \tilde{t}_{1,2}$ for the choice $m_t = 150\,\mathrm{GeV}$. The variability on the $\tilde{\tau}_{1,2}$ and $\tilde{b}_{1,2}$ masses is due to the $\tan\beta$-dependence in the off-diagonal element of the corresponding 2×2 mass matrices ($\propto m_{\tau,b}(A_{\tau,b} + \mu\tan\beta)$). The off-diagonal element in the stop-squark mass matrix ($\propto m_t(A_t + \mu/\tan\beta)$) is rather insensitive to $\tan\beta$ but still effects a large $\tilde{t}_1 - \tilde{t}_2$ mass splitting because of the significant A_t contribution. Note that both these effects are more pronounced for the $\langle F_D \rangle$ case since there $|A_{t,b,\tau}|$ are larger than in the $\langle F_M \rangle_{m_0=0}$ case. The lowest values of the $\tilde{t}_1$ mass go up with m_t and can be as low as

$$m_{\tilde{t}_1} \gtrsim \begin{cases} 160, 170, 190\,(155, 150, 170)\,\mathrm{GeV}; & \langle F_M \rangle_{m_0=0} \\ 88, 112, 150\,(92, 106, 150)\,\mathrm{GeV}; & \langle F_D \rangle \end{cases} \tag{31}$$

for $m_t = 130, 150, 170\,\mathrm{GeV}$ and $\mu > 0\,(\mu < 0)$.

The one-loop corrected lightest CP-even (h) and CP-odd (A) Higgs boson masses are shown in Fig. 4 as functions of $m_{\tilde{g}}$ for $m_t = 150\,\mathrm{GeV}$. Following the methods of Ref. [61] we have determined that the LEP lower bound on m_h becomes $m_h \gtrsim 60\,\mathrm{GeV}$, as the figure shows. The largest value of m_h depends on m_t; we find

$$m_h < \begin{cases} 106, 115, 125\,\mathrm{GeV}; & \langle F_M \rangle_{m_0=0} \\ 107, 117, 125\,\mathrm{GeV}; & \langle F_D \rangle \end{cases} \tag{32}$$

for $m_t = 130, 150, 170\,\mathrm{GeV}$. It is interesting to note that the one-loop corrected values of m_h for $\tan\beta = 2$ are quite dependent on the sign of μ. This phenomenon can be traced back to the $\tilde{t}_1 - \tilde{t}_2$ mass splitting which enhances the dominant $\tilde{t}$ one-loop corrections to m_h [71], an effect which is usually neglected in phenomenological analyses. The $\tilde{t}_{1,2}$ masses for $\tan\beta = 2$ and are drawn closer together than the rest. The opposite effect occurs for $\mu < 0$ and therefore the one-loop correction is larger in this case. The sign-of-μ dependence appears in the off-diagonal entries in the $\tilde{t}$

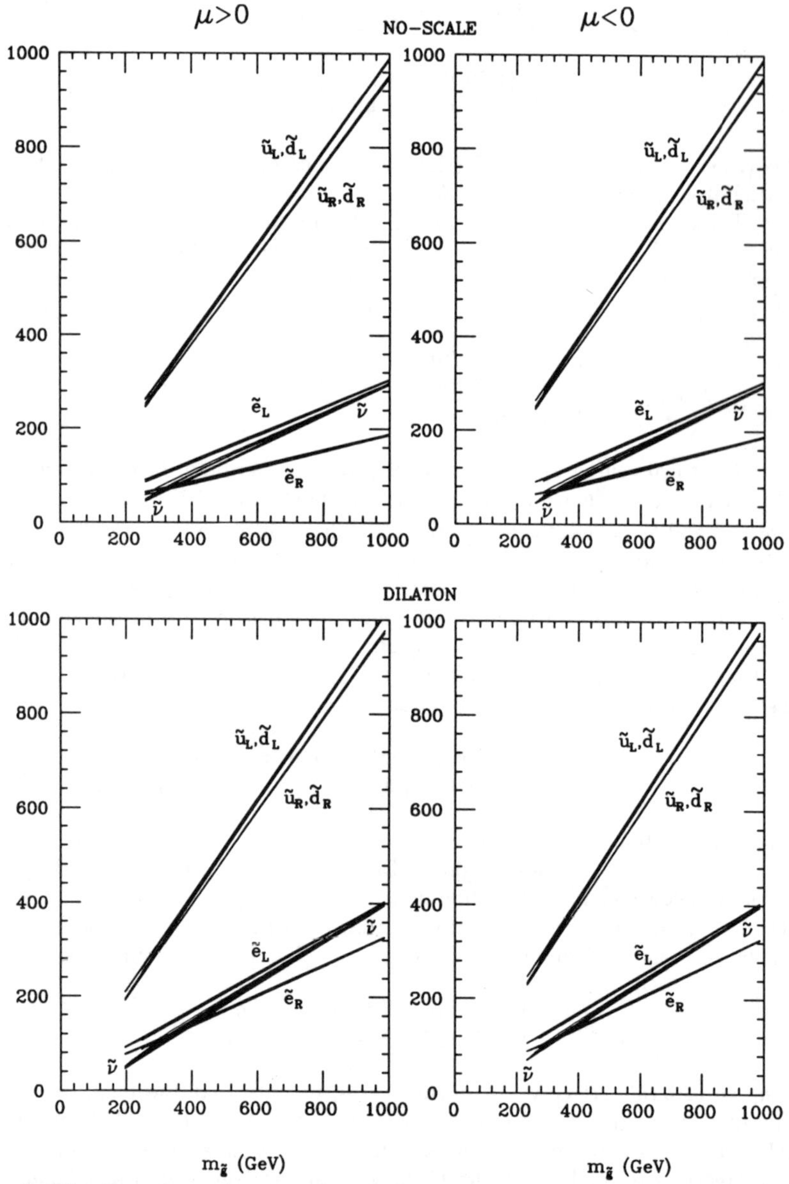

Figure 2: The first-generation squark and slepton masses as a function of the gluino mass, for both signs of μ, $m_t = 150\,\mathrm{GeV}$, and both supersymmetry breaking scenaria under consideration. The same values apply to the second generation. The thickness of the lines and their deviation from linearity are because of the small $\tan\beta$ dependence.

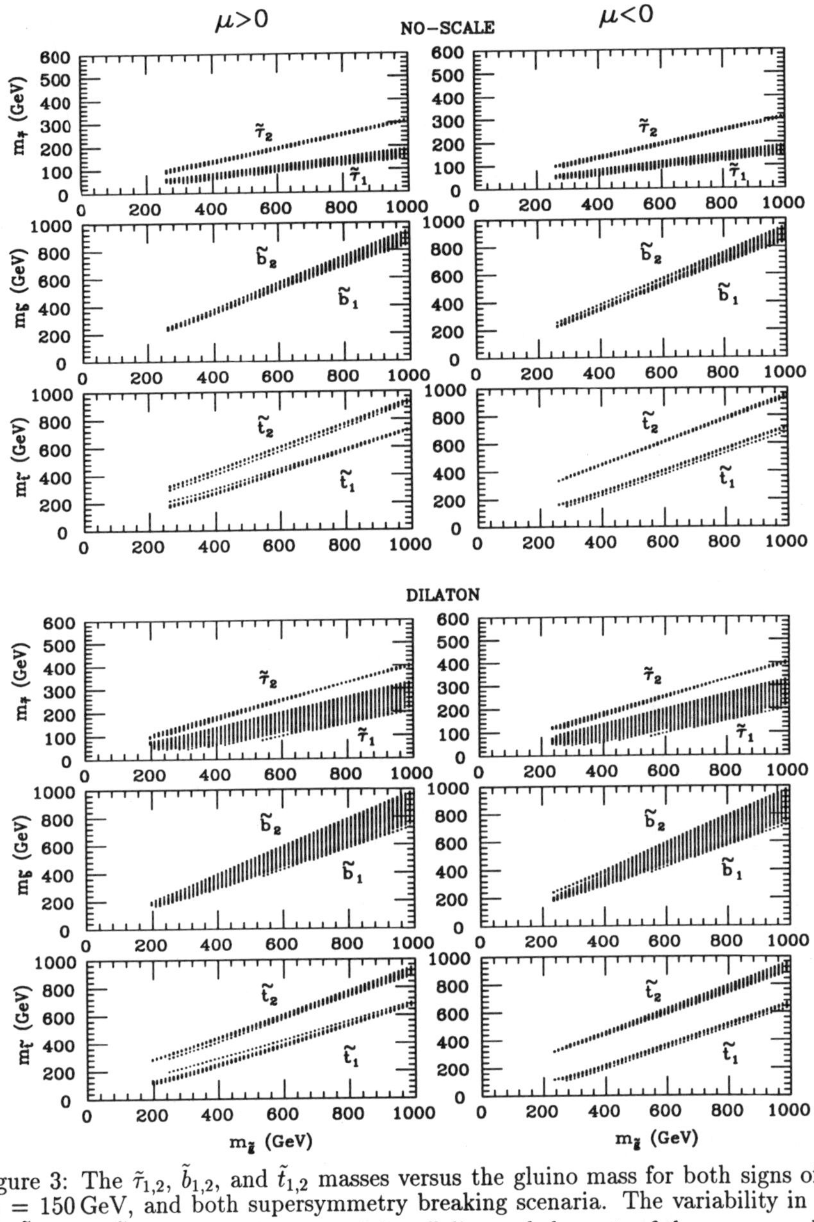

Figure 3: The $\tilde{\tau}_{1,2}$, $\tilde{b}_{1,2}$, and $\tilde{t}_{1,2}$ masses versus the gluino mass for both signs of μ, $m_t = 150\,\mathrm{GeV}$, and both supersymmetry breaking scenaria. The variability in the $\tilde{\tau}_{1,2}$, $\tilde{b}_{1,2}$, and $\tilde{t}_{1,2}$ masses is because of the off-diagonal elements of the corresponding mass matrices.

mass matrix $\propto m_t(A_t + \mu/\tan\beta)$, with $A_t < 0$ in this case. Clearly only small $\tan\beta$ matters, and $\mu < 0$ enhances the splitting. The A-mass grows fairly linearly with $m_{\tilde{g}}$ with a $\tan\beta$-dependent slope which decreases for increasing $\tan\beta$, as shown in Fig. 4. Note that even though m_A can be fairly light, we always get $m_A > m_h$, in agreement with a general theorem to this effect in supergravity theories [72]. This result also implies that the channel $e^+e^- \to hA$ at LEPI is not kinematically allowed in this model.

5.3 Neutralino relic density

The computation of the neutralino relic density (following the methods of Refs. [73, 74]) shows that $\Omega_\chi h_0^2 \lesssim 0.25\,(0.90)$ in the no-scale (dilaton) model. This implies that in these models the cosmologically interesting values $\Omega_\chi h_0^2 \lesssim 1$ occur quite naturally. These results are in good agreement with the observational upper bound on $\Omega_\chi h_0^2$ [75]. Moreover, fits to the COBE data and the small and large scale structure of the Universe suggest [76] a mixture of $\approx 70\%$ cold dark matter and $\approx 30\%$ hot dark matter together with $h_0 \approx 0.5$. The hot dark matter component in the form of massive tau neutrinos has already been shown to be compatible with the flipped $SU(5)$ model we consider here [44, 45], whereas the cold dark matter component implies $\Omega_\chi h_0^2 \approx 0.17$ which is reachable in these models.

6 Flipped Phenomenology: Special Cases

6.1 The strict no-scale case

We now impose the additional constraint $B(M_U) = 0$ to be added to Eq. (4), and obtain the so-called strict no-scale case. Since $B(M_Z)$ is determined by the radiative electroweak symmetry breaking conditions, this added constraint needs to beimposed in a rather indirect way. That is, for given $m_{\tilde{g}}$ and m_t values, we scan the possible values of $\tan\beta$ looking for cases where $B(M_U) = 0$. The most striking result is that solutions exist *only* for $m_t \lesssim 135\,\text{GeV}$ if $\mu > 0$ and for $m_t \gtrsim 140\,\text{GeV}$ if $\mu < 0$. That is, the value of m_t *determines* the sign of μ. Furthermore, for $\mu < 0$ the value of $\tan\beta$ is determined uniquely as a function of m_t and $m_{\tilde{g}}$, whereas for $\mu > 0$, $\tan\beta$ can be double-valued for some m_t range which includes $m_t = 130\,\text{GeV}$ (but does not include $m_t = 100\,\text{GeV}$). In Fig. 5 (top row) we plot the solutions found in this manner for the indicated m_t values.

All the mass relationships deduced in the previous section apply here as well. The $\tan\beta$-spread that some of them have will be much reduced though. The most noticeable changes occur for the quantities which depend most sensitively on $\tan\beta$. In Fig. 5 (bottom row) we plot the one-loop corrected lightest Higgs boson mass versus $m_{\tilde{g}}$. The result is that m_h is basically determined by m_t; only a weak dependence on $m_{\tilde{g}}$ exists. Moreover, for $m_t \lesssim 135\,\text{GeV} \Leftrightarrow \mu > 0$, $m_h \lesssim 105\,\text{GeV}$; whereas for $m_t \gtrsim 140\,\text{GeV} \Leftrightarrow \mu < 0$, $m_h \gtrsim 100\,\text{GeV}$. Therefore, in the strict no-scale case, once

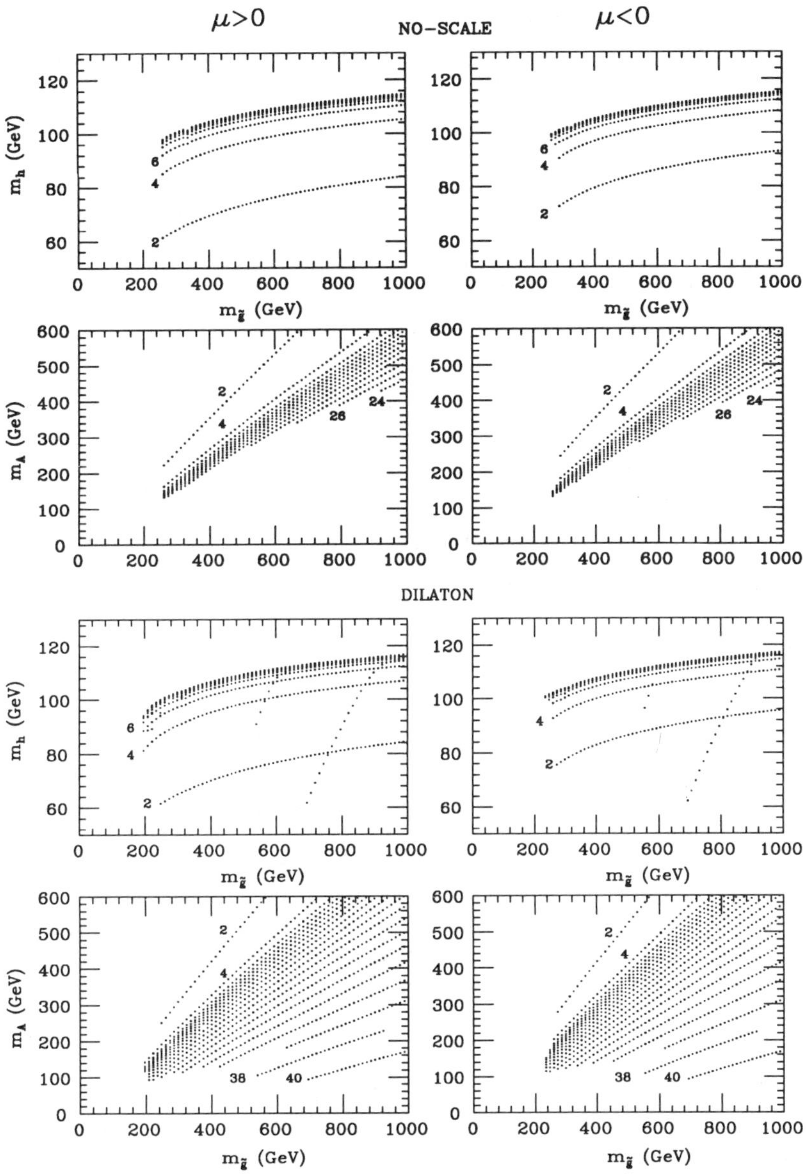

Figure 4: The one-loop corrected h and A Higgs masses versus the gluino mass for both signs of μ, $m_t = 150\,\mathrm{GeV}$, and the two supersymmetry breaking scenaria. Representative values of $\tan\beta$ are indicated.

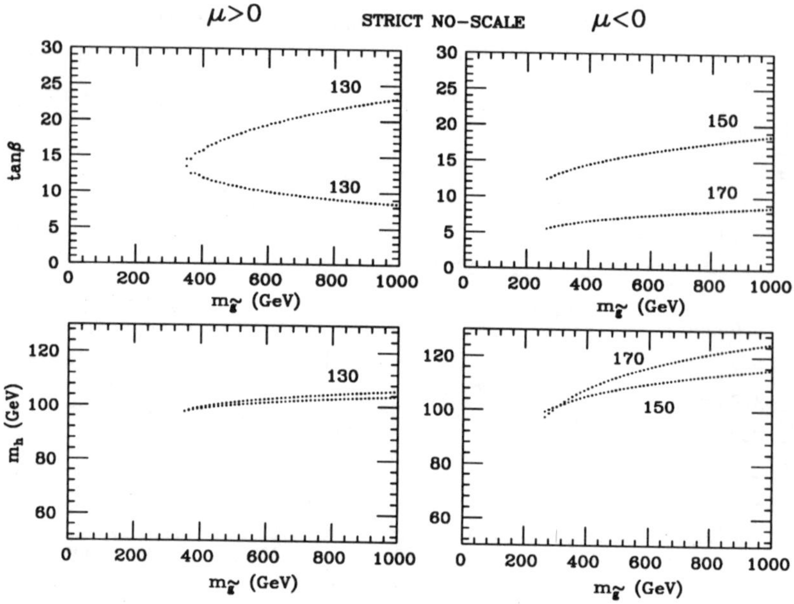

Figure 5: The value of $\tan\beta$ versus $m_{\tilde{g}}$ in the strict no-scale case (where $B(M_U) = 0$) for the indicated values of m_t. Note that the sign of μ is *determined* by m_t and that $\tan\beta$ can be double-valued for $\mu > 0$. Also shown is the one-loop corrected lightest Higgs boson mass. Note that if $\mu > 0$ (for $m_t < 135\,\mathrm{GeV}$) then $m_h < 105\,\mathrm{GeV}$; whereas if $\mu < 0$ (for $m_t > 140\,\mathrm{GeV}$) then $m_h > 100\,\mathrm{GeV}$.

the top-quark mass is measured, we will know the sign of μ and whether m_h is above or below 100 GeV.

For $\mu > 0$, we just showed that the strict no-scale constraint requires $m_t \lesssim 135\,\mathrm{GeV}$. This implies that μ cannot grow as large as it did previously in the general case. In fact, for $\mu > 0$, $\mu_{max} \approx 745\,\mathrm{GeV}$ before and $\mu_{max} \approx 440\,\mathrm{GeV}$ now. This smaller value of μ_{max} has the effect of cutting off the growth of the $\chi^0_{3,4}, \chi^\pm_2$ masses at $\approx \mu_{max} \approx 440\,\mathrm{GeV}$ (c.f. $\approx 750\,\mathrm{GeV}$) and of the heavy Higgs masses at $\approx 530\,\mathrm{GeV}$ (c.f. $\approx 940\,\mathrm{GeV}$).

6.2 The special dilaton scenario case

In our analysis above, the radiative electroweak breaking conditions were used to determine the magnitude of the Higgs mixing term μ at the electroweak scale. This quantity is ensured to remain light as long as the supersymmetry breaking parameters remain light. In a fundamental theory this parameter should be calculable and its value used to determine the Z-boson mass. From this point of view it is not clear that the natural value of μ should be light. In specific models on can obtain such values by invoking non-renormalizable interactions [34, 35]. Another contribution to this quantity is generically present in string supergravity models [36, 35, 37]. The general case with contributions from both sources has been effectively dealt with in the previous section. If one assumes that only supergravity-induced contributions to μ exist, then it can be shown that the B-parameter at the unification scale is also determined [37],

$$B(M_U) = 2m_0 = \tfrac{2}{\sqrt{3}}m_{1/2}, \tag{33}$$

which is to be added to the set of relations in Eq. (5). This new constraint effectively determines $\tan\beta$ for given m_t and $m_{\tilde{g}}$ values and makes this restricted version of the model highly predictive.

From the outset we note that only solutions with $\mu < 0$ exist. This is not a completely obvious result, but it can be partially understood as follows. In tree-level approximation, $m_A^2 > 0 \Rightarrow \mu B < 0$ at the electroweak scale. Since $B(M_U)$ is required to be positive and not small, $B(M_Z)$ will likely be positive also, thus forcing μ to be negative. A sufficiently small value of $B(M_U)$ and/or one-loop corrections to m_A^2 could alter this result, although in practice this does not happen. A numerical iterative procedure allows us to determine the value of $\tan\beta$ which satisfies Eq. (33), from the calculated value of $B(M_Z)$. We find that

$$\tan\beta \approx 1.57-1.63, 1.37-1.45, 1.38-1.40 \quad \text{for } m_t = 130, 150, 155\,\text{GeV} \tag{34}$$

is required. Since $\tan\beta$ is so small ($m_h^{tree} \approx 28 - 41\,\text{GeV}$), a significant one-loop correction to m_h is required to increase it above its experimental lower bound of $\approx 60\,\text{GeV}$ [61]. This requires the largest possible top-quark masses (and a not-too-small squark mass). However, perturbative unification imposes an upper bound on m_t for a given $\tan\beta$ [78], which in this case implies [69]

$$m_t \lesssim 155\,\text{GeV}, \tag{35}$$

which limits the magnitude of m_h

$$m_h \lesssim 74, 87, 91\,\text{GeV} \qquad \text{for} \qquad m_t = 130, 150, 155\,\text{GeV}. \tag{36}$$

Lower values of m_t are experimentally disfavored.

In Table 3 we give the range of sparticle and Higgs masses that are allowed in this case. Clearly, continuing top-quark searches at the Tevatron and Higgs searches at LEPI,II should probe this restricted scenario completely.

7 The Minimal SU(5) Supergravity Model

The minimal $SU(5)$ supergravity model [79] needs to be specified clearly in order to avoid the common misconception that it is simply the so-called MSSM with the low-energy gauge couplings meeting at very high energies. Two of its elements are

Table 3: The range of allowed sparticle and Higgs masses in the restricted dilaton scenario. The top-quark mass is restricted to be $m_t < 155\,\mathrm{GeV}$. All masses in GeV.

m_t	130	150	155
$\tilde{g}$	$335 - 1000$	$260 - 1000$	$640 - 1000$
χ_1^0	$38 - 140$	$24 - 140$	$90 - 140$
$\chi_2^0, \chi_1^\pm$	$75 - 270$	$50 - 270$	$170 - 270$
$\tan\beta$	$1.57 - 1.63$	$1.37 - 1.45$	$1.38 - 1.40$
h	$61 - 74$	$64 - 87$	$84 - 91$
$\tilde{l}$	$110 - 400$	$90 - 400$	$210 - 400$
$\tilde{q}$	$335 - 1000$	$260 - 1000$	$640 - 1000$
A, H, H^+	> 400	> 400	> 970

particularly important: (i) it is a supergravity model [80] and as such the soft supersymmetry breaking masses which allow unification are in principle calculable and are assumed to be parametrized in terms of $m_{1/2}, m_0, A$; and (ii) there exist dimension-five proton decay operators [81], which are much larger than the usual dimension-six operators, and require either a tuning of the supersymmetry breaking parameters or a large Higgs triplet mass scale, to obtain a sufficiently long proton lifetime [82, 83, 84, 85, 47, 48, 56, 49].

The $SU(5)$ symmetry is broken down to $SU(3) \times SU(2) \times U(1)$ via a vev of the neutral component of the adjoint **24** of Higgs. The low-energy pair of Higgs doublets are contained in the $\mathbf{5},\overline{\mathbf{5}}$ Higgs representations. Of the various proposals to split the proton-decay-mediating Higgs triplets from the light Higgs doublets, perhaps the most appealing one is the so-called "missing partner mechanism" [23], whereby a **75** of Higgs breaks the gauge symmetry and the $\mathbf{5},\overline{\mathbf{5}}$ pentaplets are coupled to $\mathbf{50},\overline{\mathbf{50}}$ representations ($\mathbf{50}\cdot\mathbf{75}\cdot\mathbf{5}$, $\overline{\mathbf{50}}\cdot\mathbf{75}\cdot\overline{\mathbf{5}}$). The doublets remain massless, while the triplets acquire $\sim M_U$ masses. We should remark that this non-minimal symmetry breaking mechanism is *not* the one that is usually considered in studies of high-energy threshold effects in gauge coupling unification, where one usually assumes that it is the **24** which effects the breaking.

7.1 Gauge and Yukawa coupling unification

This problem can be tackled at several levels of sophistication, which entail an increasing number of additional assumptions. The most elementary approach consists of running the one-loop supersymmetric gauge coupling RGEs starting with the precisely measured values of $\alpha_e, \alpha_3, \sin^2\theta_w$ at the scale M_Z and discovering that the three gauge couplings meet at the scale $M_U \sim 10^{16}\,\mathrm{GeV}$ [86]. More interesting from the theoretical standpoint is to assume that unification must occur, as is the case in the minimal $SU(5)$ supergravity model, and use this constraint to predict the low-energy value of $\sin^2\theta_w$ in terms of α_e and α_3. The next level of sophistication consists of increasing the accuracy of the RGEs to two-loop level and parametrize the supersymmetric threshold by a single mass parameter between $\sim M_Z$ and $\sim$ few TeV [26, 87, 88, 89]. More realistically, one specifies the whole light supersymmetric spectrum in detail [32, 90, 91, 92, 48, 93, 94, 27, 95, 96], as well as some subtle effects such as the evolution of the gaugino masses (EGM) effect [93, 91], and the effect of the Yukawa couplings on the two-loop gauge coupling RGEs [97]. A final step of sophistication attempts to model the transition from the $SU(3)\times SU(2)\times U(1)$ theory onto the $SU(5)$ theory by means of high-energy threshold effects which depend on the masses of the various GUT fields [90, 91, 98, 48, 27, 95] as well as on coefficients of possible non-renormalizable operators [95, 99]. This last step does away completely with the concept of a single "unification mass". In fact, until this last step is actually accounted for somehow, one is not dealing with a true unified theory since otherwise the gauge couplings would diverge again past the unification scale, *i.e.*, "physics is not euclidean geometry".

It is interesting to note that the original hope that precise knowledge of the low energy gauge couplings would constrain the scale of the low-energy supersymmetric particles, has not bear fruit [98, 93, 91], mainly because of the largely unknown GUT threshold effects. More precisely, the supersymmetric particle masses can lie anywhere up to $\sim$ few TeV provided the parameters of the GUT theory are adjusted accordingly.

Another consequence of the $SU(5)$ symmetry is the relation $\lambda_b(M_U)=\lambda_\tau(M_U)$ which when renormalized down to low energies gives a ratio m_b/m_τ in fairly good agreement with experiment [100]. This problem can also be tackled with improving degree of sophistication [26, 88, 90, 101, 102, 103, 97, 104, 105] and even postulating some high-energy threshold effects [97, 105]. In practice, the $\lambda_b(M_U)=\lambda_\tau(M_U)$ constraint entails a relationship between m_t and $\tan\beta$, *i.e.*, $\tan\beta=\tan\beta(m_t,m_b,\alpha_3)$, as follows: (i) the values of m_b and m_τ, together with $\tan\beta$ determine the low-energy values of λ_b and λ_τ; (ii) the input value of m_t determines the low-energy value of λ_t; (iii) running these three Yukawa couplings up to the unification scale one discovers the above relation between $\tan\beta$ and m_t if the Yukawa unification constraint is satisfied. In actuality, the dependence on m_b and α_3 is quite important. We note that for arbitrary choices of m_t and $\tan\beta$, one obtains values of m_b typically close to or above 5 GeV, whereas popular belief would like to see values below 4.5 GeV. Strict adherence to this prediction for m_b requires that one be in a rather constrained region of the $(m_t,\tan\beta)$ plane, where $\tan\beta\sim 1$ or $\gtrsim 40$ [91, 102, 97, 105], or that m_t be large (above 180 GeV). We do not impose this stringent constraint on the parameter space, hoping that further contributions to the quark masses (as required in $SU(5)$ GUTs to fit the lighter generations also [106]) will relax it somehow.

As noted above, the parameter space of this model can be described in terms of five parameters: $m_t, \tan\beta, m_{1/2}, m_0, A$. In Ref. [56] we performed an exploration of the following hypercube of the parameter space: $\mu>0, \mu<0$, $\tan\beta=2-10\,(2)$, $m_t=100{-}160\,(5)$, $\xi_0=0{-}10\,(1)$, $\xi_A=-\xi_0,0,+\xi_0$, and $m_{1/2}=50{-}300\,(6)$, where the numbers in parenthesis represent the size of the step taken in that particular direction.

Table 4: The value of the one-loop unification mass $M_U^{(0)}$, the two-loop and supersymmetric threshold corrected unification mass range $M_U^{(1)}$, the ratio of the two, and the range of the calculated $\sin^2\theta_w$, for the indicated values of α_3 (the superscript $+\,(-)$ denotes $\mu > 0\,(< 0)$) and $\alpha_e^{-1} = 127.9$. The $\sin^2\theta_w$ values should be compared with the current experimental $\pm 1\sigma$ range $\sin^2\theta_w = 0.2324 \pm 0.0006$ [95]. Lower values of α_3 drive $\sin^2\theta_w$ to values even higher than for $\alpha_3 = 0.118$. All masses in units of 10^{16} GeV.

	$\alpha_3 = 0.126^+$	$\alpha_3 = 0.126^-$	$\alpha_3 = 0.118^+$	$\alpha_3 = 0.118^-$
$M_U^{(0)}$	3.33	3.33	2.12	2.12
$M_U^{(1)}$	$1.60 - 2.13$	$1.60 - 2.05$	$1.02 - 1.35$	$1.02 - 1.30$
$M_U^{(1)}/M_U^{(0)}$	$0.48 - 0.64$	$0.48 - 0.61$	$0.48 - 0.64$	$0.48 - 0.61$
$\sin^2\theta_w$	$0.2315 - 0.2332$	$0.2313 - 0.2326$	$0.2335 - 0.2351$	$0.2332 - 0.2345$

(Points outside these ranges have little (a posteriori) likelihood of being acceptable.) Of these $92,235 \times 2 = 184,470$ points, $\approx 25\%$ passed all the standard constraints, *i.e.*, radiative electroweak symmetry breaking and all low-energy phenomenology as described in Ref. [69]. The most important constraint on this parameter space is proton decay, as discussed below. First we discuss some aspects of the gauge coupling unification calculation.

As a first step we used one-loop gauge coupling RGEs and a common supersymmetric threshold at M_Z, to determine M_U, α_U, and $\sin^2\theta_w$, once $\alpha_3(M_Z) = 0.113, 0.120$ and $\alpha_e^{-1}(M_Z) = 127.9$ were given. In Ref. [49] we refined our study including several important features: (i) recalculation of M_U using two-loop gauge coupling RGEs including light supersymmetric thresholds, (ii) exploration of values of α_3 throughout its $\pm 1\sigma$ allowed range, and (iii) exploration of low values of $\tan\beta\,(< 2)$ (which maximize the proton lifetime). We used the analytical approximations to the solution of the two-loop gauge coupling RGEs in Ref. [93] to obtain $M_U, \alpha_U, \sin^2\theta_w$. The supersymmetric threshold was treated in great detail [93] with all the sparticle masses obtained from our procedure [56]. Since, the sparticle masses vary as one explores the parameter space, one obtains *ranges* for the calculated values. In Table 4 we show the one-loop value for M_U ($M_U^{(0)}$), the two-loop plus supersymmetric threshold corrected unification mass range ($M_U^{(1)}$) [as expected [93] M_U is reduced by both effects], the ratio of the two, and the calculated range of $\sin^2\theta_w$.[11] Note that for $\alpha_3 = 0.118$ (and lower), $\sin^2\theta_w$ is outside the experimental $\pm 1\sigma$ range ($\sin^2\theta_w = 0.2324 \pm 0.0006$ [95]), whereas $\alpha_3 = 0.126$ gives quite acceptable values.

We do not specify the details of the GUT thresholds and in practice take two

[11] We should note that these ranges are obtained after all constraints discussed below have been satisfied, the proton decay being the most important one.

of the GUT mass parameters (the masses of the X, Y gauge bosons M_V, and the mass of the adjoint Higgs multiplet M_Σ) to be degenerate with M_U. Since below we allow $M_H < 3M_U$, Table 4 indicates that in our calculations $M_H < 6.4 \times 10^{16}\,\mathrm{GeV}$. In Ref. [48] it is argued that a more proper upper bound is $M_H < 2M_V$, but M_V cannot be calculated directly, only $(M_V^2 M_\Sigma)^{1/3} < 3.3 \times 10^{16}\,\mathrm{GeV}$ is known from low-energy data [48]. If we take $M_\Sigma = M_V$, this would give $M_H < 2M_V < 6.6 \times 10^{16}\,\mathrm{GeV}$, which agrees with our present requirement. Below we comment on the case $M_\Sigma < M_V$.

7.2 Proton decay

In the minimal $SU(5)$ supergravity model only the dimension-five–mediated proton decay operators are constraining. In calculating the proton lifetime we consider the typically dominant decay modes $p \to \bar{\nu}_{\mu,\tau} K^+$ and neglect all other possible modes. Schematically the lifetime is given by[12]

$$\tau_p \equiv \tau(p \to \bar{\nu}_{\mu,\tau} K^+) \sim \left| M_H \sin 2\beta \frac{1}{f} \frac{1}{1+y^{tK}} \right|^2 . \tag{37}$$

Here M_H is the mass of the exchanged GUT Higgs triplet which on perturbative grounds is assumed to be bounded above by $M_H < 3M_U$ [13] [83, 47, 48]; $\sin 2\beta = 2\tan\beta/(1+\tan^2\beta)$, thus τ_p 'likes' small $\tan\beta$ (we find that only $\tan\beta \lesssim 6$ is allowed); y^{tK} represents the calculable ratio of the third- to the second-generation contributions to the dressing one-loop diagrams. An unkown phase appears in this ratio (which has generally $|y^{tK}| \ll 1$) and we always consider the weakest possible case of destructive interference. Finally f represents the sparticle-mass–dependent dressing one-loop function which decreases asymptotically with large sparticle masses.

In Fig. 6 (top row) we show a scatter plot of $(\tau_p, m_{\tilde{g}})$. The various 'branches' correspond to fixed values of ξ_0. Note that for $\xi_0 < 3$, $\tau_p < \tau_p^{exp} = 1 \times 10^{32}\,\mathrm{y}$ (at 90% C.L. [107]). Also, for a given value of ξ_0, there is a corresponding allowed interval in $m_{\tilde{g}}$. The lower end of this interval is determined by the fact that $\tau_p \propto 1/f^2$, and $f \approx m_{\chi_1^+}/m_{\tilde{q}}^2 \propto 1/m_{\tilde{g}}(c+\xi_0^2)$, in the proton-decay–favored limit of $\mu \gg M_W$; thus $m_{\tilde{g}}(c+\xi_0^2) >$ constant. The upper end of the interval follows from our requirement $m_{\tilde{q}}(\propto m_{\tilde{g}}\sqrt{c+\xi_0^2}) < 1\,\mathrm{TeV}$. Statistically speaking, the proton decay cut is quite severe, allowing only about $\sim 1/10$ of the points which passed all the standard constraints, independently of the sign of μ.

Note that if we take $M_H = M_U$ (instead of $M_H = 3M_U$), then $\tau_p \to \frac{1}{9}\tau_p$ and all points in Fig. 6 would become excluded. To obtain a rigorous lower bound on M_H, we would need to explore the lowest possible allowed values of $\tan\beta$ (in Fig. 6, $\tan\beta \geq 2$). Roughly, since the dominant $\tan\beta$ dependence of τ_p is through the explicit $\sin 2\beta$ factor, the upper bound $\tau_p \lesssim 8 \times 10^{32}\,\mathrm{y}$ for $\tan\beta = 2$, would become

[12] Throughout our calculations we have used the explicit proton decay formulas in Ref. [84].

[13] This relation assumes implicitly that all the components of the **24** superfield are nearly degenerate in mass [48].

$\tau_p \lesssim 1 \times 10^{33}$ y for $\tan\beta = 1$. Therefore, the current experimental lower bound on τ_p would imply $M_H \gtrsim M_U$. Note also that SuperKamiokande ($\tau_p^{exp} \approx 2 \times 10^{33}$ y) should be able to probe the whole allowed range of τ_p values.

The actual value of $\alpha_3(M_Z)$ used in the calculations ($\alpha_3 = 0.120$ in Fig. 6) has a non-negligible effect of some of the final results, mostly due to its effect on the value of $M_H = 3M_U$: larger values of α_3 increase M_U and therefore τ_p, and thus open up the parameter space, and viceversa. For example, for $\alpha_3 = 0.113\,(0.120)$ we get $m_{\tilde{g}} \lesssim 550\,(800)$ GeV, $\xi_0 \geq 5\,(3)$, and $\tau_p \lesssim 4\,(8) \times 10^{32}$ y.

The refinement on the calculation of the unification mass described above, to include two-loop effects and light supersymmetric thresholds, has a significant effect on the calculated value of the proton lifetime [49], since we take $M_H = 3M_U$. With the more accurate value of M_U we simply rescale our previously calculated τ_p values which satisfied $\tau_p^{(0)} > \tau_p^{exp}$, and find that $\tau_p^{(1)} = \tau_p^{(0)}[M_U^{(1)}/M_U^{(0)}]^2 > \tau_p^{exp}$ for only $\lesssim 25\%$ of the previously allowed points. The value of α_3 has a significant influence on the results since (see Table 4) larger (smaller) values of α_3 increase (decrease) M_U, although the effect is more pronounced for low values of α_3. To quote the most conservative values of the observables, in what follows we take α_3 at its $+1\sigma$ value ($\alpha_3 = 0.126$). As discussed above, this choice of α_3 also gives $\sin^2\theta_w$ values consistent with the $\pm1\sigma$ experimental range. Finally, in the search of the parameter space above, we considered only $\tan\beta = 2, 4, 6, 8, 10$ and found that $\tan\beta \lesssim 6$ was required. Our present analysis indicates that this upper bound is reduced down to $\tan\beta \lesssim 3.5$. Here we consider also $\tan\beta = 1.5, 1.75$ since low $\tan\beta$ maximizes $\tau_p \propto \sin^2 2\beta$. These add new allowed points (*i.e.*, $\tau_p^{(0)} > \tau_p^{exp}$) to our previous set, although most of them ($\gtrsim 75\%$) do not survive the stricter proton decay constraint ($\tau_p^{(1)} > \tau_p^{exp}$) imposed here.

In Fig. 7 we show the re-scaled values of τ_p versus the lightest chargino mass $m_{\chi_1^\pm}$. All points satisfy $\xi_0 \equiv m_0/m_{1/2} \gtrsim 6$ and $m_{\chi_1^\pm} \lesssim 150$ GeV, which are to be contrasted with $\xi_0 \gtrsim 3$ and $m_{\chi_1^\pm} \lesssim 225$ GeV derived using the *weaker* proton decay constraint [56]. The upper bound on $m_{\chi_1^\pm}$ derives from its near proportionality to $m_{\tilde{g}}$, $m_{\chi_1^\pm} \approx 0.3 m_{\tilde{g}}$ [47, 56], and the result $m_{\tilde{g}} \lesssim 500$ GeV. The latter follows from the proton decay constraint $\xi_0 \gtrsim 6$ and the naturalness requirement $m_{\tilde{q}} \approx \sqrt{m_0^2 + 6m_{1/2}^2} \approx \frac{1}{3} m_{\tilde{g}} \sqrt{6 + \xi_0^2} < 1$ TeV. Within our naturalness and H_3 mass assumptions, we then obtain [14]

$$\tau_p < 3.1\,(3.4) \times 10^{32}\,\mathrm{y} \quad \text{for} \quad \mu > 0\,(\mu < 0). \tag{38}$$

The $p \to \bar{\nu} K^+$ mode should then be readily observable at SuperKamiokande and Gran Sasso since these experiments should be sensitive up to $\tau_p \approx 2 \times 10^{33}$ y. Note that if M_H is relaxed up to its largest possible value consistent with low-energy physics, $M_H = 2.3 \times 10^{17}$ GeV [48], then in Eq. (38) $\tau_p \to \tau_p < 4.0\,(4.8) \times 10^{33}$ y, and only part

[14]Note that in general, $\tau_p \propto M_{H_3}^2 [m_{\tilde{q}}^2/m_{\chi_1^\pm}]^2 \propto M_{H_3}^2 [m_{\tilde{g}}(6 + \xi_0^2)]^2$ and thus τ_p can be made as large as desired by increasing sufficiently either the supersymmetric spectrum or M_H.

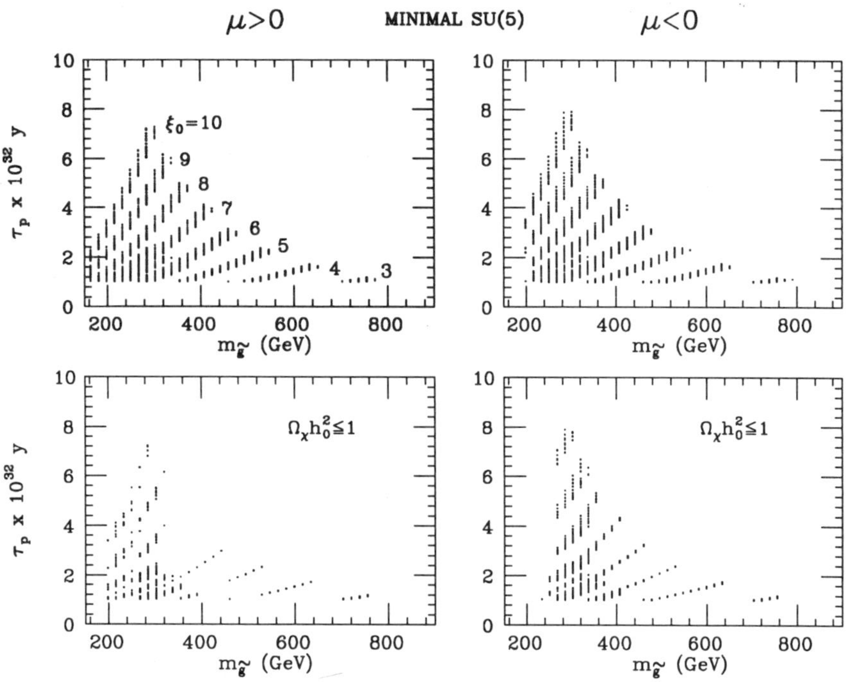

Figure 6: Scatter plot of the proton lifetime $\tau_p \equiv \tau(p \to \bar{\nu}_{\mu,\tau} K^+)$ versus the gluino mass for the hypercube of the parameter space explored. The unification mass is calculated in one-loop approximation assuming a common supersymmetric threshold at M_Z, and $M_H = 3M_U$ is assumed. The current experimental lower bound is $\tau_p^{exp} = 1 \times 10^{32}$ y. The various 'branches' correspond to fixed values of ξ_0 as indicated (the labelling applies to all four windows). The bottom row includes the cosmological constraint. The upper bound on $m_{\tilde{g}}$ follows from the requirement $m_{\tilde{q}} < 1$ TeV.

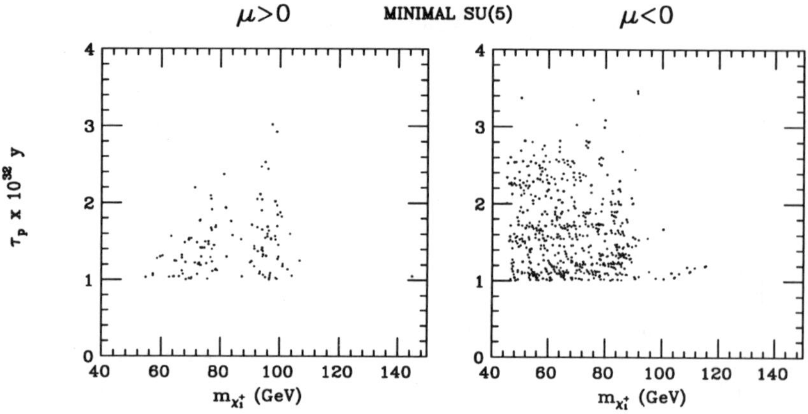

Figure 7: The calculated values of the proton lifetime into $p \to \bar{\nu} K^{+}$ versus the lightest chargino (or second-to-lightest neutralino) mass for both signs of μ, using the more accurate value of the unification mass (which includes two-loop and low-energy supersymmetric threshold effects). Note that we have taken $\alpha_3 + 1\sigma$ in order to maximize τ_p. Note also that future proton decay experiments should be sensitive up to $\tau_p \approx 20 \times 10^{32}\,\mathrm{y}$.

of the parameter space of the model would be experimentally accessible. However, to make this choice of M_H consistent with high-energy physics (*i.e.*, $M_H < 2M_V$) one must have $M_V/M_\Sigma > 42$.

7.3 Neutralino relic density

The study of the relic density of neutralinos requires the knowledge of the total annihilation amplitude $\chi\chi \to all$. The latter depends on the model parameters to determine all masses and couplings. Previously [73, 74] we have advocated the study of this problem in the context of supergravity models with radiative electroweak symmetry breaking, since then only a few parameters (five or less) are needed to specify the model completely. In particular, one can explore the whole parameter space and draw conclusions about a complete class of models. The ensuing relationships among the various masses and couplings have been found to yield results which depart from the conventional minimal supersymmetric standard model (MSSM) lore, where no such relations exist. In the minimal $SU(5)$ supergravity model we have just shown that its five-dimensional parameter space is strongly constrained by the proton lifetime. It was first noticed in Ref. [57] that the neutralino relic density for the proton-decay allowed points in parameter space is large, *i.e.*, $\Omega_\chi h_0^2 \gg 1$, and therefore in conflict with current cosmological expectations: requiring that the Universe be older than the oldest known stars implies $\Omega_0 h_0^2 \leq 1$ [75].

In Refs. [57, 56, 49] the neutralino relic density has been computed following

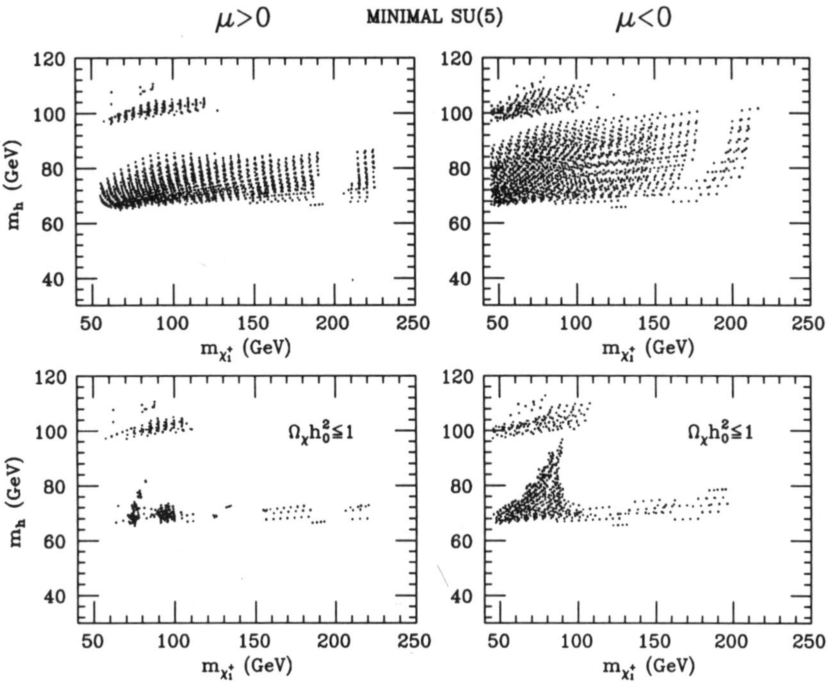

Figure 8: The allowed region in parameter space which satisfies the *weaker* proton decay constraint, before and after the imposition of the cosmological constraint. Note that when the cosmological constraint is imposed (bottom row), an interesting correlation between the two particle masses arises.

the methods of Refs. [73, 74]. In Fig. 6 (bottom row) we show the effect of the cosmological constraint on the parameter space allowed by proton decay. Only $\sim 1/6$ of the points satisfy $\Omega_\chi h_0^2 \leq 1$. This result is not unexpected since proton decay is suppressed by heavy sparticle masses, whereas $\Omega_\chi h_0^2$ is enhanced. Therefore, a delicate balance needs to be attained to satisfy both constraints simultaneously. Note that the subset of cosmologically allowed points does not change the range of possible τ_p values, although it depletes the constant-ξ_0 'branches'.

The effect of the cosmological constraint is perhaps more manifest when one considers the correlation between m_h and $m_{\chi_1^\pm}$ after imposing the (*e.g.*, *weaker*) proton decay constraint, but with and without imposing the cosmological constraint. This contrast is shown in Fig. 8.

The main conclusion is that the relic density can be small only near the h- and Z-pole resonances, *i.e.*, for $m_\chi \approx \frac{1}{2}m_{h,Z}$ [56, 58], since in this case the annihilation cross section is enhanced. It is important to note that in this type of calculations the thermal average of the annihilation cross section is usually computed using an expansion around threshold (*i.e.*, $\sqrt{s} = 2m_\chi$) [108]. In Ref. [109] however, it has been pointed out that the resulting thermal average can be quite inaccurate near poles and thresholds of the annihilation cross section, which is precisely the case for the points of interest in the minimal $SU(5)$ model. In Ref. [58, 59] the relic density calculation has been redone following the more accurate methods of Ref. [109]. The result is that the poles are broader and shallower, and thus the cosmological constraint is weakened with respect to the standard (using the expansion) procedure of performing the thermal average. However, qualitatively the cosmologically allowed region of parameter space is not changed. This result is shown in Fig. 9, where the points in parameter space allowed by the stricter proton decay constraint and cosmology are shown in the $(m_{\chi_1^\pm}, m_h)$ plane.

7.4 Mass ranges and relations

Since the proton decay constraint generally requires $|\mu| \gg M_W$ (and to a somewhat lesser extent also $|\mu| \gg M_2$), the lightest chargino will have mass $m_{\chi_1^+} \approx M_2 \approx 0.3m_{\tilde{g}}$, whereas the two lightest neutralinos will have masses $m_\chi \approx M_1 \approx \frac{1}{2}M_2$ and $m_{\chi_2^0} \approx M_2$ [47, 70]. Thus, within some approximation we expect the relation in Eq. (27) to be satisfied in this model also. Inclusion of the cosmological constraint does not affect significantly the range of sparticle masses in Sec. 7.2. The value of α_3 does not affect these mass relations either, although the particle mass ranges do change

$$m_\chi < 85\,(115)\,\text{GeV}, \qquad m_{\chi_2^0,\chi_1^+} < 165\,(225)\,\text{GeV}, \qquad \text{for } \alpha_3 = 0.113\,(0.120). \quad (39)$$

The reason is simple: higher values of α_3 increase M_U and therefore $M_H\,(= 3M_U)$, which in turn weakens the proton decay constraint. We also find that the one-loop corrected lightest Higgs boson mass (m_h) is bounded above by

$$m_h \lesssim 110\,(100)\,\text{GeV}, \quad (40)$$

independently of the sign of μ, the value of α_3, or the cosmological constraint; the stronger bound holds when the stricter proton decay constraint is enforced. In Fig. 9 we have shown m_h versus $m_{\chi_1^+}$ for $\tan\beta = 1.5, 1.75, 2$; for the maximum allowed $\tan\beta$ value (≈ 3.5), $m_h \lesssim 100\,\text{GeV}$. Note that for $\mu > 0$, $m_h \approx 50\,\text{GeV}$, and $m_{\chi_1^\pm} \gtrsim 100\,\text{GeV}$, there is a sparsely populated area with highly fine-tuned points in parameter space ($m_t \approx 100\,\text{GeV}$, $\tan\beta \approx 1.5$, $\xi_A \equiv A/m_{1/2} \approx \xi_0 \approx 6$). This figure shows an experimentally interesting correlation when the cosmological constraints are imposed,

$$m_h \gtrsim 72\,\text{GeV} \Rightarrow m_{\chi_1^+} \lesssim 100\,\text{GeV}. \quad (41)$$

The bands of points towards low values of m_h represent the discrete choices of $\tan\beta = 1.5, 1.75$. The voids between these bands are to be understood as filled by points with $1.5 \lesssim \tan\beta \lesssim 1.75$. For $m_{\chi_1^\pm} > 106\,(92)\,\text{GeV}$ (for $\mu > 0\,(\mu < 0)$), we

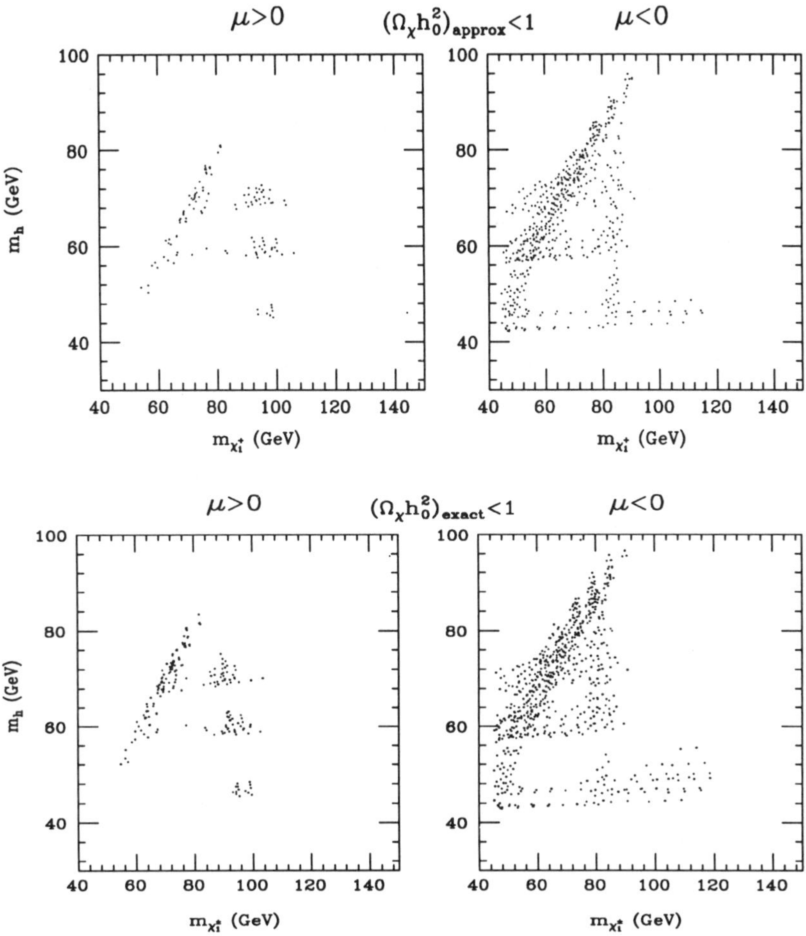

Figure 9: The points in parameter space of the minimal $SU(5)$ supergravity model which satisfy the stricter proton decay constraint and the cosmological constraint with the relic density computed in approximate and accurate way. Note the little qualitative difference between the two sets of plots.

obtain $m_h \lesssim 50\,(56)$ GeV and Higgs detection at LEP should be immediate. The correlations among the lightest chargino and neutralino masses imply analogous results for $(m_h, m_{\chi_2^0})$ and (m_h, m_χ),

$$m_h \gtrsim 80\,\text{GeV} \Rightarrow m_{\chi_2^0} \lesssim 90\,(110)\,\text{GeV}, \quad m_\chi \lesssim 48\,(60)\,\text{GeV}, \tag{42}$$

for $\alpha_3 = 0.113\,(0.120)$. These correlations can be understood in the following way: since we find that $m_A \gg M_Z$, then $m_h \approx |\cos 2\beta| M_Z +$(rad. corr.). In the situation we consider here, we have determined that all of the allowed points for $m_{\tilde{g}} > 400$ GeV correspond to $\tan\beta = 2$. This implies that the tree-level contribution to m_h is ≈ 55 GeV. We also find that the cosmology cut restricts $m_t < 130(140)$ GeV for $\mu > 0 (\mu < 0)$ in this range of $m_{\tilde{g}}$. Therefore, the radiative correction contribution to m_h^2 ($\propto m_t^4$) will be modest in this range of $m_{\tilde{g}}$. This explains the depletion of points for $m_h \gtrsim 80$ GeV in Fig. 9 and leads to the mass relationships in Eqs. (41,42).

In this model the only light particles are the lightest Higgs boson ($m_h \lesssim 100$ GeV), the two lightest neutralinos ($m_{\chi_1^0} \approx \frac{1}{2} m_{\chi_2^0} \lesssim 75$ GeV), and the lightest chargino ($m_{\chi_1^\pm} \approx m_{\chi_2^0} \lesssim 150$ GeV). The gluino and the lightest stop can be light ($m_{\tilde{g}} \approx 160 - 460$ GeV, $m_{\tilde{t}_1} \approx 170 - 825$ GeV), but for most of the parameter space are not within the reach of Fermilab.

In Ref. [61] it has been shown that the actual LEP lower bound on the ligthest Higgs boson mass is improved in the class of supergravity models with radiative electroweak symmetry breaking which we consider here, one gets $m_h \gtrsim 60$ GeV. In Sec. 8.2 below we discuss the details of this procedure. For now it suffices to note that the improved bound $m_h \gtrsim 60$ GeV mostly restricts low values of $\tan\beta$ and therefore the minimal $SU(5)$ supergravity model where $\tan\beta \lesssim 3.5$ [49]. Above we obtained upper bounds on the light particle masses in this model ($\tilde{g}, h, \chi_{1,2}^0, \chi_1^\pm$) for $m_h > 43$ GeV. In particular, it was found that $m_{\chi_1^\pm} \gtrsim 100$ GeV was only possible for $m_h \lesssim 50$ GeV. The improved bound on m_h immediately implies the following considerably stronger upper bounds

$$\begin{aligned} m_{\chi_1^0} &\lesssim 52(50)\,\text{GeV}, & (43)\\ m_{\chi_2^0} &\lesssim 103(94)\,\text{GeV}, & (44)\\ m_{\chi_1^\pm} &\lesssim 104(92)\,\text{GeV}, & (45)\\ m_{\tilde{g}} &\lesssim 320\,(405)\,\text{GeV}, & (46) \end{aligned}$$

for $\mu > 0$ ($\mu < 0$).

A related consequence is that the mass relation $m_{\chi_2^0} > m_{\chi_1^0} + m_h$ is not satisfied for any of the remaining points in parameter space and therefore the $\chi_2^0 \to \chi_1^0 h$ decay mode is not kinematically allowed. Points where such mode was previously allowed led to a vanishing trilepton signal in the reaction $p\bar{p} \to \chi_1^\mp \chi_2^0$ at Fermilab (thus the name 'spoiler mode') [60]. The improved situation now implies at least one event per $100\,\text{pb}^{-1}$ for all remaining points in parameter space (see Sec. 8.1).

8 Prospects for Direct Experimental Detection

The sparticle and Higgs spectrum discussed in Secs. 5,6,7 can be directly explored partially at present and near future collider facilities, as we now discuss for each model considered above.

8.1 Tevatron

(a) The search and eventual discovery of the top quark will narrow down the parameter space of these models considerably. Moreover, in the two special flipped $SU(5)$ cases discussed in Sec. 6 this measurement will be very important: (i) in the strict no-scale case (Sec. 6.1) it will determine the sign of μ ($\mu > 0$ if $m_t \lesssim 135\,\mathrm{GeV}$; $\mu < 0$ if $m_t \gtrsim 140\,\mathrm{GeV}$) and whether the Higgs mass is above or below $\approx 100\,\mathrm{GeV}$, and (ii) it may rule out the restricted dilaton scenario (Sec. 6.2) if $m_t > 150\,\mathrm{GeV}$.

(b) The trilepton signal in $p\bar{p} \to \chi_2^0 \chi_1^\pm X$, where χ_2^0 and $\chi_1^\pm$ both decay leptonically, is a clean test of supersymmetry [110] and in particular of this class of models [60]. The trilepton rates in the no-scale flipped $SU(5)$ and in the minimal $SU(5)$ models have been given in Ref. [60]; in Fig. 10 we show these (in the no-scale case $m_t = 130\,\mathrm{GeV}$ has been chosen). One can show that with $\mathcal{L} = 100\,\mathrm{pb}^{-1}$ of integrated luminosity basically all of the parameter space of the minimal $SU(5)$ model should be explorable. Also, chargino masses as high as $\approx 175\,\mathrm{GeV}$ in the no-scale model could be explored, although some regions of parameter space for lighter chargino masses would remain unexplored. We expect that somewhat weaker results will hold for the dilaton model, since the sparticle masses are heavier in that model, especially the sleptons which enhance the leptonic branching ratios when they are light enough [60].

(c) The relation $m_{\tilde{q}} \approx m_{\tilde{g}}$ for the $\tilde{u}_{L,R}, \tilde{d}_{L,R}$ squark masses in the flipped $SU(5)$ models should allow to probe the low end of the squark and gluino allowed mass ranges, although the outlook is more promising for the dilaton model since the allowed range starts off at lower values of $m_{\tilde{g},\tilde{q}}$ (see Eq. (26)). An important point distinguishing the two models is that the average squark mass is slightly below (above) the gluino mass in the no-scale (dilaton) model, which should have an important bearing on the experimental signatures and rates [111]. In the dilaton case the $\tilde{t}_1$ mass can be below 100 GeV for sufficiently low m_t, and thus may be detectable. As the lower bound on m_t rises, this signal becomes less accessible. The actual reach of the Tevatron for the above processes depends on its ultimate integrated luminosity. The squark masses in the minimal $SU(5)$ model ($m_{\tilde{q}} \gtrsim 500\,\mathrm{GeV}$) are beyond the reach of the Tevatron.

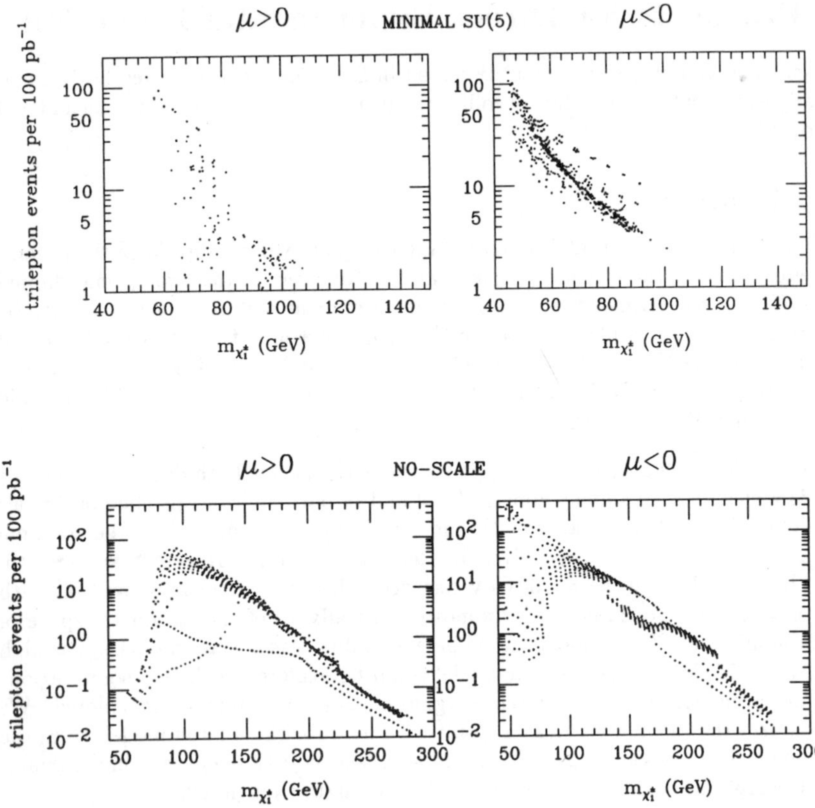

Figure 10: The number of trilepton events at the Tevatron per $100\,\mathrm{pb}^{-1}$ in the minimal $SU(5)$ model and the no-scale flipped $SU(5)$ model (for $m_t = 130\,\mathrm{GeV}$). Note that with $200\,\mathrm{pb}^{-1}$ and 60% detection efficiency it should be possible to probe basically all of the parameter space of the minimal $SU(5)$ model, and probe chargino masses as high as $175\,\mathrm{GeV}$ in the no-scale model.

8.2 LEPI

The current LEPI lower bound on the Standard Model (SM) Higgs boson mass ($m_H >$ 61.6 GeV [112]) is obtained by studying the process $e^+e^- \to Z^*H$ with subsequent Higgs decay into two jets. The MSSM analog of this production process leads to a cross section differing just by a factor of $\sin^2(\alpha-\beta)$, where α is the SUSY Higgs mixing angle and $\tan\beta = v_2/v_1$ is the ratio of the Higgs vacuum expectation values [113]. The published LEPI lower bound on the lightest SUSY Higgs boson mass ($m_h > 43\,\mathrm{GeV}$) is the result of allowing $\sin^2(\alpha-\beta)$ to vary throughout the MSSM parameter space and by considering the $e^+e^- \to Z^*h, hA$ cross sections. It is therefore possible that in specific models (which embed the MSSM), where $\sin^2(\alpha-\beta)$ is naturally restricted to be near unity, the lower bound on m_h could rise, and even reach the SM lower bound if $\mathrm{BR}(h \to 2\,\mathrm{jets})$ is SM-like as well. This has been shown to be the case for the supergravity models we discuss here, and more generally for supergravity models which enforce radiative electroweak symmetry breaking [61].

Non-observation of a SM Higgs signal puts the following upper bound in the number of expected 2-jet events.

$$\#\mathrm{events}_{\mathrm{SM}} = \sigma(e^+e^- \to Z^*H)_{\mathrm{SM}} \times \mathrm{BR}(H \to 2\,\mathrm{jets})_{\mathrm{SM}} \times \int \mathcal{L}dt < 3. \qquad (47)$$

The SM value for $\mathrm{BR}(H \to 2\,\mathrm{jets})_{\mathrm{SM}} \approx \mathrm{BR}(H \to b\bar{b} + c\bar{c} + gg)_{\mathrm{SM}} \approx 0.92$ [113] corresponds to an upper bound on $\sigma(e^+e^- \to Z^*H)_{\mathrm{SM}}$. Since this is a monotonically decreasing function of m_H, a lower bound on m_H follows, *i.e.*, $m_H > 61.6\,\mathrm{GeV}$ as noted above. We denote by $\sigma_{\mathrm{SM}}(61.6)$ the corresponding value for $\sigma(e^+e^- \to Z^*H)_{\mathrm{SM}}$. For the MSSM the following relations hold

$$\sigma(e^+e^- \to Z^*h)_{\mathrm{SUSY}} = \sin^2(\alpha-\beta)\sigma(e^+e^- \to Z^*H)_{\mathrm{SM}}, \qquad (48)$$

$$\mathrm{BR}(h \to 2\,\mathrm{jets})_{\mathrm{SUSY}} = f \cdot \mathrm{BR}(H \to 2\,\mathrm{jets})_{\mathrm{SM}}. \qquad (49)$$

From Eq. (47) one can deduce the integrated luminosity achieved, $\int \mathcal{L}dt = 3/(\sigma_{\mathrm{SM}}(61.6)\mathrm{BR}_{\mathrm{SM}})$. In analogy with Eq. (47), we can write

$$\#\mathrm{events}_{\mathrm{SUSY}} = \sigma_{\mathrm{SUSY}}(m_h) \times \mathrm{BR}_{\mathrm{SUSY}} \times \int \mathcal{L}dt = 3f \cdot \sigma_{\mathrm{SUSY}}(m_h)/\sigma_{\mathrm{SM}}(61.6) < 3. \quad (50)$$

This immediately implies the following condition for *allowed* points in parameter space [61, 114]

$$f \cdot \sin^2(\alpha-\beta) < P(61.6/M_Z)/P(m_h/M_Z), \qquad (51)$$

where we have used the fact that the cross sections differ simply by the coupling factor $\sin^2(\alpha-\beta)$ and the Higgs mass dependence which enters through a function P [113]

$$P(y) = \frac{3y(y^4-8y^2+20)}{\sqrt{4-y^2}}\cos^{-1}\left(\frac{y(3-y^2)}{2}\right) - 3(y^4-6y^2+4)\ln y - \tfrac{1}{2}(1-y^2)(2y^4-13y+47). \qquad (52)$$

The cross section $\sigma_{\rm SUSY}(m_h)$ for the minimal $SU(5)$ model also corresponds to the SM result since one can verify that $\sin^2(\alpha-\beta)>0.9999$ in this case. For the flipped model there is a small deviation ($\sin^2(\alpha-\beta)>0.95$) relative to the SM result for some points [61]. In the calculation of BR($h\to 2\,{\rm jets})_{\rm SUSY}$ which enters in the ratio f, we have included *all* contributing modes, in particular the invisible $h\to\chi^0_1\chi^0_1$ decays. The conclusion is that these models differ little from the SM and in fact the proper lower bound on m_h is very near 60 GeV, although it varies from point to point in the parameter space.

In Ref. [61] it was also shown that this phenomenon is due to a decoupling effect of the Higgs sector as the supersymmetry scale rises, and it is communicated to the Higgs sector through the radiative electroweak symmetry breaking mechanism. The point to be stressed is that if the supersymmetric Higgs sector is found to be SM-like, this could be taken as *indirect* evidence for an underlying radiative electroweak breaking mechanism, since no insight could be garnered from the MSSM itself.

Note that since the lower bound on the SM Higgs boson mass could still be pushed up several GeV at LEPI, the strict dilaton scenario in Sec. 6.2 (which requires $m_h\approx 61-91$ GeV) could be further constrained at LEPI.

8.3 LEPII

(a) At LEPII the SM Higgs mass could be explored up to roughly the beam energy minus 100 GeV [115]. This will allow exploration of almost al of the Higgs parameter space in the minimal $SU(5)$ model [62]. In the flipped $SU(5)$ models, only low $\tan\beta$ values could be explored, although the strict no-scale case will probably be out of reach (see Figs. 4,5). The $e^+e^-\to hA$ channel will be open in the flipped $SU(5)$ models for large $\tan\beta$ and low $m_{\tilde g}$. This channel is always closed in the minimal $SU(5)$ case (since $m_A\gtrsim 1$ TeV). It is important to point out that the preferred $h\to b\bar b, c\bar c, gg$ detection modes may be suppressed because of invisible Higgs decays ($h\to\chi^0_1\chi^0_1$) for $m_h\lesssim 80$ GeV ($m_h\gtrsim 80$ GeV) by as much as 30%/15% (40%/40%) in the minimal/flipped $SU(5)$ model [62].

(b) Chargino masses below the kinematical limit ($m_{\chi^\pm_1}\lesssim 100$ GeV) should not be a problem to detect through the "mixed" mode with one chargino decaying leptonically and the other one hadronically [62], *i.e.*, $e^+e^-\to\chi^+_1\chi^-_1$, $\chi^+_1\to\chi^0_1 q\bar q'$, $\chi^-_1\to\chi^0_1 l^-\bar\nu_l$. In Fig. 11 and Fig. 12 (top row) we show the correponding event rates in the minimal $SU(5)$ and no-scale flipped $SU(5)$ models. Recall that $m_{\chi^\pm_1}$ can be as high as ≈ 290 GeV in the flipped models, whereas $m_{\chi^\pm_1}\lesssim$ 100 GeV in the minimal $SU(5)$ model. Interestingly enough, the number of mixed events do not overlap (they are much higher in the minimal $SU(5)$ model) and therefore, if $m_{\chi^\pm_1}<100$ GeV then LEPII should be able to exclude at least of the models.

(c) Selectron, smuon, and stau pair production is partially accessible for both the no-scale and dilaton models (although more so in the no-scale case), and completely

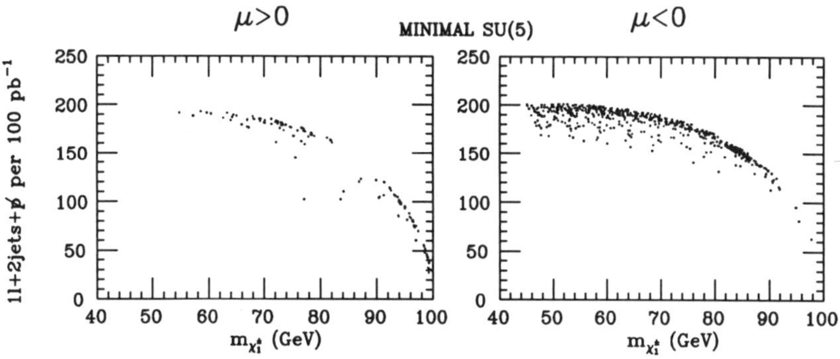

Figure 11: The number of "mixed" events (1-lepton+2jets+$\not{p}$) events per $\mathcal{L} = 100\,\text{pb}^{-1}$ at LEPII versus the chargino mass in the minimal $SU(5)$ model.

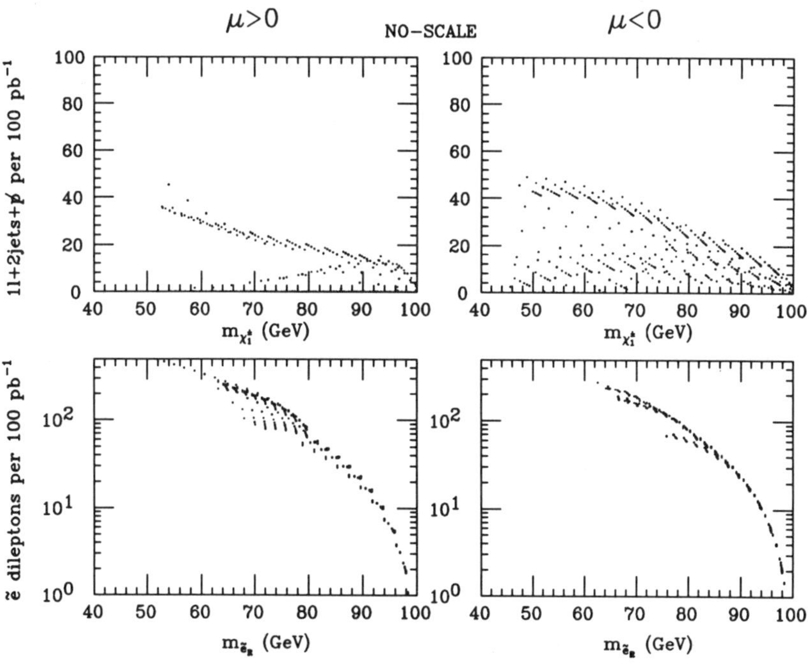

Figure 12: The number of "mixed" events (1-lepton+2jets+$\not{p}$) events per $\mathcal{L} = 100\,\text{pb}^{-1}$ at LEPII versus the chargino mass in the no-scale model (top row). Also shown (bottom row) are the number of di-electron events per $\mathcal{L} = 100\,\text{pb}^{-1}$ from selectron pair production versus the lightest selectron mass.

inaccessible in the minimal $SU(5)$ case. In Fig. 12 (bottom row) we show the rates for the most promising (dilepton) mode in $e^+e^- \to \tilde{e}_R^+\tilde{e}_R^-$ production in the no-scale model.

8.4 HERA

The elastic and deep-inelastic contributions to $e^-p \to \tilde{e}_R^-\chi_1^0$ and $e^-p \to \tilde{\nu}\chi_1^-$ at HERA in the no-scale flipped $SU(5)$ model should push the LEPI lower bounds on the lightest selectron, the lightest neutralino, and the sneutrino masses by $\approx 25\,\mathrm{GeV}$ with $\mathcal{L} = 100\,\mathrm{pb}^{-1}$ [63]. In Fig. 13 we show the elastic plus deep-inelastic contributions to the total supersymmetric signal ($ep \to$ susy $\to eX + \not{p}$) versus the lightest selectron mass ($m_{\tilde{e}_R}$) and the sneutrino mass ($m_{\tilde{\nu}}$) in the no-scale model. These figures show the "reach" of HERA in each of these variables. With $\mathcal{L} = 1000\,\mathrm{pb}^{-1}$ HERA should be competitive with LEPII as far as the no-scale model is concerned. In the dilaton scenario, because of the somewhat heavier sparticle masses, the effectiveness of HERA is reduced, although probably both channels may be accessible. HERA is not sensitive to the minimal $SU(5)$ model spectrum.

9 Prospects for Indirect Experimental Detection

9.1 $\mathbf{b \to s\gamma}$

There has recently been a renewed surge of interest on the flavor-changing-neutral-current (FCNC) $b \to s\gamma$ decay, prompted by the CLEO 95% CL allowed range[116]

$$\mathrm{BR}(b \to s\gamma) = (0.6 - 5.4) \times 10^{-4}. \tag{53}$$

Since the Standard Model (SM) prediction looms around $(2-5) \times 10^{-4}$ depending on the top-quark mass (m_t), a reappraisal of beyond the SM contributions has become topical [117, 118, 64, 119]. We use the following expression for the branching ratio $b \to s\gamma$ [118]

$$\frac{\mathrm{BR}(b \to s\gamma)}{\mathrm{BR}(b \to ce\bar{\nu})} = \frac{6\alpha}{\pi}\frac{\left[\eta^{16/23}A_\gamma + \frac{8}{3}(\eta^{14/23} - \eta^{16/23})A_g + C\right]^2}{I(m_c/m_b)\left[1 - \frac{2}{3\pi}\alpha_s(m_b)f(m_c/m_b)\right]}, \tag{54}$$

where $\eta = \alpha_s(M_Z)/\alpha_s(m_b)$, I is the phase-space factor $I(x) = 1 - 8x^2 + 8x^6 - x^8 - 24x^4\ln x$, and $f(m_c/m_b) = 2.41$ the QCD correction factor for the semileptonic decay. The A_γ, A_g are the coefficients of the effective $bs\gamma$ and bsg penguin operators evaluated at the scale M_Z. Their simplified expressions are given in the Appendix of Ref. [118], where the gluino and neutralino contributions have been justifiably neglected [120] and the squarks are considered degenerate in mass, except for the $\tilde{t}_{1,2}$ which are significantly split by m_t.

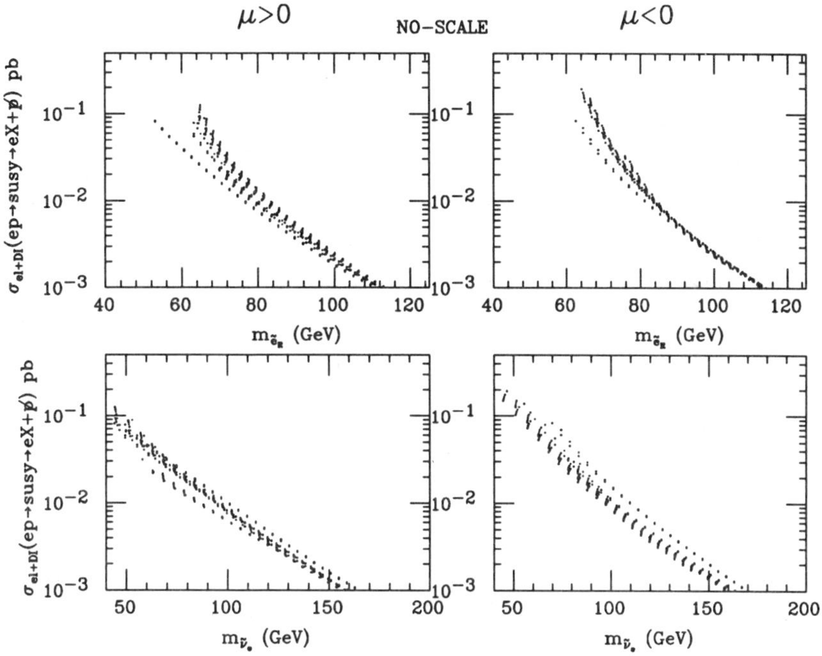

Figure 13: The elastic plus deep-inelastic total supersymmetric cross section at HERA ($ep \to$ susy $\to eX + \not{p}$) versus the lightest selectron mass ($m_{\tilde{e}_R}$) and the sneutrino mass ($m_{\tilde{\nu}}$). The short- and long-term limits of sensitivity are expected to be 10^{-2} pb and 10^{-3} pb respectively.

For the minimal $SU(5)$ supergravity model we find [64]

$$2.3\,(2.6) \times 10^{-4} < \mathrm{BR}(b \to s\gamma)_{minimal} < 3.6\,(3.3) \times 10^{-4}, \tag{55}$$

for $\mu > 0\,(\mu < 0)$, which are all within the CLEO limits. One can show that BR($b \to s\gamma$) would need to be measured with better than 20% accuracy to start disentangling the minimal $SU(5)$ supergravity model from the SM. In Fig. 14 we present the analogous results for the no-scale (top row) and strict no-scale (bottom row) flipped $SU(5)$ supergravity models. The results are strikingly different than in the prior case. One observes that part of the parameter space is actually excluded by the new CLEO bound, for a range of sparticle masses. Perhaps the most surprising feature of the results is the strong suppression of BR($b \to s\gamma$) which occurs for a good portion of the parameter space for $\mu > 0$. It has proven to be non-trivial to find a simple explanation for the observed cancellation. See Ref. [119] for the implications of this indirect constraint on the prospects for direct experimental detection of these models.

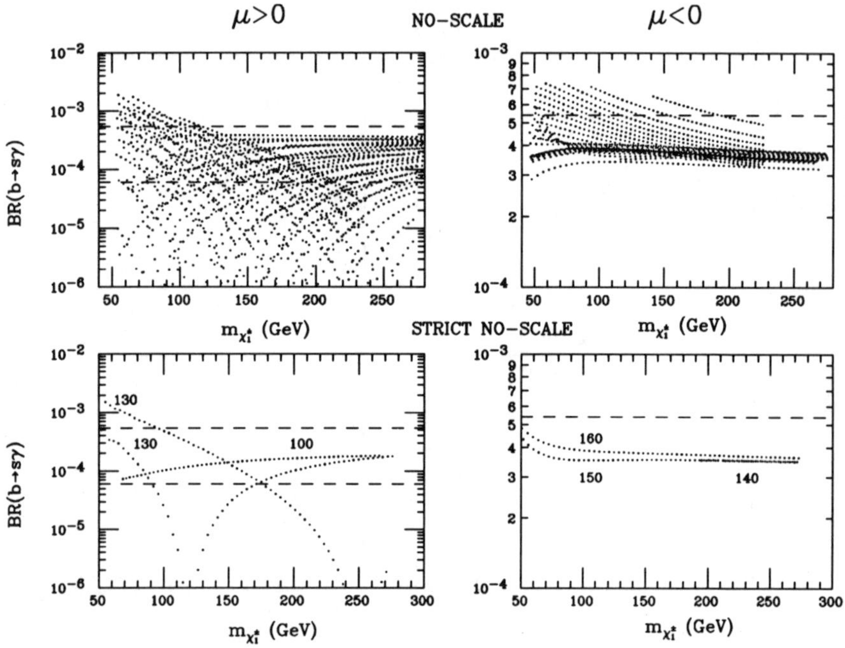

Figure 14: The calculated values of BR($b \to s\gamma$) versus the chargino mass in the no-scale (top row) and strict no-scale (bottom row, for the indicated values of m_t) flipped $SU(5)$ supergravity models. Note the fraction of parameter space excluded by the CLEO allowed range (in between the dashed lines).

9.2 $\epsilon_{1,2,3}$

A complete study of one-loop electroweak radiative corrections in the supergravity models we consider here has been made in Ref. [65]. These calculations include some recently discovered important q^2-dependent effects, which occur when light charginos ($m_{\chi^\pm} \lesssim 60 - 70\,\text{GeV}$) are present [121], and lead to strong correlations between the chargino and the top-quark mass. Specifically, one finds that at present the 90% CL upper limit on the top-quark mass is $m_t \lesssim 175\,\text{GeV}$ in the no-scale flipped $SU(5)$ supergravity model. These bounds can be strengthened for increasing chargino masses in the $50 - 100\,\text{GeV}$ interval. For example, in the flipped model for $m_{\chi_1^\pm} \gtrsim 60\,(70)\,\text{GeV}$, one finds $m_t \lesssim 165\,(160)\,\text{GeV}$. As expected, the heavy sector of both models decouples quite rapidly with increasing sparticle masses, and at present, only ϵ_1 leads to constraints on the parameter spaces of these models. For future reference, it is important to note that global SM fits to all of the low-energy and electroweak data constrain $m_t < 194, 178, 165\,\text{GeV}$ for $m_{H_{SM}} = 1000, 250, 50\,\text{GeV}$ at the 90% CL respectively [107].

One can show that an expansion of the vacuum polarization tensors to order q^2, results in three independent physical parameters. In the first scheme introduced to study these effects [122], namely the (S,T,U) scheme, a SM reference value for $m_t, m_{H_{SM}}$ is used, and the deviation from this reference is calculated and is considered to be "new" physics. This scheme is only valid to lowest order in q^2, and is therefore not applicable to a theory with new, light ($\sim M_Z$) particles. In the supergravity models we consider here, each point in parameter space is actually a distinct model, and a SM reference point is not meaningful. For these reasons, in Ref. [65] the scheme of Refs. [123, 121] was chosen, where the contributions are *absolute* and valid to higher order in q^2. This is the so-called $\epsilon_{1,2,3}$ scheme. Regardless of the scheme used, all of the global fits to the three physical parameters are *entirely consistent* with the SM at 90% CL.

With the assumption that the dominant "new" contributions arise from the process-independent (*i.e.*, "oblique") vacuum polarization amplitudes, one can combine several observables in suitable ways such that they are most sensitive to new effects. It is important to note that not all observables can be included in the experimental fits which determine the $\epsilon_{1,2,3}$ paremeters, if only the oblique contributions are kept [65].

It is well known in the MSSM that the largest contributions to ϵ_1 (*i.e.*, $\delta\rho$ if q^2-dependent effects are neglected) are expected to arise from the $\tilde{t}$-$\tilde{b}$ sector, and in the limiting case of a very light stop, the contribution is comparable to that of the t-b sector [124]. The remaining squark, slepton, chargino, neutralino, and Higgs sectors all typically contribute considerably less. For increasing sparticle masses, the heavy sector of the theory decouples, and only SM effects with a *light* Higgs survive. However, for very light chargino, a Z-wavefunction renormalization threshold effect can introduce a substantial q^2-dependence in the calculation, thus modifying significantly the standard $\delta\rho$ results. For completeness, in Ref. [65] the complete vacuum polarization contributions from the Higgs sector, the supersymmetric chargino-neutralino and sfermion sectors, and also the corresponding contributions in the SM were included.

In Fig. 15 we show the calculated values of ϵ_1 versus the lightest chargino mass ($m_{\chi_1^\pm}$) for the sampled points in the minimal (no-scale flipped) $SU(5)$ supergravity model. In the no-scale flipped $SU(5)$ case three representative values of m_t were used, $m_t = 100, 130, 160\,\mathrm{GeV}$, whereas in the minimal $SU(5)$ case several other values for m_t in the range $90\,\mathrm{GeV} \le m_t \le 160\,\mathrm{GeV}$ were sampled. In both models, but most clearly in the no-scale model one can see how quickly the sparticle spectrum decouples as $m_{\chi_1^\pm}$ increases, and the value of ϵ_1 asymptotes to the SM value appropriate to each value of m_t and for a *light* ($\sim 100\,\mathrm{GeV}$) Higgs mass. The threshold effect of $\chi_1^\pm$ is manifest as $m_{\chi_1^\pm} \to \frac{1}{2}M_Z$ and is especially visible for $\mu < 0$ in both models. This effect is not expected to be very accurate as $m_{\chi_1^\pm} \to \frac{1}{2}M_Z$. However, according to Ref. [121], for $m_{\chi_1^\pm} > 50\,\mathrm{GeV}$, this correction agrees to better than 10% with the one obtained in a more accurate way.

Recent values for $\epsilon_{1,2,3}$ obtained from a global fit to the LEP (*i.e.*, $\Gamma_l, A_{FB}^{l,b}, A_{pol}^{\tau}$)

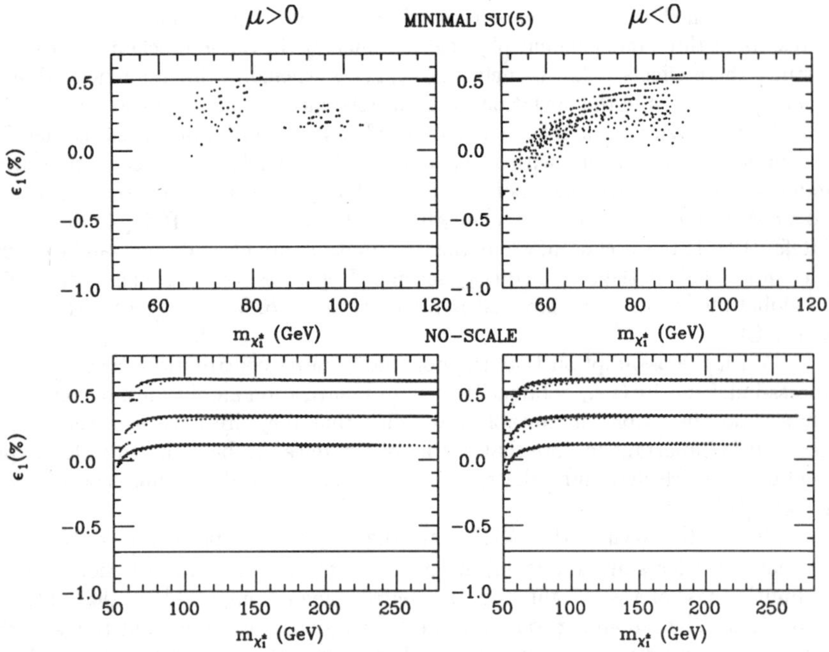

Figure 15: The total contribution to ϵ_1 as a function of the lightest chargino mass $m_{\chi_1^\pm}$ for the minimal $SU(5)$ model (upper row) and the no-scale flipped $SU(5)$ model (bottom row). Points between the two horizontal solid lines are allowed at 90% CL. The three distinct curves (from lowest to highest) in then no-scale case correspond to $m_t = 100, 130, 160\,\mathrm{GeV}$.

and M_W/M_Z measurements are [125],

$$\epsilon_1 = (-0.9 \pm 3.7)10^{-3}, \quad \epsilon_2 = (9.9 \pm 8.0)10^{-3}, \quad \epsilon_3 = (-0.9 \pm 4.1)10^{-3}. \tag{56}$$

For ϵ_1 it is clear that virtually all the sampled points in the minimal $SU(5)$ supergravity model are within the $\pm 1.64\sigma$ (90% CL) bounds (denoted by the two horizontal solid lines in the figures). Since several values for $90 \leq m_t \leq 160\,\mathrm{GeV}$ were sampled, the trends for fixed m_t are not very clear from the figure. Nonetheless, the points just outside the 1.64σ line correspond to $m_t = 160\,\mathrm{GeV}$, which are therefore excluded at the 90% CL. In the no-scale model, the upper bound on m_t depends sensitively on the chargino mass. For example, for $m_t = 160\,\mathrm{GeV}$, only light chargino masses would be acceptable at 90% CL. In fact, in Ref. [65] the region $130\,\mathrm{GeV} \leq m_t \leq 190\,\mathrm{GeV}$ was scanned in increments of 5 GeV and obtained the maximum values for $m_{\chi_1^\pm}$ allowed by the experimental value for ϵ_1 at 90% CL. These are given in the Table 5. One can immediately see the strong correlation between m_t and $m_{\chi_1^\pm}$: as m_t rises, the upper

Table 5: Maximum allowed chargino mass ($m_{\chi_1^\pm}$) for different m_t (in GeV) at 90%CL in the no-scale flipped $SU(5)$ model. In the entries Y(N) means all points are within (outside) the LEP bounds at 90%CL.

m_t	$\mu > 0$	$\mu < 0$
145	Y	Y
150	Y	Y
155	68	95
160	66	72
165	N	63
170	N	58
175	N	53
180	N	N

limit to $m_{\chi_1^\pm}$ falls, and vice versa. In particular, for $m_t \le 150\,\mathrm{GeV}$ all values of $m_{\chi_1^\pm}$ are allowed, while one could have m_t as large as $160\,(175)\,\mathrm{GeV}$ for $\mu > 0\,(\mu < 0)$ if the chargino mass were light enough.

10 Conclusions

The recent surge of interest in supersymmetric models, spurred by the precise LEP measurements of the gauge couplings and their unification at very high energies [26], has made it clear that some sort of organizing principle is needed to tame the zoo of supersymmetric particles at low energies, as encompassed by the minimal supersymmetric standard model (MSSM). Even though it is usually not acknowledged that this model needs at least *twenty-one parameters* for its full description, the fact remains that this is the case. One reason for disregarding this fact is that many phenomenological calculations of interest have been performed and numerical results obtained by making ad-hoc assumptions about the values of these parameters, giving the erroneous impression that the results so-obtained are largely insensitive to changes in these assumptions. This mindset has been explicitly exposed by now on a variety of calculations. These have been performed in the context of the class of unified supergravity models which we consider here, and have been found to yield quite different results than previously expected. A few examples include: the calculation of the cosmological relic density of neutralinos in the MSSM [126] and in supergravity models without [127, 73, 77] and with [128, 74] radiative electroweak symmetry breaking; the enhancement of the leptonic branching ratios of the chargino in the presence of light sleptons [60]; the large sparticle mass dependence of the $b \to s\gamma$ branching ratio [64] which can completely wash out the charged Higgs contribution [117]; etc. It then becomes apparent that one must make further well motivated theoretical assumptions to make further headway into this problem. As discussed above, supergravity models with radiative electroweak symmetry breaking accomplish this goal very economically, needing just three supersymmetry breaking parameters, the ratio $\tan\beta$, and the soon-to-be-measured top-quark mass to be fully described. We have discussed two classes of such models, the class of string-inspired flipped $SU(5)$ models and the traditional minimal $SU(5)$ supergravity model.

The simplest, string-derivable, supergravity model has as gauge group flipped $SU(5)$ with supplementary matter representations to ensure unification at the string scale ($\sim 10^{18}\,\text{GeV}$). This basic structure is complemented by two possible string supersymmetry breaking scenaria: $SU(N,1)$ no-scale supergravity and dilaton-induced supersymmetry breaking. These two variants should be considered to be idealizations of what their string-derived incarnation should be. The specification of the hidden sector is crucial to the determination of the supersymmetry breaking scenario at work. A thorough exploration of the parameter spaces of the two models yields interesting results for direct and indirect experimental detection at present or near future colliders. In this regard, the no-scale model is more within reach than the dilaton model, because of its generally lighter spectrum. In both supersymmetry breaking scenaria considered, there ia a more constrained special case which allows $\tan\beta$ to be determined in terms of m_t and $m_{\tilde{g}}$. In the strict no-scale case we find a striking result: if $\mu > 0$, $m_t \lesssim 135\,\text{GeV}$, whereas if $\mu < 0$, $m_t \gtrsim 140\,\text{GeV}$. Therefore the value of m_t determines the sign of μ. Furthermore, we found that the value of m_t also determines whether the lightest Higgs boson is above or below 100 GeV. In the restricted dilaton case there is an upper bound on the top-quark mass ($m_t \lesssim 155\,\text{GeV}$) and the lightest Higgs boson mass ($m_h \lesssim 91\,\text{GeV}$). Thus, continuing Tevatron top-quark searches and LEPI,II Higgs searches could probe this restricted scenario completely. In Table 6 we give a summary of the general properties of these models and a comparison of their spectra.

The minimal $SU(5)$ supergravity model is strongly constrained by the proton lifetime and the cosmological neutralino relic density. The former constraint implies light chargino masses and heavy scalar masses (beyond the reach of present or near future colliders), while the second constraint nearly enforces the mass relations $m_\chi \approx \frac{1}{2}m_{h,Z}$. The prospects for detection of the light particles in this model ($\chi^0_{1,2}, \chi^\pm_1, h$) is very promising in the near future. It is interesting to remark that the resulting allowed choices for the supersymmetry breaking parameters bear close resemblance to those predicted in moduli-induced string-inspired supersymmetry breaking scenaria. In Table 7 we give a summary of the general properties of this models and its spectrum.

We conclude that these well motivated supergravity models (especially the strict versions of the string-inspired/derived models) could soon be probed experimentally. The various ingredients making up the flipped $SU(5)$ models are likely to be present in actual fully string-derived models which yield the set of supersymmetry breaking parameters in Eqs. (4,5). The search for such models is imperative, although it may not be an easy task since in traditional gaugino condensation scenaria Eqs. (4,5) are usually not reproduced (see however Refs. [129, 130]). Moreover, the requirement of vanishing vacuum energy may be difficult to fulfill, as a model with

Table 6: Major features of the $SU(5) \times U(1)$ string-inspired/derived model and a comparison of the two supersymmetry breaking scenaria considered. (All masses in GeV).

$\mathbf{SU(5) \times U(1)}$
• Easily string-derivable, several known examples • Symmetry breaking to Standard Model due to vevs of $\mathbf{10},\overline{\mathbf{10}}$ and tied to onset of supersymmetry breaking • Natural doublet-triplet splitting mechanism • Proton decay: $d = 5$ operators very small • Baryon asymmetry through lepton number asymmetry (induced by the decay of flipped neutrinos) as processed by non-perturbative electroweak interactions

$\langle F_M \rangle_{m_0=0}$ (no-scale)	$\langle F_D \rangle$ (dilaton)
• Parameters 3: $m_{1/2}, \tan\beta, m_t$ • Universal soft-supersymmetry breaking automatic • $m_0 = 0$, $A = 0$ • Dark matter: $\Omega_\chi h_0^2 < 0.25$ • $m_{1/2} < 475\,\text{GeV}$, $\tan\beta < 32$ • $m_{\tilde{g}} > 245\,\text{GeV}$, $m_{\tilde{q}} > 240\,\text{GeV}$ • $m_{\tilde{q}} \approx 0.97 m_{\tilde{g}}$ • $m_{\tilde{t}_1} > 155\,\text{GeV}$ • $m_{\tilde{e}_R} \approx 0.18 m_{\tilde{g}}$, $m_{\tilde{e}_L} \approx 0.30 m_{\tilde{g}}$ $m_{\tilde{e}_R}/m_{\tilde{e}_L} \approx 0.61$ • $60\,\text{GeV} < m_h < 125\,\text{GeV}$ • $2m_{\chi_1^0} \approx m_{\chi_2^0} \approx m_{\chi_1^\pm} \approx 0.28 m_{\tilde{g}} \lesssim 290$ • $m_{\chi_3^0} \sim m_{\chi_4^0} \sim m_{\chi_2^\pm} \sim \lvert\mu\rvert$ • Spectrum easily accessible soon	• Parameters 3: $m_{1/2}, \tan\beta, m_t$ • Universal soft-supersymmetry breaking automatic • $m_0 = \frac{1}{\sqrt{3}} m_{1/2}$, $A = -m_{1/2}$ • Dark matter: $\Omega_\chi h_0^2 < 0.90$ • $m_{1/2} < 465\,\text{GeV}$, $\tan\beta < 46$ • $m_{\tilde{g}} > 195\,\text{GeV}$, $m_{\tilde{q}} > 195\,\text{GeV}$ • $m_{\tilde{q}} \approx 1.01 m_{\tilde{g}}$ • $m_{\tilde{t}_1} > 90\,\text{GeV}$ • $m_{\tilde{e}_R} \approx 0.33 m_{\tilde{g}}$, $m_{\tilde{e}_L} \approx 0.41 m_{\tilde{g}}$ $m_{\tilde{e}_R}/m_{\tilde{e}_L} \approx 0.81$ • $60\,\text{GeV} < m_h < 125\,\text{GeV}$ • $2m_{\chi_1^0} \approx m_{\chi_2^0} \approx m_{\chi_1^\pm} \approx 0.28 m_{\tilde{g}} \lesssim 285$ • $m_{\chi_3^0} \sim m_{\chi_4^0} \sim m_{\chi_2^\pm} \sim \lvert\mu\rvert$ • Spectrum accessible soon
• Strict no-scale: $B(M_U) = 0$ $\tan\beta = \tan\beta(m_t, m_{\tilde{g}})$ $m_t \lesssim 135\,\text{GeV} \Rightarrow \mu > 0, m_h \lesssim 100\,\text{GeV}$ $m_t \gtrsim 140\,\text{GeV} \Rightarrow \mu < 0, m_h \gtrsim 100\,\text{GeV}$	• Special dilaton: $B(M_U) = 2m_0$ $\tan\beta = \tan\beta(m_t, m_{\tilde{g}})$ $\tan\beta \approx 1.4 - 1.6$, $m_t < 155\,\text{GeV}$ $m_h \approx 61 - 91\,\text{GeV}$

Table 7: Major features of the minimal $SU(5)$ supergravity model and its spectrum (All masses in GeV).

SU(5)
• Not easily string-derivable, no known examples
• Symmetry breaking to Standard Model due to vevs of **24** and independent of supersymmetry breaking
• No simple doublet-triplet splitting mechanism
• Proton decay: $d = 5$ operators large, strong constraints needed
• Baryon asymmetry ?

Spectrum
• Parameters 5: $m_{1/2}, m_0, A, \tan\beta, m_t$
• Universal soft-supersymmetry breaking automatic
• $m_0/m_{1/2} > 3$, $\tan\beta \lesssim 3.5$
• Dark matter: $\Omega_\chi h_0^2 \gg 1$, 1/6 of points excluded
• $m_{\tilde{g}} < 400\,\mathrm{GeV}$, $m_{\tilde{q}} > m_{\tilde{l}} > 2m_{\tilde{g}} \gtrsim 500\,\mathrm{GeV}$
• $m_{\tilde{t}_1} > 45\,\mathrm{GeV}$
• $60\,\mathrm{GeV} < m_h < 100\,\mathrm{GeV}$
• $2m_{\chi_1^0} \approx m_{\chi_2^0} \approx m_{\chi_1^\pm} \approx 0.28 m_{\tilde{g}} \lesssim 100$
• $m_{\chi_3^0} \sim m_{\chi_4^0} \sim m_{\chi_2^\pm} \sim \lvert\mu\rvert$
• Chargino and Higgs easily accessible soon

these properties and all the other ones outlined in Sec. 2 is yet to be found. This should not be taken as a discouragement since the harder it is to find the correct model, the more likely it is to be in some sense unique.

Acknowledgements

This work has been supported in part by DOE grant DE-FG05-91-ER-40633. The work of J.L. has been supported by an SSC Fellowship.

References

[1] See *e.g.*, L. Rolandi in Proceedings of the XXVI International Conference on High Energy Physics, Dallas, August 1992, ed. by J. R. Sanford (AIP Conference Proceedings No. 272), p. 56.

[2] See *e.g.*, S. Ting, CERN-PPE/93-34, to appear in Proceedings of the 1992 DPF Meeting, Fermilab, November 1992.

[3] See *e.g.*, J. Ellis, G. L. Fogli, and E. Lisi, Phys. Lett. B **274** (1992) 456; F. del Aguila, M. Martinez, and M. Quiros, Nucl. Phys. B **381** (1992) 451.

[4] M. Veltmann, Acta Phys. Pol. B8 (1977) 475; C. Vayonakis, Lett. Nuovo Cimento, **17** (1976) 383; B. W. Lee, C. Quigg, and H. B. Thacker, Phys. Rev. D **16** (1977) 1519; M. Lüscher and P. Weisz, Phys. Lett. B **212** (1989) 472; W. Marciano, G. Valencia, and S. Willenbrock, Phys. Rev. D **40** (1989) 1725

[5] L. Durand, J. Johnson, and J. L. Lopez, Phys. Rev. Lett. **64** (1990) 1215 and Phys. Rev. D **45** (1992) 3112.

[6] M. Lindner, M. Sher, and H. W. Zagluer, Phys. Lett. B **228** (1989) 139. For a review see, M. Sher, Phys. Rep. **179** (1989) 273.

[7] L. Ibáñez and G. Ross, Phys. Lett. B **110** (1982) 215; K. Inoue, *et. al.*, Prog. Theor. Phys. 68 (1982) 927; L. Ibáñez, Nucl. Phys. B **218** (1983) 514 and Phys. Lett. B **118** (1982) 73; H. P. Nilles, Nucl. Phys. B **217** (1983) 366; J. Ellis, D. V. Nanopoulos, and K. Tamvakis, Phys. Lett. B **121** (1983) 123; J. Ellis, J. Hagelin, D. V. Nanopoulos, and K. Tamvakis, Phys. Lett. B **125** (1983) 275; L. Alvarez-Gaumé, J. Polchinski, and M. Wise, Nucl. Phys. B **221** (1983) 495; L. Ibañéz and C. López, Phys. Lett. B **126** (1983) 54 and Nucl. Phys. B **233** (1984) 545; C. Kounnas, A. Lahanas, D. V. Nanopoulos, and M. Quirós, Phys. Lett. B **132** (1983) 95 and C. Kounnas, A. Lahanas, D. V. Nanopoulos, and M. Quirós, Nucl. Phys. B **236** (1984) 438.

[8] G. Gamberini, G. Ridolfi, and F. Zwirner, Nucl. Phys. B **331** (1990) 331.

[9] See *e.g.*, *String theory in four dimensions*, ed. by M. Dine (North-Holland, Amsterdam, 1988); *Superstring construction*, ed. by A. N. Schellekens (North-Holland, Amsterdam, 1989).

[10] P. Ginsparg, Phys. Lett. B **197** (1987) 139.

[11] V. Kaplunovsky, Nucl. Phys. B **307** (1988) 145 and Nucl. Phys. B **382** (1992) 436(E).

[12] I. Antoniadis, J. Ellis, R. Lacaze, and D. V. Nanopoulos, Phys. Lett. B **268** (1991) 188; S. Kalara, J. L. Lopez, and D. V. Nanopoulos, Phys. Lett. B **269** (1991) 84.

[13] L. Ibáñez, D. Lüst, and G. Ross, Phys. Lett. B **272** (1991) 251.

[14] D. Lewellen, Nucl. Phys. B **337** (1990) 61; J. A. Schwarz, Phys. Rev. D **42** (1990) 1777.

[15] L. Ibáñez, H. Nilles, and F. Quevedo, Phys. Lett. B **187** (1987) 25; L. Ibáñez, J. Mas, H. Nilles, and F. Quevedo, Nucl. Phys. B **301** (1988) 157; A. Font, L. Ibáñez, F. Quevedo, and A. Sierra, Nucl. Phys. B **331** (1990) 421.

[16] A. Faraggi, D. V. Nanopoulos, and K. Yuan, Nucl. Phys. B **335** (1990) 347; A. Faraggi, Phys. Lett. B **278** (1992) 131, Phys. Lett. B **274** (1992) 47, Nucl. Phys. B **387** (1992) 239.

[17] I. Antoniadis, G. Leontaris, and J. Rizos, Phys. Lett. B **245** (1990) 161.

[18] S. Barr, Phys. Lett. B **112** (1982) 219, Phys. Rev. D **40** (1989) 2457; J. Derendinger, J. Kim, and D. V. Nanopoulos, Phys. Lett. B **139** (1984) 170.

[19] I. Antoniadis, J. Ellis, J. Hagelin, and D. V. Nanopoulos, Phys. Lett. B **194** (1987) 231.

[20] I. Antoniadis, J. Ellis, J. Hagelin, and D. V. Nanopoulos, Phys. Lett. B **231** (1989) 65.

[21] J. Ellis, J. Hagelin, S. Kelley, and D. V. Nanopoulos, Nucl. Phys. B **311** (1988/89) 1.

[22] See *e.g.*, J. Ellis, J. L. Lopez, and D. V. Nanopoulos, Phys. Lett. B **245** (1990) 375; A. Font, L. Ibáñez, and F. Quevedo, Nucl. Phys. B **345** (1990) 389.

[23] A. Masiero, D. V. Nanopoulos, K. Tamvakis, and T. Yanagida, Phys. Lett. B **115** (1982) 380; B. Grinstein, Nucl. Phys. B **206** (1982) 387.

[24] J. L. Lopez, D. V. Nanopoulos, and A. Zichichi, Texas A & M University preprint CTP-TAMU-68/92, CERN-TH.6667/92, and CERN-PPE/92-188.

[25] J. L. Lopez, D. V. Nanopoulos, and K. Yuan, Nucl. Phys. B **399** (1993) 654.

[26] J. Ellis, S. Kelley, and D. V. Nanopoulos, Phys. Lett. B **249** (1990) 441.

[27] F. Anselmo, L. Cifarelli, A. Peterman, and A. Zichichi, Nuovo Cimento **105A** (1992) 1179; F. Anselmo, L. Cifarelli, and A. Zichichi, Nuovo Cimento **105A** (1992) 1335.

[28] I. Antoniadis, J. Ellis, S. Kelley, and D. V. Nanopoulos, Phys. Lett. B **272** (1991) 31; D. Bailin and A. Love, Phys. Lett. B **280** (1992) 26.

[29] S. Kelley, J. L. Lopez, and D. V. Nanopoulos, Phys. Lett. B **278** (1992) 140.

[30] G. Leontaris, Phys. Lett. B **281** (1992) 54.

[31] For a review see *e.g.*, J. L. Lopez and D. V. Nanopoulos, in Proceedings of the 15th Johns Hopkins Workshop on Current Problems in Particle Theory, August 1991, p. 277; ed. by G. Domokos and S. Kovesi-Domokos (World Scientific, Singapore 1992).

[32] J. Ellis, S. Kelley, and D. V. Nanopoulos, Phys. Lett. B **260** (1991) 131.

[33] J. L. Lopez and D. V. Nanopoulos, Phys. Lett. B **251** (1990) 73, Phys. Lett. B **256** (1991) 150, and Phys. Lett. B **268** (1991) 359.

[34] J. E. Kim and H. P. Nilles, Phys. Lett. B **138** (1984) 150 and Phys. Lett. B **263** (1991) 79; E. J. Chun, J. E. Kim, and H. P Nilles, Nucl. Phys. B **370** (1992) 105.

[35] J. Casas and C. Muñoz, Phys. Lett. B **306** (1993) 288.

[36] G. Giudice and A. Masiero, Phys. Lett. B **206** (1988) 480.

[37] V. Kaplunovsky and J. Louis, Phys. Lett. B **306** (1993) 269.

[38] For a review see A. B. Lahanas and D. V. Nanopoulos, Phys. Rep. **145** (1987) 1.

[39] E. Witten, Phys. Lett. B **155** (1985) 151.

[40] J. Ellis, C. Kounnas, and D. V. Nanopoulos, Nucl. Phys. B **241** (1984) 406.

[41] J. Ellis, C. Kounnas, and D. V. Nanopoulos, Nucl. Phys. B **247** (1984) 373.

[42] S. Kalara, J. Lopez, and D. V. Nanopoulos, Phys. Lett. B **245** (1990) 421, Nucl. Phys. B **353** (1991) 650.

[43] S. Kelley, J. L. Lopez, and D. V. Nanopoulos, Phys. Lett. B **261** (1991) 424.

[44] J. Ellis, J. L. Lopez, and D. V. Nanopoulos, Phys. Lett. B **292** (1992) 189.

[45] J. Ellis, J. L. Lopez, D. V. Nanopoulos, and K. Olive, Phys. Lett. B **308** (1993) 70.

[46] J. Ellis, D. V. Nanopoulos, and K. Olive, Phys. Lett. B **300** (1993) 121.

[47] R. Arnowitt and P. Nath, Phys. Rev. Lett. **69** (1992) 725; P. Nath and R. Arnowitt, Phys. Lett. B **287** (1992) 89.

[48] J. Hisano, H. Murayama, and T. Yanagida, Phys. Rev. Lett. **69** (1992) 1014 and Tohoku University preprint TU–400 (July 1992).

[49] J. L. Lopez, D. V. Nanopoulos, H. Pois, and A. Zichichi, Phys. Lett. B **299** (1993) 262.

[50] S. Bethke, in Proceedings of the XXVI International Conference on High Energy Physics, Dallas, August 1992, ed. by J. R. Sanford (AIP Conference Proceedings No. 272), p. 81.

[51] J. Derendinger, L. Ibáñez, and H. Nilles, Phys. Lett. B **155** (1985) 65; M. Dine, N. Seiberg, and E. Witten, Phys. Lett. B **156** (1985) 55; A. Font, L. Ibáñez, D. Lüst, and F. Quevedo, Phys. Lett. B **245** (1990) 401; S. Ferrara, N. Magnoli, T. Taylor, and G. Veneziano, Phys. Lett. B **245** (1990) 409; H. Nilles and M. Olechowski, Phys. Lett. B **248** (1990) 268; P. Binetruy and M. Gaillard, Nucl. Phys. B **358** (1991) 121.

[52] I. Antoniadis, J. Ellis, A. Lahanas, and D. V. Nanopoulos, Phys. Lett. B **241** (1990) 24.

[53] J. Ellis and D. V. Nanopoulos, Phys. Lett. B **110** (1982) 44.

[54] L. Ibáñez and D. Lüst, Nucl. Phys. B **382** (1992) 305.

[55] B. de Carlos, J. Casas, and C. Muñoz, Nucl. Phys. B **399** (1993) 623 and Phys. Lett. B **299** (1993) 234.

[56] J. L. Lopez, D. V. Nanopoulos, and H. Pois, Phys. Rev. D **47** (1993) 2468.

[57] J. L. Lopez, D. V. Nanopoulos, and A. Zichichi, Phys. Lett. B **291** (1992) 255.

[58] R. Arnowitt and P. Nath, Phys. Lett. B **299** (1993) 58 and **307** (1993) 403(E); P. Nath and R. Arnowitt, Phys. Rev. Lett. **70** (1993) 3696.

[59] J. L. Lopez, D. V. Nanopoulos, and K. Yuan, Texas A & M University preprint CTP-TAMU-14/93 (to appear in Phys. Rev. D).

[60] J. L. Lopez, D. V. Nanopoulos, X. Wang, and A. Zichichi, Texas A & M University preprint CTP-TAMU-76/92 and CERN-PPE/92-194 (to appear in Phys. Rev. D).

[61] J. L. Lopez, D. V. Nanopoulos, H. Pois, X. Wang, and A. Zichichi, Phys. Lett. B **306** (1993) 73.

[62] J. L. Lopez, D. V. Nanopoulos, H. Pois, X. Wang, and A. Zichichi, Texas A & M University preprint CTP-TAMU-89/92 and CERN-PPE/93-16.

[63] J. L. Lopez, D. V. Nanopoulos, X. Wang, and A. Zichichi, Texas A & M University preprint CTP-TAMU-15/93 and CERN-PPE/93-64.

[64] J. L. Lopez, D. V. Nanopoulos, and G. Park, Texas A & M University preprint CTP-TAMU-16/93 (to appear in Phys. Rev. D).

[65] J. L. Lopez, D. V. Nanopoulos, G. Park, H. Pois, and K. Yuan, Texas A & M University preprint CTP-TAMU-19/93.

[66] R. Gandhi, J. L. Lopez, D. V. Nanopoulos, K. Yuan, and A. Zichichi (in preparation).

[67] J. L. Lopez, D. V. Nanopoulos, and A. Zichichi, Texas A & M University preprint CTP-TAMU-31/93 and CERN-TH.6903/93.

[68] R. Barbieri, J. Louis, and M. Moretti, CERN-TH.6856/93.

[69] S. Kelley, J. L. Lopez, D. V. Nanopoulos, H. Pois, and K. Yuan, Nucl. Phys. B **398** (1993) 3.

[70] P. Nath and R. Arnowitt, Phys. Lett. B **289** (1992) 368.

[71] J. Ellis, G. Ridolfi, and F. Zwirner, Phys. Lett. B **262** (1991) 477.

[72] M. Drees and M. Nojiri, Phys. Rev. D **45** (1992) 2482.

[73] J. L. Lopez, D. V. Nanopoulos, K. Yuan, Nucl. Phys. B **370** (1992) 445.

[74] S. Kelley, J. L. Lopez, D. V. Nanopoulos, H. Pois, and K. Yuan, Phys. Rev. D **47** (1993) 2461.

[75] See *e.g.*, E. Kolb and M. Turner, *The Early Universe* (Addison-Wesley, 1990).

[76] See *e.g.*, R. Schaefer and Q. Shafi, Nature **359** (1992) 199; A. N. Taylor and M. Rowan-Robinson, Nature **359** (1992) 393.

[77] J. L. Lopez, D. V. Nanopoulos, and K. Yuan, Phys. Lett. B **267** (1991) 219.

[78] L. Durand and J. L. Lopez, Phys. Lett. B **217** (1989) 463, Phys. Rev. D **40** (1989) 207.

[79] For reviews see *e.g.*, R. Arnowitt and P. Nath, *Applied N=1 Supergravity* (World Scientific, Singapore 1983); H. P. Nilles, Phys. Rep. **110** (1984) 1.

[80] A. Chamseddine, R. Arnowitt, and P. Nath, Phys. Rev. Lett. **49** (1982) 970.

[81] S. Weinberg, Phys. Rev. D **26** (1982) 287; N. Sakai and T. Yanagida, Nucl. Phys. B **197** (1982) 533.

[82] J. Ellis, D. V. Nanopoulos, and S. Rudaz, Nucl. Phys. B **202** (1982) 43; B. Campbell, J. Ellis, and D. V. Nanopoulos, Phys. Lett. B **141** (1984) 229.

[83] K. Enqvist, A. Masiero, and D. V. Nanopoulos, Phys. Lett. B **156** (1985) 209.

[84] P. Nath, A. Chamseddine, and R. Arnowitt, Phys. Rev. D **32** (1985) 2348; P. Nath and R. Arnowitt, Phys. Rev. D **38** (1988) 1479.

[85] M. Matsumoto, J. Arafune, H. Tanaka, and K. Shiraishi, Phys. Rev. D **46** (1992) 3966.

[86] See *e.g.*, U. Amaldi, *et. al.*, Phys. Rev. D **36** (1987) 1385; G. Costa, *et. al.*, Nucl. Phys. B **297** (1988) 244 and references therein.

[87] P. Langacker and M.-X. Luo, Phys. Rev. D **44** (1991) 817; U. Amaldi, W. de Boer, and H. Fürstenau, Phys. Lett. B **260** (1991) 447; L. Clavelli, Phys. Rev. D **45** (1992) 3276.

[88] H. Arason, *et. al.*, Phys. Rev. Lett. **67** (1991) 2933, Phys. Rev. D **46** (1992) 3945, Phys. Rev. D **47** (1992) 232.

[89] F. Anselmo, L. Cifarelli, A. Peterman, and A. Zichichi, Nuovo Cim. **104A** (1991) 1817 and **105A** (1992) 1025.

[90] J. Ellis, S. Kelley and D. V. Nanopoulos, Nucl. Phys. B **373** (1992) 55.

[91] J. Ellis, S. Kelley and D. V. Nanopoulos, Phys. Lett. B **287** (1992) 95.

[92] G. Ross and R. Roberts, Nucl. Phys. B **377** (1992) 571.

[93] F. Anselmo, L. Cifarelli, A. Peterman, and A. Zichichi, Nuovo Cim. **105A** (1992) 581; F. Anselmo, L. Cifarelli, and A. Zichichi, Nuovo Cim. **105A** (1992) 1357.

[94] F. Anselmo, L. Cifarelli, A. Peterman, and A. Zichichi, Nuovo Cim. **105A** (1992) 1201.

[95] P. Langacker and N. Polonsky, Phys. Rev. D **47** (1993) 4028.

[96] M. Carena, S. Pokorski, and C. Wagner, MPI-Ph/93-10.

[97] V. Barger, M. Berger, and P. Ohman, Phys. Rev. D **47** (1993) 1093.

[98] R. Barbieri and L. Hall, Phys. Rev. Lett. **68** (1992) 752; K. Hagiwara and Y. Yamada, Phys. Rev. Lett. **70** (1993) 709; A. Faraggi, B. Grinstein, and S. Meshkov, Phys. Rev. D **47** (1993) 5018.

[99] L. Hall and U. Sarid, Phys. Rev. Lett. **70** (1993) 2673.

[100] M. Chanowitz, J. Ellis, and M. Gaillard, Nucl. Phys. B **128** (1977) 506; A. J. Buras, J. Ellis, M. K. Gaillard, D. V. Nanopoulos, Nucl. Phys. B **135** (1978) 66; D. V. Nanopoulos and D. A. Ross, Nucl. Phys. B **157** (1979) 273, Phys. Lett. B **108** (1982) 351, and Phys. Lett. B **118** (1982) 99.

[101] A. Giveon, L. Hall, and U. Sarid, Phys. Lett. B **271** (1991) 138.

[102] S. Kelley, J. L. Lopez, and D. V. Nanopoulos, Phys. Lett. B **274** (1992) 387.

[103] S. Dimopoulos, L. Hall, and S. Raby, Phys. Rev. Lett. **68** (1992) 1984, Phys. Rev. D **45** (1992) 4192; G. Anderson, S. Dimopoulos, L. Hall, and S. Raby, Phys. Rev. D **47** (1992) R3702.

[104] S. Naculich, JHU-TIPAC-930002 (January 1993).

[105] P. Langacker and N. Polonsky, UPR-0556T (May 1993).

[106] J. Ellis and M. Gaillard, Phys. Lett. B **88** (1979) 315; D. V. Nanopoulos and M. Srednicki, Phys. Lett. B **124** (1983) 37.

[107] Particle Data Group, Phys. Rev. D **45** (1992) S1.

[108] M. Srednicki, R. Watkins, and K. Olive, Nucl. Phys. B **310** (1988) 693.

[109] P. Gondolo and G. Gelmini, Nucl. Phys. B **360** (1991) 145; K. Griest and D. Seckel, Phys. Rev. D **43** (1991) 3191.

[110] P. Nath and R. Arnowitt, Mod. Phys. Lett. A **2** (1987) 331; R. Barbieri, F. Caravaglios, M. Frigeni, and M. Mangano, Nucl. Phys. B **367** (1991) 28; H. Baer and X. Tata, Phys. Rev. D **47** (1993) 2739.

[111] See *e.g.*, J. White, talk presented at the "Recent Advances in the Superworld" Workshop, The Woodlands, Texas, April, 1993; R. M. Barnett, J. Gunion, and H. Haber, LBL-34106; H. Baer, C. Kao, and X. Tata, FSU-HEP-930527.

[112] J. Hilgart, Talk presented at the 1993 Aspen Winter Conference.

[113] See *e.g.*, *The Higgs Hunter's Guide*, J. Gunion, H. Haber, G. Kane, and S. Dawson (Addisson-Wesley, Redwood City, 1990).

[114] J. L. Lopez, D. V. Nanopoulos, and X. Wang, CERN-TH.6890/93 (to appear in Phys. Lett. B).

[115] See *e.g.*, J. Alcaraz, M. Felcini, M. Pieri, and B. Zhou, CERN-PPE/93-28.

[116] E. Thorndike, talk given at the 1993 Meeting of the American Physical Society, Washington D. C., April 1993.

[117] V. Barger, M. Berger, and R. J. N. Phillips, Phys. Rev. Lett. **70** (1993) 1368; J. Hewett, Phys. Rev. Lett. **70** (1993) 1045.

[118] R. Barbieri and G. Giudice, CERN-TH.6830/93.

[119] J. L. Lopez, D. V. Nanopoulos, G. Park, and A. Zichichi (in preparation).

[120] S. Bertolini, F. Borzumati, A. Masiero, and G. Ridolfi, Nucl. Phys. B **353** (1991) 591.

[121] R. Barbieri, M. Frigeni, and F. Caravaglios, Phys. Lett. B **279** (1992) 169.

[122] M. Peskin and T. Takeuchi, Phys. Rev. Lett. **65** (1990) 964; W. Marciano and J. Rosner, Phys. Rev. Lett. **65** (1990) 2963; D. Kennedy and P. Langacker, Phys. Rev. Lett. **65** (1990) 2967.

[123] G. Altarelli and R. Barbieri, Phys. Lett. B **253** (1990) 161; G. Altarelli, R. Barbieri, and S. Jadach, Nucl. Phys. B **369** (1992) 3.

[124] M. Dress and K. Hagiwara, Phys. Rev. D **42** (1990) 1709.

[125] R. Barbieri, CERN Report No. CERN-TH.6659/92 (unpublished).

[126] J. Ellis, J. Hagelin, and D. V. Nanopoulos, Phys. Lett. B **159** (1985) 26; K. Griest, Phys. Rev. D **38** (1988) 2357; Phys. Rev. D **39** (1989) 3802(E); K. A. Olive and M. Srednicki, Phys. Lett. B **230** (1989) 78 and Nucl. Phys. B **355** (1991) 208; K. Griest, M. Kamionkowski, and M. S. Turner, Phys. Rev. D **41** (1990) 3565; J. Ellis, L. Roszkowski, and Z. Lalak, Phys. Lett. B **245** (1990) 545; L. Roszkowski, Phys. Lett. B **262** (1991) 59; K. Griest and L. Roszkowski, Phys. Rev. D **46** (1992) 3309; L. Roszkowski, Phys. Lett. B **278** (1992) 147; J. McDonald, K. A. Olive, and M. Srednicki, Phys. Lett. B **283** (1992) 80; S. Mizuta and M. Yamaguchi, Phys. Lett. B **298** (1993) 120; S. Mizuta, D. Ng, M. Yamaguchi, TU-410 (August 1992).

[127] M.M. Nojiri, Phys. Lett. B **261** (1991) 76; J. Ellis and L. Roszkowski, Phys. Lett. B **283** (1992) 252.

[128] J. Ellis and F. Zwirner, Nucl. Phys. B **338** (1990) 317; S. Kelley, J. L. Lopez, D. V. Nanopoulos, H. Pois, and K. Yuan, Phys. Lett. B **273** (1991) 423; M. Kawasaki and S. Mizuta, Phys. Rev. D **46** (1992) 1634; M. Dress and M. M. Nojiri, Phys. Rev. D **47** (1993) 376; R. Roberts and L. Roszkowski, RAL-93-003 and UM-TH-33/92.

[129] See *e.g.*, J. Casas, Z. Lalak, C. Muñoz, and G. Ross, Nucl. Phys. B **347** (1990) 243; A. de la Macorra and G. Ross, OUTP-92-14P.

[130] I. Antoniadis, J. Ellis, E. Floratos, D. V. Nanopoulos, and T. Tomaras, Phys. Lett. B **191** (1987) 96; S. Ferrara, C. Kounnas, M. Porrati, and F. Zwirner, Nucl. Phys. B **318** (1989) 75; M. Porrati and F. Zwirner, Nucl. Phys. B **326** (1989) 162.

CHAIRMAN: D.V. Nanopoulos

Scientific Secretaries: N.E. Moulai, R. Sehgal

DISCUSSION

– *do Couto e Silva:*

If neutrinos have masses $\neq 0$ something has to be changed in the "straight SU(5)" model. Can you explain again what is this change? How is the flipped SU(5) model affected by a massive neutrino?

– *Nanopoulos:*

1) We have to introduce a SU(5) singlet beyond the normal $\underline{10}$ and $\underline{\overline{5}}$. This looks like an ad-hoc procedure;

2) In the case of flipped SU(5), because the ν^c has to be inside the $\underline{10}$, its introduction is natural. Furthermore there is the coupling $F\overline{H}Q$ that provides ν^c with a superheavy mass ($\sim< \overline{H} >$) in a natural way. In other words flipped SU(5) automatically provides a see-saw mechanism for neutrino mass generation.

– *Tao:*

Can you give your comments on why people need GUT with gravity, since in my view gravity is quite different from other interactions?

– *Nanopoulos:*

Well, the dream of unification is to put gravity together with strong, electromagnetic and weak interaction. The problem of unification has gone a little bit further than the original unification. We have not only tried to understand the so-called internal symmetry (electromagnetic, weak, strong etc.), but at the same time we also try to understand standard space-time with these internal symmetries. The beauty of the string theory is that somehow it puts on an equal footing space-time and normal internal symmetries.

– *Hanany:*

Explain the problem of quantum coherence in B.H. and your solution to this problem. How do you relate $g_{\mu\nu} = 0$ to Topological Field Theory since we know that there is no dependence on the metric in such theories?

– *Nanopoulos:*

Let us go back to classical theory. If Hawking was right and if a system goes from a pure state to a mixed state then this is in contradiction with the Liouville theorem. In order to do this you have to go from a closed system to an open system. What Hawking had proposed was this extra term. This is a well known equation in statistical mechanics for open systems. Topological field theory: When I talked about topological field theory I did not mean summing over the different topologies, but a topological field theory on the world-sheet.

– *Benakli:*

What is the prediction of the flipped SU(5) for the proton lifetime?

– *Nanopoulos:*

The flipped SU(5) model predicts the proton lifetime to be of the order of 10^{35} years for $p \to e^+\pi^0$. This is the same as given by non-supersymmetric Grand Unification Theories. But if we have to use string theory we have to be very careful because we can have extra type of operators that can mediate proton decay.

– *Cho:*

What are your thoughts about the decision of the US Congress regarding the SSC?

– *Nanopoulos:*

It is very sad because this has set a bad precedent for the entire world.

– *Leike:*

In your flipped SU(5) theory you made a SU(5) × U(1) gauge group. That leads to one additional gauge boson. Are there some predictions for the properties of this additional gauge boson?

– *Nanopoulos:*

Yes, the theory predicts that this gauge boson is of the order of 10^{16} GeV. When one breaks the SU(5) × U(1) down to SU(3) × SU(2) × U(1) then the moment one breaks SU(5) one also breaks U(1).

– *Raciti:*

What does the superstring solution predict for the merging of two black holes? Is the entropy of the newly formed black hole greater or equal to the sum of the entropy of the two initial black holes?

– *Nanopoulos:*

String black holes behave like normal massive particles. We do not need to resort to thermodynamic properties in order to describe them. If we "ignore" the W-hair then we get back to the classic Hawking-Beckenstein thermodynamic description of black holes.

– *Sopczak:*

The modified uncertainty principle you presented changes the uncertainty principle for high momenta. Has anyone investigated its experimental implications in astrophysics?

– *Nanopoulos:*

This modification of the uncertainty principle is a well accepted result from string theory developed by Gross and Veneziano. I have basically used it here to explain the collapse of space-time at very very small distances.

CHAIRMAN: E. Fisk

Scientific Secretaries: M.T. Dova, M. Wadhwa

DISCUSSION

– *do Couto e Silva:*

In the experiment to determine Structure Functions from neutrino scattering it was said that the ν flux is twice the $\bar{\nu}$ flux. Why?

– *Fisk:*

This is important because it tells you the relative numbers of neutrino and antineutrino events you expect. One likes to have as much antineutrino flux as can be achieved. Since the number of antineutrino events is small and the Structure Function F_3 is determined from the differences in the number of neutrino and antineutrino events, the statistical accuracy on F_3 is compromised if the observed number of antineutrinos is too small.

– *do Couto e Silva:*

Last year there was a talk at CERN from CDF on Silicon Detectors. They showed a signal/noise of about 8 and they were still using DC coupled devices. Has the signal to noise ratio changed ever since? Are you already using AC coupled devices?

– *Fisk:*

The signal to noise ratio is not much different from what you have mentioned. They have radiation hard and AC coupled devices in production which they will use in the second half of the first run which will begin in about one year.

– *Rind:*

You talked about the measurement of $\sin^2\theta_W$ from neutrino scattering. What is the biggest error we can expect from these measurements and how do these measurements compare with the value obtained at LEP?

– *Fisk:*

The error is larger than the number measured by LEP. There are a number of reasons for this, some of which are systematic and others that are statistical. You should refer to the list of systematic and statistical errors to see this, even the neutrino scattering experiments have larger errors than those of LEP it is important to see that all measurements of $\sin^2\theta_W$ converge to the same value. We should appreciate the fact that different methods of measuring $\sin^2\theta_W$ contribute to rigorous tests of the Standard Model. The neutrino experiment errors on $\sin^2\theta_W$ can be reduced if the experiment can accumulate more events and if systematic errors can be reduced. There is a new proposal to continue the Fermilab neutrino experiment with a better beam. It may be true, however, that the errors on $\sin^2\theta_W$ will be dominated by LEP.

– *Rind:*

Is there any reason why the value of $\sin^2\theta_W$ is lower than the LEP value?

– *Fisk:*

It is a preliminary value which is 1.5σ below the world average value. The group is continuing to make checks on a number of systematic errors before publication of the final number.

– *Genovese:*

I would like to make a remark about the determination of $\sin^2\theta_W$ considering the ratio of neutral current and charge current. Usually this determination is done assuming universality in parton distribution in the two kinds of scattering, but as we have shown last year one cannot consider at this level of precision the universality of neutral current and charge current scattering. So, at least the error that contains the theoretical uncertainty should be bigger as claimed or one has to consider properly the non-universality.

– *Hambye:*

Concerning the E731 experiment, the results you have shown for the $\mathrm{Br}(K_L \to \pi^0\gamma\gamma)$ is a final result?

– *Fisk:*

No, it is not a final result. The analysis is still in progress.

– *Hambye:*

Did they do their analysis properly because there was some controversy about this decay?

– *Zichichi:*

What about this controversy?

– *Hambye:*

The controversy is among different models: chiral perturbation theory and others. We need more results in the low energy of two gamma region.

– *Wadhwa:*

You gave us nice results on E760 Collaboration on J, χ_c states, ψ etc. Then you mentioned CDF has observed χ_c states. Are there any results on χ_c branching ratios, production rates etc.?

– *Fisk:*

CDF uses branching ratios from CLEO and ARGUS to estimate production rates for χ states.

– *Block:*

If I recall correctly, when speaking about CP violation you showed the results obtained at Fermilab (E731) as being $(6.0 \pm 6.9) \times 10^{-4}$ and at CERN (NA31) as being $(23 \pm 7) \times 10^{-4}$. Can you comment on how much we can trust the first mentioned value?

– *Fisk:*

The result from E731 has remained stable for the last year and now they are trying to increase their data sample by loosening cuts etc. So they may be able to increase the number of events and decrease statistical errors at the expense of some slight increase in systematic errors. They expect to reduce the total error to 6×10^{-4}.

– *Baillie:*

In the CDF measurement of the B total cross-section they look at $B \rightarrow K\psi$ but not $B \rightarrow K\psi'$. Do you know why?

– *Fisk:*

The statistics for ψ' is low.

– *Glashow:*

I am astonished that you did not mention the new exciting development: the challange to the Standard Model offered by the inexplicable Double Vertex Events seen in the CCFR neutrino experiment. These have been presented at Harvard and at Granada by S. Misra. Could you comment on that now?

– *Fisk:*

If you look at the neutrino events in the CCFR detector you find cases where there is a downstream vertex "secondary vertex." There is a difference between those which come from charged currents and the neutral current and the issue is what are these events. There is a small number of neutral current events upstream and a hadronic energy deposition downstream in their detector. Present preliminary estimates of overlay events (two independent simultaneous neutral current interactions) and hadronic punchthrough show a possible small excess of unexplained events.

Note added in proof: The CCFR group have decided not to publish these preliminary data.

– *Tao:*

Is CDF going to measure $D^0 - \overline{D}^0$ mixing because if there is any signal then it might give us new physics?

– *Martin:*

CDF is not trying to measure lepton pair events. It is much better to do this kind of measurement at e^+e^- experiments.

– *Zichichi:*

I was very interested in the time-like proton form factor and since this is a school, I would like to show a plot to let our fellow students know how tough it is to measure something. This game started 30 years ago. At that time the EMFF of the proton was known in the space-like region. There was a theory by Yang and co-workers predicting that in the time-like region the form factor would be 1. This is the first evidence for the strong point-like and non point-like structure of the proton as shown in fig. 1.

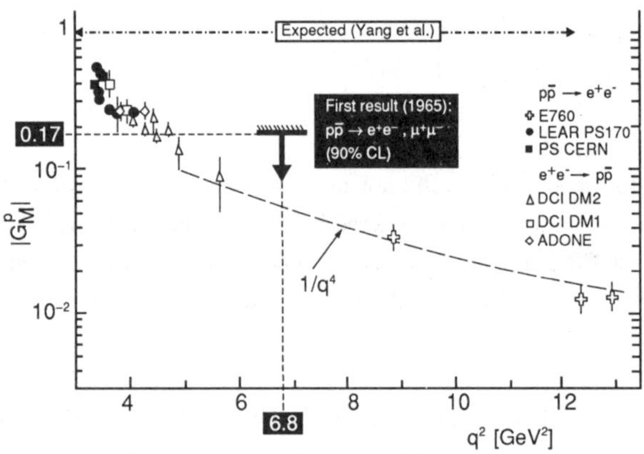

Fig. 1. Proton time-like form factor.

– *do Couto e Silva:*

In the CDF search for the $\Lambda_b \to \Lambda J/\psi$, are there kinematic ambiguities in identifying the Λ? For example, do they uniquely identify the $\Lambda \to \rho\pi^-$ or could the decay be $K^0 \to \pi^+\pi^-$?

– *Fisk:*

First of all there is no Č identification of the decay products. However the proton carries most of the momentum so the kinematics of the K^0 and Λ decays are rather different. Furthermore the CDF momentum determination for charged particles is quite good, i.e. $\Delta p_t/p_t = 0.0002\ p_t$ where p_t is in GeV/c. With the p_t cuts they make in selecting events the p_t determination is quite good and it follows the momentum is well determined. This results in no ambiguity in selecting K^0's and Λ's.

– *Zichichi:*

Could you please review the UA1 and CDF evidence for producing Λ_b and the branching ratio for its decay? There is published evidence from our group (reference: Nuovo Cimento Letters **31**, 97 (1981, 23 May, M. Basile et al.) that could be relevant.

– *Fisk:*

UA1 (references: CERN-PPE91-002, Phys. Lett. B273 (1991) 540), finds a signal for $\bar{p}p \to \Lambda_b X$ where the Λ_b decays to $\Lambda + J/\psi$ with the Λ_b mass at 5.64 GeV/c^2. They postulate a value of f_b, the probability the b quark hadronizes to Λ_b, is 10% and they find the BR is $(1.8 \pm 0.9) \times 10^{-2}$. CDF on the other hand, finds no peak in their $\Lambda J/\psi$ mass spectrum. In a bin by bin analysis they quote the 95% confidence limit for $f_b * BR(\Lambda_b \to \Lambda J/\psi) < 0.55 \times 10^{-3}$ to be compared with the equivalent UA1 result of $(1.8 \pm 1.0) \times 10^{-3}$. This result is a several standard deviation disagreement.

– *Zichichi:*

Λ_b^0 has been observed in 1981 by the BCF group, as shown in fig. 2. It is published in Nuovo Cimento Letters **31**, 97 (1981), 23 May, M. Basile et al. The decay mode is $pD^0\pi^-$ where $D^0 \to K^-\pi^+$. The experiment has been repeated in 1983 and Λ_b has been observed in the decay mode $pD^0\pi$ and in a new channel which is $\Lambda_c^+\pi^+\pi^-\pi^-$. The observed masses are shown in figs. 3, 4, 5.

– *Liu:*

The most worrisome problem for the silicon vertex detector is the stability of the silicon material under significant amounts of radioactivity. Could you tell us the test result?

– *Fisk:*

CDF had in the previous run, a piece of silicon detector in the colliding beam line and they measured the radiation damage to the detectors that are identical to what they now use. They are able to measure the leakage current and signal-noise ratio as a function of the dose. At 20 krad the signal to noise ratio decreases from 8 or 10 to 1 down to 2:1, where it would make it difficult to use it as a vertex detector in their experiment.

– *Mainusch:*

Is the D0 calorimeter compensating for signal electrons and hadrons? What are the electromagnetic and hadron energy resolutions?

– *Fisk:*

It is basically compensating. $e/\pi \sim 1.05$; the electron resolution is $16\%/\sqrt{E}$ and the hadron resolution is about $45\%/\sqrt{E}$ where E is in GeV.

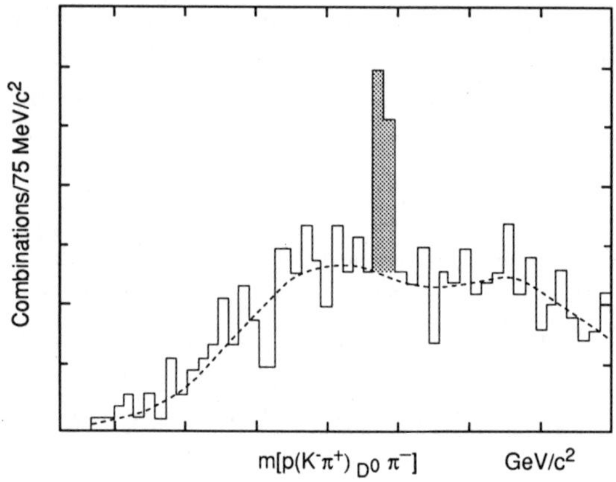

Fig. 2. ($pK^-\pi^+\pi^-$) invariant mass spectrum in events selected with the set-α conditions plus a "sharp" D^0 trigger.

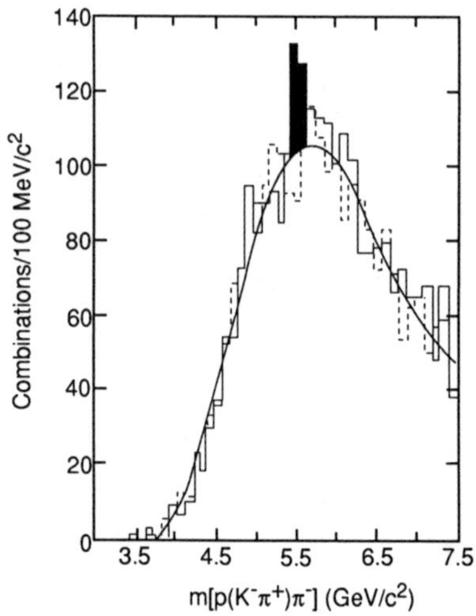

Fig. 3. The peak [$p(K^-\pi^+)\pi^-$] invariant-mass spectra of fig. 4a) and b), corresponding to the e^+ triggers, superimposed after normalization. The full-line histogram is obtained for $m(K^-\pi^+)$ in the interval $(1.90 \div 2.05)$ GeV/c^2, the dashed-line histogram for $m(K^-\pi^+)$ in the intervals $(1.75 \div 1.90)$ GeV/c^2 and $(2.05 \div 2.20)$ GeV/c^2. A fit to the latter is also shown.

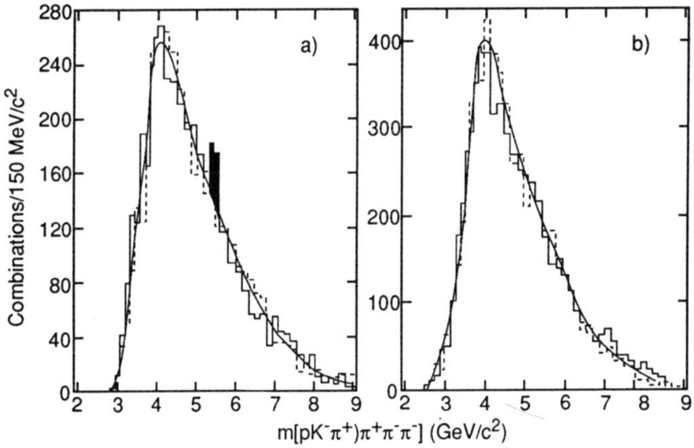

Fig. 4. The $[p(K^-\pi^+)\pi^+\pi^-\pi^-]$ invariant-mass spectra obtained for e^+ triggers (full-line histogram); a) when $2.23 \leq m(pK^-\pi^+) < 2.43$ GeV/c^2; b) when $2.03 \leq m(pK^-\pi^+) < 2.23$ GeV/c^2 or $2.43 \leq m(pK^-\pi^+) < 2.63$ GeV/c^2. The histograms corresponding to e^- are normalized to those corresponding to e^+. The curves superimposed are polynominal fits of the dashed-line histograms.

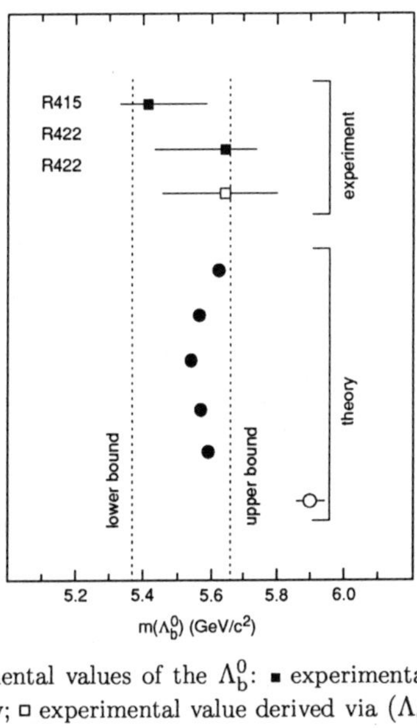

Fig. 5. The experimental values of the Λ_b^0: ▪ experimental value derived via $(pD\pi)$ decay; ▫ experimental value derived via $(\Lambda_c\pi\pi\pi)$ decay; • potential mode estimate; ∘ scalar lattice QCD estimate. The theoretical lower and upper bounds for potential model estimates are indicated as vertical lines.

WHY THE PROTON IS GETTING BIGGER

Jay Orear

Laboratory of Nuclear Studies
Cornell University
Ithaca, NY

TOTAL CROSS SECTION

As I shall show, the proton is getting bigger when measured at very high energies. So far it is almost twice as big as its usual size of 40 mb. The rise in cross section at beam momenta above 400 GeV/c was a big surprise, and so far no one has a convincing answer to the question "Why is the proton getting bigger?" My title which was suggested by A. Zichichi implies that I have the answer. So for the sake of this paper I will propose an answer.

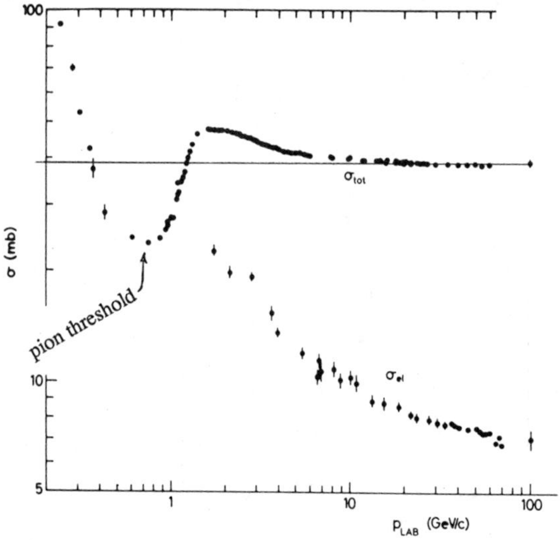

Fig. 1 Total and elastic pp cross section data.

My answer becomes apparent if we compare the very low energy nucleon-nucleon total cross section with the recent very high energy results. At low energies it has been known for ~60 years that there is a strong nucleon-nucleon potential - strong enough in the triplet state to bind them. Even in the singlet state this attractive potential is about half as strong. At energies below pion threshold it is this potential which causes the nuclear part of pp elastic scattering. As the energy increases, quantum mechanics causes this cross section to decrease as seen in Fig. 1. But at beam momentum of ~700 MeV/c there is a rise in σ_t because this is the threshold for single pion production. As the beam momentum increases beyond 700 MeV/c more inelastic channels open up such as 2 pions, eta production, rho production, etc. There is a rise in σ_t until these new channels saturate at beam momentum ~2 GeV/c. Then there is a long stretch of beam momentum from ~10 GeV/c to 400 GeV/c where the cross section stays fixed at about 40 mb. In the days of those measurements most physicists believed in a version of the Pomeromchuk Theorem which proposed that all hadron-hadron cross sections would approach an asymptotic limit. As seen in Fig. 1, the pp system appeared to reach this limit at 10 GeV/c. However in 1973 the ISR started getting pp cross section results which showed a small increase of 2 or 3 mb at equivalent beam momenta ~1000 GeV/c; also the pp and $p\bar{p}$ total cross sections had become the same. (See Fig. 2.) Then came the UA4 SPS results with the $p\bar{p}$ total cross section equal to 51 mb at $\sqrt{s} = 540$ GeV. The latest result which is the highest that can be obtained with existing accelerators ($\sqrt{s} = 1800$ GeV at the Tevatron) gives a $p\bar{p}$ total cross section of 73 mb.[(1)] In Fig. 2 the world data are fit by a log(s) dependence (the solid curve). I have also drawn a dashed curve which repeats the rise of cross section behavior shown just above the pion threshold.

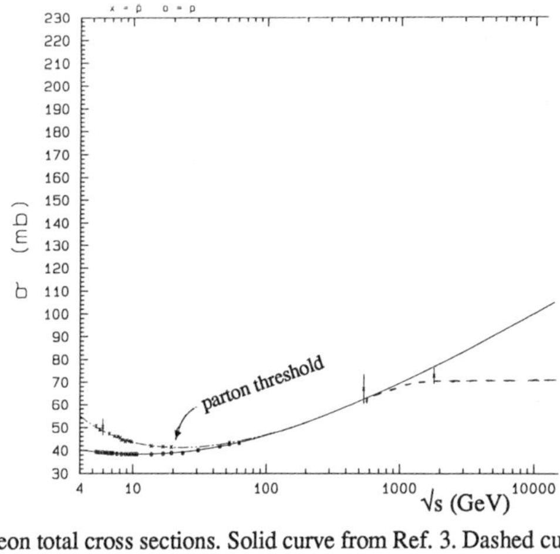

Fig. 2 Nucleon-nucleon total cross sections. Solid curve from Ref. 3. Dashed curve repeats the shape in Fig. 1.

At the ISR energy region a new kind of threshold appears: production of quark jets. So

I am suggesting that the new rise in cross section is due to new parton channels opening up with a saturation around $\sqrt{s}$ = 1800 GeV (the dashed curve). This is my simple- minded explanation of the increase in size of the proton. Gaisser and Halzen(2) have speculated that the observed cross section for jet production is enough to account for the rise in σ_t . Their speculation requires an arbitrary low energy jet cutoff near the threshold for jet production.

The ρ-value

Fortunately it is possible to distinguish between the solid and dashed line curves in Fig. 2 without the need to go to higher energies. One merely does an accurate measurement of ρ (ratio of real to imaginary parts of the forward scattering amplitude) at the Tevatron energy. If the pp and pp cross sections are the same (as is observed so far) there is a simple relation between ρ and the slope in Fig. 3:(3)

$$\rho = \frac{\pi}{2\sigma_t} \frac{d\sigma}{d\ln(s)} \qquad (1)$$

If the slope at $\sqrt{s}$ = 1800 GeV is near zero (dashed line), then ρ must be close to zero. If the slope is 7.0 mb as given by the solid line, the above equation gives $\rho = 0.15$. So far the experiments on ρ are not accurate enough to make a definitive statement (see Fig. 3). The UA4 point of 0.24±0.04 has been redone and the new data are presently under analysis. The E710 point will be redone with a new, more accurate detector in 1993 in Fermilab experiment E811.

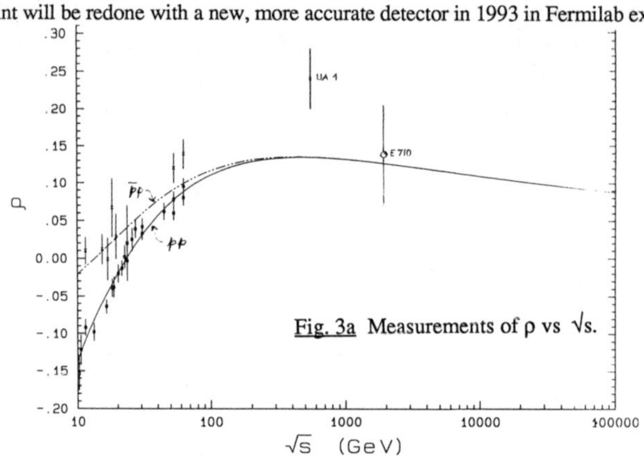

Fig. 3a Measurements of ρ vs $\sqrt{s}$.

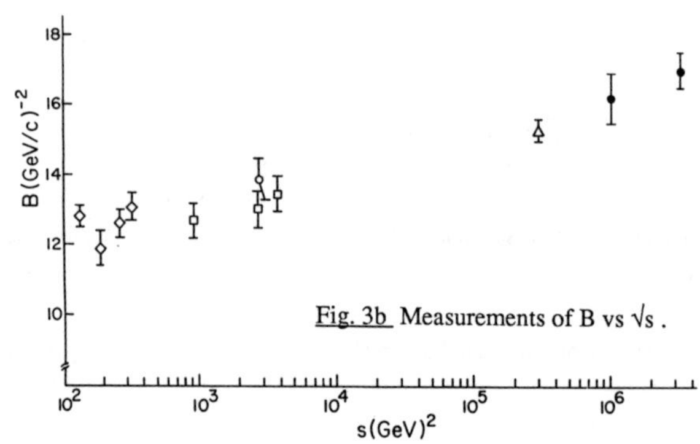

Fig. 3b Measurements of B vs $\sqrt{s}$.

The Froissart Bound

There is a theoretical upper limit on just how fast σ_t can increase with energy. If one invokes analyticity and unitarity, one can show that σ_t must be bounded by some constant times $\log^2 s$. Unfortunately this constant is so large (~60 mb) that the upper limit is orders of magnitude larger than what is observed. So far the world data seem to fit a log(s) term better than $\log^2 s$.(3)

The Optical Theorem

It is almost impossible to build a truly 4π detector and to know accurately the luminosity in order to directly measure σ_t. But the total cross section can be obtained indirectly by use of the Optical Theorem even without any knowledge of the luminosity. A version of the Optical Theorem first appears in classical physics. If a plane electromagnetic wave impinges on a black disk of radius R, the probability for scattering in a direction θ is(4):

$$\frac{d\sigma}{d\omega} = \left(\frac{k^2R^4}{4}\right)\left[\frac{2J_1(kR\sin\theta)}{kR\sin\theta}\right]^2$$

Converting to $-t = p^2\sin^2\theta$, this becomes

$$\frac{d\sigma}{dt} = \frac{\sigma_t^2}{16\pi}\left[\frac{2J_1(\sqrt{t})}{\sqrt{t}}\right]^2 \qquad (1)$$

As $t \to 0$, the square bracket equals one and

$$\left[\frac{d\sigma}{dt}\right]_{t=0} = \frac{\sigma_t^2}{16\pi}$$

Taking the square root of both sides:

$$\sqrt{\left[\frac{d\sigma}{dt}\right]_{t=0}} = \frac{\sigma_t}{\sqrt{16\pi}}$$

In converting from classical physics to quantum mechanics, one might expect that the left hand side is the forward scattering amplitude. But unlike classical physics, one must deal with complex amplitudes in quantum mechanics. It turns out that by invoking unitarity, the left hand side is only the diffractive part; i.e., the imaginary part of f(t) where $|f(t)|^2 = d\sigma_{el}/dt$. So the most general form for the Optical Theorem is

$$\mathrm{Im}[f(0)] = \frac{\sigma_t}{\sqrt{16\pi}}$$

By definition, $\mathrm{Re}[f(0)] = \rho \times \mathrm{Im}[f(0)]$, so then

$$f(0) = (\rho + i)\,\frac{\sigma_t}{\sqrt{16\pi}}$$

and $|f(0)|^2 = \dfrac{1+\rho^2}{16\pi}\,\sigma_t^2$ (2)

The experimentalist measures $\left.\dfrac{dN_{el}}{dt}\right|_0 = (Lt)\,\dfrac{1+\rho^2}{16\pi}\,\sigma_t^2$ (3)

The left hand side is measured by extrapolating the nuclear part to zero degrees. The quantity ρ is determined from the amount of coulomb-nuclear interference. The quantity (Lt) is the integrated luminosity which at the Tevatron is not too well known. But there is a way of using Eq. 3 to get σ_t without knowledge of the luminosity. A large angle detector which is triggered by any secondary that is produced will have a total number of triggers $N_{inel} = (Lt)\sigma_{inel}$. The sum $(N_{el}+N_{inel}) = (Lt)\sigma_t$. Dividing Eq. 3 by this equation gives

$$\frac{\left[\frac{dN_{el}}{dt}\right]_0}{N_{el}+N_{inel}} = \frac{1+\rho^2}{16\pi}\,\sigma_t$$

This luminosity independent method is called the combined method.

If the detector gets to small enough angles such that the coulomb cross section dominates at the smallest angle then the equation below can be used to determine (Lt) and that value of (Lt) can be substituted into Eq. 3:

$$\frac{dN_{coul}}{dt} = (Lt)\,\frac{4\pi\alpha^2}{t^2}$$

This method is called the coulomb normalization method.

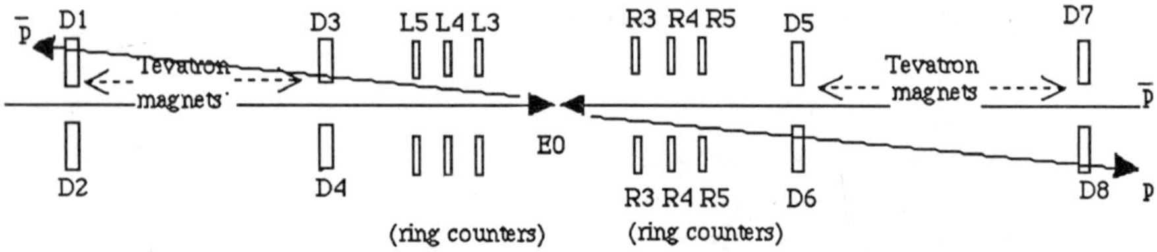

Fig. 4 Schematic representation of E710 layout. The elastic scattering shown gives rise to a (1,8) coincidence.

Experimental Details

After the beam is stored, track detectors 1,2,7, and 8 are pushed to within 2 mm of the beam. A schematic view of the detectors is shown in Fig. 4. Elastic scatterings will give a (1,8) or a (2,7) coincidence. N_{inel} is measured using the ring counters. The original detectors used in E710 were small drift chambers. In the next run a new type of scintillating fiber detector will be using which will give much improved measuring resolution (see Fig. 5). A bundle of ~15,000 scintillating fibers oriented parallel to the beam is viewed by 2 image intensifiers in cascade. The final phosphor is viewed by a CCD and the "exposed" pixel coordinates written on data tape. Test beam runs have shown that each fiber can be indentified and is 100% efficient (each track gives ~70 photoelectrons). So this new detector is equivalent to a wall of 15,000 independent scintillation counters all at 100% efficiency with no gaps or cracks.

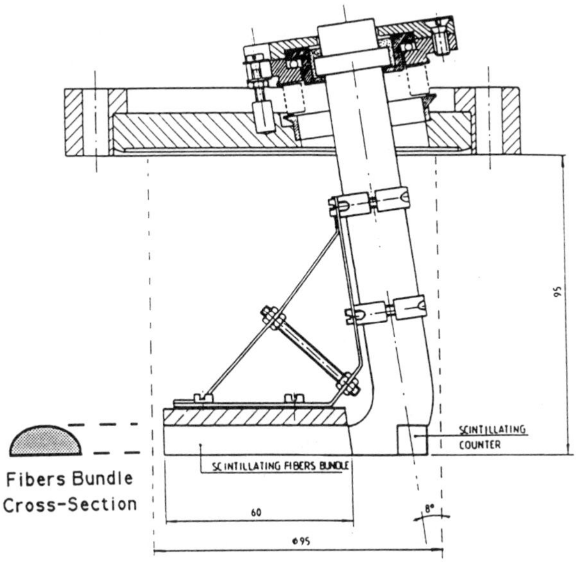

Fig.5 Side view of scintillating fiber detector showing fiber bundle and fiber optics light guide.

Shape of $d\sigma/dt$

If Eq.1 for a black disk is plotted on semilog paper, the initial slope is $B = R^2/4$ and the slope gets steeper with increasing $|t|$. But experimentally at energies below 1,800 GeV the slope gets less steep with increasing $|t|$. This is what is expected of a partially opaque scattering center. But as the scattering center becomes more opaque the scattering pattern should approach that of the black disk. This transition from a "concave" curve to a "convex" curve is verified by the 1,800 GeV results of E710 where the semilog plot is straight all the way out to $|t| \approx 0.7$ GeV^2.

Let $R = \sqrt{(4B)}$ which we call the optical radius. The opaqueness or opacity is equal to the true σ_t divided by $2\pi R^2$ (the black disk value for σ_t). So

opacity = $\sigma_t/(8\pi B)$.

As seen by comparing Fig. 2 and 3b, σ_t is increasing faster with energy than is B, therefore the opacity is increasing with energy which is in agreement with the change in shape observed by E710. We also note that the optical theorem when combined with $\sigma_{el} = (d\sigma/dt)_0/B$ gives

$$\frac{\sigma_{el}}{\sigma_t} = \frac{1+\rho^2}{2} \times \text{(opacity)}$$

LEADING PARTICLE PHYSICS

In addition to elastic scattering, there are other processes where a leading particle goes "down the beam pipe" and would be lost by most so-called 4π detectors. However, the detector described above can measure angles within 30 μrad of the beam at the Tevatron and 1 μrad at the SSC. Of the inelastic processes which give rise to a leading particle, the most prominent is single diffraction scattering: $\bar{p} + p \rightarrow \bar{p}' + p^*$ where p^* is an excited state of the proton having mass M. In E710 some running was done with a leading antiproton trigger on the antiproton side in coincidence with any secondary on the proton side. Almost as many single diffractives were seen as elastics in the same amount of time. The single diffractives were fit to the form

$$\frac{d^2\sigma}{dt\,dM^2} = A\,\frac{1}{M^{2\alpha}}\,e^{bt}$$

The experimental results were $\alpha = 1.13\pm0.07$, $b = 10.5\pm\mathrm{GeV}^{-2}$, and the ratio of total single diffractive cross section to elastic was 0.49 ± 0.09.

LARGE ANGLE ELASTIC SCATTERING

Since the J_1 Bessel function has zeros, one might expect to see dips in $d\sigma/dt$ as one goes to larger $|t|$. The first dip is seen in nucleon-nucleon and pion-nucleon elastic scattering and the positions of the dip are consistent with the radii that are measured in form factor experiments. Beyond the first dip, $d\sigma/dt$ does not drop as fast with $|t|$ as predicted by the black disk optical model. So it is natural to expect that for $|t| > 2|t_{dip}|$ most of the elastic scattering is due to another process. Fortunately perturbative QCD does make a prediction at high momentum transfer. The prediction is that at fixed angle $d\sigma/dt$ drops off as s^{-10} for nucleon-nucleon and s^{-8} for meson-nucleon elastic scattering. And in nucleon-nucleon the lowest order term in momentum transfer is t^{-8}. So the complete prediction for pp elastic scattering is

$$\frac{d\sigma}{dt} \propto \frac{1}{s^2 t^8} \quad \text{for } |t| > 3\ \mathrm{GeV}^2 .$$

The world's data which cover a range of 21 powers of ten are shown in Fig. 6. The agreement with the above s and t-dependence is remarkably good over a range of 10^{21}.

Pion proton large angle elastic scattering has been observed up to beam momenta of 200 GeV/c. Here the dip is at $-t \sim 4\ \mathrm{GeV}^2$. Since $2|t_{dip}|$ is about 8 GeV^2, we would expect for $|t| <$ 8 GeV^2 that $d\sigma/dt$ would be mainly diffractive and have little s-dependence. And that is what is observed in high energy pion-proton elastic scattering. See Fig. 7. Above $|t| \sim 8\ \mathrm{GeV}^2$ one might expect to see the QCD prediction of s^{-8} dependence (at fixed angle), but so far there are no pion-proton data in this region.

References

1. Amos, et al, Phys. Rev. Letters **68** (1992) 2433-2436
2. Gaisser and Halzen, Phys. Rev. Letters **54** (1985) 1754.
3. Block and White, Nucl. Phys. B (Proc. Suppl.) **25B** (1992) 59-69.
4. Jackson, *Classical Electromagnetism*, p. 297.
5.Amos, et al, Phys. Lett. B 247 (1990) 127.

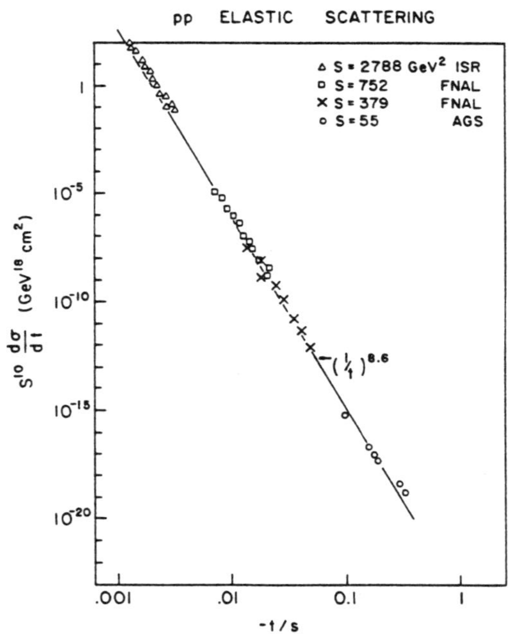

Fig. 6 pp elastic dσ/dt times s^{10} plotted vs. (-t/s). The best fit t-dependence is $(1/t)^{-8.6}$.

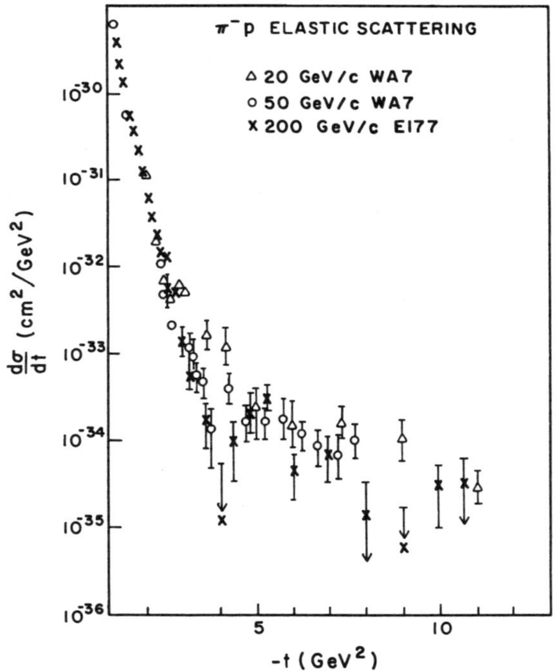

Fig. 7 dσ/dt for π^-p elastic scattering vs -t. Note lack of s-dependence.

CHAIRMAN: J. Orear

Scientific Secretaries: U. Paz, G. Word

DISCUSSION

– *Ren:*

1) To deal with a scattering problem with energy more than 1 TeV using classic wave optics approximation is not suitable.
2) When proton and anti-proton move together to collide, there are photon clouds around each of them. Maybe the increase of the total cross-section is the result of photon-photon interactions (the two photon cross-section increases with beam energy).

– *Orear:*

Actually the measured scattering pattern gets *closer* to the classical black disk with increasing energy. The shape of $d\sigma/dt$ is approaching the $(J_1(x)/x)^2$ shape and the measured σ_T is getting closer to $2\pi R^2$ where R^2 is measured by the width of the diffraction peak. This scattering is due to strong interaction, not electromagnetic. (No photons.)

– *Nanopoulos (remark):*

I would like to stress that we who are working at super high energies, take QCD "for granted." We would like to have as much experimental proof as we can get.

STATUS OF HERA AND THE FIRST RESULTS

B.H.Wiik
II.Institut für Experimentalphysik, Univ. Hamburg, and
Deutsches Elektronen-Synchrotron DESY,
2000 Hamburg 52 - Notkestrasse 85 - Germany

1. INTRODUCTION

During the past few weeks the electron-proton collider HERA[1-5] and its large, multipurpose detectors H1[6,7] and ZEUS[8,9] have made the transition from a virtual to a real source of data obtained in a new kinematic region.

HERA is the first electron-proton collider ever and it has been constructed within the framework of a novel form of international collaboration where institutions in Canada, China, Czechoslovakia, France, Israel, Italy, Netherlands, Poland, United Kingdom and USA have contributed either components built at home or delegated skilled manpower to work on the project at DESY.

Also the large detectors H1 and ZEUS have been built and will be exploited by an international collaboration. Only some 25% of the 750 physicists presently involved in the programme are from DESY or German universities while the remainder come from 64 institutions in 15 countries.

Thus the construction of HERA and its detectors has been, and the exploitation of the physics will be a truly international effort.

The HERA project was authorized in April 1984 and its construction completed in November 1990 in accordance with the original time schedule. The first electron beam was stored in August 1988 and the first proton beam in April 1991. Electron-proton collisions were observed for the first time in October 1991.

After extensive detector testing with cosmic rays both H1 and ZEUS were installed in the ring during the 1991/92 winter shutdown. The commissioning of HERA resumed in April 1992 and by the end of June luminosity was made available to the experiments.

In the first part of my talk I'll briefly remind you of the HERA layout and parameters. After a short outline of the preaccelerators, I'll discuss the performance of the HERA electron and proton rings before reviewing its performance as an ep collider.

In the second part I'll first focus on the detectors, then discuss the experimental conditions and finally present first physics results. Both experiments report[10,11] deep inelastic neutral current data for small values of x, and a measurement of the total photoproduction cross section around 200 GeV in center of mass. Preliminary lower limits on the mass of leptoquarks and on excited states of the electron have also been extracted from the data.

I'll close my talk with a brief summary and a look at the future.

2. HERA

2.1 Overview

HERA is made of two independent accelerators designed to store respectively 820 GeV protons and 30 GeV electrons and to collide the two counterrotating beams

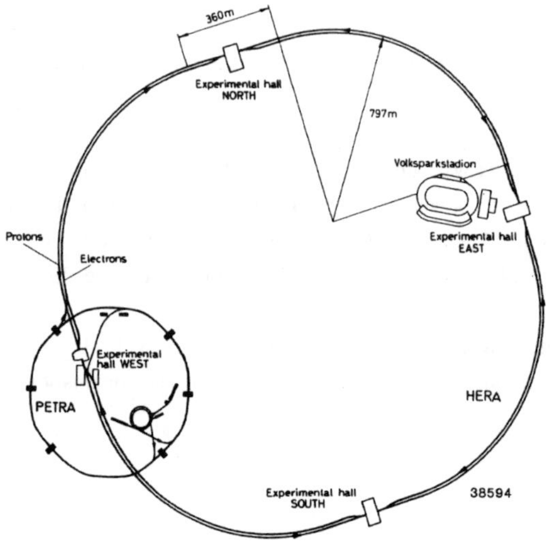

Fig. 1 - The layout of the HERA accelerator complex.

	p-ring	e-ring	units
Nominal energy	820	30	GeV
Polarization time		28	min
Luminosity	1.5×10^{31}		$cm^{-2}s^{-1}$
Space between IR Quad	15		m
Interaction points	4		
Crossing angle	0		mrad
Circumference	6336		
Magnetic field	4.68	0.165	T
Number of particles	2.1	0.8	10^{13}
Number of bunches	210		
Injection energy	40	14	GeV
Filling time	20	15	min
σ_x/σ_y at I.P.	0.29/0.07	0.26/0.07	mm
σ_z at I.P.	110	8.0	mm
Energy loss/turn	6.24×10^{-6}	127	MeV
Circumferential RF voltage	0.2/2.4	260	MV
RF-frequency	52.033/208.13	499.776	MHz
RF-power	1	13.2	MW
Refrigerator	21.0 kW (isothermal at 4.3K) 60 g/s Liq.He 60 kW (40 K - 80 K)		

42047

Table 1 - Main HERA Parameters.

head on in four interactions points spaced uniformly around its 6.3 km long circumference.

The main parameters of the HERA rings are listed in Table 1 and the layout of the accelerator complex is shown in Fig. 1. The installed accelerators in the arcs are depicted in Fig. 2. The proton ring is installed 81 cm above the electron ring in the arcs. In the straight sections the proton beam is deflected into the plane of the electron ring. In the region adjacent to the interaction point the protons are guided by normal conducting magnets.

The H1 and ZEUS detectors are installed in straight section North, respectively South. Straight section East has been reserved for new experiments. Straight section West is presently being used as a general utility section incorporating the electron and proton injection system, the proton beam dump, proton collimators and RF system.

Fig. 2 - The installed HERA electron and proton rings in the arc.

2.2 The Preaccelerators

The layout of the HERA preaccelerators is shown in Fig. 3. The electron (positron) injection complex is based on available accelerators. Electrons (positrons) from a 500 MeV linear accelerator are injected into a small storage ring, accumulated into a 60 mA single bunch, injected on axis into DESY II,

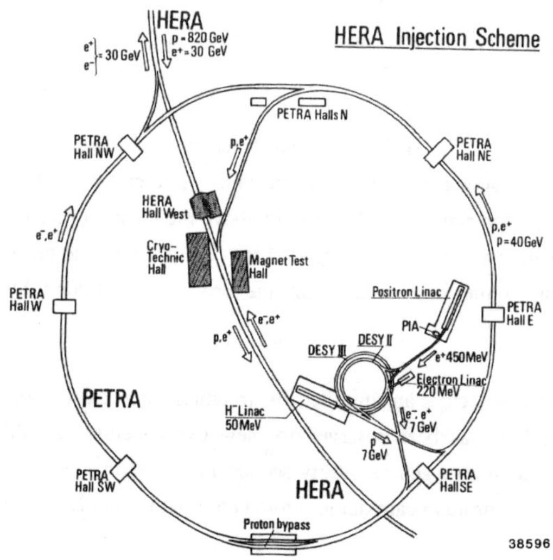

Fig. 3 - The layout of the HERA preaccelerators.

accelerated to 7 GeV and transferred to the modified PETRA II ring. This is repeated at a rate of 12.5 Hz until PETRA II has been filled with 70 bunches each with $0.4 \cdot 10^{11}$ electrons (positrons) spaced 28.8 m apart as in HERA, which are then injected into the HERA main ring. This is repeated 3 times until HERA has been filled with 210 bunches. The expected 15 minutes filling time of the electron ring is dominated by the cycle time of PETRA II. The threshold for multibunch instabilities in PETRA II has been raised from 2.6 mA to the design current of 58 mA by using a new longitudinal and transverse feedback system[12].

A whole new chain of preaccelerators has been built in order to inject protons into HERA.

The 50 MeV linear accelerator[13] for negatively charged hydrogen ions is in routine operation and delivers a 6 mA beam with a normalized 95% emittance of 3.3π mm mrad horizontally, 5.7π mm mrad vertically, and a momentum spread of ±0.1%. The negatively charged hydrogen ions are stripped upon their injection into DESY III[14]. After a multiturn injection, protons are captured into 11 buckets, spaced 28.8 m apart as in HERA, accelerated to 7.5 GeV, and transferred to PETRA II. Roughly $1.2 \cdot 10^{12}$ protons in 11 bunches reach 7.5 GeV compared to the design intensity of $1.1 \cdot 10^{12}$ protons. The normalized 1 σ transverse emittance at ejection is 3.6 πmm mrad horizontally, 5.4 πmm mrad vertically and the longitudinal emittance is 0.1 eVs.

The maximum number of 70 bunches has been accumulated in PETRA II and accelerated to the final energy of 40 GeV. The lifetime was 1 hour at 7.5 GeV and increased to more than 10 hours at 40 GeV. The cycle time of 5 min is limited by the tune shift caused by eddy currents in the thick aluminium vacuum chamber. The tune shift is measured in real time using a fast Q-measuring system. It is compensated by adjusting the current in two distributed quadrupole circuits.

A 52 MHz RF system is used in HERA to accept the 2.2 m long bunches delivered from PETRA II. The HERA frequency is not phase locked to that of PETRA II during injection, a difference frequency of 1.3 kHz is necessary to make up for a small difference in orbit length in PETRA II. The correct RF phase at injection is found by a digital counter technique triggering the kickers at the right moment. The same procedure[15] is used between DESY III and PETRA II. This "flying injection" could be used because synthesized RF frequencies are used only. Also during the fast ramping of DESY III with a frequency ratio of 3 between injection and extraction, a quartz controlled preprogrammed frequency synthesizer is used. The system worked well from the beginning and is extremely flexible. A similar system is also in use for the electron ring.

2.3 Status of the Electron Ring

While the first run of the electron ring in August 1988 was mainly used to commission technical components, a second run in September 1989 focussed on the performance of the ring. In this run beam currents of 3 mA could be injected, accelerated to 27.5 GeV and stored with a life-time of several hours. The maximum electron energy was limited by the available RF voltage. The commissioning was greatly eased by the reproducibility and stability of the electron ring.

At present the conventional RF system has been augmented by 16 four cell superconducting 500 MHz cavities[16,17], assembled pairwise into 8 cryostats. These cavities have been industrially produced from high purity Niobium (RRR = 300). At the design current of 58 mA the gradient is limited to 2.05 MV/m due to the 100 kW power rating of the input coupler. New couplers rated at 200 kW are now being tested. Without beam all cavities reached 5 MV/m. Initially the design Q-value of $2 \cdot 10^9$ could not be reached in all cases, due to the presence of Nb_xH_y precipitation on the Nb surface.

The beam has been accelerated to 30.3 GeV with both the normal and superconducting RF systems working in parallel. The superconducting system was stable and reliable and it provided roughly one quarter of the circumferential RF voltage of 160 MV.

The maximum single bunch current was 2.5 mA nearly one order of magnitude above its design value. The maximum current which can be stored with lifetimes on the order of a few hours is presently limited to roughly 4 mA at 27 GeV. This limit may be due to dust particles trapped in the strong field of the circulating electron beam. A run with a positron beam is planned for the autumn.

The electron beam polarization[18] is determined by measuring the up-down asymmetry of backscattered polarized laser light using a tungsten scintillator sandwich counter. The observed transverse beam polarization is shown in Fig. 4 as a function of synchrotron tune and beam energy. The buildup time of 25.8 min is consistent with the measured[18] maximum beam polarization of 58 ± 5%.

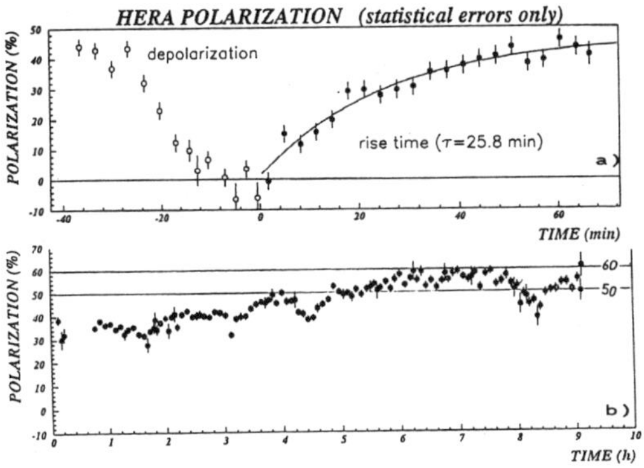

Fig. 4 - a) The transverse electron beam polarization as a function of time. The build up of transverse polarization is clearly seen.
b) Beam polarization during a long storage.

A pair of 60 m long magnetic spin rotators designed to turn the vertical spin direction into a state of positive or negative helicity have been built and are ready for installation in straight section East.

2.4 Status of the Proton Ring

The cryogenic system and the superconducting magnets are the most challenging components of the HERA proton ring. I will first describe the status and the performance of these components and then discuss the results obtained during the first commissioning runs.

2.4.1 The Refrigeration System - The central refrigerator is located on the DESY site. It is subdivided into three identical plants each providing 6.6 kW isothermally at 4.3 K, 20.4 g liquid helium per second and 20 kW at 40 K to 80 K. The cryogenic plant is very reliable and has been available 98.5% of the scheduled time. Each of the three coldboxes has now run for about 20000 hours.

Liquid helium and 40 K helium gas are supplied by a fourfold transferline to precoolers and feedboxes which are installed at the ends of each octant. The same transferline is used to return helium gas of 4.6 K and of 80 K to the refrigerator. In the case of a quench the warm gas from the quenched magnets is fed through a safety valve to a ring line which returns the gas to the storage vessels at a pressure up to 20 bar. A detailed description of the cryogenic system and its performance can be found elsewhere[19]. Here we only summarize the main results.

The cooldown time of 140 hours for a HERA octant was determined by the condition that mechanical stresses caused by the temperature gradient within a magnet should be limited to 100 MPa which is less than 50% of the critical value. Note that all octants can be cooled down in parallel.

Semiconductor temperature sensors are installed in the single-phase helium volume of each dipole, allowing to monitor the temperature with a precision of 0.02 K. Using these monitors at a known mass flow, the heat loads of the whole ring, including transfer lines and feed boxes, were measured to be 5.1 kW at 4.4 K and 28.5 kW at the shield level. These values compare favourably to the proposal values. The heat leak of the magnets alone is 3.6 kW at 4.4 K and 19.6 kW at the shield level, in excellent agreement with heat load measurements on single magnets. At a current of 5020 A an additional heat load of 21 W at 4.4 K per octant was observed. The maximum single-phase helium temperature occurs in the magnets installed immediately downstream of the main current leads. In these magnets the temperature may reach 4.5 K compared to the critical temperature of the superconductor of 6 K at the nominal field of 4.65 T.

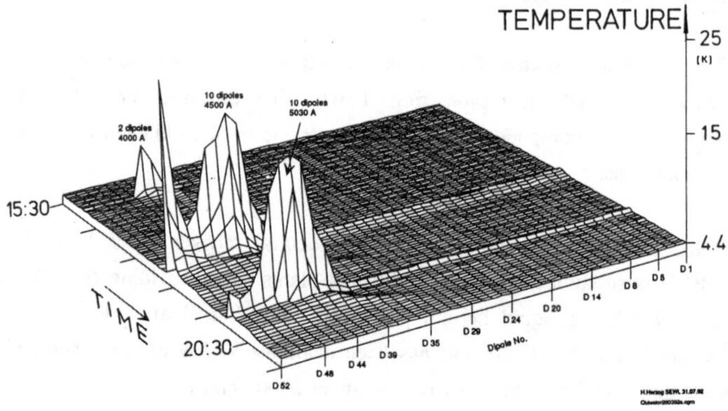

Fig. 5 - Recovery of the superconducting magnet chain after an induced quench. Groups of magnets - 2 dipoles at a current of 4000 A and 10 dipoles at 4500 A and 5030 A - were quenched by means of electric heaters. Plotted is the helium temperature as a function of time and magnet number.

The magnet recovery time after a quench is quite short as shown in Fig. 5. After an induced quench of 10 magnets at the nominal current of 5030 A it takes 20 min to cool the magnets back down from 20 K to 5 K and another 60 min to reach the operating temperature of 4.4 K. These quenches were induced by firing heater strips mounted close to the superconducting coils. An important observation is that quenches do not propagate into non triggered magnets.

2.4.2 The Superconducting Magnets - A total of 2156 superconducting magnets and correction coils have been installed in the HERA proton ring. In the arcs the superconducting magnets are arranged in 104 cells with an ordering shown in Fig. 6.

The design[20] and the performance[21] of the magnets have been reported elsewhere. Here we just summarize the main results. The industrial production of superconducting magnets was a success. During series production, DESY received an average of 8 dipoles and 6 quadrupoles per week, exceeding the contractual rates. Out of 449 dipoles and 246 quadrupoles, only 5 magnets were rejected, four of which had shortened windings and one a bad spot in the superconductor.

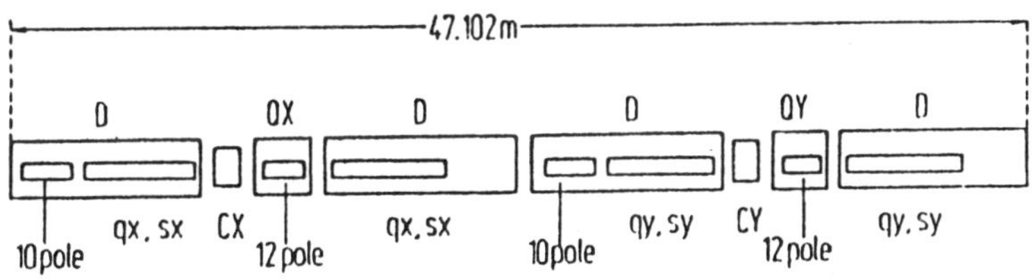

Fig. 6 - A unit cell of the proton ring. D: main dipoles; QX, QY: main quadrupoles, qx, qy: quadrupole correction coils, sx, sy: sextupole correction coils, CX, CY: correction dipoles. In addition, there are 10-pole and 12-pole correction coils.

All magnets were tested at liquid helium temperature and the results can be summarized as follows:

Nearly 93% of all magnets reached the short sample current at the first or second excitation step. None of the magnets quenched below the nominal operating current.

Adjusted to an operating temperature of 4.4 K the average quench current was (6900±130) A for the dipoles and (7840±160) A for the quadrupoles. Only magnets with a quench current above 6500 A were installed in HERA.

The field quality of both the dipole and the quadrupole magnets is better than specified.

The superconducting magnet system has now been in use for more than a year. The magnets in the ring are very reliable and so far faults have not occurred in the cold part of the system.

The energy stored in the magnets at 820 GeV exceeds 270 MJ. A sophisticated quench detection and protection system[22] is installed to dump this energy in a safe way if a quench occurs. We have observed only few beam induced quenches during the first year of operation. This is presumably due to the large safety margin which exist both in field (30%) and in temperature (1.5K). The quench protection system has also been reliable and only few spurious quenches have occurred.

2.4.3 Proton Ring Commissioning - During HERA commissioning PETRA II could deliver a single proton bunch every 5 to 10 minutes with intensities varying between $1 \cdot 10^9$ to $3 \cdot 10^{10}$ protons per bunch. Also the commissioning of the HERA proton accelerator has been eased by the reliability and stability of the accelerator and by the excellent performance of the beam instrumentation.

The field quality of the superconducting magnets is seriously affected by persistent magnetization currents. Any change of the field in a magnet induces magnetization currents within the superconducting filaments whose strength is proportional to the filament diameter and to the critical current density at ambient temperature and field strength. The persistent current changes the value of all allowed multipoles. At HERA, with a ratio of peak magnetic field to injection field of 20 and a filament diameter of 14 μm, the persistent current effects are rather strong as shown in Fig. 7. For example the persistent current sextupole enhances the uncompensated chromaticity at HERA by more than a factor of five. However, the persistent current multipoles vary little from magnet to magnet and are well reproducible and can hence be compensated by correction coils wound directly on the dipole and quadrupole beam pipes. The arrangement of the compensation magnets is shown in Fig. 6. Caused by flux creep in the superconductor and other effects, the magnetization currents decay with a nearly logarithmic time dependence[23]. This drift is also compensated using the correction coils.

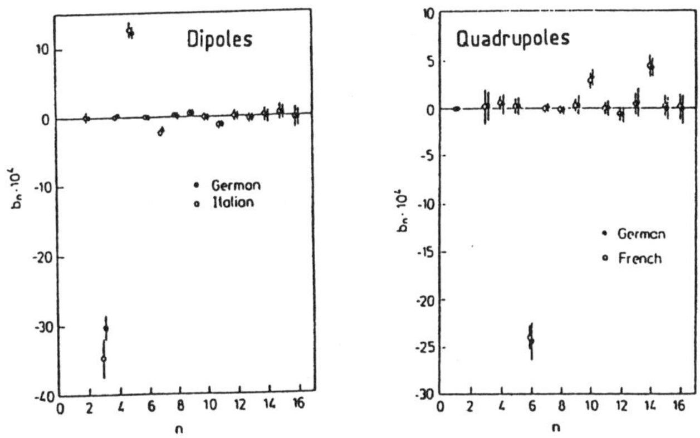

Fig. 7 The average value of the normal multipole coefficients for 436 dipole magnets and 246 quadrupole magnets at injection field. The strength of the integrated multipole field at a radius of 2.5 cm is normalized to the strength of the integrated dipole, respectively the integrated quadrupole field. The large values of b_3 and b_5 in the dipoles and of b_6 in the quadrupoles are caused by persistent currents.

In order to determine the required strength of the correction elements at injection and during acceleration, the dipole and sextupole fields are measured continously in two superconducting reference magnets[24], powered in series with the ring magnets.

The proton beam lifetime at injection is on the order of 10 hours after a careful correction of the persistent-current multipoles using the scheme outlined above. This lifetime is long compared to the 30 min filling time. The geometric acceptance is larger than 2π mm mrad and the dynamic acceptance is larger than 0.8π mm mrad. The design value of the unnormalized 2σ transverse emittance is 0.5π mm mrad at injection.

The acceleration cycle proceeds in two steps. The beam is injected at 40 GeV and then accelerated to 300 GeV. At 300 GeV the optics of the ring is changed from injection to luminosity conditions and then the beam is accelerated to 820 GeV and stored. In the transition from injection to luminosity conditions the maximum value of the β functions changes from 270 m to 1000 m. Only very small losses occur during the acceleration cycle as shown in Fig. 8.

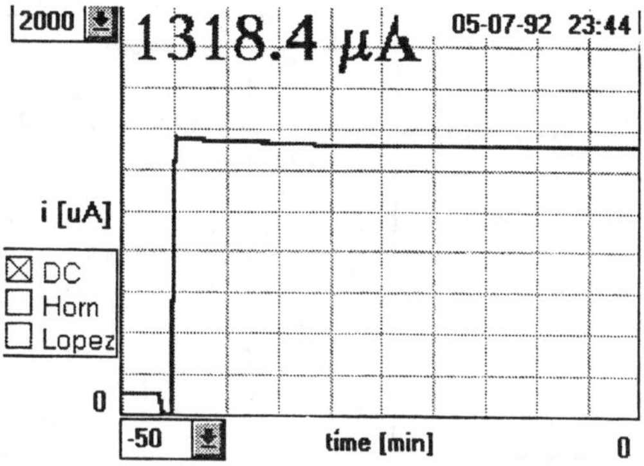

Fig. 8 - The circulating proton current in HERA is plotted as a function of time from injection to storage at 820 GeV.

Because of the long time constant of the main superconducting magnet circuit, the instantaneous value of the magnetic field in the reference dipole magnet, powered in series with the main circuit, is used to derive a reference signal for all other energy dependent systems like correction coils, normal-conducting magnets in the straight sections and the RF system.

The measured proton beam life time at 820 GeV for 10 bunches each with 20% of its design intensity is several weeks. The normalized 2σ transverse emittance is about 35π mm mrad in both planes, nearly a factor of two above the design values. For a well tuned machine we do not observe a significant emittance blow-up, neither during acceleration nor during storage.

During the main part of the data taking period the protons were accelerated and stored using only the 52 MHz RF system. The resulting bunch length was reduced from 35 cm (1σ) at the beginning of the run to its design value of 18 cm by improving the bucket matching between PETRA II and HERA and the beam loading compensation of the idling 208 MHz system. Towards the end of the run the 208 MHz system was turned on, yielding a bunch length of 11.6 cm in agreement with the design value. The bunch length before and after turning on the 208 MHz system is shown in Fig. 9.

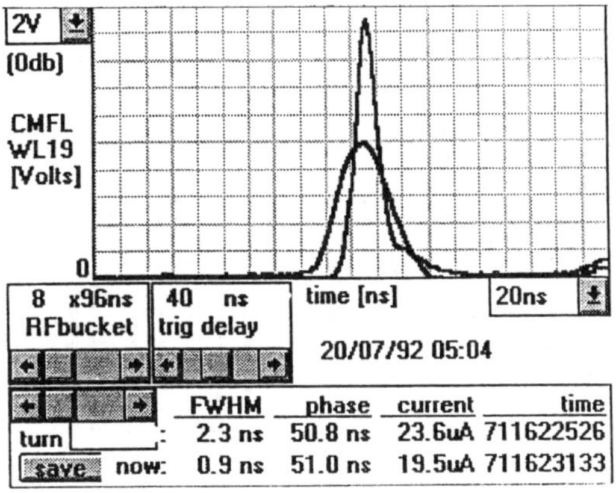

Fig. 9 The proton bunch length at 820 GeV. The long bunch corresponds to the protons being stored in a 52.4 MHz bucket. Note the bunch shortening which results from the turn on of the 208 MHz system.

2.5 Colliding Beams

The electron and proton ring were operated simultaneously for the first time in October 1991. Note that magnets near the interaction points are common to both rings. A correction was applied to compensate for the small magnetic coupling between the rings.

The first collisions between a 12 GeV electron bunch and a 480 GeV proton bunch was observed on October 19th. The following strategy was adopted to bring the two counter-rotating bunches into collisions. The beams were first centered transversely using beam position monitors located ±7 m from the interaction point. Then, to insure proper collisions in the interaction points North and South, the proton RF frequency was locked to the electron RF frequency and the arrival time of the bunches at a monitor located in interaction point East was used to adjust the phase such that the bunches collided in the interaction points North and South.

The luminosity is measured using the reaction $e + p \rightarrow e + \gamma + p$ with the electron and the photon detected in coincidence[25,26]. The arrangement of the luminosity monitors is shown in Fig. 10.

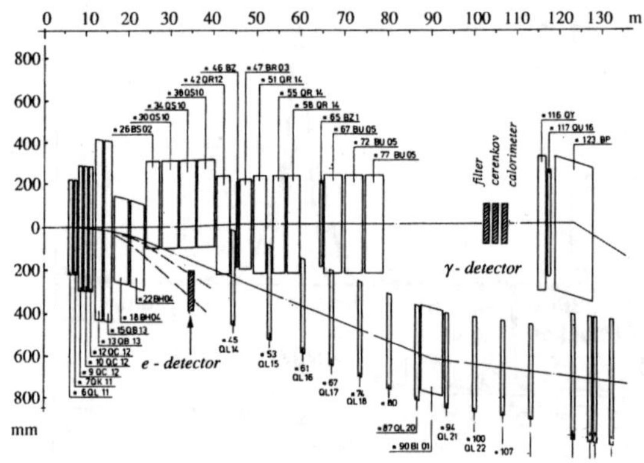

Fig. 10 The layout of the magnets in the straight section and the luminosity detectors designed to monitor the reaction e+p→e+γ+p.

The electron-photon coincidence rate measured by the ZEUS luminosity monitor during collisions of 26 GeV electrons and 480 GeV protons is plotted in Fig. 11.

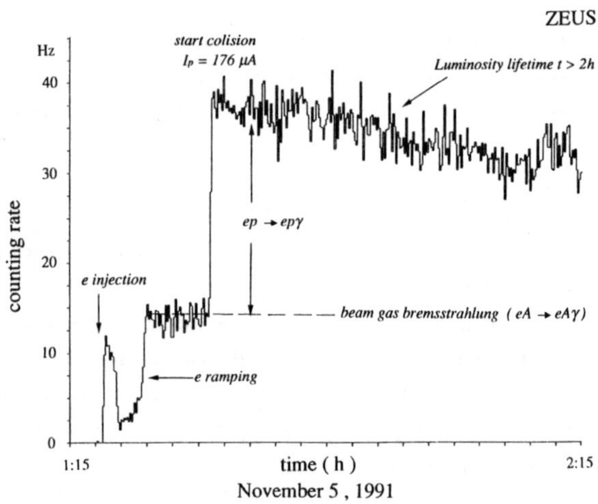

Fig. 11 The electron-photon coincidence rate as measured with the ZEUS detector as a function of time.

A clear step in the coincidence rate is observed when the timing between the electron and the proton bunch is adjusted for collisions. The observed peak luminosity is about $7 \cdot 10^{27}$ cm^{-2}s^{-1} in good agreement with the value predicted from the beam parameters.

Note that in multibunch operation the counter-rotating bunches are first adjusted to have the same revolution time and the correct phase. The final step is then to make the beams overlap transversely at the interaction point.

A new method[27] has been successfully tested, aimed at finding the optimum beam alignment at the interaction point. The electron beam is excited very weakly by transverse kickers near the betatron frequencies for the protons. A central beam collision maximizes the signal on the Schottky monitor of the protons. Even when the beams are missing each other by some standard deviations the signal is clearly visible and may be used for automatic alignment.

The electron-proton luminosity is given by:

$$L = \frac{N_p \cdot N_e \cdot n_b \cdot f_R}{(\sigma_{x,e}^2 + \sigma_{x,p}^2)^{1/2}\,(\sigma_{y,e}^2 + \sigma_{y,p}^2)^{1/2}} .$$

N_p and N_e denotes the number of protons, respectively electrons per bunch, n_b is the number of colliding bunch pairs and f_R is the revolution frequency. The transverse electron and proton beam size is denoted by $\sigma_{x,e}$, $\sigma_{y,e}$, $\sigma_{x,p}$, $\sigma_{y,p}$.

A maximum luminosity of $1.3 \cdot 10^{29}$ cm^{-2}s^{-1} with 9 bunches in each beam has been observed corresponding to 20% of the design luminosity per bunch crossing. A typical set of beam parameters yielding a luminosity of $0.8 \cdot 10^{29}$ cm^{-2}s^{-1} is listed in Table 2.

Parameter	Present Run	Design
$\sigma_{x,p}/\sigma_{y,p}$ (mm)	0.27/0.09	0.29/0.07
$\sigma_{x,e}/\sigma_{y,e}$ (mm)	0.33/0.09	0.26/0.07
$\beta_{x,p}/\beta_{y,p}$ (m)	7/0.7	10/1
$\beta_{x,e}/\beta_{y,e}$ (m)	2.2/1.4	2/0.7
n_b^p	10	210
n_b^e	11	210
Bunch spacing (ns)	96	96
N_p	$0.2 \cdot 10^{11}$	$1.0 \cdot 10^{11}$
N_e	$0.3 \cdot 10^{11}$	$0.35 \cdot 10^{11}$
$\Delta\nu_{x,p}/\Delta\nu_{y,p}$	0.001/0.0005	0.001/0.0005
$\Delta\nu_{x,e}/\Delta\nu_{y,e}$	0.009/0.015	0.016/0.02
Lifetime		
Proton beam (h)	50	-
Electron beam (h)	4	-
Luminosity ($cm^{-2}s^{-1}$)	$0.8 \cdot 10^{29}$	$1.5 \cdot 10^{31}$

Table 2 - Typical luminosity beam parameter.

The transverse beam dimensions are derived from the measured values of the β functions at the interaction point and the measured beam emittance. The proton beam lifetime is strongly dependent on the transverse dimensions of the electron beam and is maximized when the cross sections of the two beams match. The values of $\beta_{x,e}$ and $\beta_{y,e}$ have been adjusted to meet this condition.

The number of electron bunches n_b^e has been limited by the maximum electron current which can be stored in the ring. The eleventh electron bunch is used to measure the background in the luminosity monitors due to interactions of the electron beam with the rest gas.

The number of proton bunches n_b^p corresponds to one DESY III bunch train and is chosen to match the number of electron bunches minus 1. The number of protons per bunch has so far been limited by the proton injectors and by losses during beam transfer to 10% to 30% of the design value. Recently the proton intensity in DESY III has been raised by a factor of 2 to its design value.

The maximum proton beam tune shifts are now at their design values and have so far not limited the performance. The tune shifts suffered by the electron beam are still smaller than the proposal value.

The stored proton and electron currents versus time for a well centered proton beam colliding with an electron beam of same transverse size is plotted in Fig. 12.

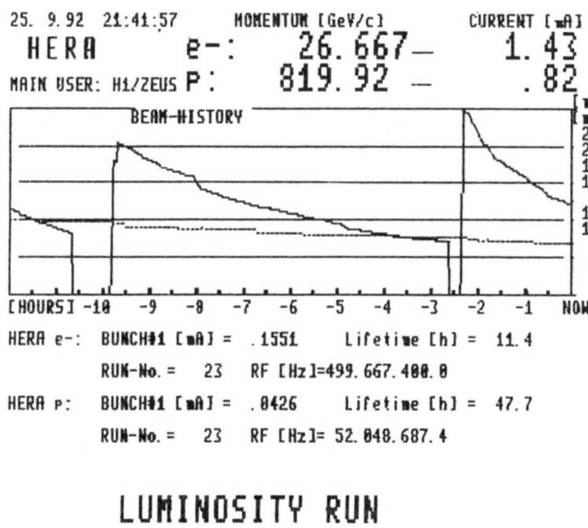

Fig. 12 The stored proton and electron currents as a function of time during luminosity conditions.

Note that under these conditions the proton beam lifetime is of order 50 hours compared to a typical 4 h lifetime of the electron beam. Thus in general protons are filled once every 24 hours whereas the electrons are dumped and reinjected every 4 to 5 hours.

The specific luminosity, defined as $L/I_p \cdot I_e$, is a measure of beam overlap and transverse beam dimensions. The measured luminosity and the derived specific luminosity are plotted in Fig. 13 as a function of time. Whereas the luminosity drops by a factor of 3.5 during the 5 hour long storage time the specific luminosity remains constant. This demonstrates the stability of the two rings and that the proton beam emittance remains constant over the fill. It demonstrates also that the proton beam is not strongly influenced by noise in the electron beam - i.e. there is negligible cross talk between the two beams. This remains true even when the amplitude of the electron feedback system is turned up to its maximum value.

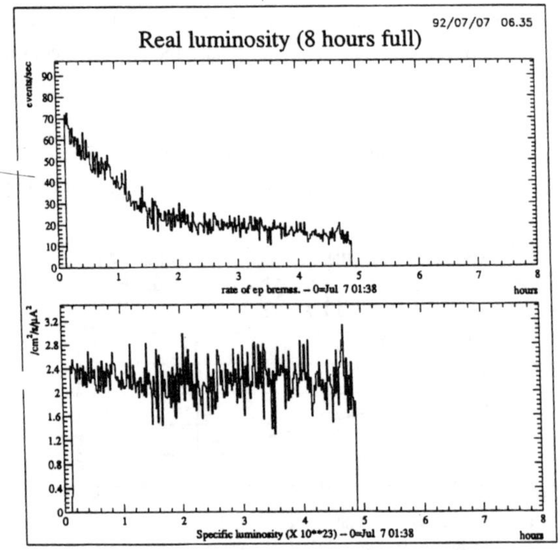

Fig. 13 - The luminosity and specific luminosity as a function of time (ZEUS).

Recently the specific luminosity has been raised to $3.7 \cdot 10^{28} cm^{-2} s^{-1} mA^{-2}$ compared to its design value of $5.4 \cdot 10^{28} cm^{-2} s^{-1} mA^{-2}$.

The beam orbits are very reproducible. By following a well defined magnet "massage" procedure, ep collisions are obtained for the same setting of the correction elements.

3. THE PHYSICS

In a deep inelastic electron-proton collision as depicted in Fig. 14 the incoming electron interacts directly with one of the quarks in the proton by means of a spacelike current, charged or neutral. The kinematic variables Q^2 and ν or the scaled variables x and y, used to describe this process, are also defined in Fig. 14.

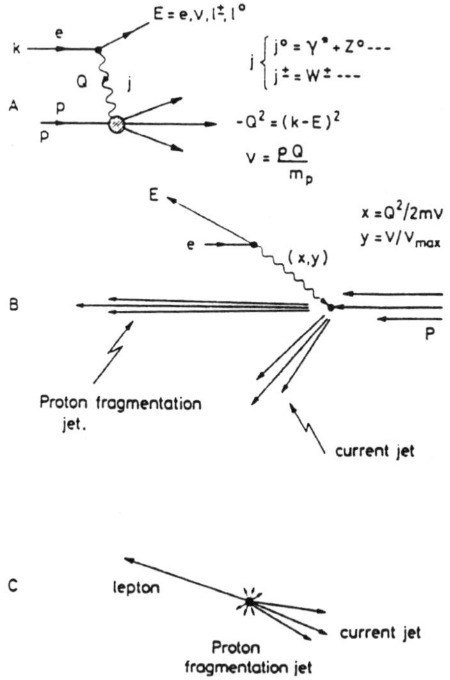

Fig. 14 - Deep inelastic electron-proton interactions at HERA and the definition of kinematic variables.

An electron-proton collision in HERA yields a very simple final state topology. The struck quark will materialize as one or several jets of hadrons whose momentum components transverse to the beam axis are balanced by the transverse momentum of the final state electron (or neutrino). The remainder of the proton will appear as a sharply collimated jet of hadrons travelling along the initial proton direction (Fig. 14).

The expected event topology is indeed observed at HERA as shown in Fig. 15.

The kinematic region in Q^2 and ν which has been liberated by the turn on of HERA is shown in Fig. 16 and compared to the region which can be investigated with a 600 GeV muon beam interacting with protons at rest.

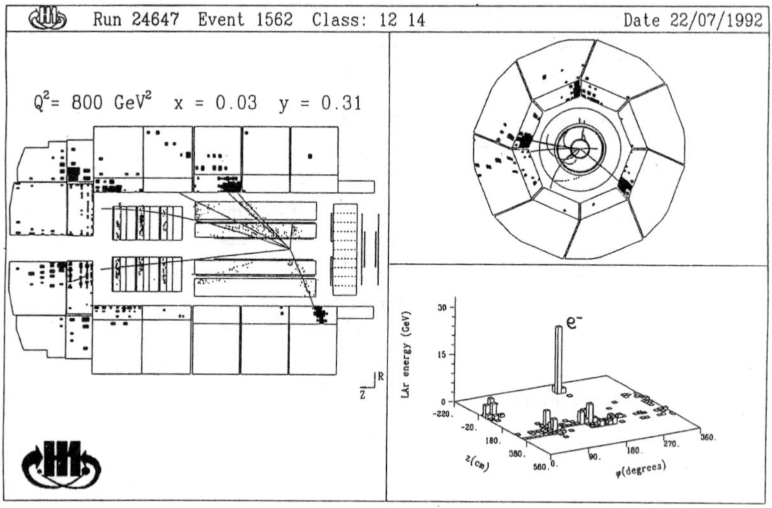

Fig. 15 - Deep inelastic ep event observed by the H1 detector at HERA.

The present maximum values of Q^2 = 600 GeV^2 and ν = 400 GeV can be extended by nearly two orders of magnitude to Q^2 and ν values of order 30000 GeV^2 and 40000 GeV. Furthermore, measurements of the cross section in the scaling region can be extended down to x-values on the order of a few 10^{-5} compared to the present lower limit of roughly 10^{-2}.

Photoproduction and other low Q^2 processes can be investigated up to c.m. energies of 250 GeV. Up to now, data have been available only up to 20 GeV.

Also the range available to search for new physics in electron-proton interactions is extended considerably. Leptoquarks and electron-like new leptons can be searched for nearly up to the kinematic limit of 250-300 GeV. Right-handed currents can be observed for a propagator mass of order 1 TeV.

A complete up-to-date discussion on HERA physics can be found in reference 28.

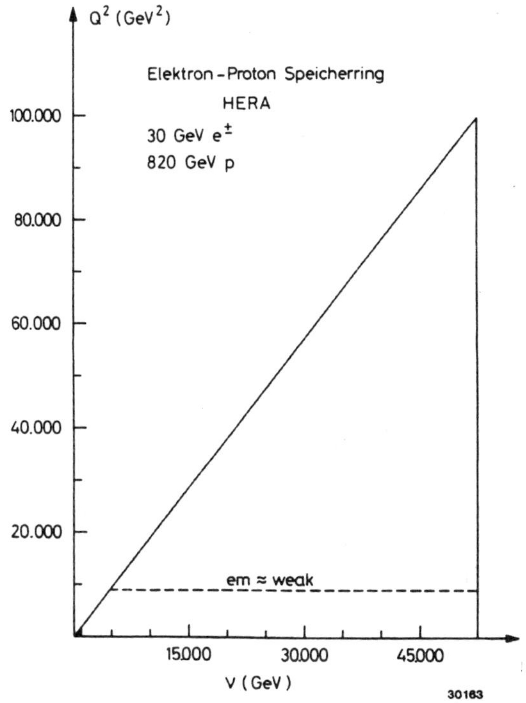

Fig. 16 - The kinematic region in Q^2 and ν at HERA. The black area at the left hand corner corresponds to a 600 GeV muon incident on protons at rest.

By the end of July a total integrated luminosity of 5 nb^{-1} had been delivered to the experiments. Despite the fact that luminosity became routinely available only in July, both experiments have been able to extract data on neutral current deep inelastic scattering for $x < 0.01$. They also report preliminary values of the total photoproduction cross section at a c.m. energy of 200 GeV and preliminary limits on the mass of lepto-quarks.

3.1 The Detectors

A simulated electron-proton event is shown in Fig. 17.

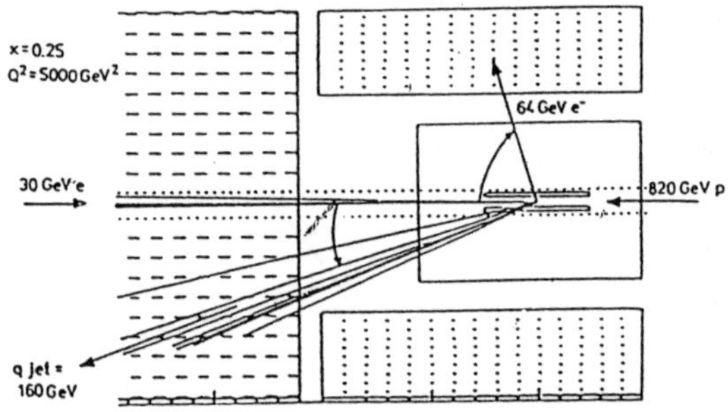

Fig. 17 - The topology of a typical deep inelastic electron-proton event.

The kinematic variables Q and ν or x and y as defined in Fig. 14 can be determined from a measurement of the energy and direction of either the final state electron or of the current jet. Thus the basic requirement is that the detector can measure single electrons and hadron jets with good resolution, both in space and energy, over nearly 4π.

The energies of the hadron jets vary between 820 GeV in the forward direction (incident proton direction) to roughly 30 GeV in the backward direction. It is thus clear that a high resolution, finely grained 4π calorimeter is the cornerstone of a general purpose detector at HERA. Since the ratio of neutral to charged pions varies from event to event, it is crucial that the calorimeter has the same response to showering as to non-showering particles.

New particles produced in the collision may reveal themselves through their semileptonic decay modes. The detector must thus be able to identify leptons travelling within the hadron jet and to determine the charge sign of the lepton.

The properties of various detector elements of H1 and ZEUS are listed in Table 3.

3.1.1 The H1 Detector

An isometric view of the H1 detector is shown in Fig. 18.

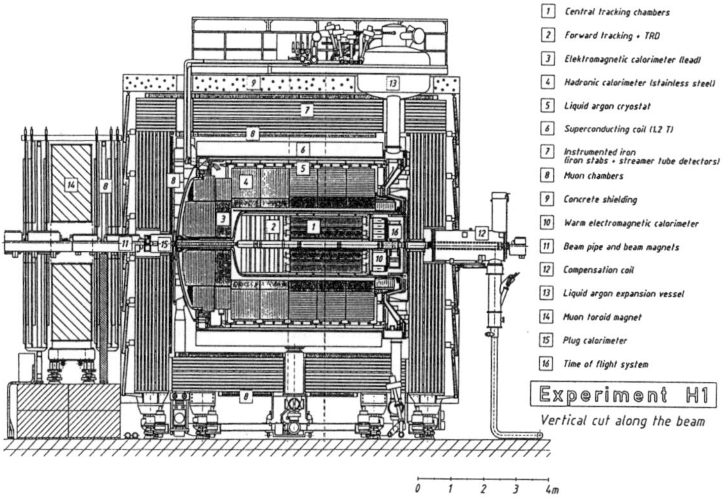

Fig. 18 - An isometric view of the H1 detector

The H1 electromagnetic calorimeter is made of 2.4 mm thick lead plates spaced 2.5 mm apart and the hadronic calorimeter is made of 16 mm thick stainless steel plates with gaps of 2 x 2.5 mm. Both calorimeters are read out using liquid Argon and are installed in a common cryostat. The liquid Argon read-out allows naturally for a fine segmentation both transversely and longitudinally. Transversely, the tower size varies between 3 x 3 and 8 x 8 cm^2 in the electromagnetic calorimeter and between 8 x 8 and 20 x 20 cm^2 in the hadronic calorimeter. Longitudinally, the electromagnetic shower is sampled 3 to 4 times, the hadronic shower 4 to 6 times. There is a total of 45000 independent segments in the detector.

A liquid Argon calorimeter has inherently a different response to showering and non-showering particles as shown in Fig. 19a. However, an electron or neutral pion will deposit its energy over a very short distance compared to the 6 to 8 absorption lengths needed to contain the energy of a charged pion. Since the liquid Argon calorimeter is finely segmented also in the longitudinal direction, the neutral pion content in a jet can be identified by looking for "hot" spots" in the longitudinal energy loss profile. By weighting the measured longitudinal energy loss profile, the H1 group has achieved the

Table 3 - Properties of H1 and ZEUS

	H1	ZEUS
Size	12 x 10 x 15 m^3	11.6 x 10.8 x 20 m^3
Polar angle	0.7^o-175°	10 σ_{beam} - 176°
Solenoid		
Field	1.2 T	1.8 T
Diameter	5.6 m	1.72 m
Length	5.75 m	2.8 m
Central Tracking	11<r<85 cm	10<r<85 cm
Resolution $\sigma(p)/p$	0.003 p	0.002 p $\oplus$ 0.002
Electromagnetic calorimeter	lead plates 2.4 mm liquid Argon 2.5 mm	Uranium plates 3.3 mm Scintillator 2.6 mm
Thickness	20 - 30 X_o	27 - 32 X_o
Tower size	3 x 3 - 8 x 8 cm^2	5 x 20 cm^2
Long.segmentation	3 - 4	1
Resolution	$(0.10 / \sqrt{E}) \oplus 0.01$	$(0.18 / \sqrt{E}) \oplus 0.01$
Hadronic calorimeter	steel plates 16mm liquid Argon 2x2.5mm	Uranium plates 3.3mm Scintillator 2.6mm
Response e/h	1, by weighting	1 inherent
Thickness	$(4.7 - 8)\lambda_{abs}$	$(4 - 7.1)\lambda_{abs}$
Size	8 x 8 - 20 x 20 cm^2	20 x 20 cm^2
Long.segmentation	4 - 6	2
Resolution	$(0.45 / \sqrt{E}) \oplus 0.01$	$(0.35 / \sqrt{E}) \oplus 0.01$
Tail catcher	4.5 λ_{abs}	4 λ_{abs}
Electron-pion separation	10^4 - 10^5	10^4 - 10^5 with silicon pads
Myon-pion separation	10^3	10^3

$\oplus$: errors added in quadrature

same response to electrons and charged pions and hence an improved energy resolution. The energy resolution for electrons and hadrons (after weighting) is plotted in Fig. 19b as a function of energy.

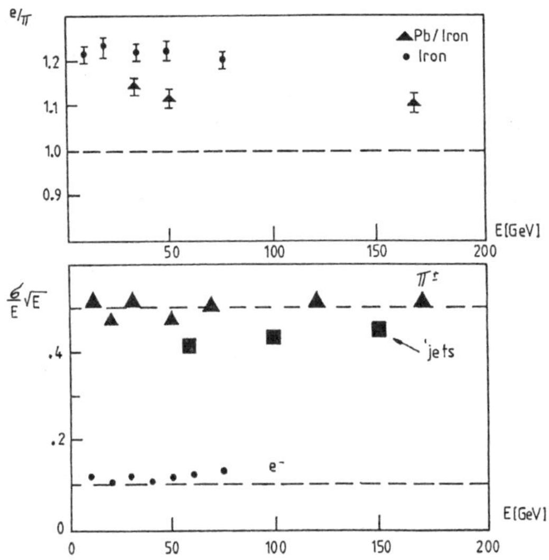

Fig. 19 - a) The measured signal ratio of e to π for the lead and iron calorimeter modules as a function of energy

b) The electron and hadron energy resolution after weighting.

The calorimetry is completed by a forward hadron calorimeter made of copper absorber plates with silicon read-out and a backward electromagnetic lead scintillator sandwich calorimeter.

The calorimeter has been operated since April 1991. The energy calibration has been stable to better than 0.2% and less than 10^{-3} of the read-out channels are not operational. Also the relative energy calibration between various units in the central calorimeter is quite good. However, more ep data are needed to reduce the uncertainty on the absolute energy calibration to a value below 5%. The measured energy resolution of the backward calorimeter is between 2% and 3% for electrons between 10 GeV and 30 GeV. An ultimate resolution of 1% can be reached using ep data.

The liquid Argon read-out can be operated in a magnetic field and the H1 group has chosen to install the calorimeter inside the large superconducting coil, which produces a 1.2 T field over a length of 5.75 m and a diameter of 5.2 m.

In the central region charged particles are being tracked using two jet chambers, augmented by two z-drift chambers and two multiwire proportional chambers. In the forward direction the trajectories are measured with three sets of radial and planar drift chambers. Electrons are identified using transition radiation detectors interleaved with tracking chambers. Trajectories of particles travelling in the backward direction are measured using multiwire proportional chambers. In the forward direction, a measured resolution (σ) of 150 μm has been achieved compared to the central chamber values of 190 μm in r, Θ and 350 μm in z.

The calorimeters and the tracking detectors are enclosed and supported by an iron structure of 2200 tons which serves both as the return yoke of the magnet and as a hadronic backing calorimeter and muon-filter. The iron yoke is made of 7.5 cm thick iron plates interleaved with streamer tube chambers. Muons are identified and measured using three layers of chambers which are installed on both sides and in the middle of the iron structure. In the forward direction muons are detected in a muon spectrometer made of a magnetized iron toroid and six layers of drift chambers. The forward spectrometer covers production angles between 3° and 17° and is designed to detect muons in the range 5 GeV to 200 GeV.

3.1.2 The ZEUS Detector

A vertical cut of the ZEUS detector along the beam is shown in Fig. 20. It was the aim of the ZEUS Collaboration to construct an inherently compensating calorimeter with the best possible hadronic energy resolution. Compensation can be achieved by using a high-Z material as an absorber and judiciously choosing the thickness of the absorber plates and the scintillator plates. The ZEUS calorimeter is made of 3.3 mm thick, 20 cm wide and up to 4.6 m long depleted uranium plates, cladded with stainless steel, which are interleaved with 2.6 mm thick scintillator plates. The tower size, defined by the scintillator plates, is 5 x 20 cm^2 in the electromagnetic part and 20x20 cm^2 in the hadronic calorimeter. The scintillator plates are read out by means of electromagnetic wavelength shifters, light guides and photo-multipliers. Longitudinally the calorimeter is segmented into an electromagnetic and two hadronic sections. The calorimeter is made of three parts; the central region is surrounded by the barrel (BCAL) calorimeter, the forward (FCAL) and the rear (RCAL) calorimeters cover the proton, respectively the electron direction. The total coverage is 99.8% of 4π.

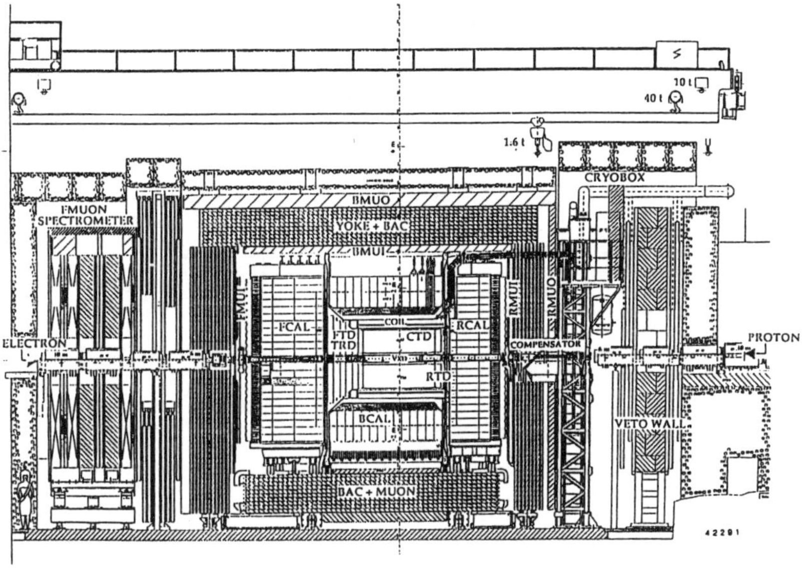

Fig. 20 - Schematic view of the ZEUS detector. The component labels are defined in the text.

A prototype calorimeter module has been extensively tested and some of the results are shown in Fig. 21. Note that the compensation indeed works over the whole energy range tested and that hadrons are measured with excellent resolution.

The natural radioactivity of the uranium is used to monitor the calibration of the calorimeter to better than 0.2%. The noise caused by the uranium decay is on the order of 12-30 MeV depending on cell size. The total noise, integrated over an area of 80x80 cm^2 which corresponds to the area needed to contain a normal jet, is of order 200 MeV. The calorimeter has a time resolution of better than 1 ns for particle energies above 3 GeV. In order to be able to identify electrons within a jet, the ZEUS Collaboration has proposed to insert two layers of silicon pads at depths of 3 and 5 radiation lengths.

Photomultipliers are sensitive to magnetic fields and the calorimeters are hence installed outside the superconducting coil. The solenoid, 1.72 m in diameter and 2.8 m long, is 0.9 radiation lengths thick and produces a field of 1.8 T.

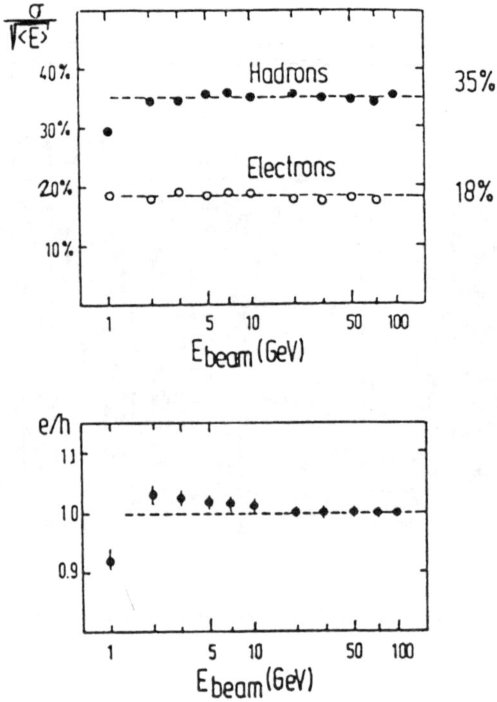

Fig. 21 - a) The energy resolution observed with the uranium calorimeter module for incident hadrons and electrons as a function of energy.
b) The ratio of energy deposits by electrons and charged pions as a function of the particle energy.

The inner tracking detector is located in the magnetic volume. In the central part, particles are tracked with a time expansion vertex chamber (VXD) and a vector drift chamber (CTD). The vertex chamber has a measured resolution of 35 μm. The drift chamber has 9 superlayers each with 8 layers of sense wires, yielding a total of 4608 sense wires. A resolution of 120 μm was achieved in a test beam. In the superlayers 1, 3 and 5 a z read-out is available with a resolution of 40 mm. The forward tracking detector (FDET) contains three planar drift chambers and four transition radiation detectors to identify particles within the jet. Particles emitted along the incident electron direction are measured in the rear tracking detector (RTD) which is made of planar drift chambers. The planar drift chambers have an expected spatial resolution of order 120 μm and the energy loss of a minimum ionizing particle is measured with an uncertainty of 6-7%.

The iron yoke of 2000 tons is segmented into 7.5 cm thick iron plates, interleaved with proportional tube chambers (BAC) which serve to measure the energy leaking out of the hadronic calorimeter and to identify muons. Four layers of limited streamer tube chambers are installed on both sides of the barrel part (BMUI, BMUO) and the rear part (RMIZU, RMUO) of the iron yoke. In the forward direction high energy muons are detected using a toroidal spectrometer equipped with planar drift chambers and limited streamer tube chambers.

Protons scattered under small angles (~ 0.2 mrad) can be measured to within 10 σ of the proton beam size by an elaborate set of six Roman pots, installed between 23 m and 90 m downstream of the interaction point.

3.2 Experimental Conditions

The HERA design parameters impose rather stringent requirements on the data acquisition system and its ability to recognize and separate physics events from the much more abundant background events in real time.

The most serious source of background is beam halo protons traversing material upstream and then showering the detector with secondary tracks. An example of this class of events is shown in Fig. 22. Several methods are used to reduce this source of background to a manageable level.

A set of precise aperture-limiting collimators, reducing this background by one to two orders of magnitude, has been installed in straight section West 1.6 km upstream of the ZEUS detector. Both experiments have installed shielding walls and veto counters just upstream of the detectors. The remaining events of this type can be removed by making use of the time-of-flight information.

The secondaries of genuine ep events are produced close to the center of the detector and have nearly equal flight times t_F and t_R to the front respectively rear calorimeter whereas background particles, originating upstream of the detector, will cluster around negative values of t_R and positive values of t_F-t_R (see Fig. 23).

A second class of background events is caused by proton interactions with the residual gas in the beam pipe within the detector. An example of such an event observed with the H1 detector is shown in Fig. 24. Using dE/dx information, 21 proton tracks were identified. Obviously this class of events can be eliminated by improving

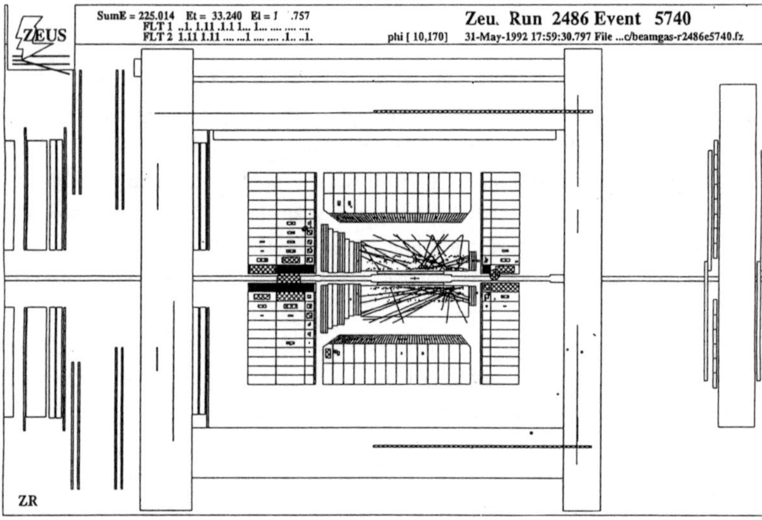

Fig. 22 - Computer reconstructed background event caused by a proton interacting with material upstream of the ZEUS detector.

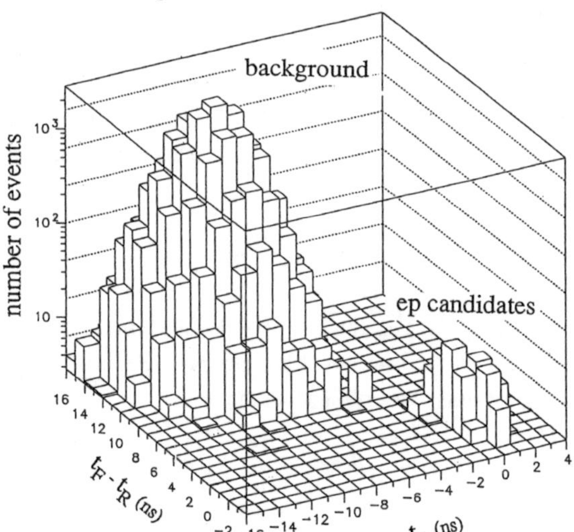

Fig. 23 - Difference in arrival times t_F-t_R of signals from the front and the rear calorimeters versus the signal arrival time t_R in the rear calorimeter for events which meet the selection criteria (ZEUS).

the beam pipe vacuum, by a more stringent cut on transverse momentum or by excluding events with many slow protons.

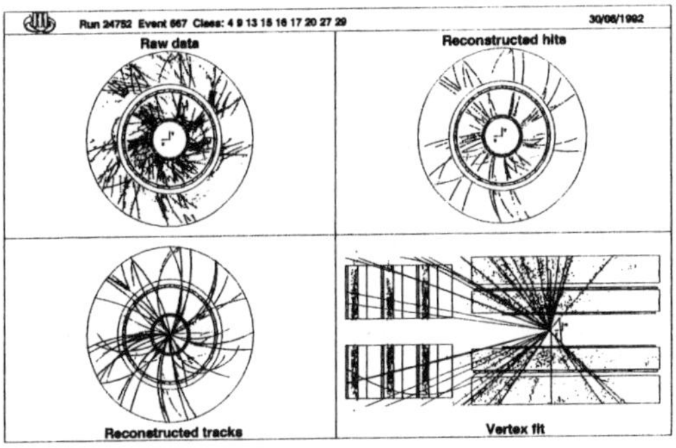

Fig. 24 - Computer reconstructed background event caused by a proton interacting with the rest gas in the beam pipe.

Up to now, typically 5% of the design electron current could be stored. At these current levels the synchrotron radiation was effectively masked out and did not cause problems. However, electrons suffering beam gas interactions or scattering on the synchrotron masks do lead to a high single rate in the electron luminosity counter. This background was determined from the counting rate measured during the passage of the electron pilot bunch.

Cosmic rays are another potential source of background events. They are removed using time of flight information from the muon counters.

The z-distribution of neutral-current events which meet the selection criteria is shown in Fig. 25. The event distribution is well fitted by a Gaussian with σ_z = 35 cm in accordance with the observed proton bunch length. This demonstrates that neutral-current events can be well separated from background events.

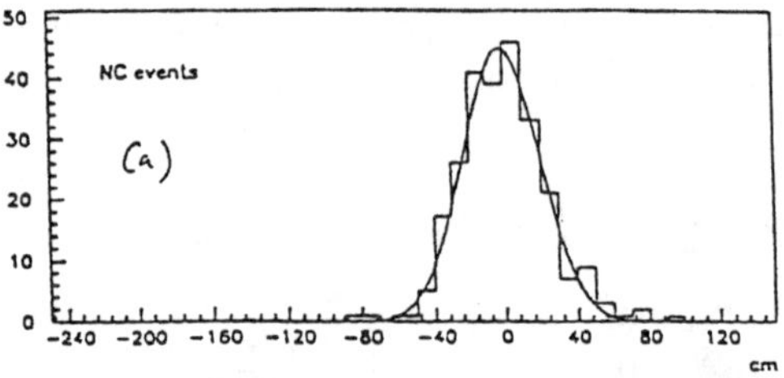

Fig. 25 - The z-distribution of neutral-current events which meet the selection criteria. The data were obtained by the ZEUS group.

The trigger and data acquisition system must cope with the collision rate of roughly 10 MHz, corresponding to the 96 ns bunch spacing, and with the large background to signal ratio in a minimum bias trigger configuration. Both experiments select the interesting events using a series of trigger levels. The first-level trigger decision needs 2.2 μs (H1) respectively 5 μs (ZEUS). Until then the data from 200000 to 300000 detector elements are stored in deadtime-less analog or digital pipelines for 23 (H1) or 52 (ZEUS) successive beam crossings. The final event selection is based on a partial event reconstruction using fast processors farms. During the first experimental period both experiments were using rather loose event selection criteria yielding a final event rate of 3-5 Hz corresponding some 600 kbyte/s.

The block diagramme of the ZEUS data acquisition and trigger system is shown in Fig. 26 as an example. Each detector component generates a first level trigger (FLT) and passes the information on to a general first level trigger (GFLT). Based on this information the first level trigger makes a decision to accept or reject the event after 5 μs. In the case of a positive decision the information stored in the pipeline is digitized, stored in digital buffers and passed on to a component second level trigger (SLT). If accepted, the data are then passed on to the global second level trigger (GSLT). The second level trigger can handle a 1 kHz input rate. Events accepted by the second level trigger are passed on to the third level trigger (TLT). This level is made of a 1000 mips processor farm and here the event is reconstructed and analyzed. The event builder and the third level trigger are designed to accept an input rate of 100 Hz. By applying further selection criteria the event rate is reduced to a few Hz which are transferred via an optical link to the DESY central computing center where they are stored on disks or on cassettes.

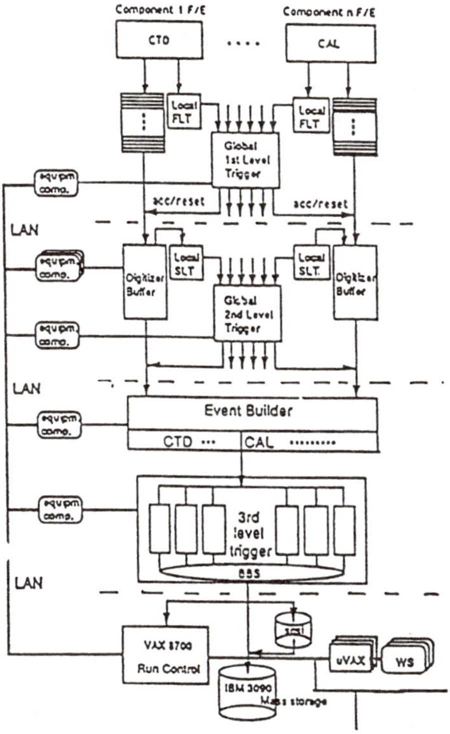

Fig. 26 - The ZEUS data acquisition and trigger system.

3.3 Physics Results

3.3.1 Deep inelastic neutral current events

The differential neutral-current cross section e p → e'X can be written as:

$$\frac{d^2\sigma(e^{\mp})}{dx\ dQ^2} = \frac{4\pi\ \alpha^2}{x\ Q^4}[(1-y+y^2/2)F_2(x,Q^2) - (y^2/2)F_L(x,Q^2) \pm (y-y^2/2)xF_3(x,Q^2)]$$

The longitudinal structure function

$$F_L(x,Q^2) = F_2(x,Q^2) - 2x\ F_1(x,Q^2)$$

does not receive contributions from scattering off spin 1/2 constituents if transverse momenta are neglected.

The parity and charge conjugation violating "weak" structure function $x F_3(x,Q^2)$ is to first order proportional to the ratio of the Z and γ propagators

$$\frac{Q^2}{Q^2 + M_Z^2}$$

and can safely be neglected in the present analysis.

The kinematic region in 1/x and Q^2 available to HERA and to a 600 GeV muon beam incident on a proton at rest is plotted in Fig. 27.

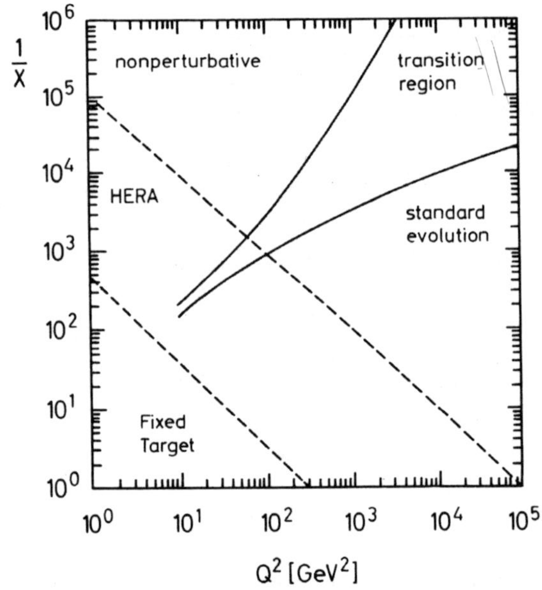

Fig. 27 - The kinematic region in 1/x and Q^2 available at HERA and with a 600 GeV muon beam incident on protons at rest. The perturbative and non-perturbative domains are separated by a transition region.

Thus HERA data on e p $\rightarrow$ e'X will allow us to investigate the behaviour of the structure functions down to x values of 10^{-4}, a factor of 100 below the region available in fixed target experiments. This is the last frontier of QCD and has been the focus of intense theoretical work[29] during the past few years. The main features are summarized below.

The kinematic region can be divided into three main areas: a QCD perturbative region located in the lower right hand corner at moderate x and large Q^2, a non-perturbative region in the left hand upper corner and a transition region in between.

In the classic region of QCD, the Q^2 evolution of the structure functions can be computed from the Gribov-Lipatov-Altarelli-Parisi equations[30,31]. These equations are defined in terms of the parton splitting functions as indicated in Fig. 28.

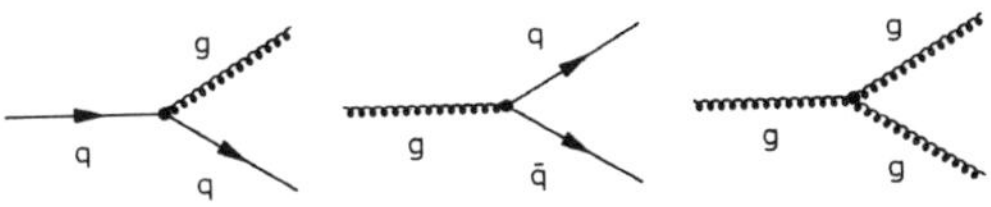

Fig. 28 - The relevant parton splitting diagrams.

At low x-values the structure functions are dominated by the gluon distribution function $xG(x,Q^2)$. Extrapolating $xG(x,Q^2)$, as determined in the perturbative region, towards small x for constant Q^2 leads to a steeply increasing function that violates unitarity. Since the gluon density increases with 1/x, the gluon-gluon interaction can no longer be neglected although the gluon-gluon coupling is still small. In this transition region one may be able to use the parton language and the behaviour of the structure functions may be described[32] by adding a recombination term onto the perturbative evolution equations. The strength of this non-linear recombination term depends on an unknown parameter R. If the gluons are distributed uniformly inside the proton, the parameter R may be identified with the proton radius i.e. $R \simeq 5\ GeV^{-1}$. In some models, however, the gluons will cluster around the valence quarks leading to local "hot spots". In this case $R \simeq 2\ GeV^{-1}$ and the onset of the nonlinear region will start earlier.

A further extrapolation in 1/x yields a very dense partonic system. Although $\alpha_s(Q^2)$ is still small the effective interactions are strong due to the high parton densities. In this region the perturbative approach breaks down and the parton picture may not be applicable.

With HERA we should be able to explore the transition region and perhaps also the non-perturbative region.

Both experiments report data on $e\,p \to e'X$. The data are based on a useful luminosity of 1.5 nb^{-1} (H1) and 2.5 nb^{-1} (ZEUS). The details of the experiment and the data reduction can be found elsewhere[10,11]. Note that the results are still preliminary and have not yet been corrected for radiative effects.

The data analysis resulted in a clean sample of events. Computer reconstructed events are shown in Fig 15 and 29.

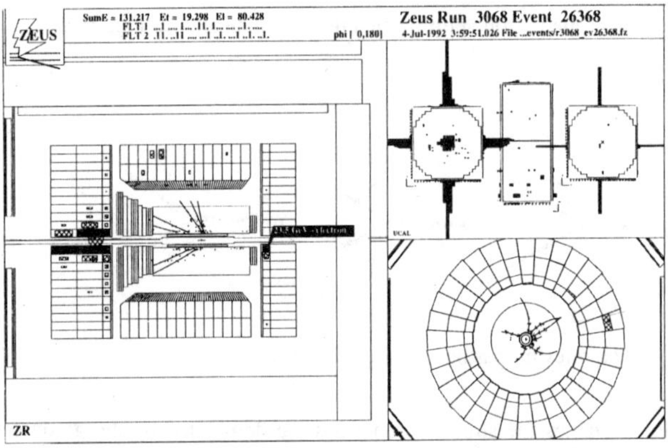

Low X, low Q^2 event
($x = 6.3 \cdot 10^{-3}$, $Q^2 = 35$ GeV^2)

Fig. 29 - Computer reconstructed neutral-current events (ZEUS):
a) Event with $x = 0.065$, $Q^2 = 380$ GeV^2
b) Event with $x = 6.3 \cdot 10^{-3}$, $Q^2 = 35$ GeV^2

Note that the events are clean with only a few spurious tracks. The ZEUS collaboration determines the Q^2 and x of an event by combining the information obtained from a measurement of the final state electron and the hadrons in the current jet, whereas H1 relies on only the measured angle and energy of the scattered electron. Scatter plots of the events which meet the selection criteria are shown in Fig. 30.

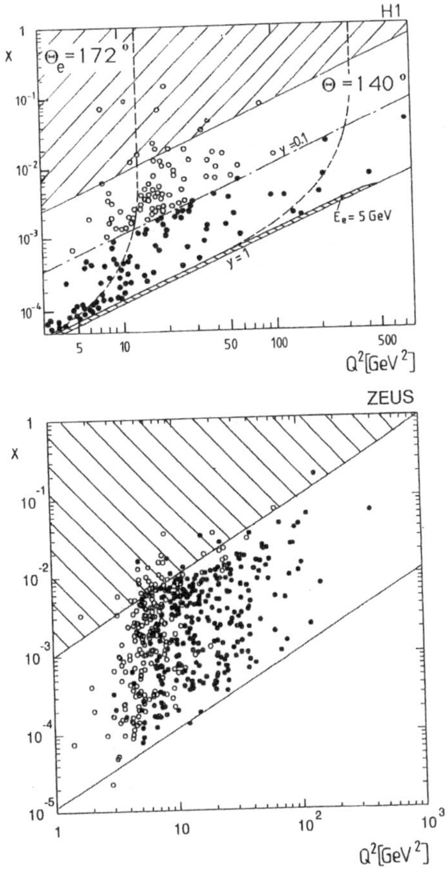

Fig. 30 - A scatter plot of deep inelastic (ep → e'X) events in the x, Q^2 plane measured by the H1 and ZEUS experiments. The kinematic region explored by fixed target experiments is indicated by the hatched area. Well measured events are indicated by the full circles.

A firm prediction of the quark model is that the transverse momentum of the scattered electron should be balanced by the transverse momentum of the hadrons in the current jet. That HERA events have this property is clearly demonstrated in Fig. 31 where the transverse energy flow of the hadrons is plotted as a function of the opening angle $\varphi_e - \varphi_{hadron}$ between the hadron and the electron. This distribution is shown for various cuts in x. There is a clear peaking around $\varphi_e - \varphi_{hadron} \sim 180^o$ for events with $x > 0.003$, consistent with the jet picture.

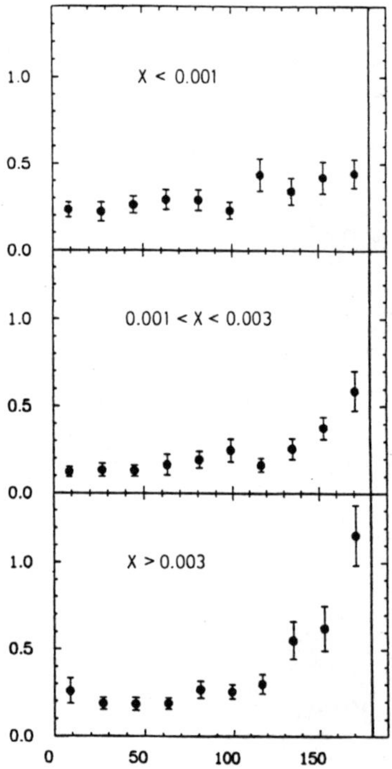

Fig. 31 - Transverse energy flow of hadrons as a function of the opening angle between the electron and the hadrons φ_e - φ_{hadron} for various cuts in x. The data were obtained by the H1 Collaboration.

The Q^2 distributions as measured by the two collaborations, are plotted in Fig. 32a,b and compared to model calculations[33,34]. Within the rather large error bars the data agree well with the predictions.

The scattered electron energy distribution at a fixed scattering angle is very sensitive to the gluon distribution at small x as shown in Fig. 33.

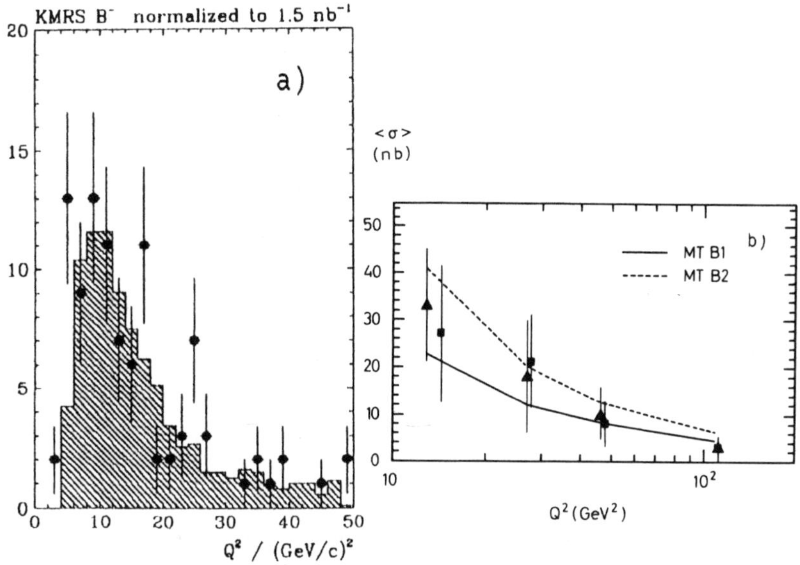

Fig. 32 - The measured Q^2-distribution for events meeting the selection criteria.
a) Data reported by the H1 Collaboration for $0.05 < y < o.55$ normalized to 1.5 nb^{-1}. The hatched area shows the absolute prediction based on the KMRS parametrization[33] of the structure functions.
b) The Q^2-distribution reported by the ZEUS Collaboration for events with $Q^2 > 10$ GeV^2 and $8 \cdot 10^{-4} < x < 8 \cdot 10^{-3}$. The results of two different assumptions in the data analysis are presented and compared to theoretical predictions[34].

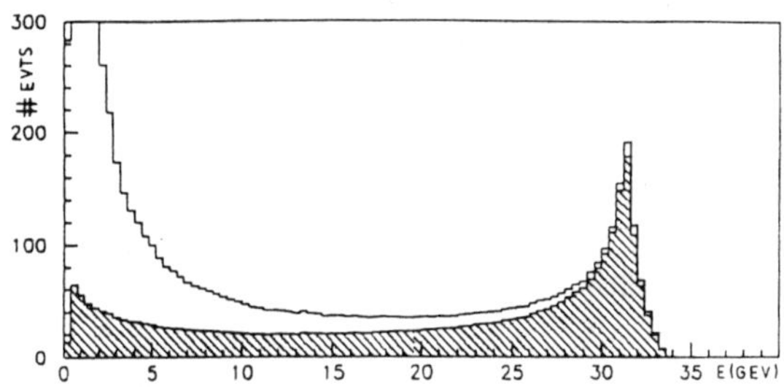

Fig. 33 - The simulated energy spectrum (Ref. 35) of the scattered electron at a fixed scattering angle for a flat (batched) and a steeply increasing gluon distribution.

The prominent peak at the incident electron energy is simply a kinematic effect. Note that the scattered electron energy distribution must be measured down to quite low energy in order to be able to discriminate between different shapes of the gluon form factor. Preliminary energy distributions obtained by the two experiments are shown in Fig. 34 and compared to the theoretical predictions.

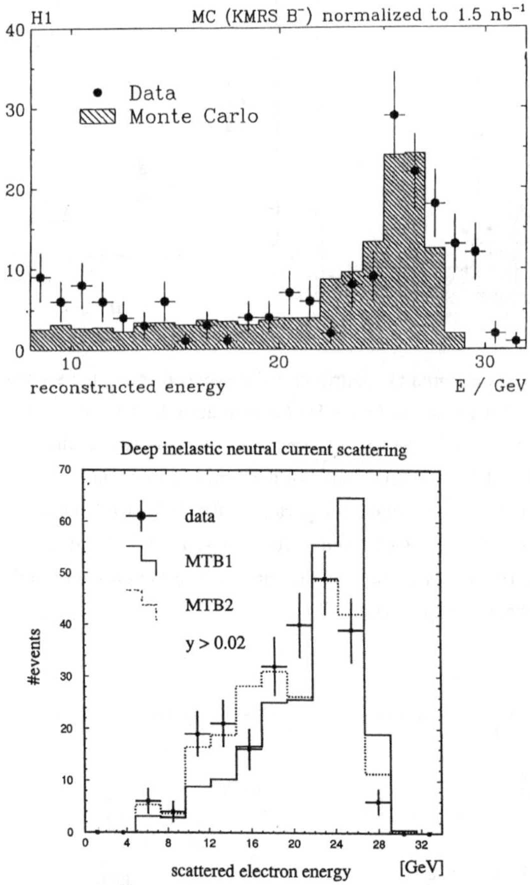

Fig. 34 - The measured energy distribution of the scattered electron in neutral current events.

a) Data obtained by the H1 Collaboration and compared to Monte Carlo predictions[33]

b) Data obtained by the ZEUS Collaboration and compared to Monte Carlo predictions[34].

The early data on the neutral-current cross section e p → e'X are not yet precise enough to allow a determination of the gluon structure function at small x. However, the quality of the data demonstrates that such detailed comparisons will soon become feasible with improved statistics.

3.3.2 Photoproduction

The photon is a particle with unique properties. On the one hand it is a fundamental gauge boson with well defined couplings to basic fermions and gauge bosons as shown in Fig. 35a. On the other hand, part of the time the photon behaves like a strongly interacting vector boson (Fig. 35b). For larger values of $p_\perp$, the photon interacts indeed dominantly via its hadronic constituents, quarks and gluons, yielding resolved photon events as shown in Fig. 35c.

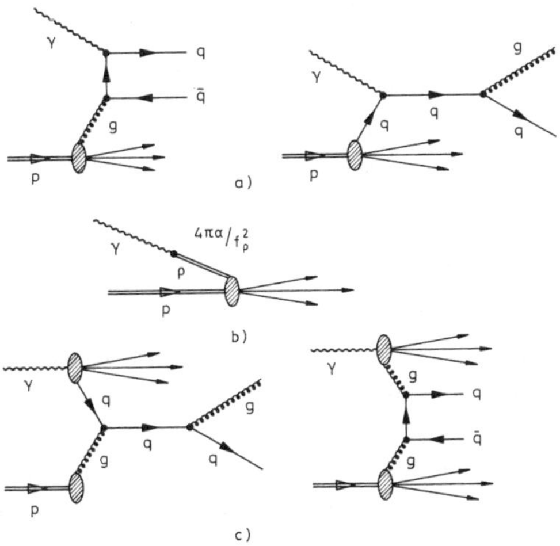

Fig. 35 - Photon - hadron interactions:
a) Examples of the pointlike coupling of photons
b) The vector meson dominance model
c) "Resolved Photon" events induced by the gluon and quark content of the photon.

The total photon-proton cross section has previously been studied for center of mass energies up to 18 GeV. A measurement of photon-proton interactions at higher energies allows us to address the following questions:

- What is the connection between high energy non-perturbative QCD and the Pomeron?

- What is the density of quarks and gluons in the photon ?
- Does the total photon-proton cross section rise gently with energy as observed in $p\bar{p}$ interactions or is the photon-proton cross section anomalously large[36] at HERA energies as first conjectured to explain the high muon yield[37] in very high energy cosmic showers?

HERA is well suited to study photon induced processes for c.m. energies up to 250 GeV. For very small values of Q^2, the incident electron beam is equivalent to a well collimated beam of quasi real photons. In this case the electron scattering cross section is related to the photoproduction cross section as follows:

$$\frac{d\sigma(ep)}{dy} = \frac{\alpha}{2\pi}\frac{1+(1-y)^2}{y}\sigma(\gamma p)\,\ell n\frac{Q^2_{max}(y)}{Q^2_{min}(y)}$$

The photon energy and Q^2 are determined from the energy and angle of the scattered electron. Both experiments use the luminosity electron counter for this purpose. An electron which has emitted a photon at a small angle is deflected away from the circulating electron beam in the off-axis, low β quadrupoles and in the downstream magnets, exits the accelerator vacuum through a thin window, and impinges on the luminosity electron counter.

At ZEUS electrons with scattering angles between 0 and 6 mrad and energies E' in the range $0.2E < E' < 0.9E$ are accepted. This corresponds to a range in Q^2 of the quasi-real photons from 10^{-7} GeV^2 to $2 \cdot 10^{-2}$ and center of mass energies between 186 GeV and 220 GeV. The bulk of the ZEUS data result from a dedicated photoproduction run with an integrated luminosity of 0.23 nb^{-1}.

The H1 tagger accepts electrons scattered at angles below 5 mrad and energies in the range of 6 GeV to 19 GeV corresponding to a Q^2 range between 10^{-7} GeV^2 and 0.015 GeV^2 and a mean c.m. energy of 200 GeV. H1 data were collected concurrent with the deep inelastic data and are based on an integrated luminosity of 1 nb^{-1}.

The data analysis and the event selection criteria are discussed in detail in references 10 and 11. Both experiments report data on the total photoproduction cross section:

H1 quotes at a c.m. energy of 200 GeV: $\sigma_T(\gamma p) = (150 \pm 15(\text{stat.}) \pm 19(\text{syst.}))\mu b$

ZEUS reports at a c.m. energy of 210 GeV: $\sigma_T(\gamma p) = (154 \pm 16(\text{stat.}) \pm 32(\text{syst.}))\mu b$.

The quoted systematic errors result from the errors in the measurement of the photon flux, the ep luminosity and the detection efficiency, all added in quadrature.

The detection efficiency depends on the relative abundance of elastic and diffractive events to low transverse-energy inelastic events and hard processes. The total photoproduction cross section has been evaluated under various assumptions on the ratio of these processes, consistent with the observed final state topology. The measured cross sections are plotted in Fig. 36 versus energy together with data at lower energies and various predictions.

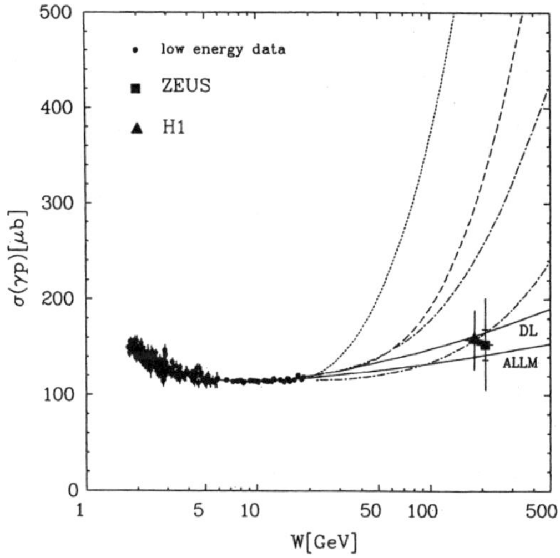

Fig. 36 - The total photoproduction cross sections measured by H1 and ZEUS are plotted versus energy together with low energy data and compared to theoretical models.

The data are in good agreement with predictions based on Regge models[39,40] and show an increase with energy similar to that observed in the $\bar{p}$-p total cross section.

Mini-jet models[38], which simply add the perturbative (QCD) and non-perturbative (soft) contributions, predict steeply rising cross sections with energy for smallish values of $p_{\perp}^{min}$, the lowest momentum below which perturbative QCD cannot be used. With the choice of $p_{\perp}^{min}$ made in the mini-jet calculation, there is disagreement with the data. Including[41] multiple scattering corrections brings also the minijet model into agreement with the observed data.

Hard photon processes, leading to large $p_\perp$ jet events, result either from direct photon processes (Fig. 35a) or from resolved photon processes (Fig. 35c). Direct photon processes will yield 2-jet final states in lowest order, whereas resolved final states will populate the 3-jet final state.

Large-$p_\perp$ photoproduction events at HERA energies show a clear jet structure as seen in Fig. 37.

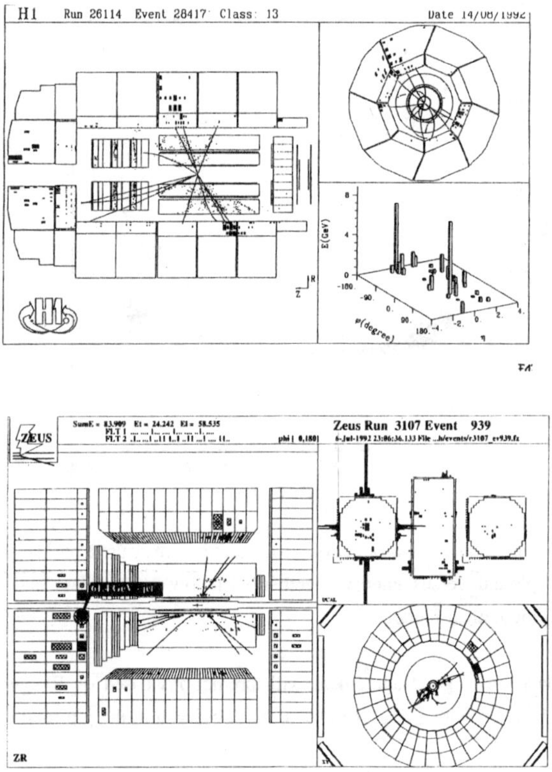

Resolved photon event
(2 jets + remnant of low Q^2 photon)

Fig. 37 - Photoproduction events observed in the H1 and ZEUS detectors. Both events show a clear jet structure.

Based on a jet cone algorithm the H1 group has searched for final states with 2 or more jets. They are left with 34 events of which 7 have a tagged electron. The transverse energy distribution of these events is plotted in Fig. 38 and compared to predictions based on a complete Monte Carlo simulation and on a simulation of the direct process only. The comparison of the data with the Monte Carlo results suggest that the bulk of the observed photoproduced jet events are due to the "resolved photon" process.

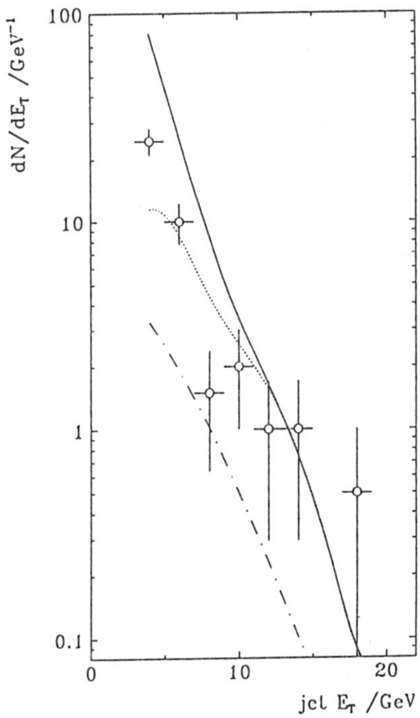

Fig. 38 - The jet transverse energy distribution measured by H1. The solid (dotted) curve is the absolute prediction of the PYTHIA MC programme for $p_{\perp}^{min}$ = 1 GeV ($p_{\perp}^{min}$ = 5 GeV). The prediction for the direct process with $p_{\perp}^{min}$ = 1 GeV is shown as the dash-dotted line.

3.4 Search for Leptoquarks

HERA is ideally suited to search for exotic particles with electron quantum numbers. Both experiments report limits on leptoquarks[42], massive particles with both electron and baryon quantum numbers which occur naturally in various unified models and also in composite models of leptons and quarks.

The cross section for producing a scalar leptoquark in ep collisionss is given by

$$\sigma_{LQ} = \frac{\pi}{4s} \lambda^2_{L,R} \; q\,(M^2_{LQ}/s)$$

The coupling constants $\lambda_{L,R}$ are defined in units of $\sqrt{4\pi\alpha_{em}}$ and the subscripts L,R denote the chirality of the electron. $q(M^2_{LQ}/s)$ is the probability of finding a quark (antiquark) with a fraction $x = M^2_{LQ}/s$ of the proton momentum.

The favoured leptoquark decay mode is into an electron and a quark jet. Leptoquark production will thus lead to a neutral-current type final state and will indeed show up in the x-distribution of neutral-current events as a bump at $x = M^2_{LQ}/s$.

Fig. 39 shows the x-distribution of neutral-current events, which meet the leptoquark selection criteria. One event was found for $x > 0.008$ compared to the expected background of 4.2 events.

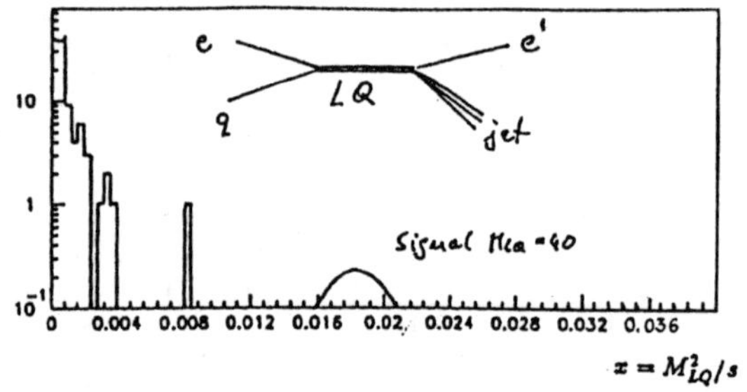

Fig. 39 - The x distribution of neutral-current events which meet the leptoquark selection criteria. The expected signal for a leptoquark with M_{LQ} = 40 GeV at 90% confidence level is also shown (ZEUS).

From these data the ZEUS group derives the following 90% confidence limits:

$\lambda_L = 0.3 \qquad \lambda_R = 0$

$29.5\ \text{GeV} < M_{LQ} < 47\ \text{GeV}$

$\lambda_L = 0.0 \qquad \lambda_R = 0.3$

$29.5\ \text{GeV} < M_{LQ} < 67\ \text{GeV}$

The mass limits set by the H1 group as a function of the coupling strength are shown in Fig. 40 and compared to mass limits obtained in $p\bar{p}$ collisions by UA2 and CDF. The mass limits which can be reached at HERA for integrated luminosities of 1 pb^{-1} and 100 pb^{-1} are also shown.

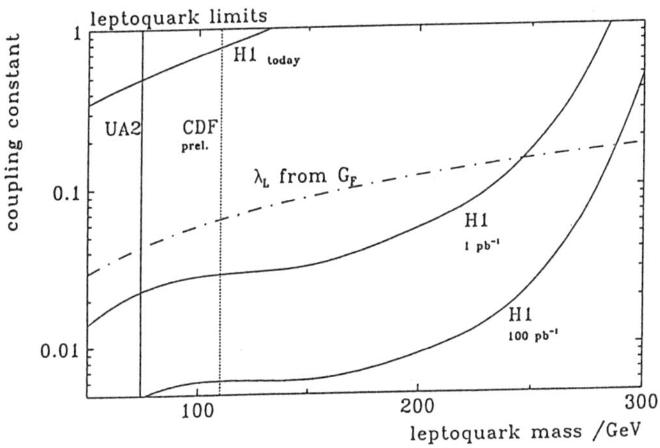

Fig. 40 - Limits on leptoquark mass obtained by the H1 group in ep collisions and by UA2 and CDF in $p\bar{p}$ collisions.

4. SUMMARY and OUTLOOK

During the past months the HERA proton-electron collider and the general purpose detectors H1 and ZEUS have been successfully transformed from a potential to a real source of physics.

The observed ep luminosity per crossing has reached 20% of its design value and we have so far not discovered any fundamental reason why HERA should not be able to reach its design value. The next step will be to overcome the current limitation in the electron ring and to increase the number of bunches in the two rings from 10 to the design value of 210.

Both experiments have been able to extend the measurement of the deep inelastic neutral-current cross section down to x values of order 10^{-4}, a factor of 100 smaller than before. Although present statistical errors do not yet allow precise statements on the behaviour of a weakly coupled, densely packed parton system it is clear that experiments in this region can be carried out and that data will be forthcoming in the near future.

The experiments have demonstrated that photoproduction can be investigated at HERA for center of mass energies a factor of 10 above the range accessible in fixed target experiments. At these energies the final state jet structure for either direct of "resolved photon" events should become clearly visible.

The experiments have also shown that they will soon be able to carry out significant searches for new objects like leptoquarks or excited electrons.

After a brief shutdown the HERA accelerator has resumed operation with the time split between physics runs and accelerator development. The increase in peak luminosity by a factor of two and the fact that many operational aspects have now become routine has raised the average integrated luminosity per day by almost a factor of 10 compared to the July run.

In addition to the general ep detectors H1 and ZEUS, HERA also offers the intriguing possibility of carrying out high luminosity, high duty cycle fixed target experiments by using internal targets. The proposed HERMES experiment[43] plans to use the longitudinally polarized electron beam impinging on polarized p, d and He^3 gas jet targets to investigate the proton spin structure. Another group has proposed[44] to position thin wire targets in the tails of the proton beam. With this arrangement and the full circulating proton current they estimate a production rate of 60 $b\bar{b}$ pairs every second. Such a rate would be sufficient to measure the CP-violating parameters in the $J/\psi K\bar{K}$ channel.

Both experiments are now under active considerations and - if approved - they will be a new and fascinating aspect of the HERA experimental programme.

References

1. HERA, A Proposal for a Large Electron-Proton colliding beam facility at DESY, DESY HERA 81-10
2. G.-A.Voss, Proceedings of the First European Accelerator Conference, Rome, June 1988, p. 7
3. B.H.Wiik, Proceedings of the 1991 IEEE Particle Accelerator Conference, San Francisco, May 1991
4. D.Degèle, Proceedings of the Third European Accelerator Conference, Berlin, March 1992
5. F.Willeke, Proceedings of the XII International Accelerator Conference, Hamburg, July 1992
6. Technical Proposal for the H1 Detector, March 1992 for the H1 Collaboration
7. F.Brasse - for the H1 Collaboration - contribution to this conference
8. Technical Proposal for the ZEUS detector, March 1986 and updates
9. D.Caldwell - for the H1 Collaboration - contribution to this conference
10. B.Löhr - for the ZEUS Collaboration - contribution to this conference
11. F.Eisele - for the H1 Collaboration - contribution to this conference
12. M.Ebert et al., DESY 91-036
13. U.Timm et al., Rev.Sci.Instr. 62 (4), April 1991, 687
14. K.Balewski et al., Part.Acc. Conf. 1990, Vol. 27, 39
15. W.Kriens, contribution to the XII International Conference on High Energy Acc., Hamburg, July 1992
16. R.Byrns et al., Proc. 2nd European Acc. Conf., Nice 1990, 13
17. A.Matheisen et al., Proc. IEEE Part.Acc. Conf., San Francisco 1991, 2429
18. HERA Polarimeter Group, contribution to HEACC'92, Hamburg, July 1992
    HERA Polarimeter Group, Internal Report
19. G.Horlitz, Proceedings of the 1991 IEEE Particle Acc.Conf., San Francisco, May 1991, and DESY HERA 91-10
    M.Clausen et al., Int. Cryo.Eng.Mat.Conf. 91, Huntsville, Alabama, and DESY HERA 91-17
20. P.Schmüser, Proc.Oregon APS Meeting (1985), p.968
    H.Kaiser, Proceedings of the 13th Intern.Conf. on High Energy Acclerators, Novosibirsk, August 1986, Vol.2, p.49
    S.Wolff, Proceedings of the 10th Int. Conf. on Magnet Technology , Boston, September 1987, p. 729
    R.Auzolle et al., IEEE Trans. on Magnetics, Vol.25, p.1660, 1989

21. R.Meinke, Proceedings of the 1990 Applied Superconductivity Conference, Snowmass 1990 and DESY HERA 90-17
22. R.Bacher et al., Contribution to HEACC'92, Hamburg, July 1992
23. P.Schmüser, Proceedings of the 1991 IEEE Particle Accelerator Conference, San Francisco, May 1991, and Proceedings of the Third European Accelerator Conference, Berlin, March 1992
24. H.Brück et al., Contribution to HEACC'92, Hamburg, July 1992
25. D.Kisielewska et al., DESY HERA 85-25
26. S.V.Levonian et al., DESY H1-TR113, 1987
27. S.Herb, Contribution to HEACC'92, Hamburg, July 1992
28. Physics at HERA, Proceedings of the Workshop Hamburg, October 29-30, 1991 Edited by W.Buchmüller and G.Ingelmann
29. For a discussion on the low x-region including references see: J.Bartels and J.Feltesse in Ref. 10
30. V.N.Gribov and L.N.Lipatov, Sov.Journ.Nucl.Phys.15, 438 and 675 (1972)
31. G.Altarelli and G.Parisi, Nucl.Phys. 126, 297 (1977)
32. L.V.Gribov, E.M.Levin, and M.G.Ryskin, Phys.Rep. 1001, 1982
33. J.Kwiencisnki, A.D.Martin, R.G.Roberts and W.J.Stirling, Phys.Rev. D42, 3645, 1990
34. J.G.Morfin and W.K.Tung, Z.Phys. C52, 13, 1992
35. A.De Roeck and M.Klein, Zeuthen PHE91-014, 1991
36. M.Drees and F.Hatzen, Phys.Rev.Lett. 61, 275, 1988
37. G.Yodh, Nucl.Phys. B.12, 277 (1999)
38. For a discussion on low Q^2 Physics at HERA including references see G.A.Schuler in Ref. 10
39. A.Donnachie and P.V.Landshoff, Nucl.Phys. B231, 189, 1984
40. H.Abramowicz, E.M.Levin, A.Levy and U.Maor, Phys.Lett. B269, 465, 1991
41. J.R.Forshaw and J.K.Storrow, Phys.Lett. B268, 116, 1991
42. W.Buchmüller, R.Rückl, D.Wyler, Phys.Lett. B191, 442, 1987
43. The HERMES Collaboration, K.Coulter et a., DESY/PRC/90/1 (1990)
44. Private Communication - W.Hoffmann, W.Schmidt-Parzefall.

CHAIRMAN: B. Wiik

Scientific Secretaries: Z. Ren, S. Shotkin

DISCUSSION

– *Dreiner:*

Please comment on the discovery potential for SUSY at HERA. The discovery potential for the MSSM is close to zero. If one allows for the violation of R-parity then there is the possibility of resonant quark production, similar to lepton-quark production. The discovery potential allows for squark masses up to close to the kinematic limit. This region (200 GeV) for squarks is "expected" and would thus not be a lucky discovery of SUSY.

– *Wiik:*

I agree with your comment. My strong motivation for building this machine was supersymmetry, but you know as time goes, other machines develop and I think we are more or less running out of kinematic space for finding supersymmetric particles, but, you know, we could be lucky.

– *Dreiner:*

Could I just add one more comment, for R-parity violation you would not have to be lucky, it would stand out like a sore thumb.

– *do Couto e Silva:*

1) How crucial is the scheme to correct the persistent current in the superconductivity cavities? Since these cavities were commissioned to industry, how accurate were they in reproducing the devices? What about the yield?

2) What is the difference in using proton target and the He^3 target in the proton spin experiment?

3) Can you give me more information on ZEUS's microvertex detector?

– *Wiik:*

(1) The persistent current is of course a problem related to the superconducting magnets, and has nothing to do with cavities. The properties of the persistent current depend upon two things: it depends upon the thickness of the filaments and upon the quality of the conductor, that is, the current-carrying capability at very low field and very low temperatures. And if the thickness and the critical

current is reproduced then also the persistent current multipoles will vary little from magnet to magnet. At HERA the persistent current effects are constant to within 5 to 10% from magnet to magnet and that is the reason that we can compensate their effects with correction coils as I showed you.

Superconducting cavities are even more delicate than superconducting magnets. All the activity in a superconducting cavity takes place in a 80 nm thick layer on the surface. If a tiny spot of the surface is not superconducting then the Q-value is bad and the cavity useless. You need extreme cleanliness in producing superconducting cavities and so far there is not too much industrial experience of producing completely finished cavities including all the processing.

2) The He^3 target has a high polarization and also allows you to measure the neutron form factors.

3) For detailed information, please contact the ZEUS group directly.

– *Baillie:*

How many of your events will be due to W-exchange, as compared to photon exchange?

– *Wiik:*

Given the present integrated luminosity we should not see any charged current events.

– *Fisk:*

In the Argus Collaboration an experiment is being proposed for studying $B^0 \to \Psi K_s^0$ using an internal target. Do you detect both B's?

– *Wiik:*

I believe so.

– *Fisk:*

How do you expect to deal with the charm background?

– *Wiik:*

I do not know. The ARGUS group at DESY which is centered around Schmidt-Parsifal and Hoffman has not yet submitted a proposal.

– *Liu:*

How many events do you expect of the type $\gamma p \to p + g$?

– *Leike:*

You got nice predictions for the quark mixing matrix. Is there room in your theory for possible neutrino mixing?

– *Ross:*

I gave an example where there are predictions for mixing angles involving slepton number violation. But I do not have predictions for neutrino masses. Dimopoulos, Hall and Raby have another good candidate mass matrix which they have extended to give predictions for neutrino masses.

– *Hanany:*

What about the high temperature phase structure of this theory?

– *Ross:*

The phase structure of this theory is just like the phase structure of the supersymmetric standard model. At high temperature ($\simeq 10^{10}$ GeV) we expect supersymmetry restoration. Below this temperature some supersymmetry is broken and below approximately 10^2 GeV it will trigger a vacuum expectation value for the Higgs and thereafter the quarks and leptons will get masses as well as the W.

– *Benakli:*

How do you think the disagreement between the string predicted value $E_{SU} \sim 3.73 \times 10^{17}$ GeV and the MSSM value $M_{GUT} \sim 10^6$ GeV can be reconciled?

– *Ross:*

The question is: is your prediction true for all the string theories? There are some orbifolds in which this scale can go down. Much more plausible for me is the idea that we have a string theory which at the compactification scale gives you some group, say $[SU(3)]^3$. This group could persist down to the 10^{16} GeV. Other people have suggestions of additional multiplets at the intermediate scales which affect the unification scale and push it to 10^{18} GeV.

– *Wiik:*

There should be quite a few of those and at HERA energies they should be visible as two jet events. Perhaps the experiments will be able to see them by the time of the Dallas meeting.

– *Ball:*

I would like to ask you why the magnetic fields in your detectors are so low, it is about 1 or 2 Tesla, whereas at LEP, as far as I remember, it is about 4 Tesla.

– *Wiik:*

In the detector magnets?

– *Ball:*

Yes the magnetic field in H1 and ZEUS.

– *Wiik:*

The H1 is a copy of OPAL and it is run at the same magnetic field as OPAL. The magnetic field in ZEUS is 50% bigger, it is nearly 2 Tesla, and I do not know of any LEP experiment with a higher magnetic field than that. These are big volume magnets so the stored energy and the forces are pretty impressive. In an accelerator magnet you can go up to 10 Tesla now, but then the field is confined to a very small volume and hence the forces and the stored energy are much less.

– *Ren:*

I have a question. I notice that your H1 detector is perfect covering the theta angle from 0.7° to 175°. I also notice that you have got dominant beam gas events, so are there any connections between them?

– *Wiik:*

Actually the ZEUS with its forward proton spectrometer is covering even smaller angles with the counters just 10σ of the transverse beam width from the beam i.e. just a few mm. The main background is due to protons traversing the beam pipe upstream of the detector and spraying the whole detector with secondaries. This affects the trigger rate but real events are clean with only a few spurious hits. It simply means that:

a) we have to close the aperture defining collimators more;
b) the experiments must increase the trigger thresholds.

– *Shotkin:*

In determination of $\gamma + \mathrm{p}$ cross-section how are the events tagged to identify the photon?

– *Wiik:*

The experiments are using the luminosity monitor. When the electron loses energy in the interaction, it goes through the electron low β-quadrupole magnets off-axis. A quadrupole magnet acts like a dipole magnet, so the electron, since it has a lower energy, gets bent away from the main beam into the luminosity monitor. So then we measure the energy of the scattered electron and since we know the energy of the incident electron, we know the photon energy.

– *Shotkin:*

So you do not attempt to measure any energy through the electromagnetic calorimeter?

– *Wiik:*

Sure we do, but the electron that has lost a photon basically comes in at zero angle so it goes straight forward; we can only separate it using the magnetic field, and with the calorimeter we look at the rest of the photoproduction events, but the electron ends up in this tagging counter sitting 40 or 50 metres downstream, after the magnetic elements.

– *Cho:*

What will the signals and scale for quark and gluon compositeness be at HERA? Will HERA be able to set a tighter limit on the quark compositeness scale than CDF?

– *Wiik:*

I do not know. This is model-dependent. What HERA can do very nicely is the case where the electron and the quark have a common component. In this case the quark and the electron may exchange constituents during the collision and this will modify the neutral current cross-section. This is unique to HERA. The experiments, in the usual definition of the compositeness scale, may be sensitive to a constituent mass of 5 to 10 TeV.

Where is Top?*

R.H.Dalitz[α],Gary R.Goldstein[β]

[α]Department of Physics(Theoretical Physics),University of Oxford, 1 Keble Rd,Oxford OX1 3NP,U.K.
[β]Department of Physics,Tufts University,Medford,Massachusetts 02155,USA

ABSTRACT

Possibilities are discussed for determining the top quark mass m_t from observations on the decay processes for top-antitop pairs produced in antiproton-proton collisions at Tevatron energies, assuming that the $t \to bW^+$ decay channel is dominant and much faster than hadronization. The final states $t\bar{t} \to b\bar{b}\mu^\pm e^\mp$ provide the most striking signal, with little background, but they are rare (≈2/81). If all three candidate events prove to be from $t\bar{t}$, an estimate follows for $P(m_t|\text{rate})$, the probability distribution for m_t. The one reported configuration allows an independent estimate for $P(m_t|\mu^\pm e^\mp 2\text{jets})$. These two distributions are compatible and can be combined to give an m_t estimate of about 122 GeV.

Decay events "1 energetic lepton(ℓ) + 4jets" should appear twelve times as often as "$\mu^\pm e^\mp$2jets" events and can be analysed to give estimates for $P(m_t|\ell 4\text{jets})$; this is illustrated for a fictitious event. There may be background from non-top events but suitable cuts on the data and our analysis procedure together reduce this to a low level. The rate observed for these events does not appear to be as large as this factor 12.

Identification of either or both of the $(b,\bar{b})$ jets would be a great step forward, separating out the "$\mu^\pm e^\mp b\bar{b}$" and "$\ell b\bar{b}$2jets" events of importance with negligible background. We advocate an energetic approach to the analysis of individual events (whether 2,1 or no (b or $\bar{b}$) jets identified) on an event by event basis, with the hope of finding a subgroup of events with a common mass estimate.

*This paper was given by the first author (R.H.D.) on 19 July 1992 at the International School of Subnuclear Physics at ERICE. Its text has been updated to the end of 1992.

1. INTRODUCTION

1.1. Existence of the top quark.

There is little doubt today about the existence of a top quark t, the partner to the well-known and much studied bottom quark b. The Standard Model, with SU(2)xU(1) symmetry for the Electroweak Interactions has had remarkable success, many of its parameters now being known to high accuracy. In particular, the weak-isospin component T^{wk} for the b quark has now been determined empirically in a very direct way, from measurements of the forward-backward asymmetry in the process $e^+e^- \to b\bar{b}$ and of the decay width $\Gamma(Z^0 \to b\bar{b})$. From these data, Kane and Kolda[1] have deduced the following T^{wk} values for the b-quarks:

$$T_3^{wk}(b_L) = -0.49(+0.05,-0.02), \quad T_3^{wk}(b_R) = 0.00(+0.10,-0.08), \qquad (1.1)$$

which are in accord with the Standard Model assignments of the b quark to the third quark-lepton family, for which the leptonic member τ^+ is already known, as shown in Table 1. The fact that the predictions of the (spontaneously broken) SU(2)xU(1) symmetry have been so successful provides a powerful argument that the third (quark-lepton) family must be complete. The absence of a $T_3^{wk} = +1/2$ partner to the b_L quark would be such a gross violation of this symmetry that it would no longer be possible to understand the detailed and widespread agreement of the data with the predictions based on this electroweak symmetry. In short, there must exist a $T_3^{wk} = +1/2$ partner to the bottom quark b_L, and it has been natural for this partner to be named the top quark t.

Empirically, little else is known about the top quark, although its interactions are prescribed in form and magnitude from its place in the third quark-lepton family. However, we do know that its mass m_t is very much greater than the mass of the b quark, $m_b \approx 5$ GeV. A firm limit on m_t is provided by the fact that top-antitop pair creation is not observed in e^+e^- annihilation in the energy region of Z^0 excitation; in other words, the threshold for $\bar{t}t$ production must be above M_Z, giving the lower limit

$$m_t \geq M_Z/2 \approx 46 \text{ GeV}.$$

Indeed, the top mass is now believed to be above 91 GeV[2]. If top were lighter than this, much $\bar{t}t$ production would have been observed already at the Tevatron, where antiproton-proton interactions are studied at c.m. energy 1800 GeV, an energy far more than adequate for any reasonable value for m_t. At the Tevatron, the limiting factor for top mass determination is the rate of events, rather than the available energy. We shall discuss this further in Sec.2..

The purpose of current top quark research is not just to demonstrate directly the existence of the top quark (which we believe already) nor to check that its interactions are in accord with SU(2)xU(1) symmetry, although these latter studies will provide a very fruitful field for research later on.

Table 1. The three families in the SU(2)xU(1) Standard Model.

T_3^{wk}	Y^{wk}	Quark and Lepton states for each Family		
+1/2	+1/3	u_L	c_L	(t_L)
-1/2	+1/3	d_L	s_L	b_L
0	{+4/3, -2/3}	$\{u_R, d_L\}$	$\{c_R, s_R\}$	$\{(t_R), b_R\}$
+1/2	+1	$\bar{\nu}_{eL}$	$\bar{\nu}_{\mu L}$	$(\bar{\nu}_{\tau L})$
-1/2	+1	e_L	μ_L	τ_L
0	+2	e_R	μ_R	τ_R

The top quark will not lose its polarization by hadronization, as do the other heavy quarks, charm and bottom (cf.Sec.1.2), so that it will be possible to test all of the detailed spin dependences of its interactions[3-5]. It may even turn out to be possible to use the top quark decay as a spin analyser for heavy particle and/or high energy processes which happen to give rise to top quarks. No, the paramount purpose of present top research is to determine the mass m_t of the top quark; how this may be achieved will be discussed in Sec.3 below. Only after this mass has been determined can we move on with the Standard Model, e.g. with the determination of the Higgs particle mass and with the testing of more extensive symmetry models which contain SU(2)xU(1) as a subgroup. Only then will we know what kind of accelerator we shall need for all these detailed studies.

1.2.The top quark lifetime.

When m_t exceeds $(M_W+m_b)\approx 85$ GeV, top quark physics becomes qualitatively different from the physics we have become accustomed to for the c and b quarks, as was first pointed out by Bigi[6]. The dominant decay mode for the top quark is then the two-body mode

$$\text{(a)} \quad t \to W^+ + b, \qquad\qquad \text{(b)} \quad \bar{t} \to W^- + \bar{b}. \tag{1.2}$$

As the mass m_t increases, this decay process becomes faster and faster until it becomes dominant over hadronization. The top decay lifetime calculated for this mode[7] is plotted as function of m_t on Fig.1. For m_t appreciably lower than (M_W+m_b), the W-boson is virtual and its leptonic decay interaction generates the overall mode $t\to b\ell^+\nu_\ell$,whose partial lifetime falls like m_t^{-5}. In the transition region, as m_t approaches and then exceeds (M_W+m_b), the decay lifetime falls even more rapidly with m_t. For m_t well above this threshold, as holds for the physical situation, the decay lifetime falls more slowly, ultimately like m_t^{-3}. The calculated partial width is given as function of m_t in Table 2.

Table 2. Partial decay width for $t\to bW^+$, as function of m_t[29].

m_t(GeV)	=	100	120	140	160	180	200
$\Gamma(t\to bW^+)$(MeV)	=	88	298	612	1033	1572	2242

[By way of contrast, we comment briefly on the other heavy quarks Q = c and b. Their decay lifetimes are of order 10^{-12}s. In c and b jets, a polarized quark Q forms a meson $(Q\bar{q})$, where q denotes a light quark q = (u,d,s), by picking up a light antiquark from the vacuum. If this meson is in the $^1S_0$ ground state, all polarization carried by the quark Q is lost, since this state is spherically symmetric and cannot carry spin information. If the meson state $(Q\bar{q})$ which is formed has non-zero spin,e.g.the $^3S_1$ state or orbitally-excited states like $^3P_2$, it undergoes fast hadronic or electromagnetic transitions (since these conserve parity, the angular distributions do not depend on any initial polarization, although they can transfer polarization from the initial to the final state) until it reaches the $^1S_0$ state, which is the lowest state for the $(Q\bar{q})$ system. Thus, all of the polarization information carried initially by the quark Q is lost, for the cases Q = (c,b)[8].]

2. TOP MASS ESTIMATES FROM EMPIRICAL DATA

2.1.Virtual top radiative corrections.

It was first pointed out by Veltman[9] that, although the e^+e^- energy at LEP is too low for the production of real top-antitop pairs, the existence of top quark can still have a substantial effect on the predictions of the Standard Model through the "radiative corrections" generated by virtual top-quark loops and exchanges. Since the Standard Model is a renormalizable theory, these corrections are computable, at least in a perturbative approximation. Now that LEP experiments have measured with high accuracy many quantities related with the electroweak interactions, these measurements can be compared with the corrected theoretical predictions in order to draw some conclusions concerning the top quark and any other particles of high mass. Some of these quantities are the masses M_W and M_Z, the total width Γ_Z and some partial widths for Z^0 decay, and the forward-backward asymmetries for $e^+e^- \to \bar{b}b$ and $e^+e^- \to$leptons in an energy range covering the Z^0 peak, which arise from γ-Z^0 interference. In the minimal Standard Model, there are several other parameters also relevant, namely M_H, the Higgs particle mass, and $\alpha_s(M_Z^2)$, the QCD coupling strength evaluated at the Z^0 mass. The latter can be deduced with fair accuracy from purely hadronic processes as well as from electroweak studies in the Z^0 mass range. Given the recent LEP data and the theoretically computed expressions, it is then possible to lay out on an (m_t, M_H) plane, the regions consistent with these data. Quite a number of such analyses have been carried out along these lines recently [10,11]. With the LEP data updated to July 1992, Ellis et al.[10] have given the value:

$$\text{all data, with } M_H \text{ free:} \qquad m_t = 124(27)\text{GeV}, \tag{2.1}$$

using $\alpha_s(M_Z^2) = 0.118(8)$. Their analysis has some sensitivity to M_H;adopting by choice a rather low value for M_H, their analysis gives

$$\text{all data, with } M_H = M_Z: \qquad m_t = 132(25)\text{GeV}, \tag{2.2}$$

requiring a small increase for the optimum m_t.

More elaborate models generally have more free parameters. The Minimal Supersymmetric Standard Model(MSSM) has aroused much interest recently, owing to its success in extrapolating the three coupling strengths, $\alpha_i(\mu^2)$, at scale μ^2 = (momentum transfer)^2 and with i = (1,2,3) appropriate to the strong and electroweak sectors of the SU(3)xSU(2)xU(1) symmetry contained within a Minimal SU(5)SUSY, back to a common value at a GUT scale μ of order 10^{16}GeV [12], a test that an earlier attempt based on SU(5)⊃SU(3)xSU(2)xU(1) had failed[13]. Besides M_H and m_t, this MSSM introduces five new parameters, m_A and $\tan\beta$ for the Higgs sector, (m_g, m_0) the gluino mass and a common sfermion mass, respectively, and Λ_t, a supersymmetry-breaking parameter. The discussion of the allowed regions for all of these parameters is naturally rather complicated. Ellis et al.[14] have discussed these dependences only for m_t = 130GeV, mainly to illustrate how an empirical value for m_t would give much-needed information on the bounds which constrain the MSSM parameters. Most probably, given the number of new parameters introduced, the limits placed on m_t in the MSSM will be much less restrictive than those for the minimal SM without supersymmetry.

2.2.Total cross section for top-antitop production.

At the Tevatron energy of 900GeV for proton and antiproton, the (900+900)GeV $\bar{p}p$ interactions are rather like (300+300)GeV interactions between a valence quark q and a valence antiquark $\bar{q}$. Creating a $\bar{t}t$ pair, each with (say) m_t = 150GeV, through the interaction process,

$$\bar{q} + q \to \bar{t} + t \tag{2.3}$$

absorbs half of the initial energy, not leaving much energy for the creation of more particles. The energies of the residual (qq) and $(\bar{q}\bar{q})$ quarks from the proton and antiproton, respectively, lead through hadronization to particles which mostly go down the beam pipe. Thus, after the production process,

$$\bar{p} + p \to \bar{t} + t + X, \tag{2.4}$$

the $\bar{t}$ and t quarks decay accordingto (1.4a,b), giving

$$\bar{p} + p \to \bar{b} + W^- + b + W^+ + X, \tag{2.5}$$

where X consists of (i) X_{inv} consisting of hadrons which go down the beam pipe, and (ii) X_{vis}, the other hadrons recorded by the detector. The $W^\pm$ bosons then decay, with lifetime about 3×10^{-25}s. Their simplest and most visible decay processes are

$$\text{(a) } W^+ \to \bar{\ell}^+ + \nu_\ell, \qquad \text{(b) } W^- \to \ell^- + \bar{\nu}_\ell, \tag{2.6}$$

for the three leptons ℓ = e, μ and τ.

The most striking events are those where the W^+ and W^- decays lead to two charged leptons from different families, without any X_{vis}, a typical state being

$$\bar{b} + b + \mu^\pm + e^\mp + (\nu_\mu + \bar{\nu}_e + X), \tag{2.7}$$

where the $\bar{b}$ and b quarks hadronize to give corresponding jets j($\bar{b}$) and j(b). There may also be some secondary jets, emitted from the initial quarks or by the quarks heading for the beam-pipe, or arising from the development of the $\bar{b}$ and b jets, but these will generally have relatively low energy, so that these ($\mu^\pm e^\mp \bar{b}b$) final states may be expected to have rather simple structure. Another attractive feature of these ($\mu^\pm e^\mp \bar{b}b$) events is that they may be expected to be rather free of background, since μ and e belong to different families. There is no particle Y known whose decay leads to the $\bar{\mu}e$ or $\mu\bar{e}$ configuration without corresponding neutrinos ν_e(or $\bar{\nu}_e$) and $\bar{\nu}_\mu$(or ν_μ). Indeed, lepton conservation is believed to hold separately for muonic and electronic numbers: strong limits have been placed on any violation of this conservation, through searches for reactions of the type $\mu^- p \to e^- p$ and for decays of the type $Y \to \mu^\pm e^\mp X$, especially for $Y^0 = K_L{}^0$ and $Y^+ = K^+$.

The simplest $\bar{p}p$ process leading to final states with ($\mu^\pm e^\mp$) is W-pair production

$$\bar{p} + p \to W^- + W^+ + n \text{ jets}, \tag{2.8}$$

followed by the decays (2.6). This process involves two weak vertices whereas the process (2.4) has none; provided that the energy available is reasonably large compared with $2(M_W + m_b)$, the process (2.8) with n = 2 may be expected to have a rate smaller by two factors of $(\alpha_{em}/2\sin^2\theta_W)$ than that for (2.5), if the n = 2 jets were b-jets. However, the process (2.8) will be in fact dominated by events with light quark jets in large numbers, and it is difficult to estimate how much background (2.8) will produce for the $\bar{t}t$ events (2.4) leading to final states (2.5). It is a question of what fraction of these non-$\bar{t}t$ events (2.8) having n = 2 light quarks and surviving the transvers momentum cuts (cf.Secs.3 and4) can be adequately fitted to the interpretation (2.7). The only further remark we can make is that, if a microvertex detector is used, so that it is possible to add the requirement that two of the jets in (2.8) be identified as b jets, the background rate from (2.8) would be of order $\{\alpha_{em}/(2\alpha_s\sin^2\theta_W)\}^2 \approx 10^{-2}$ times the signal for (2.7), following the top-pair production (2.4). In our further discussion here and in Sec.3, we shall neglect background to the process (2.5) arising from the direct production of W pairs. However, the validity of this neglect is under discussion at present.

The cross section for $\bar{t}t$ production in $\bar{p}p$ annihilations at the Tevatron energy of 900GeV through reaction (2.4) has been calculated by Eichten[15], as function of m_t. At Tevatron energies and for such large m_t values (≥120GeV) as we are led to consider here, $\bar{t}t$ production is dominantly due to the quark-antiquark process (2.3); production by gluon-gluon processes is small in comparison.

The $W^\pm$ bosons have hadronic modes, in addition to the leptonic modes (2.6). These are due to the W-coupling with the quark weak-isospin charged current. Since we are interested here only in total rate, we limit our discussion to the couplings

$$W^+ \rightarrow \quad (a)\ u + \bar{d}, \qquad\qquad (b)\ c + \bar{s}, \tag{2.9}$$

given by diagonal terms of the Cabibbo-Kobayashi-Maskawa matrix. Each of these couplings leads to two final quark jets. The third coupling is $W^+ \rightarrow (t + \bar{b})$, whose threshold lies far above M_W, so that it contributes only to virtual processes, such as the radiative corrections mentioned above.

Recognising colour, there are 6 couplings in all, each with the same strength as each $W^+ \rightarrow \ell^+\nu_\ell$. Empirically, the branching ratio for each $(\ell\nu_\ell)$ final state is 10.5(9)%, which agrees with the naive expectation of 1/9, from the channels just enumerated. If both W^+ and W^- in (2.5) decay hadronically, the final state will have 6 energetic jets, a configuration which it may be difficult to disentangle. If only one W decays hadronically, the final state will generally have one energetic lepton and four quark jets. The analysis of this final state is generally possible and we shall discuss this in Sec.3. The dilepton events have been described briefly just above. The decays (2.6) include the possibility of $\tau^\pm$ emission; this is far more difficult to deal with experimentally than is $e^\pm$ or $\mu^\pm$. For this reason, we shall not refer explicitly to the possibility of $\tau^\pm$ emission again. On the other hand, final electrons or muons can be easily recognised and distinguished. Among the dilepton final states, the (e^+e^-) and $(\mu^+\mu^-)$ cases need special attention and much caution, since there are many other mechanisms which can give rise to electron or muon pairs, one being the production and decay of a Z^0 boson, and which will still require special cuts for their exclusion. In fact, no candidate event of this type has yet been reported, although such events $(e^+e^-$ and $\mu^+\mu^-)$ must have the same rate as the $(\mu^\pm e^\mp)$ events.

2.3. Top mass estimates from production data.

In this Sec., we confine attention to the $(\mu^\pm e^\mp)$ events. The expected number $N_{\mu e}$ of these events is plotted as function of m_t in Fig.2 for an integrated luminosity IL = 30pb^{-1}, based on the estimates by Crane[16] who used Eichten's calculation of the cross section for top-antitop pair production [15] and took due account of the efficiency κ for the CDF detector as function of the event location and configuration. For mass m_t, we have

$$N_{\mu e}(m_t) = \kappa.\text{IL}.\sigma_{\mu e}(m_t), \tag{2.10}$$

where $\sigma_{\mu e})$ denotes the cross section for $\mu^\pm e^\mp$ events in $\bar{p}p$ collisions at 1800 GeV c.m. energy. The probability of producing n such events, given m_t, is

$$P(n|m_t) = (N_{\mu e}(m_t)^n/n!)\exp(-N_{\mu e}(m_t)). \tag{2.11}$$

Given IL, $\sigma_{\mu e}(m_t)$ and the observation of n events, the Bayesian probability distribution for m_t is then

$$P(m_t|n) = N_{\mu e}(m_t)^n.\exp(-N_{\mu e}(m_t))/\{\int dm.N_{\mu e}(m)^n).\exp(-N_{\mu e}(m)\}, \tag{2.12}$$

where the integral is taken over all possible values m allowed for the top quark t by all the conservation laws. This probability may be reduced to the form

$$P(m_{t|n}) = c(n).(\kappa.IL.\sigma_{\mu e}(m_t))^n.\exp(-\kappa.IL.\sigma_{\mu e}(m_t)), \tag{2.13}$$

One good ($\mu^- e^+$) candidate event has already been published by the CDF collaboration[2] and has been discussed[17] in some detail (cf.Sec.3 below). A second (μe) candidate was shown by the CDF collaboration in their report given at the November 1992 Chicago Meeting of the Division of Particles and Fields of the American Physical Society, although no measurement details were released. It was generally believed at that meeting that the D0 collaboration also had their first (μe) candidate[1]. Although the integrated luminosities IL are not known to us precisely, a value of about 20 pb^{-1} for CDF (including IL = 4.7 pb^{-1} from their 1989 paper) and 10 pb^{-1} for D0 would appear to be plausible estimates,up to the end of 1992, at least of the right order of magnitude. The mean detector efficiency can be deduced from a comparison of Crane's rates with Eichten's total $\bar{t}t$ production cross sections.

On the assumption that these three (μe) candidates do stem from top-antitop production, and that the integrated luminosity up to November 1992 was about 30 pb^{-1}, the probability distribution for m_t is shown on Fig.3. Its peak is at 120GeV, the one standard deviation limits being 109 and 135 GeV. Since the curve for $N_{\mu e}(m_t)$ shown in Fig.2 is a steeply falling function of m_t, the peak value thus determined for m_t is not strongly dependent of our estimate for IL, nor on the number of μe events. For (IL,n) = (40pb^{-1},3), the peak value is m_t = 127.5GeV; for (30pb^{-1},4), it is at 114GeV, and for (30pb^{-1},2), it is at 129GeV.

3. ANALYSIS OF DILEPTON EVENTS

3.1.Kinematics of top decay $t \to bW^+ \to b\bar{l}^+\nu_l$.

Consider first the kinematics of top decay to bW^+, followed by W^+ decay to $\bar{l}^+ \nu_l$, in any frame. Energy-momentum conservation gives

$$t = b + \bar{l} + \nu \tag{3.1}$$

where a denotes the energy-momentum four-vector of the particle a. Since $\bar{l}$ and ν_l are decay products of W, we have

$$M_W^2 = (\bar{l} + \nu_l)^2 = (t - b)^2, \tag{3.2}$$

and since the neutrino has zero mass,

$$0 = (t-b-\bar{l})^2 = (t-b)^2 - 2\bar{l}\cdot(t+b) + \bar{l}^2. \tag{3.3}$$

where $a \cdot b$ denotes the scalar product of the four-vectors a and b and $a^2 = a \cdot a$. For l = e or μ, we can neglect the lepton mass,i.e. $l^2 = m_l^2 = 0$. Using (3.2) in (3.3b), we obtain the result

$$\bar{l}\cdot t = \bar{l}\cdot b + M_W^2/2. \tag{3.4}$$

Evaluating the r.h.s. in the lab. frame and l.h.s. in the top rest-frame, we deduce that

$$E_{\bar{l}t} = (\bar{l}\cdot b + M_W^2/2)/m_t, \tag{3.5}$$

giving $E_{\bar{l}t}$, the lepton energy in the top rest-frame, in terms of lab. measurements for $\bar{l}$ and b.

It is useful to derive here an inequality for m_t. From (3.1) and (3.2), we have

---

[1] This event was shown at a Fermilab seminar in February 1993, but without full details.

$$m_t^2 = (b + \bar{l} + \nu_l)^2 \qquad (3.6a)$$
$$= m_b^2 + M_W^2 + 2b\cdot\bar{l} + 2b\cdot\nu_l. \qquad (3.6b)$$

Evaluate the invariant products $b\cdot l$ and $b\cdot\nu_l$ in the W rest-frame, where $E_\ell = \ell$ and E_ν are each $M_W/2$, and the momenta $\underline{l}$ and $\underline{\nu}_\ell$ are opposite. Denoting by θ the angle between $\underline{l}$ and $\underline{b}$ in this frame, their product takes the form

$$b\cdot\bar{l}b\cdot\nu_l = (E_b - b\cos\theta)(E_b + b\cos\theta)(M_W^2/2)^2 \qquad (3.7a)$$
$$= (m_b^2 + b^2\sin^2\theta)(M_W^2/4) \qquad (3.7b)$$
$$\geq m_b^2 M_W^2/4. \qquad (3.7c)$$

Using this inequality for $b\cdot\nu_l$ in eq.(3.6b), leads to the inequality

$$m_t^2 \geq m_b^2 + M_W^2 + 2b\cdot\bar{l} + m_b^2 M_W^2/2b\cdot\bar{l} \qquad (3.8a)$$

$$= (m_b^2 + 2b\cdot\bar{l})(M_W^2 + 2b\cdot\bar{l})/2b\cdot\bar{l}, \qquad (3.8b)$$

as stated in the Appendix of ref.[5].

3.2.A geometric construction.

Now consider the kinematics in the lab frame, starting with the 3-momenta $\underline{b}$ and $\underline{l}$, depicted in Fig.4. The top momentum $\underline{t}$ is constrained by two relations,

$$(\underline{t} - \underline{b})^2 = (E - E_b)^2 - M_W^2, \qquad (3.9)$$

from eq.(3.2) and

$$(\underline{t} - \underline{b} - \underline{l})^2 = (E - E_b - E_l)^2, \qquad (3.10)$$

from eq.(3.3a), where E has not been constrained to the value E_t because the value of m_t is still left open. For given E, eqs.(3.9) and (3.10) constrain $\underline{t}$ to lie on the intersection of two spheres, one centred on the point $\underline{b}$ and the other centred on $(\underline{b} + \underline{l})$. Their intersection is a circle on a plane perpendicular to the line BL joining these two centres and centred on this line. As E varies, the centre of this circle moves along the BL at a rate linear in E, and the square of its radius also increases linearly with E; in short, as E varies, this circle traces out a paraboloid with axis BL, as described in refs.[8,17]. All points $\underline{t}$ satisfying the constraints (3.2) and (3.3a) for given $\underline{b}$ and $\underline{\ell}$, for all possible m_t values, lie on this paraboloid. The points of interest to us are those for a definite m_t, still to be determined. These points lie on an ellipse formed by the intersection of the paraboloid by a plane whose normal lies in the plane OBL and makes an angle σ with the axis BL, where

$$\tan\sigma = b_1/(E_b - b_3), \qquad (3.11)$$

is independent of m_t, and b_1 and b_3 are defined on Fig.4. For a specified m_t, the possible vectors $\underline{t}$ are given by OT, as T moves around this ellipse, and the vector LT gives the corresponding neutrino momentum $\underline{\nu}$. It is apparent that the top quark energy $E_t = \sqrt{(m_t^2 + \underline{t}^2)}$ must lie between two limits $E_{min}(m_t) \leq E_t(m_t) \leq E_{max}(m_t)$.

As m_t increases, the plane PTQ moves upwards, with a constant normal since the angle σ (given by (3.11)) does not depend on m_t. The ellipse retains the same eccentricity but increases in linear dimension. As m_t decreases, the plane moves downward and the ellipse shrinks until for a limiting value $m_t = m_*$ the plane becomes tangent to the paraboloid and the ellipse is reduced to a point. For $m_t < m_*$, there is no solution for $\underline{t}$. This value m_* is the lowest limit for an m_t consistent with the momenta $(\underline{b},\underline{l})$; m_* is, of course, the value given by the square root of the expression (3.8b). The projection of the ellipse for m_t on a plane perpendicular to the vector

$\underline{l}$ is a circle, and the centre of the ellipse projects onto the centre of the circle. As m_t varies, the centre of the ellipse moves parallel to the vector BL $= \underline{l}$ and distant from it by amount $(M_W^2/2E_l)\tan\sigma$ in the plane of $\underline{b}$ and $\underline{l}$.

It is worth noting here that any configuration $(\underline{b},\underline{l})$ can be fitted by a sufficiently large m_t. However, a very large m_t means that LT is very large, but this is the neutrino momentum $\underline{\nu}_l$, of which we have no direct knowledge. It means also that OT must be correspondingly large, and this is the top momentum $\underline{t}$. These two very large momenta would be compensating each other, differing by the observed momentum $(\underline{l} + \underline{b})$, and this is not generally a plausible interpretation of the top decay event observed.

For top-antitop production, with lepton emission in both decays, two paraboloids are to be constructed, one from $(\underline{b},\underline{l})$ for t and one from $(\underline{\bar{b}},\underline{\ell})$ for $\bar{t}$, both in the same laboratory frame. One dilepton event has already been described in detail in the literature [17] from the CDF experiment in 1988 as a possible $t\bar{t}$ production event. We shall refer to it as CDF-1. Its measurements are given here in Table 3, the z-axis being along the beam. There is a third lepton in CDF-1, a μ^+ with quite high energy but with rather small transverse energy. It allows interpretation as a secondary muon, emitted from $\bar{B}$ meson decay in the $\bar{b}$-jet following the $\bar{t} \to \bar{b}W^-$ decay of the hypothetical antitop quark produced in this $\bar{p}p$ interaction. It has the right charge sign and a low transverse energy compatible with the sequence $\bar{b} \to \bar{c}W^+$, $W^+ \to \mu^+\nu_\mu$, and it travels in the same general direction as an energetic hadronic jet. These features strongly suggest that the μ^+ and the hadronic jet are to be taken together, as the components of a $\bar{b}$ jet. We expect that the other jet is a b quark jet but there is no evidence to demonstrate this.

3.3. Top mass estimation from dileptonic decays $\bar{t}t \to \mu^\pm e^\mp + 2$jets.

If we now <u>assume that CDF-1 represents top-antitop production</u>, with the assignments given in Table I, the two paraboloids then constructed are those shown in Fig.5, projected onto the (y,z) plane. The ellipses representing t decay are clearly seen; the $\bar{t}$-ellipses happen to be seen edge-on in this projection, since the momentum vectors $\underline{\bar{b}}$ and $\underline{\ell}$ almost lie in this plane, their components $\bar{b}_x$ and ℓ_x being small. These two sets of ellipses are shown for m_t ranging from 115 to 305 GeV, in 10 GeV intervals. The quark-antiquark interaction which gave rise to this event was strongly asymmetric, since the two paraboloids lie almost entirely within the same hemisphere.

It had been assumed generally, I think (but see ref.[19]), that it would not be possible to analyse such an event in terms of top-antitop pair creation since the final state would then include at least two energetic neutrinos, ν_e and $\bar{\nu}_\mu$. [CDF-1 actually has a third neutrino, of type ν_μ and with unknown energy, from the $\bar{b}$ decay mentioned above. It is assumed here that this ν_μ energy is small although its associated μ^+ has quite high energy. This assumption needs checking for internal consistency]. Of course, we do not know whether this $\bar{t}t$ assumption for event CDF-1 is correct, but if it is, then it is possible to deduce something probabilistic about m_t.

Table 3. Input data for two $\bar{t}t$ candidates, one real (CDF-1) and the other fictitious ($F_{\mu e}$), both of the form ($\mu^\pm e^\mp b\bar{b}$).

	Event CDF-1				Event $F_{\mu e}$ (fictitious)			
GeV/c	e^+	jet($\bar{b}$)	e^-	jet($\bar{b}$)*	e^-	jet($\bar{b}$)	μ^+	jet(b)
p_x	-21.2	18.7	0.6	-0.4	22.5	66.3	8.0	-47.7
p_y	23.6	-6.25	-43.7	16.8	-5.45	86.4	20.0	9.5
p_z	-28.55	25.25	16.8	-98.9	-21.5	-12.0	94.0	27.5
	*includes a low energy μ^+.							

There is one further input. In the simplest parton picture of the $\bar{p}p$ collision, [20], each of the partons q and $\bar{q}$ has no transverse component of momentum. It follows that the net transverse momentum of the final t and $\bar{t}$ quarks must be zero, so that we have the two further constraints:

$$\underline{\bar{t}}_T + \underline{t}_T = 0. \tag{3.12}$$

(3.12)where the suffix T denotes the component of the vector transverse to the incident proton beam. What is the consequence of this condition? We can see this most simply by rotating the $\bar{t}$ paraboloid by 180° about the z-axis. If we view the two resulting paraboloids along the z-axis, i.e. project them on to the x0y plane, the constraint equations (3.12) for an assumed m_t correspond to a crossing of the two ellipses for this m_t, since (3.12) then reads

$$-\bar{t}_T = t_T. \tag{3.13}$$

These crossings are shown for a number of m_t values on Fig.6, for the event CDF-1. For $m_t < 110.2$ GeV, the ellipses do not cross (for $m_t < 100$ GeV, they do not even exist) nor do they cross for $m_t > 410$ GeV (which is an unreasonably large m_t, as is explained below). For m_t between these limits, there are two or four crossing points, each giving a solution for $\underline{t}$ and $\underline{\bar{t}}$ (and hence for $\underline{\nu}_e$ and $\underline{\bar{\nu}}_\mu$ separately). The next step is to assign a probability to each configuration.

From the z-components of a solution for the momenta $(\underline{t},\underline{\bar{t}})$, the momenta xP and $\bar{x}$P of the parton and antiparton, where P denotes the total momentum of p and $\bar{p}$ in the Lab. frame, can be deduced, giving

$$x = \{E_t - E_{\bar{t}} + (t_L + \bar{t}_L)\}/2P, \tag{3.14a}$$

$$\bar{x} = \{E_t - E_{\bar{t}} - (t_L + \bar{t}_L)\}/2P, \tag{3.14b}$$

where the suffix L denotes the component of the vector along the incident proton beam. The structure functions for the proton and the antiproton are the same and are well-known [21]. For the proton there are three of these, F(x) for the quark partons, $\bar{F}(x)$ for the antiquark partons and $F_g(x)$ for the gluons, and they depend on (momentum transfer)^2, for which we have taken the value $(m_t)^2$. For the antiproton, the functions F(x) and $\bar{F}(x)$ are to be interchanged and x replaced by $\bar{x}$. For the values of $(\underline{t},\underline{\bar{t}})$ relevant to the Tevatron experiments, the relevant values x and $\bar{x}$ are quite large, being typically 0.1 or larger. For such values of x and $\bar{x}$, $\bar{F}$ is small and F_g is even smaller. We should add here a note that, some way below the upper limit allowed for m_t by the kinematics, the values of x and $\bar{x}$ required by eqs.(3.14) may exceed unity, as m_t increases; of course, the expressions (3.14) are then not physically meaningful solutions. Thus, the condition that x and $\bar{x}$ must be less than unity imposes an upper limit on the magnitude of m_t.

Thus the $\bar{t}t$ production rate at the Tevatron is dominated by the quark-antiquark collisions, which are proportional to $F(x)F(\bar{x})$. We have used for F set 2 of the Duke and Owens [22] structure functions for (u+d), neglecting $\bar{F}$ and F_g. Even F(x) falls to zero quite rapidly, like $(1-x)^3$, as x increases above 0.5. The rate is also proportional to the differential cross section for parton-antiparton→$t\bar{t}$, summed over all of the parton-antiparton collisions. Including for completeness the i = {1,2,3} = {($\bar{q}$q),(q$\bar{q}$),(gg)} initial states, we find for the rate factor,

$$R(x,\bar{x}) = \sum_i F_i(x)F_i(\bar{x}).d\sigma(\hat{s},\hat{t})_i/d\hat{t}, \tag{3.15}$$

where $F_1(x)$ now stands for F(x), $F_2(x)$ for $\bar{F}(x)$ and $F_3(x)$ for $F_g(x)$, and $\hat{s} = x\bar{x}s$, and $\hat{t} = m_t^2-(E_t - t_L)x\sqrt{s}$ are the differential cross section variables.

Finally, we have to consider the spectrum of the lepton in the top rest-frame (and of the antilepton in the antitop rest frame). With the Standard Model couplings $W\cdot(\bar{t}\gamma(1+\gamma_5)b)$ and $W\cdot(\bar{l}\gamma(1+\gamma_5)\nu_l)$, there is a large forward-backward asymmetry in the $W \to \bar{l}\nu_l$ decay relative to the b-momentum in the W rest-frame. Since there is a one-to-one relationship between this angle and the energy $E_{\bar{l}t}$, we can express this angular distribution as an energy distribution:

$$d\Gamma/dE_{\bar{l}t} = G_F{}^2M_W{}^3E_{\bar{l}t}(m_t{}^2- m_b{}^2-2m_tE_{\bar{l}t})/(4\pi^2m_t\Gamma_W). \qquad (3.16)$$

If we neglect $(m_b/m_t)\ll 1$, and normalize the distribution to unity, we have the probability distribution

$$P(E_{\bar{l}t}) = (24/m_t{}^2)E_{\bar{l}t}(1 - 2E_{\bar{l}t}/m_t)dE_{\bar{l}t} \qquad (3.17)$$

where $E_{\bar{l}t}$ runs from 0 to $m_t/2$ in this approximation.

Taking all these factors together, the total probability for reaching the observed configuration is proportional to

$$P(\underline{t},\underline{\bar{t}}|m_t) = \sum_j\sum_i F_j(x_i)F_j(\bar{x}_i).d\sigma(\hat{s}_i,\hat{t}_i)_j/d\hat{t}_i.P(E_{\bar{l}t}|m_t)P(E_{l\bar{t}}|m_t), \qquad (3.18)$$

where the sum i is over $(\bar{q}q)$,$(q\bar{q})$ and (gg), and the sum j is over all of the ellipse-crossing points for the m_t value under consideration. As discussed here and in ref.[18], $P(\underline{t},\underline{\bar{t}}|m_t)$ is a discontinuous function of m_t, the ellipses being lines of zero width which either cross or don't cross. In reality, these lines have finite width, partly because top quarks and W-bosons have finite widths, but more importantly because of statistical uncertainties in jet development and in measurements of the particles in the event, and because of a distribution of transverse momenta for the partons and the antipartons in the incident proton and antiproton. In order to estimate m_t from the data on CDF-1, assuming it to be due to top-antitop production and decay, we appeal to Bayes Theorem, which gives the probability distribution as:

$$P(m_t|\underline{t},\underline{\bar{t}}) = P(\underline{t},\underline{\bar{t}}|m_t)\Phi(m_t)/\{\int dm.P(\underline{t},\underline{\bar{t}}|m)\Phi(m)\}, \qquad (3.19)$$

where $\Phi(m)$ is the à priori probability that the top mass is m and the integral is taken over all possible values of m. With $\Phi(m) = 1$, i.e. for the CDF-1 event shape alone, the probability $P(m_t|\underline{t},\underline{\bar{t}})$ is plotted in Fig.7(a), as taken from ref.[18]. Its peak lies at about 122 GeV, and it is broad and flat.

3.4. More realistic top mass estimation.

Although the above analysis is geometrically simple and easy to comprehend, we have come to realise, from the analysis of the "ℓ+4jets" events in Sec.4 below, that it is really necessary to allow (i) for the transverse momenta of the initial parton-antiparton system, and also (ii) for the uncertainties of the b and $\bar{b}$ jet energies. In this Section, for simplicity, we shall consider only the former of these; the inclusion of the latter is discussed below, in Sec.4.2.

Consider first the t-ellipse for given $\underline{b}$ and $\underline{\bar{l}}$, for assumed top mass m. Divide its boundary by N points by dividing the ellipse of Fig.4 into equal azimuthal slices with respect to the axis $C_\bullet N$, where $C_\bullet$ denotes the centre of the ellipse and $C_\bullet N$ intersects BL at angle σ given by expression (3.11) and is normal to the plane of the ellipse. At each one of these N points (a set labelled by γ) there is a definite momentum $\underline{t}(\gamma)$. Next, consider the $\bar{t}$-ellipse for $\underline{\bar{b}}$ and $\underline{l}$, for the same value of m, following the same procedure for the N points (this set labelled by $\bar{\gamma}$), each point leading to a definite momentum $\underline{\bar{t}}(\bar{\gamma})$. We now soften the condition (3.12) by the introduction of a weighting factor T_ρ of finite range, thus obtaining:

$$0 \qquad P(\underline{t},\underline{\bar{t}}|m) = \sum_{\gamma\bar{\gamma}}\{P(\underline{t}_\gamma|m)\ T_\rho(\underline{t}_{\gamma T}+\underline{\bar{t}}_{\bar{\gamma}T})\ P(\underline{t}_{\bar{\gamma}}|m), \qquad (3.20)$$

where $T_\rho(\underline{x})$ is the function

$$T_\rho(\underline{r}) = (1/2\pi\rho^2)\exp[-(\underline{r})^2/(2\rho^2)], \tag{3.21}$$

which is a representation of the two-dimensional δ-function

$$T_\rho \longrightarrow \delta(\underline{r}) = \delta(\underline{t}_{\gamma T} + \underline{\bar{t}}_{\bar{\gamma}T}), \tag{3.22}$$

as $\rho \longrightarrow 0$. In practice, we will use the sum (3.20) over the set of points $(\gamma,\bar{\gamma})$, each having 180 points, so that the slices are of angle 2°. The contribution of each point to the sum is evaluated and summed. The largest individual contributions are naturally those from the vicinity of the crossing points of Fig.6. The results of this calculation are shown in Fig.7(a) for $\rho = 0.1m_t$ and $0.025m_t$. We see there that the forms obtained differ considerably from that obtained with the condition (3.12), and further, that there is no sign that the earlier result is recovered in the limit $\rho \to 0$. However, it is easy to understand the change in form. Consider the case of m = 125 GeV. We no longer have single large contributions coming only from the crossing points C and D, but find a large number of smaller contributions all the way from C to D; this situation persists as the mass falls below 125 GeV, even after there are no crossings corresponding to C and D. Since the ellipses become smaller as m falls, the finite range gains increasing relative importance and it is not surprising that the non-crossing contributions become dominant there, giving greater weight to low m values and thus distorting the m-distribution away from the form which we presented in ref.[18].

The dashed line on Fig.7(a) shows the m-probability of the single point having the greatest probability for the m considered. This point naturally lies close to the cross-over point for that m, if there is one, so that this entry must be closely related with the value obtained with the procedure based on (3.12). This curve, based on these entries of maxima, does have the slow fall-off for increasingly large m which we noted in ref.[18] and show here in Fig.7(a). The change introduced by the use of (3.20) arises from the fact that there are a large number of points $(\gamma,\bar{\gamma})$ which cannot contribute to the sum $P(\underline{t},\underline{\bar{t}}|m)$ when (3.12) is required but which dominate its sum for low m when (3.20) is used, so depressing the value of the most likely value for m. The probability distribution we now adopt is that shown in Fig.7(a) for $\rho = 0.1m_t$.

3.5. Compatibility and combining of top mass probability distributions.

The probability functions $P(m_t|N_{\mu e})$ and $P(m_t|\underline{t},\underline{\bar{t}})$ are independent and at present compatible. They can be combined to give an overall probability function from the data publically available today:

$$P(m|\text{data}) = P(m|N_{\mu e}=3).P(m|\underline{t},\underline{\bar{t}}:\text{CDF-1}), \tag{3.23}$$

shown on Fig.7(b). Its peak is at 121 GeV, the one-standard-deviation limits being 114.5 and 130 GeV. Of course, these remarks are very tentative since they are based on the assumption that the three $(\mu^\pm e^\mp)$ events are due to top production and decay; all three might be due to background.

At present, we do not know whether or not $m_t = 121(+9,-6.5)$ GeV is the top mass. However, the hope for the future is that a substantial number of $(\mu^\pm e^\mp)$ events will be observed at the Tevatron. An m_t distribution can be determined for each of them, as done for CDF-1 here. These distributions may turn out to vary randomly from event to event; if so, then these events could not have anything to do with top-antitop production and decay. More likely, a large fraction of them (and perhaps all) may peak at a definite m_t value and we shall be able to conclude that they form a well-defined group of events due most probably to $\bar{t}t$ production and decay. The total cross section for this group of events will provide a quantitative test on this interpretation for them.

In order to illustrate one difficulty in following this line of thought, we shall close this Section by discussing a fictitious event which we shall refer to as $F_{\mu e}$. Its input momenta are given in Table 3 and it has been subjected to the same analysis, with ρ=0.1m_t, as was applied to the real event CDF-1. Evaluating the limit (3.8b), the antitop momenta lead to $(m_t)_{min}$ = 95.3 GeV, while the top momenta lead to $(m_t)_{min}$ = 134.5 GeV. Since the latter is the stronger limit, we conclude that, if $F_{\mu e}$ were data from a real $\bar{t}t$ event, then the top mass could not lie lower than 134.5 GeV. The m_t distribution for $F_{\mu e}$ is shown on Fig.8. It peaks at 154 GeV, with one-standard-error limits at 144 and 168 GeV and it is almost incompatible with the m_t distribution given in Fig.7(a) for the real event CDF-1. We would have to conclude that either one (or both) of these events is not the result of top-antitop production and decay, or that there has been a large statistical fluctuation. In either case, the situation would be unsatisfactory and we would have to await further μe events. At the moment, as mentioned in Sec. 2(b), it is known that there do exist two further events of this kind and we all look forward to the release of the detailed data from them.

4. ANALYSIS OF "LEPTON AND FOUR JETS" EVENTS

4.1. Expected rate and nature of the events.

As mentioned following eq. (2.9), the hadronic decay modes $W^+ \to u\bar{d}$ and $c\bar{s}$ have a net rate about six times greater than either of the decay modes $W^+ \to e^+\nu_e$ or $\mu^+\nu_\mu$. From this remark it follows that $\bar{t}t$ decay leading to the final charged particles

$$(\ell^+(=e^+ \text{ or } \mu^+)b) \;+\; \bar{b}(\bar{u}d \text{ or } \bar{c}s), \tag{4.1a}$$

and

$$(\ell^-(=e^- \text{ or } \mu^-)\bar{b}) \;+\; b(u\bar{d} \text{ or } c\bar{s}) \tag{4.1b}$$

are more frequent than the $(\mu^\pm e^\mp)$ dilepton modes

$$\text{(a)}\quad (\mu^+b + (e^-\bar{b}), \qquad\qquad \text{(b)}\quad (\mu^-\bar{b}) + (e^+b), \tag{4.2}$$

by a factor of twelve. Here, we have confined attention to the diagonal elements of the Cabibbo-Kobayashi-Maskawa matrix and then only for λ = 0 (Wolfenstein's parametrization [28]), since this is a satisfactory approximation for the present. If the three $(\mu^\pm e^\mp)$ events mentioned in Sec. 2(b) were all due to top-antitop pairs, then we would expect about another 36 pairs, each giving one energetic lepton with high pT accompanied by one b-jet, one $\bar{b}$-jet and two other jets, as specified by the possibilities given in (4.1). Up to the present, it has not been possible to identify which are the b and $\bar{b}$ jets, except by the observation of a low-energy muon resulting from the secondary decay $b \to \ell^-\nu_\ell c$ or $\bar{b} \to \ell^+\nu_\ell\bar{c}$, as was the case in the CDF-1 event, where there was a low-energy μ^+ (see Table 3). The secondary lepton from b or $\bar{b}$ decay has branching ratio about 21%, so that one should appear in ≈ 42% of $\bar{t}t$ events; the appearance of two secondary leptons has a rate an order of magnitude lower. In the present run, CDF is equipped with a microvertex detector, which will greatly increase its efficiency of identification of the b and $\bar{b}$ jets, because of the relatively long lifetime of the b quarks ($\approx 10^{-12}$s), and so restrict the fitting of the final state to the $\bar{t}t$ hypothesis in a very helpful way.

Table 4. Input momenta for a fictitious $\bar{t}t$ event (F_ℓ) of the form ($\ell^-b\bar{b}q'\bar{q}$), where $(q',\bar{q})$ = $(s,\bar{c})$ or $(d,\bar{u})$.

GeV/c	e^-	jet(ϵ)	jet(ψ)	jet(ϕ)	jet(η)
p_x	11.0	42.0	0.7	0.0	7.0
p_y	-19.0	-23.5	23.5	-25.5	16.0
p_z	-7.0	25.0	-56.0	23.0	72.0

4.2. Their analysis and top mass probability distributions.

Here we discuss the analysis of the (ℓ+4j) events, when there is no information as to which are the b and $\bar{b}$ jets. What we say below will assume the lepton to be the ℓ^+; if it is ℓ^-, then the terms top and antitop are to be interchanged in what follows.

We start by labelling the jets according to their momenta, using the symbols $\chi = (\phi, \psi, \epsilon$ and $\eta)$, as illustrated in Fig.9 and in Table 4. The mass M_W = 80.2(3) GeV and the width Γ_W = 2.1(1) GeV are now rather well known for the W bosons; we shall therefore impose this value of M_W on the event, while neglecting the width Γ_W for convenience, since its effect are of secondary importance. Two of the jets are selected, say ϵ and η, and identified as 1 and 2, with energies E_1 and E_2 and separation angle θ_{12}. We then have

$$M_W^2 = m_1^2 + m_2^2 + 2(E_1E_2 - p_1p_2\cos\theta_{12}) \approx 2E_1E_2(1 - \cos\theta_{12}). \tag{4.3}$$

Each jet energy E_i has a probability distribution $Q_i(E_i)$, partly because jet development is a stochastic process and partly because of uncertainties in its determination from the observations. Eq.(4.3) fixes the product E_1E_2, so that the allowed values correspond to a hyperbola on the (E_1, E_2) plane. N_{12} points are chosen on that hyperbola and the probability assigned to each of these points is deduced from the integral

$$Q_{\epsilon\eta}(1,2) = \int Q_\epsilon(E_1)Q_\eta(E_2)\delta(E_1E_2 - M_W^2/\{2(1 - \cos\theta_{12})\})dE_1dE_2. \tag{4.4}$$

A third jet, say ϕ, is selected and labelled 3; N_3 points are chosen over the energy range $(E_3 \pm \chi\sigma_3)$, where σ_3 is the assigned uncertainty at the one-standard-error level and χ is chosen suitably (usually χ = 2). Thus, we have chosen $N_3.N_{12}$ energy values (a set labelled α); for each member of this set, an antitop momentum $\bar{\underline{t}}(\alpha)$ and mass $m(\alpha)$ are deduced, with an assigned probability $Q(\alpha)$, which is the product of Q_{12}, computed from (4.4), and $Q_\phi(E_3)$.

We now turn to ℓ and the fourth jet ψ, labelled 4. This jet is tacitly (and necessarily) assumed to be a b jet, although this identification is not used here. We take N_4 energy values (a set labelled β) over the range $E_4 \pm \chi\sigma_4$, each with probability $Q_\psi(\beta)$. For each point (α, β), we determine the ellipse for mass $m(\alpha)$, based on the vectors $\underline{l}$ and $\underline{b}_\psi$, and divide its boundary by N_ℓ points (a set labelled γ), as we did in Sec.3.4 by taking equal azimuthal slices of the ellipse of Fig.4 with respect to the axis C_eN, where C_e denotes the centre of the ellipse and C_eN intersects BL at the angle σ given by expression (3.11) and is normal to the plane of the ellipse. At each one of these N_ℓ points, there is a definite momentum $\underline{t}(\alpha,\beta,\gamma)$ for the top quark. Next, we ask for a match between $\underline{t}$ and $\bar{\underline{t}}$, i.e. that

$$\underline{t}(\alpha,\beta,\gamma)_T + \bar{\underline{t}}(\alpha)_T = 0 \tag{4.5}$$

if the incident parton-antiparton system is required to have no transverse momentum. However, in general, this equation will not be satisfied, except at isolated points. As above, in Sec. 3.4, we give up the equality (4.5), allow initial transverse momentum and introduce the probability distribution (3.21), where here $\underline{r}$ is given by the l.h.s. of eq.(4.5), adopting as before the value $\rho = 0.1m_t$. Just as for the dilepton modes in Sec.3, a lepton factor (3.17) is necessary, which we denote here by $P_\psi(E_{\bar{l}t}, 4)$. For the accompanying decay sequence $t \rightarrow bW^-$, $W^- \rightarrow \bar{c}s$ or $\bar{u}d$, the light quark s or d will have the same strong backward/forward asymmetry as holds for the lepton in $W^- \rightarrow \ell\bar{\nu}_\ell$, but since we have (in general) no means to distinguish the d jet from the $\bar{u}$ jet (or the s jet from the $\bar{c}$ jet), we have to average over the two final jets and this removes the backward/forward asymmetry. For simplicity, we have taken the decays $\bar{t} \rightarrow \bar{b}\bar{c}s$ and $\bar{b}\bar{u}d$ to be isotropic. For the production of $\bar{t}$ and t quarks with momenta $\bar{t}(\alpha)$ and $\underline{t}(\alpha,\beta,\gamma)$, through the process $\bar{q}q \rightarrow \bar{t}t$, there is also a rate factor $R(x(\alpha,\beta,\gamma),\bar{x}(\alpha))$ necessary, given by (3.15) with x and $\bar{x}$ defined by eqs.(3.14). Thus, for the labelling (3,4,1,2) of the four jets $(\phi,\psi,\epsilon,\eta)$ and for each of the number $(N_\ell.N_{12}.N_3.N_4)$ of points defined above, we have calculated a mass $m(\alpha)$, the two momenta $\bar{\underline{t}}(\alpha)$ and $\underline{t}(\alpha,\beta,\gamma)$ and an associated probability for reaching this event configuration, given by the product

$$P(\text{event}|m(\alpha);3,4,1,2) = Q_{\epsilon\eta}(1,2)Q_\phi(3)Q_\psi(4)T(\alpha,\beta,\gamma)P_\psi(E_{\bar{\ell}t},4)R(x(\alpha,\beta,\gamma),\bar{x}(\alpha)). \quad (4.7)$$

Finally, for this particular assignment of the labels (1,2,3,4), the net probability for this event configuration when $m(\alpha)$ lies in the interval $(m,m+\delta m)$ is obtained by summing these probabilities (4.7) for all of these $(N_\ell.N_{12}.N_3.N_4)$ points for which their $m(\alpha)$ lies in this interval.

When none of the four jets is identified, it is necessary to consider all possible identifications for them. The final expression for the net probability that the observed event configuration could occur for top quark mass m is given by the sum of (4.7) over all permutations of the labels (1,2,3,4):

$$P(\text{event}|m) = \Sigma_{\text{Perm.}}\ P(\text{event}|m;\ 1,2,3,4). \quad (4.8)$$

As in Sec. 3 above, we now use Bayes' Theorem to obtain the desired expression for the probability that the top quark mass is m, given the data on this event and an à priori probability $\Phi(m)$ from other information (such as that given by the $\bar{t}t$ production rate observed; cf.Sec.2.2 above), as follows

$$P(m|\text{this event + earlier info.}) = c.P(\text{event}|m)\Phi(m), \quad (4.9)$$

where c is a normalization constant, independent of m.

Of course, most of the identifications in the sum (4.8) are necessarily inappropriate but we have no means, in general, to know which of them correspond to the correct physical interpretation. It is our optimistic expectation that incorrect identifications will generally lead to negligible contributions to the net probability for top mass m, although it can certainly happen that several different interpretations have comparable probabilities. The merit of this method of analysing the data on an event is that it is systematic, so that no possible interpretations can be overlooked, and that it is quantitative, assigning a numerical value for the relative probability for different interpretations. It is not an elegant procedure but it does not involve multivariate searches for the maximum likelihood.

We may illustrate the procedure by discussing a fictitious event of the type (ℓ+4jets), with final state momenta as given in Table 4, which we shall refer to as F_ℓ. The probability distribution P(m) calculated for F_ℓ is shown on Fig.8. This distribution peaks at 135 GeV, with one-standard-deviation limits at ±6 GeV; it has quite a strong overlap with those for CDF-1 and it overlaps with that for $F_{\mu e}$ no more than that for CDF-1 does.

4.3. Simulated data and its analysis.

To test the method more extensively, we have considered two sets of simulated data:

(a) A Toy Model.

This is described in ref.[23]. Adopting a mass of 140 GeV, the vectors $\bar{\underline{t}}$ and $\underline{t}$ were chosen in a random way, and their decay configurations follow the leptonic sequence for t discussed in Sec.3 and an isotropic 3 jet sequence for $\bar{t}$. These events were processed by the procedure described above, including allowance for the Gaussian probabilities appropriate to the CDF determination of jet energies from the data on each jet and for all permutations of the assignments of (1,2,3,4) to the jets. In this analysis, the same cuts were made on the simulated data as are routinely applied by CDF to their real data.

These cuts are as follows:

(i) $p_T > 15$ GeV for each jet,
(ii) $E_{\ell T} > 20$ GeV for the lepton,
(iii) missing transverse $E_T > 35$ GeV,
(iv) pseudorapidity lower than 2.44 in magnitude for all four jets.

The P(m|event) distributions for three "toy events" chosen at random from 1000 simulated events are shown in Fig.10. In two of them, Figs.10(a) and 10(b), the peak of the distribution lies close to 140 GeV, the input value although with much spreading due to wrong jet combinations. However the third, Fig.10(c), shows that individual events can deviate widely from our simple expectations, for its peak is at 150 GeV while it has a marked dip in the vicinity of 140 GeV. In Fig.11, mean probability distributions for 1000 "toy model" events generated for $m_t = 140$ GeV, are shown for two cases. For Fig.11(a), the analysis made allowance for the Gaussian probabilities appropriate to the CDF determination of jet energies and excluded events which did not satisfy the CDF cuts as specified in the last paragraph. The reader should note that this is <u>not</u> a distribution of the mean peak mass, but represents rather the form of a "typical m distribution" under the circumstances specified. Its peak is at 137 GeV with a width (FWHM) of about 18 GeV. For Fig. 11(b) the jet energies input were smeared by making random changes in their magnitudes, with a standard deviation taken from the CDF work, intended to represent the effects of jet fragmentation and soft-gluon bremsstrahlung and of the detector efficiencies. This has depressed the mean probability distribution by a factor of about 2/3; the peak location is about 1 GeV lower and the FWHM increased to about 25 GeV.

(b) ISAJET [24]

This provides a more sophisticated simulation of $\bar{t}t$ production. A full simulation of the CDF detector effects has been carried out, using the appropriate jet-finding procedures and fragmentation codes. These detector characteristics were taken account of in the code QFL, as understood through the analysis of the 1988-89 CDF run [24]. The standard ISAJET program does not take account of t decay to bW^+, the process which dominates over t-jet fragmentation for the top mass values of interest to us here; however, ISAJET does have the option of adding one additional channel for transitions of the top quark, beyond those for the usual jet development, and this has been made use of here. This channel then becomes the dominant one, so that the later jet development is entirely that of the secondary b-quark from top decay and of the tertiary (c and $\bar{s}$ or u and $\bar{d}$) quark decays. The 500 $\bar{t}t$ production and decay events generated in this way for $m_t = 140$ GeV have been processed as discussed above for systems with the (ℓ^++4jets) final state structure and the resulting mean probability distribution has been plotted in Fig.12. This distribution does show a quite definite peak at 130 GeV, a shift of 10 GeV below the input mass, with a FWHM of about 25 GeV if the secondary peak at about 108 GeV is ignored. This relatively large shift is believed to be due the gluon radiation which is taken into account in ISAJET, but this is still under investigation. The secondary peak is believed to arise from the "wrong" combinations of jets but this is not yet certain; however, it appears unlikely to interfere with the use of the upper peak to determine the top quark mass, since it may be masked by the low-mass contributions to P(m|event) from background events. We note also that the mass distributions predicted for the ISAJET model spread to much higher top-mass values than do the simpler "toy model" calculations.

These model calculations demonstrate that when sets of input data obtained from "events" generated by an algorithm representing any of the processes

$$\bar{p} + p \rightarrow \left\{ \begin{matrix} e^{+} & + & \mu^{-} \\ e^{-} & + & \mu^{+} \end{matrix} \right\} + b + \bar{b} + \left\{ \begin{matrix} \nu_e & + & \bar{\nu}_\mu \\ \bar{\nu}_e & + & \nu_\mu \end{matrix} \right\} + X, \qquad (4.10)$$

or their charge conjugates, which have proceeded through the hadronic $\bar{t}t$ production process (2.4), are considered, the method of analysis we have proposed [23] does lead to m distributions which peak strongly at a mass close to the top mass used in generating these events, at least on the average. This holds true even when we do not know the identity of the quark or gluon which has generated each of the observed jets. However, we must next enquire whether a systematic peaking of the m distributions computed from real data at some particular mass value, say m*, necessarily implies the existence of a top quark with a mass close to m*. Are there other processes leading to the final states (ℓ+4jets), which we might describe as "background", for which our analysis leads to such peaks even though there may be no top quark at all or perhaps just no top quark in the mass range explored? We shall see that the answer is "yes" because the distribution P(m) from the background must vanish at the threshold (M_W+m_b) and approach zero again as m becomes sufficiently large since the parton parameters α and $\bar{\alpha}$ cannot exceed unity. Precisely where its maximum occurs depends on the nature of the background and on what cuts are made to reduce its contribution to P(m) in the mass region accessible for the top quark.

4.4. Simulation of non-top background and its analysis as if top-antitop.

It is apparent that final states of the form

$$\bar{p} + p \rightarrow \ \ell \ + 4\text{jets} + \nu_\ell \ + X \qquad (4.11)$$

can readily arise without top production, through single W production and decay, as follows:

$$\begin{aligned} p + p \dashrightarrow W \ &+ (4 \text{ hadronic jets} = \bar{q}\bar{q}qq \text{ or } \bar{q}qgg \text{ or } gggg). \qquad (4.12) \\ \rightarrow \ell + \nu_\ell. & \end{aligned}$$

Berends et al.[26] have provided full tree-level calculations of the cross sections for all of the Standard Model processes which involve one W-boson and n ≤ 6 quarks, antiquarks and gluons. The case n=6 includes all of these processes (4.11) and therefore provides a natural and plausible "ℓ + 4jets" background to the "ℓ +4jets" states (4.1) which result from the sequence of $\bar{t}t$ production (2.3), their decays (2.6) and finally leptonic decay for one W and decay to three quarks for the other, as outlined in (4.1). In the latter sequence, all four jets are quark or antiquark jets, one being a b jet and another a $\bar{b}$ jet. The calculation by Berends et al. for the processes (4.12) at the Tevatron energy predicts, after application of the cuts specified in Sec. 4.3 above, that the set of (one quark, one antiquark and two gluon)-jets occur more often (56%) than the set of (two quarks and two antiquarks)-jets (42%), while the set of four gluon jets is quite rare (2%). It predicts also that, with these cuts, b and $\bar{b}$ jets should occur among these quark jets only rarely (3%). Berends et al. also calculated the rate for the $\bar{t}t$ production process (2.4) in the same approximation, with the decay of the consequent W^+ or W^- boson leading to the final states (4.10). Applying the same cuts to these final states, they then concluded from their calculation that, for top masses between 100 and 135 GeV, the number of these events due to $\bar{t}t$ production and decay would exceed the number of background events.

The m distributions obtained by subjecting these QCD background events to our analysis procedure may be expected to differ appreciably from those for real $\bar{t}t$ production and decay. We have examined this question by using the VECBOS Monte Carlo program[27] which implements these Standard Model calculations of Berends et al. for "W+3jets" and "W+4jets" events, to obtain a large sample of calculated events which are input to the ISAJET+QFL evolution and development. The sample of "W+3jets" used corresponds to an IL = 112 pb^{-1} and that of "W+4jets" to IL = 128 pb^{-1}. Each event was put through our analysis procedure as described above, assuming it to result from $\bar{t}t$ production and decay, and gave an m probability distribution $P(m|\ell+4j)$. The mean m distribution, normalised to IL = 4 pb^{-1}, is shown by the shaded area on Fig.13. It does show a peak at m≈112 GeV, with an appreciable tail running up as high as 140 GeV. The open area of Fig.13 corresponds to the fictitious event F_ℓ shown on Fig.8, with its peak at 135 GeV; the two peaks are quite well separated, showing that a $\bar{t}t$ event of type F_ℓ with top mass 135 GeV could readily be separated from this background, given sufficient events. However, if the top mass were much lower than 120 GeV - the value our analysis of the data on CDF-1 might suggest - the "ℓ+4jets" m distribution would overlap so strongly with the background m distribution as to make their separation more problematical, requiring a very large body of data, at the least. In this situation, the most convincing evidence on m_t would be that obtainable from the ($\mu^\pm e^\mp$) dilepton events, where no source of serious background is known.

4.5. b- and/or $\bar{b}$-quark tagging.

This situation could be much improved if we had some knowledge of the identities of the four jets in these "ℓ+4jets" events, since some permutations of the labels (1,2,3,4) would then be excluded and the spread of the $P(m_t|\ell+4j)$ distribution due to incorrect identifications would be much reduced. For example, for our fictitious event $F_{\mu e}$, if jet(ϕ) had an accompanying low-energy lepton ℓ^+ in roughly the same direction, this would suggest that jet(ϕ) in Table 4 is the $\bar{b}$ jet, since the two leptons e^- and ℓ^+ would then stem naturally from the $\bar{t}$ decay sequence $\bar{t}\text{-->}\bar{b}W^-$, followed by $\bar{b}\text{-->}\bar{c}\ell^+\nu_\ell$ and $W^-\text{-->}e^-\bar{\nu}_e$. In this case, jet($\phi$) would be assigned the label 4 and the only permutations which need be made are those within (1,2,3). On the other hand, if the secondary lepton were an ℓ^-, then jet(ϕ) would be the b jet and should be assigned the label 3 in the above discussion; only the permutations within (1,2,4) would be necessary in the sum for the net probability. In Sec.4.1, we noted that the emission of a lepton from either the b or the $\bar{b}$ quark will occur in about 42% of the $\bar{t}t$ events (2.5), so that observation of one secondary lepton should be a common occurrence. Sometimes this lepton may escape detection because of its relatively low energy, so that, in practice, its rate may not be as high as this remark would suggest, but whenever it can be observed and measured, it can be made use of.

Now that microvertex detectors are being used by CDF, it will often be possible to identify the b and $\bar{b}$ jets from the finite visible (or inferred) path up to the decay of the $B^\pm$ and $B^0(\bar{B}^0)$ mesons to which they give rise. With a magnetic field present, measurement of the charge signs of the B decay products will enable B^+ and B^- to be separated. The existence of $(B_s^0,\bar{B}_s^0)$ mixing, with quite a high rate, will complicate their assignment to b or $\bar{b}$ jet. However, the formalism to take this mixing into account is well-known and without free parameters (apart from those concerning CP violation, which makes no appreciable contribution to the hadronic phenomena under discussion here), so that it can be built into the calculation of $P(m|\ell+4j)$ from the data. Observations with the microvertex detectors should both support and supplement the observations of secondary leptons, but even together they will not make the identification of the b jet or $\bar{b}$ jet possible in all events. Nevertheless, such "b-tagging", whether from secondary leptons or from determining the flight time from source to decay for the b quark, will prove to be of great value both for reducing the number of misinterpreted background events and for identifying some of the quark and antiquark jets, thus reducing the number of irrelevant jet assignments included in the determination of $P(m|\ell+4j)$ from the data.

5. DISCUSSION

It is clear that we can do little more at present beyond discussing possible procedures for the analysis of top-antitop candidate events. There is little data and it suffers serious uncertainties, so that we are not able even to test our theoretical assumptions yet. Nevertheless, it is natural to assume that the top quark exists with some mass value yet to be determined, and that it does decay dominantly through t-->bW+, as the Standard Model indicates for the mass values we must consider today. We believe that we should launch into the discussion of all candidate events. Even if it is not possible to prove that individual events are necessarily top-antitop, we may still find that there is a substantial fraction of the ($\mu^{\pm}e^{\mp}$2jets) events give P(m|event) distributions which are compatible and can be combined, in a first step to the determination of the top mass. We have illustrated this here by combining the P(m|CDF-1) distribution with the P(m|$N_{\mu e}$) determined from the rate of observed top-antitop pair production at the Tevatron; they are compatible and combining them gives an improved estimate of mt.

We have noted above that "ℓ+4jets" events from the decay of a $\bar{t}t$ pair are expected to occur twelve times as often as "$\mu^{\pm}e^{\mp}$2jets". If the three events of the latter type are really due to $\bar{t}t$, there should also be, in the recent data, roughly 12x3 = 36 events of the former type. At present, there do not appear to be even as many candidate events as this remark suggests, whereas the majority of such candidate events are not compatible with top mass as high as the analysis of CDF-1 and the ($\mu^{\pm}e^{\mp}$) event rate would suggest. This channel calls for open discussion of all "ℓ+4jets" events. It could prove to be a very fruitful line of investigation, where the advantage of much higher statistics has to be balanced against the necessity for discriminating against background events. It is worth remarking here that candidate events are a most powerful stimulus to a theoretician's thinking; it is not good for the progress of particle physics that experimenters should hold the details of new events unnecessarily long.

Berends et al. have provided an excellent QCD model for non-top production of "ℓ+4jets" events, whose implications must continue to be studied further. Nevertheless, this is only a tree-level model and the proposal and study of other, no doubt more elaborate, models of background will merit much more attention. However, these models will not come forth in the absence of a knowledge of the nature of the events which are actually occurring. For example, there are certainly events which have more than the minimal number of jets, such as "ℓ+njets" coming from $\bar{t}t$ production as well as from its background of W-production and decay, where n≥5. We shall need quantitative estimates of the rate of these more complex final states, which compete with the simpler states having the minimal number of jets and which so reduce the rate of the latter.

The observation of the rate $N_{\mu e}$ of the production of the states ($\mu^{\pm}e^{\mp}$) for high energy leptons is a direct measure of $\bar{t}t$ production. The competing source of ($\mu^{\pm}e^{\mp}$) is W^+W^- pair production, as pointed out in Sec.2.2, which is anticipated to have a lower rate. Both of these processes,($\bar{t}t$+jets) and (W^+W^- +jets) are also calculable at tree level. The former modifies the cross section for ($\mu^{\pm}e^{\mp}$), while the latter provides background. The determination of m_t from the rate of ($\mu^{\pm}e^{\mp}$) events also needs attention to theory beyond tree level. The simplest calculations indicate that the $N_{\mu e}$ rate at energy $2E_p$ falls rapidly with increasing m_t, at the Tevatron energy, and it is reasonable to expect that the corrections just mentioned will not change this. The conclusion is that the value of m_t obtained from $N_{\mu e}$ should be relatively stable with respect to experimental errors in $N_{\mu e}$, such as the inclusion of uncertain background events, as we noted in Sec.2.2.

6. CONCLUSION

No firm conclusion can be expected from this work so far, except that we should be active in analysing all of the candidate events as they emerge, and that the "$\mu^{\pm}e^{\mp}$2jets" events have a special value since they suffer little background from non-top events. It will be possible to combine the P(m_t)

probabilities from different events and event types as the data improves. One advantage of the relatively low energy of the Tevatron is that two $\bar{t}t$ pairs will be produced in the same event only very rarely, so that this source of confusion will be essentially absent. Of course, a somewhat higher energy would have the advantage of increasing single $\bar{t}t$ production beyond the low rate possible from the present Tevatron. The proposed increase of its c.m. energy from 1800 to 2000 GeV this Summer, with other upgrades, will raise the rate by a factor of two; clearly an optimum energy would lie still somewhat higher than this. We shall not try here to estimate what would be the optimum energy, by balancing a high rate of single $\bar{t}t$ pairs against a higher rate of multiple $\bar{t}t$ pairs, since it would be a rather academic exercise at the moment. However, it is a question worthy of serious consideration to give a well-judged answer.

Finally, we must emphasize that no top candidate yet reported has been demonstrated to represent $\bar{t}t$ production and decay, and that it may still be some considerable time before we reach that stage of certainty. However, we understand that the present plans at Fermilab are to run on until the end of May 1993, by which time the integrated luminosity IL should reach about $50pb^{-1}$. The upgrade to 2000 GeV will take place in Summer 1993, and Tevatron running should begin again in the following October and run steadily all through 1994. Thus, it is hoped that the net IL reached by the end of 1994 may be about $200pb^{-1}$, a very substantial improvement over the IL up to the end of 1992, with which this paper has been concerned.

7.ACKNOWLEDGEMENTS

The first author (R.H.D.) thanks Prof.A.Zichichi for the opportunity to speak in this Summer School at ERICE, the site of so many important physics meetings, and Profs.D.Sherrington and G.G.Ross for the continuing hospitality of the Theoretical Physics Department of the University of Oxford, and the second author (G.R.G.) wishes to acknowledge the U.S.Department of Energy for partial support during this work. We have appreciated discussions with Dr.K. Sliwa and the help of Drs.J.Benlloch, M.Timko and N.Wainer of C.D.F. with various aspects of the simulations described briefly here.

REFERENCES

1. G.L.Kane, Top Quark Physics, in Proc. Workshop on High Energy Phenomenology at Mexico City,1-10 July,1991. Also issued as Univ. of Michigan preprint UM-TH-91-32 (December,1991).
2. CDF Collaboration,F.Abbe et al.,Phys.Rev.Lett.68(1992);ibid,Phys.Rev.D 45 (1992)3921.
3. C.A.Nelson, Phys.Rev.D 41(1990)2805.
4. G.L.Kane,G.A.Ladinsky and C.-P. Yuan,Phys.Rev.D 45(1992)124.
5. R.H.Dalitz and Gary R.Goldstein,Phys.Rev.D 45(1992)1531.
6. I.Bigi,Phys.Lett.B 175(1986)233;I.Bigi et al.,ibid,181(1986)157.
7. F.J.Gilman and R.Kauffman,Phys.Rev.D 37(1988)2676.
8. R.H.Dalitz,Gary R.Goldstein and R.Marshall,Z.Phys.C 42(1988)441.
9. M.Veltman,Nucl.Phys.B 123(1977)89.
10. J.Ellis and G.L.Fogli,Phys.Lett.B 232(1989);ibid, 249(1990)543;J.Ellis, G.L.Fogli and E.Lisi,Phys.Lett.B 274(1992)456.
11. F.del Aguila,M.Martinez and M.Quiros,preprint CERN-TH.6389/92.
12. U.Amaldi,W.de Boer and H.Furstenau,Phys.Lett.B 260(1991)447.
13. U.Amaldi et al.,Phys.Rev.D 36(1987)1385.
14. J.Ellis,G.L.Fogli and E.Lisi,Phys.Lett.B 287(1992)335;ibid, 292(1992)427.
15. E.Eichten,in Physics at Fermilab in the 1990's. World Scientific, Singapore(1990)p.56.
16. D.Crane, in Physics in Collision XI(ed.J.-M.Brom,D.Huss and M.-E.Michalon, Editions Frontieres,1991)p.145.
17. CDF Collaboration, F.Abe et al.Phys.Rev.Lett. 64(1990)147;K.Sliwa, in

Proc.XXVth Intl.Symp.on Heavy Flavors Physics,Orsay,France,1991(eds.M. Davier and G.Wormser,Editions Frontieres,Gif-sur-Yvette,1992)p.567.
18. R.H.Dalitz and Gary R.Goldstein,Phys.Lett.B 287(1992)225.
19. K.Kondo,J.Phys.Soc.Japan 57(1988)4126;ibid,60(1991)836.
20. R.P.Feynman,Phys.Rev.Lett.23(1969)1415.
21. R.G.Roberts, The Structure of the proton.(C.U.P.,Cambridge, England, 1990).
22. D.W.Duke and J.F.Owens,Phys.Rev.D 30(1984)49.
23. Gary R.Goldstein,K.Sliwa and R.H.Dalitz,Phys.Rev.D 47(1993)967.
24. F.Paige and S.Protopopescu,in Physics of the Superconducting Super Collider,Snowmass,1987,Proc.Summer Study,Snowmass,Colorado,1986(ed.R. Donaldson and J.Marx,Division of Particles and Fields of the American Physical Society, New York,1986)p.320.
25. C.Newman-Holmes and J.Freeman, in Proc.Workshop on Detector Simulation for the SSC,Argonne Natl.Lab.,1988)pp.190 and 285. Also issued as Rept.No. ANL-HEP-CP-88-51.
26. F.A.Berends,H.Kuijf,B.Tausk and W.T.Giele,Nucl.Phys.B 357(1991)32.
27. We are indebted to Jose Benlloch for generating the VECBOS samples used. The cuts imposed at the generator level differed from the cuts used for the Toy Model and the ISAJET calculations given here in Sec.4(a). Jets were defined by the condition $R = \sqrt{\{(\delta\phi)^2 + (\delta\eta)^2\}} > 0.6$, where ϕ denotes the azimuth angle about the direction of the jet and η denotes the pseudo-rapidity, with $p_T > 10$ GeV and $|\eta| < 2.5$; the corresponding lepton cuts were $p_{\ell T} > 10$ GeV and $|\eta| < 1.2$.
28. L.Wolfenstein, Phys.Rev.Lett.51(1983)1945.
29.

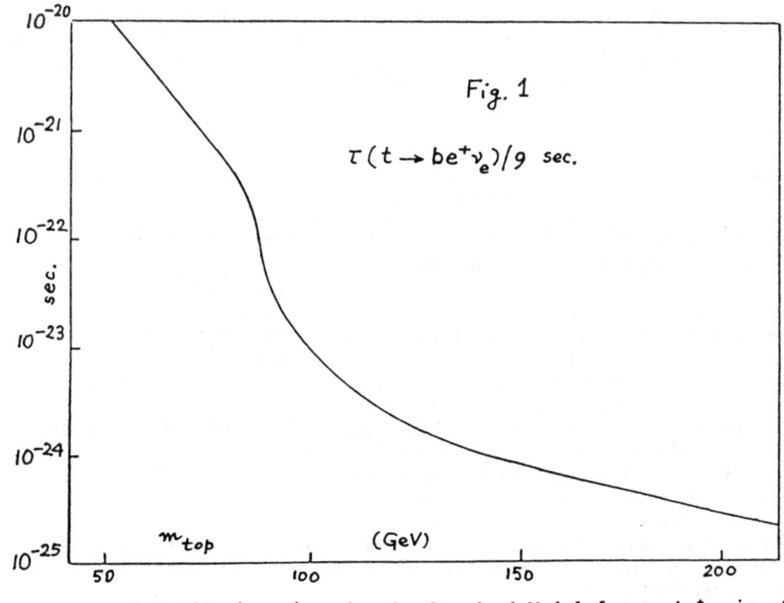

1. The partial lifetime given by the Standard Model for $t \to be^+\nu_e$ is plotted vs. top mass m_t. The factor 9 ensures that for $m_t > 100$ GeV the quantity plotted is the total top quark lifetime.

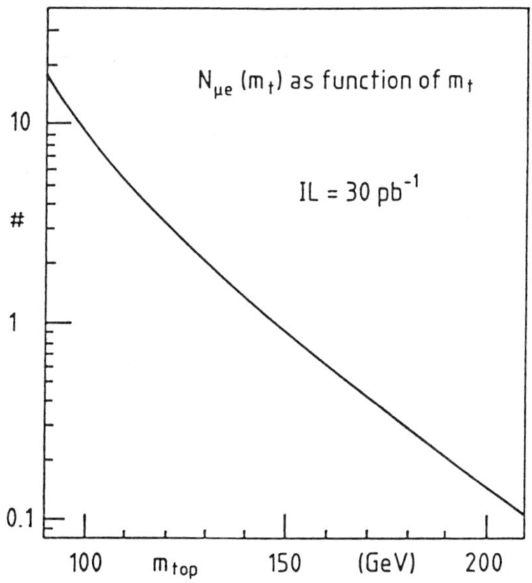

2. The mean number of ($\mu^{\pm}e^{\mp}$) events expected for integrated luminosity 30 pb^{-1} for $\bar{p}p$ collisions at total energy 1800 GeV, as calculated by Eichten[15] and Crane[16].

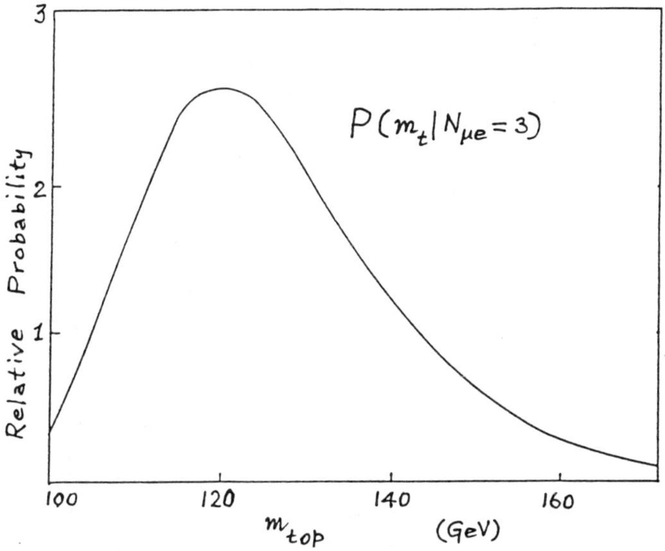

3. Shows the probability distribution for m_t, given that there are 3 $\mu^{\pm}e^{\mp}$ events observed for integrated luminosity 30 pb^{-1}.

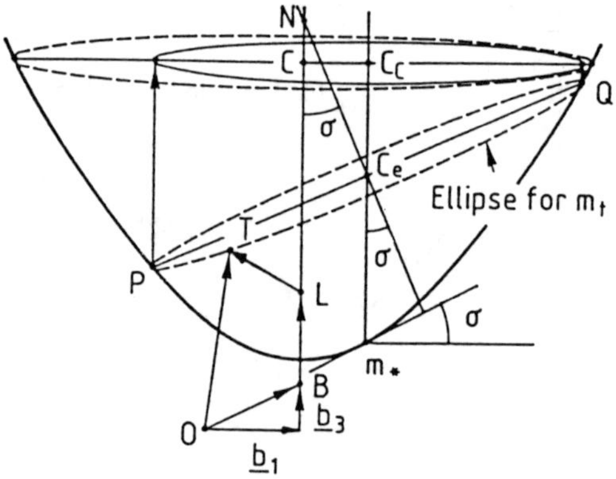

Fig. 4

4. Shows the ellipse PQ comprising all configurations for the vectors $\underline{t}$, $\underline{b}$ and $\underline{\ell}$ when the top mass is m_t.

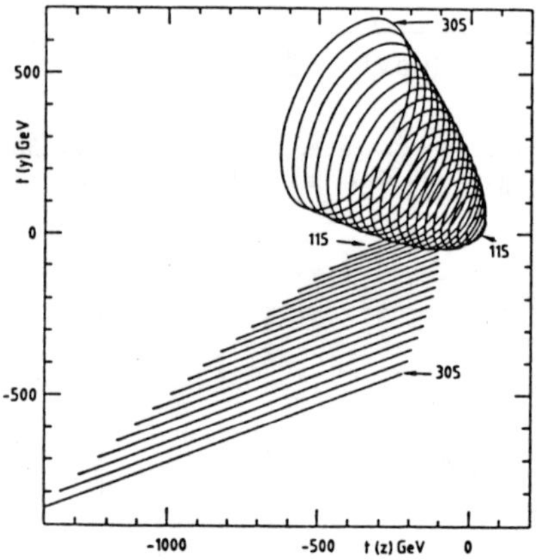

5. The t and $\bar{t}$ ellipses for the event CDF-1, as function of m_t, specified at 10 GeV intervals.

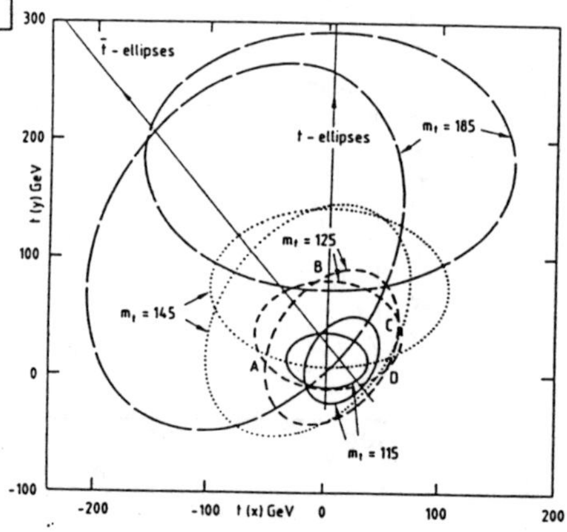

6. Projections of the t and $\bar{t}$ ellipses for the CDF-1 event onto the plane perpendicular to the beam direction for a number of m_t values.

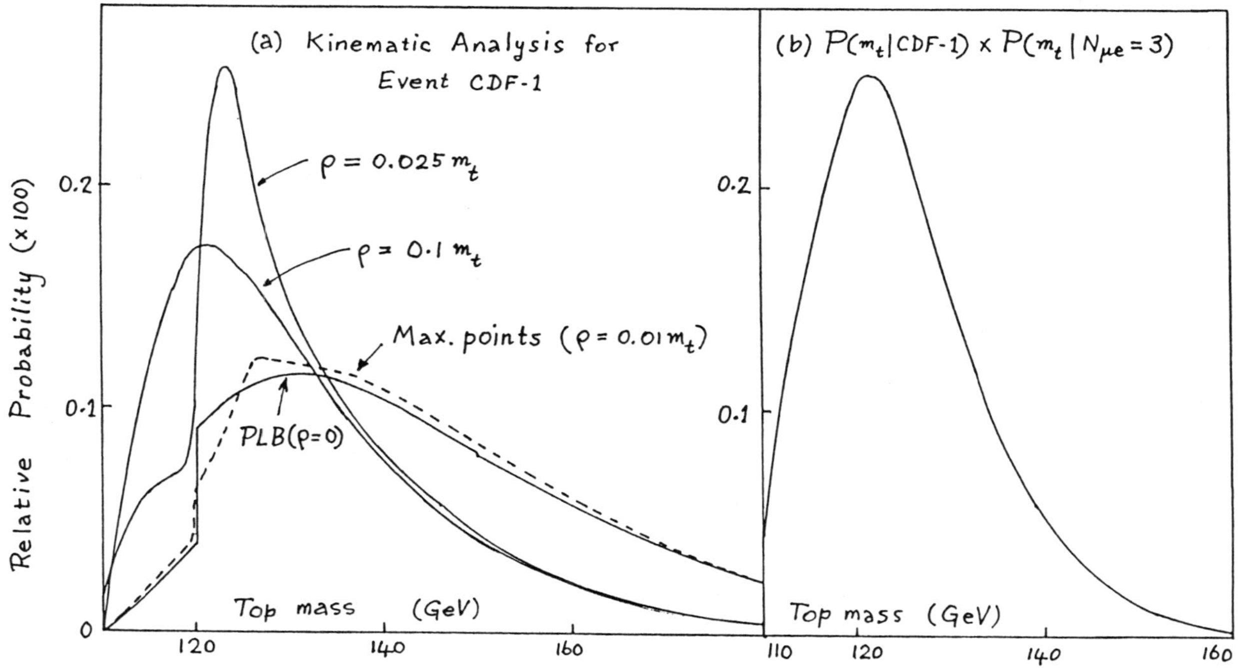

FIG. 7. Shows (a) $P(m_t|\text{CDF-1})$ for zero initial transverse momentum and for Gaussian initial transverse momentum with s.d. $\rho = 0.025m_t$ and $0.1m_t$; for the dashed curve, see text. (b) $P(m_t|N_{\mu e}=3 \text{ and CDF-1})$, for $\rho = 0.1m_t$.

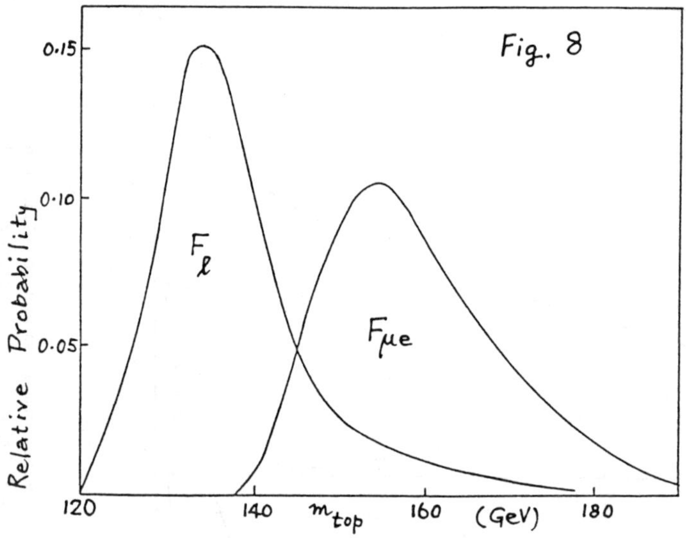

8. Shows $P(m_t|F)$ for the fictitious events $F_{\mu e}$ and F_ℓ, described in the text.

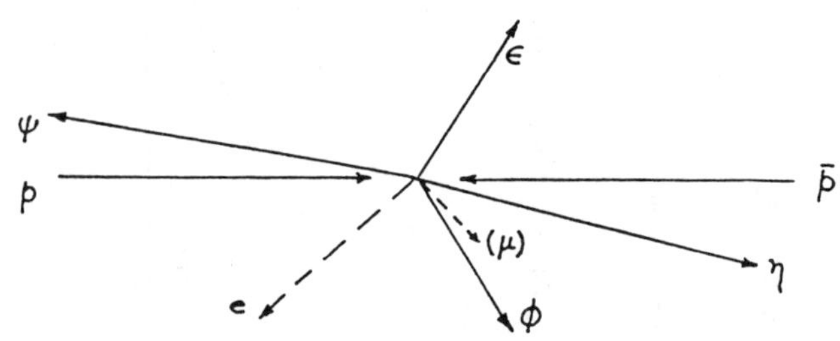

9. The leptons and jets in a typical "ℓ+4jets" configuration (actually for the fictitious event $F\ell$), following $\bar{t}t$ production and decay. A low-energy secondary muon, from b or $\bar{b}$ decay, is added, as indicated by a short dashed line.

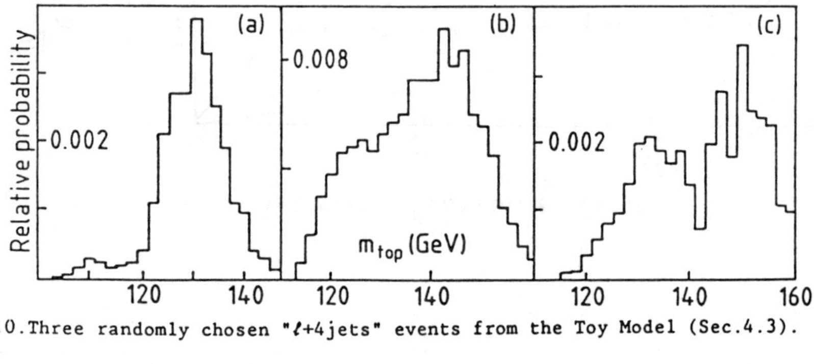

10. Three randomly chosen "ℓ+4jets" events from the Toy Model (Sec.4.3).

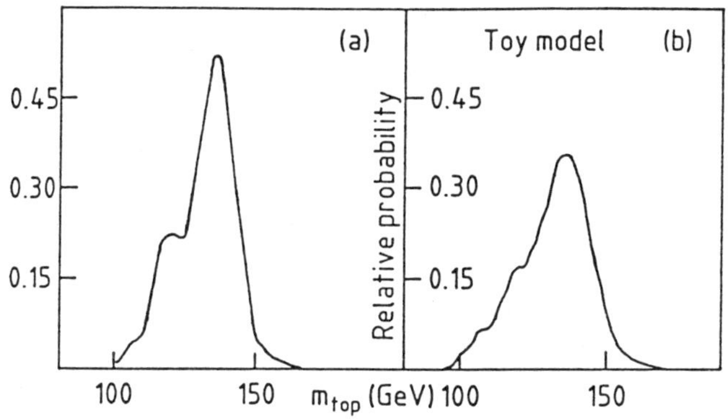

11. The mean probability distribution obtained (a) from 1000 Toy Model events and (b) from the same events after "smearing" (see text above, also ref.[23]) to simulate real events,

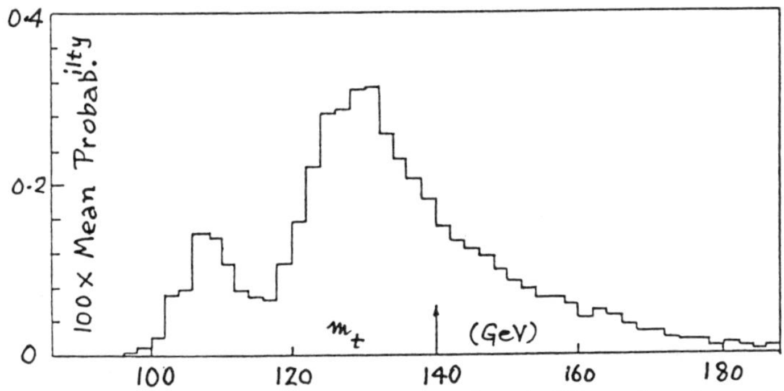

12. The mean m_t distribution from our analysis of 500 production and decay events generated by the ISAJET+QFL programs, taking into account fully the CDF detector characteristics.

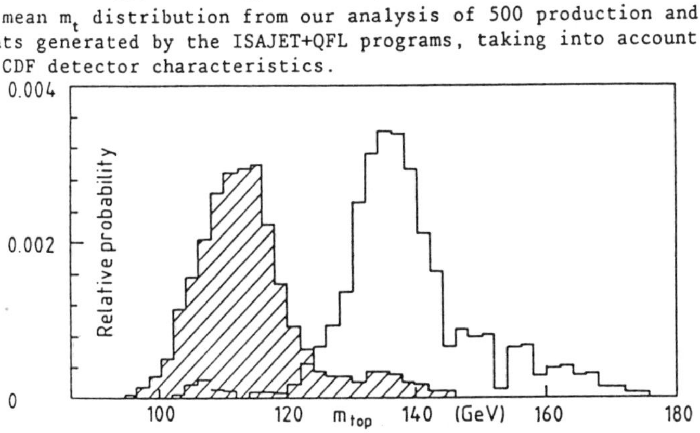

13. The mean m_t distribution from our analysis of Standard Model "ℓ+4jets" events generated by VECBOS simulating the tree-level calculations of backgrounds events by Berends et al.[26]; cuts as in Sec.4.2. The sample was about 60 times that appropriate for the 1988-89 CDF run.

CHAIRMAN: R.H. Dalitz

Scientific Secretaries: N. Moulai, R. Sehgal, N. Yunus

DISCUSSION

– *Liu:*

In your last transparency you showed a histogram which showed top quark of mass around 130 GeV. Where did you get that data?

– *Dalitz:*

No, those data were not from real events. They are all from simulated events. The only experimental data available to date are from the μ^+e^- event, which was reported in the literature two years ago.

– *Malik:*

1) What is the probability of having a toponium ($t\bar{t}$) resonance? Is it possible to study it on LEP II?

2) What does the background in your plots on the mass of top (fig. 1) correspond to?

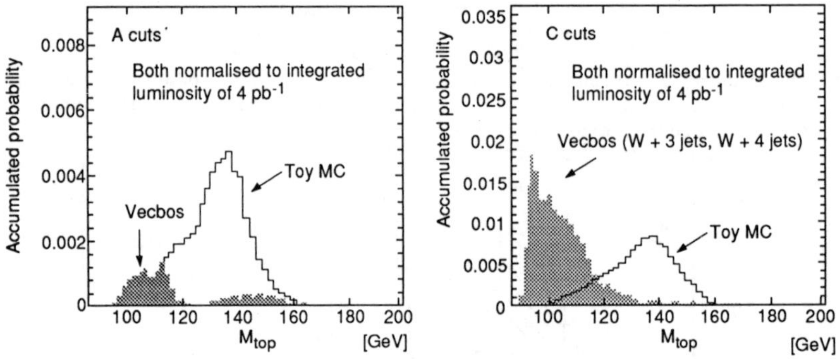

Fig. 1. See text of Lecture above

– *Dalitz:*

1) The production of toponium in antiproton (or proton) interactions on protons will have a very very low cross-section. For ($\bar{c}c$), charmonium was readily found in proton-nucleus collisions. The ($\bar{b}b$) bound system, which we might call

"bottonium", has been detected in proton-nucleus collisions but it is clear that the cross-section for production of $(\overline{Q}Q)$ is a rapidly falling function of m_Q - it appears most unlikely that toponium could be produced and detected in these collisions. A further difficulty is that the $(\bar{t}t)$ system will have only fleeting existence. Its decay rate is about $2 \times \tau_t$, where τ_t is the lifetime for a top quark. This means that its lifetime width will be of the order of 1 GeV. Toponium will not be a narrow state, not like charmonium!

Toponium may be produced in e^+e^- annihilations. Probably the ground states $^3S_1$ and $^1S_0$ for $(\bar{t}t)$ have almost complete overlap in energy but could be separated from the excited states. However, it will probably be impossible to separate any of the excited states of $(\bar{t}t)$ - they will all merge together to give a production cross-section which will be a smooth function (without peaks) of the energy of excitation. Even this process would require e^+ and e^- beams each with energy greater than $m_t c^2$. LEP II will not reach such a high energy.

2) The background consists of states of the form $(W^\pm + 4 \text{ jets})$ from the process $q + \bar{q} \rightarrow W^\pm + 4$ jets, with $W^\pm \rightarrow e^\pm$ or $\mu^\pm$. They are calculated using the QCD results of Berends et al. (Nucl. Phys. **B357**, (1991) 32) for all of the (tree level) graphs of the form Vacuum $\rightarrow W + 6$ partons, with electroweak interactions and QCD. We identify two of these partons as the incident parton in the proton and the antiparton in the incident antiproton. The four final partons may be quarks, antiquarks or gluons (but not top quarks since this is non-top background). The final states found by applying ISAJET to these 4 final partons to produce 4 final jets are then analysed (after making cuts on the momenta of the final particles) by our method to give a probability distribution vs. the parameter m_t, the top mass. Our purpose is to learn what cuts applied to this simulated data can remove configurations which lead to non-top events with apparent m_t values exceeding about 120 GeV (assuming that the top quark mass has a value appreciably higher then 120 GeV), which might be confused with events which come from top-antiquark pair production and decay. The real background might well be more complicated than that calculated by Berends et al., which represents only the simplest calculable model for the background.

– *Sopczak:*

Are detector effects, like limited acceptance ranges and detector inefficiencies included in the simulations you presented?

– *Dalitz:*

Yes, of course, all these effects have been included in our analysis of the simulated events. The acceptance and efficiencies are exactly those used by the CDF collaboration in their analysis of real events of the class "lepton + 4 jets".

– *Vassilevskaya:*

Recently many authors have given sufficiently definite values for the t-quark mass from different effects in the framework of the standard theory (decays of Z-boson, rare decays of t-mesons and so on). May this preliminary value be used to simplify your procedure?

– *Dalitz:*

The values that have been quoted range from 100 to 200 GeV. To produce a top quark heavier than 200 GeV in the proton-antiproton experiment at TEVATRON would be very difficult. A report from CDF to the "Physics in Collisions" conference of 1991 showed that if the top quark mass were 90 GeV, then they should be seeing about five "$t\bar{t} \to e\bar{e}, t\bar{t} \to e\bar{\mu}$ or $\bar{e}\mu$, and $t\bar{t} \to \mu\bar{\mu}$" events. They have not seen any $t\bar{t} \to e\bar{e}$ and $t\bar{t} \to \mu\bar{\mu}$ events and they have only one candidate $t\bar{t} \to \bar{\mu}e$ event. They argued that within 95% confidence level, this one event was not different from five. However, if one takes a top quark mass of 130 GeV, then the predicted number of such lepton-antilepton events is 0.6 and they have just one candidate, which is at least consistent.

The short answer to your question is that it does not really help, for the events should be analysed allowing all values for m_t between 90 and (say) 200 GeV. The reliability of the estimates obtained from calculating radiative corrections to standard model predictions is not very great and we must not assume them to be firm limits, as is apparent from the fact that different limits are determined on the basis of radiative corrections, from different pieces of Electroweak data.

– *Baillie:*

Assuming the top quark is discovered in the present CDF run, can you give an indication of the errors in m_t your two different methods would give?

– *Dalitz:*

The events which could be most reliably identified as top-antitop production and decay would be those where the top-antitop particles decay to give the observable particles, $e + \mu$ + two b-jets. The other mode in which (say) the top decays to three quarks and the antitop decays to an antibottom and a W boson is much more difficult to analyse uniquely. Although the events in this mode should be six times more frequent than in the $e\mu$ channel the fluctuations which can occur may be much more complicated to deal with.

It is difficult to estimate the accuracy these methods can achieve, at the present stage.

– *Hanany:*

1) What is the meaning of the intersections of the ellipses that you show in fig. 2 for different top masses?

2) What is the difference between the top and antitop ($t\bar{t}$)?

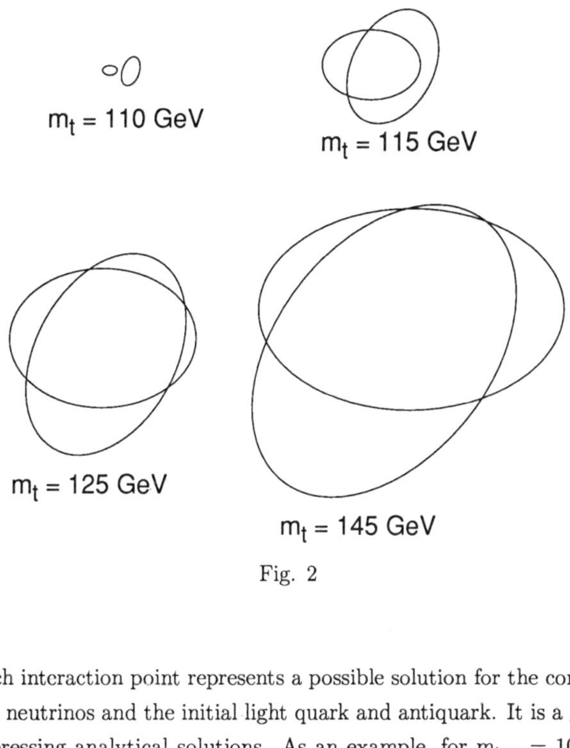

Fig. 2

– *Dalitz:*

1) Each interaction point represents a possible solution for the configurations of the final neutrinos and the initial light quark and antiquark. It is a geometrical way of expressing analytical solutions. As an example, for $m_{top} = 100$ GeV the two ellipses do not overlap. This means that there is simply no solution; there is no top mass which can be consistent with the lepton, antilepton and two-jet configuration observed for this event.

2) The observed configurations will be different for each of the two (or four) crossings. You have finally two or four sets of vectors ($\vec{b}$, $\vec{\bar{\ell}}$; $\vec{\bar{b}}$, $\vec{\ell}$) and each set gives a different configuration for t and $\bar{t}$ production and decay.

The signs of the highest energy leptons tell which are the products of the top and which are those of the antitop. For the top, $t \rightarrow W^+$ which gives e^+ or μ^+, whereas for antitop, $\bar{t} \rightarrow W^-$, giving e^- or μ^- in its decay.

– *Rind:*

Is there anything you can say about the number of six jets events expected from the hadronic decay of the W?

– *Dalitz:*

You ask about the case where both W's undergo decay to $\bar{q}q'$, in which the t-decay and the $\bar{t}$-decay each gives rise to three quark jets. These six-jet final states from top-antitop annihilation without any accompanying lepton with high transverse momentum should be 36 times as frequent as the top-antitop events which give e, μ and two (b-quark) jets. They will be very difficult events to identify, separate out and analyse. We have not yet made any attempt to deal with them. The combinatorics are enormous, of course; that is the problem.

REVIEW OF NEUTRINO MASSES

Douglas R.O. Morrison

CERN, Geneva, Switzerland.

SUMMARY OF TALK PRESENTED AT ERICE, July 1992.

1. ASTROPHYSICS and COSMOLOGY

They give two limits on the sum of the three neutrino masses - the sum must be less than (23 to 39) eV or greater than about 1.5 GeV.

Neutrinos with mass, e.g. about 7 eV, would help solve the problem of Dark Matter, but not alone as their reaction rate is so low that there would not be enough time for quasers, etc. to form after the period of recombination.

2. LEP

Results from LEP exclude heavy leptons with masses less than 45 GeV.

3. DIRECT MASS MEASUREMENT

The limits have been obtained on the masses of tau and muon neutrinos - they are 31 MeV and 250 keV resp., but it should be noted that the astrophysical constraints are more restrictive.

Some six experiments have recently measured the end-point region of tritium decay and all exclude the positive value reported by ITEP. These 6 values are all negative and there is now a suspicion that the process is not well-understood.

4. DOUBLE BETA DECAY

Best measurements can now give limits down to 1 eV. However these apply to Majorana neutrinos and not Dirac ones. Note that the value obtained depends on knowing the matrix elements.

5. 17 keV NEUTRINO

The 17 keV neutrino suggested by Simpson, should now be considered disproved, both by the weight of many careful experiments which failed to find it and by the some of the original proposers who have found mistakes - in particular Hine and Jelley did not take into fully account the effects of a badly placed baffle despite warnings.

6. CCFR DOUBLE VERTEX EVENTS

It had been suggested that an analysis of the CCFR neutrino data by Mishra showed evidence for a heavy neutrino, but this was not observed by their neighbouring experiment, E632, with their bubble chamber, who did however observe a steeply rising cross section for K0 production which may help to explain the effect. Later the CCFR Collaboration withdrew the suggestion.

7. SOLAR NEUTRINOS

Two effects have been claimed as possible evidence for New Physics; that the experimental flux of solar neutrinos is less than that calculated from models of stellar evolution e.g. the Standard Solar Model, SSM, and secondly that the solar neutrino flux fluctuates strongly with a periodicity close that of Solar sunspots but inversely in intensity.

Much of the evidence depends on choosing SSM theories that have very small errors - too small when one considers the uncertainties of the models and the disagreements of models with other experimental data such as the fact that the lithium observed is only one-hundredth of that predicted (important as including results from other stars - Boesgaard dip - this indicates meriditional circulation and turbulent diffusion which are not considered in the SSM). Taking other evolutionary models with more reasonable errors makes most experimental results not inconsistent with theoretical predictions. It was pointed out that the experiments are very difficult and that pioneering neutrino experiments have a history of requiring later correction to rates.

Data from the Kamiokande experiment does not find any indication of an 11-year fluctuation of the neutrino flux and this is considered to negate the 5 sigma effect claimed by the Chlorine experiment - it would be very unexpected that a surface phenomenon, sunspots would have any relation to production of solar neutrinos deep inside the SUN.

It is concluded that there is no compelling evidence for any Solar Neutrino problem

7. NEUTRINO OSCILLATIONS

Several theories predict that neutrinos have masses and serious theories do not exclude it. If neutrinos have masses, then it should be possible for them to oscillate, that is, switch from one form to another. A large theoretical structure has grown up as to the effects that would then be observed. Indications of anomalous effects (particularly concerning Solar Neutrino experiments) have frequently been interpreted as evidence for neutrino oscillations and many papers have been published. Unfortunately at this time there is no compelling evidence that neutrinos have mass, though it is always possible that new experiments looking at some lower mass range, or different couplings, may find effects proving that neutrinos have mass - this would have very important consequences and for this reason many experiments are exploring new regions.

CHAIRMAN: D.R.O. Morrison

Scientific Secretaries: Y. Khodyrev, H. Petrache

DISCUSSION

– *Ball:*

How difficult is it to change the input parameters of the SSM in order to change the neutrino flux by 30%?

– *Morrison:*

In general it is surprisingly difficult to change the input parameters by large amounts as there are very strong feedback mechanisms. Thus if you change a cross-section which changes a temperature, then it is found that the luminosity of the Sun does not agree - and the luminosity is very accurately measured. Thus in general it is difficult when the luminosity is involved - which is the case for gallium experiments. But for the ^8B neutrinos, this is a small side chain, and the flux is directly related to the cross-section for the $^7Be(p,\gamma)^8B$ cross-section, so that any change in it is directly related to the ^8B flux and the luminosity of the Sun is barely involved. So changing this by 30%, which is possible, is easy as the change in the luminosity is negligible. Note that I gave rates for Chlorine flux which extend from 5.8 to 8 SNU which is about 30%.

– *Sopczak:*

LEP experiments measure the $e^+e^- \rightarrow Z^0$ cross-section very precisely. For which mass range can this measurement indicate the existence of a massive neutrino?

– *Morrison:*

There are two types of high mass neutrinos - stable and unstable. The stable ones have restrictions from cosmology as they would be produced and finally become relics. Their mass would have to be less than 23 to 39 eV or more than about 1.5 GeV. Now LEP has given an upper limit of about 45 GeV for a heavy stable neutral particle. However the CCFRW Collaboration claim to have seen double vertex events which could be a new neutrino - this is difficult for LEP to test as they give it unusual properties. This is because it is unstable and that it couples very weakly to the Z^0 so that it is rarely produced at LEP. However we expect that the LEP experiments should be able to answer the question of the existence of the CFFRW "particle".

– *Leike:*

You reported that the Solar Standard Model has problems in predicting the amounts of Li and Be. Can this model be tested on other stars? Do we have same problems there?

– *Morrison:*

This is a very good question. Indeed progress has been made by considering other stars. In the Evolutionary Model there are three zones, the inner Core where the nuclear fusion takes place, the outer Convection Zone, and in between, the Intermediate Zone which is rather stable. The temperature at the Convection-Intermediate Zone boundary is about 2 million K. Now Lithium burns at 2.5 MK and Beryllium burns at 3.5 MK. The popular theory of why the Lithium and Beryllium are depleted, was that there were strong currents (downdrafts, vortices) which carried them from the Convection Zone into the Intermediate Zone where they were burnt up. However there is a recent unpublished paper from Yale which notes that in other stars the dispersion of the depletion values is very large and it is hard to understand this variation in the standard stellar model. Now there is a new theory that the Lithium and Beryllium depletions are not occurring now but happened near the transition of the protostar into a star. The protostar is a great cloud which gradually coalesces into one, two, three or more stars. The mass flowed from the surroundings to the star and this was a highly variable process depending on the local distribution of mass near the star. Hence Lithium and Beryllium were burnt up in the violent first 100 million years or so in different amounts in different stars. This theory explains the dispersion of values of Lithium depletion whereas considering the Sun only, such a dispersion would not be predicted.

– *Cho:*

Could you comment upon the possibility of using neutrino detectors as telescopes for observing astrophysical events such as supernovae explosions?

– *Morrison:*

There is great interest in using and developing large neutrino detectors. Most calculations suggest that most astrophysical events (except nearby supernovae), for example Active Galactic Nuclei, will yield few neutrinos at the Earth so the detectors have to be very large, about 100 000 m^2. Such detectors are planned:

for example DUMABD which is off the island of Hawaii, or NESTOR off the coast of Greece, both at a depth in the sea of over 4000 metres.

For supernovae the situation is much better and indeed SN1987A was outside our galaxy but was detected by Kamiokande and the IBM detector. Now a new generation of detectors is being built such as Super Kamiokande with 33 000 tons and the Sudbury Lab, SNO, with 1000 tons of heavy water, which give very high counting rates. Also there is a very imaginative detector being proposed at the South Pole with the elements buried deep in the ice. One expects one supernova explosion in our galaxy about once per 10 years but most are invisible to telescopes because of the cloud of dust etc. that hides most of our galaxy from us since we are on the outside. But neutrinos and gravity waves can penetrate the dust and should be observable by the new series of gravity wave detectors and neutrino detectors.

– *Liu:*

Even though the lepton number is the same for neutrino and antineutrino the helicity is still different. Therefore how can you see double beta decay without neutrinos?

– *Peccei:*

All you need to have are Majorana neutrinos and it will go, there is no extra suppression.

– *Liu:*

The times at which the gamma and neutrinos are generated can be very different even though we see both at the same time. So, are the parameters from luminosity good for neutrinos?

– *Morrison:*

Basically what you say is correct. The neutrinos that are observed come from the Core of the Sun and are observed directly whereas the gammas interact very many times before being absorbed and re-emitted by nuclei as well as being scattered so that the luminosity observed comes from very different gammas. If there were a significant change in the emission of gammas from the centre of the Sun, it would take about 10 000 years for this change to be observed at the Earth. However the Sun burns very stably so that the luminosity which describes what happened 10 000 years ago, is in fact a very good indicator of what is happening now when the neutrinos are observed.

– *Liu:*

The Gallex result is 83 SNU, but SSM predicts 120 SNU. Why did you say they were consistent?

– *Morrison:*

In considering whether the two numbers are consistent, the errors on both the experiment and on the theory must be considered. The experimental value is 83 +/-21 SNU and the theory value of about 120 SNU has an error of about 15 to 20 SNU. So the two values agree to less than two standard deviations. The probability of this is sufficiently high that one does not worry and consider it as evidence for New Physics.

– *Liu:*

Can you say a little bit more about the CERN neutrino beam sent to Gran Sasso?

– *Morrison:*

The protons are extracted and directed downwards by about three degrees towards the Gran Sasso Lab. They hit a target producing mainly pions and kaons plus other particles. These particles are focussed and decay in a very long tunnel giving mainly muon neutrinos plus about 2% electron neutrinos. There may be a few tau neutrinos but very few.

– *Liu:*

So what is the mean energy of the beam?

– *Morrison:*

There are two different quantities; one is the average energy of the beam and the other is the average energy of the interaction. This is because the cross-section rises linearly with energy so that higher energy neutrinos are favoured. Also the exact energy depends on how the beam is focussed, but one would expect that the average energy of the interactions from a beam from the SPS accelerator would be about 50 GeV.

– *Hanany:*

Recently there have been suggestions that the neutrinos have a magnetic moment. Can you explain that?

– *Morrison:*

Okun and the others thought of the neutrino having an appreciable magnetic moment in an attempt to explain the Solar Neutrino Problem - but as the problem is less evident now, the need for an unusual magnetic moment is less compelling. Also the magnetic moment proposed seemed to be in contradiction to other measurements.

– *do Couto e Silva:*

There was a talk at CERN by a person from KEK. He mentioned that the use of solid state devices was not a good choice, because back scattering is not very well understood. You mentioned that with these magnetic spectrometers one always has to introduce a scale factor to correct the data. Which would you consider the more reliable detector to work with?

– *Morrison:*

These are two very different techniques, so let us discuss them separately.

The silicon solid state detector has a major problem - back-scattering. If the beam falling onto it comes at all angles (said to be diffuse), then 30% of the electrons scatter so much that they emerge out again - are back-scattered. If the beam enters the detector at 90 degrees (i.e. normally), then only 15% back-scatters. So the angle of the incoming electrons is very important and in the experiments of Simpson, Hime and Jelley, this angle was not very well known. Also the absolute values (e.g. 15%) varied from experiment to experiment. Further the fraction back-scattered varied with energy even though Hime and Jelley claimed it did not (as shown in Hime's thesis). The top expert in CERN says that these detectors are very good to 1% accuracy - but Hime and Jelley were claiming to measure a value of 0.8% to an accuracy of 0.1%.

On the other hand if you are designing an experiment, you almost always want to have a magnetic field as this gives you a means of control - for example to check on momentum dependence. It is often an essential extra handle. Good and reliable experiments always have redundancy. The problem of adding a very small correction to produce a linear Curie plot, is a small correction and is a smooth one - it has never been found to produce a kink or an anti-kink.

Hence overall I believe a magnetic field is a great advantage.

– *do Couto e Silva:*

Some people are working with low temperature devices in order to detect dark matter, for instance liquid helium and superconducting devices. Can you mention the status of these techniques?

– *Morrison:*

There has been a great deal of development done on cryogenic detectors, but the work is more in a development stage and not an application stage. The idea is to make a detector which is unstable and which a neutrino interaction will trigger and release a much larger amount of stored energy. So the basic theory of these detectors is fine and on a small scale they appear to work, but to use them for very small interaction rates as with solar neutrinos, you need a very large detector - this will take some time to develop.

– *Petrache:*

In the matter oscillation mechanism the solar densities of electrons and neutrons play a major role. What can you say about the variation of these densities from the core to the solar surface?

– *Morrison:*

The basis of the evolutionary models is quite good and involves very detailed information of the densities of electrons, neutrons and all the elements. So the information is certainly easily available for people to make detailed calculations if they wish.

– *Petrache:*

Do you think that neutrino oscillations in the Earth matter could affect the Solar neutrino experiment results?

– *Morrison:*

The Earth is so much smaller than the Sun that it does not seem too important to consider matter oscillations in it now - particularly when the evidence for any such oscillation, is so weak.

– *Petrache:*

If neutrinos are the Majorana type could they annihilate and in this way, suppress the flux?

– *Morrison:*

They certainly could, but they would have to find one another to annihilate - the cross-section is so small that the effect must be negligible.

– *Shotkin:*

You said that astrophysics sets limits on the sum of the neutrino masses and that this has to be less than 23-39 eV or greater than 1.5 GeV. Could you explain the origin of the gap between the two allowed mass ranges?

– *Morrison:*

This is a theoretical question on the Big Bang and how it developed. Initially The Universe was very small and reached a stage when there were quarks and anti-quarks. These then gave matter and anti-matter which almost all annihilated except for one in a billion which was matter. The products of these annihilations were mainly gammas and neutrinos. As the Universe expanded the temperature dropped until these gammas and neutrinos could no longer find one another to interact and so they were "frozen out" at that temperature. Thus we have now a sea of electromagnetic radiation which is the 2.7 K radiation and there are about a billion gammas for each baryon. Similarly we expect a sea of neutrinos with about a billion neutrinos for each baryon existing. If neutrinos have no mass, this is no problem. But if the neutrinos do have a mass, then the sum of the masses of the three neutrinos cannot be too large or the total mass of neutrinos would be greater than the upper limit on mass of the Universe - this gives the upper bound of about 23 to 39 eV mass.

POTENTIAL MODELS: PREDICTIVE RIGOROUS RESULTS

A. MARTIN
Theoretical Physics Division, CERN
CH - 1211 Geneva 23

ABSTRACT

After presenting a "progress report" on order and spacing of energy levels and related topics, with illustrations in atomic and particle physics, we describe some old and new successes of the quark model of hadrons, in the framework of potential models.

1. Introduction

The first part of this lecture will be a "progress report" of rigorous theorems on order and spacing of energy levels of the Schrödinger equation, with applications to particle and atomic physics. The reason for this is historical. In 1977, after some discussions on the energy levels of charmonium with M.A.B. Bég, Harald Grosse and myself derived some theorems on the order of energy levels of the Schrödinger equation. Professor Zichichi happened to have heard about this, attended my seminar at CERN and asked me to give a series of lectures in Erice about quarkonium in the summer of 1977. I knew very little about the subject in general, spend sleepless nights worrying about the lectures I was supposed to give, but, in the end it all went well[1]. I continued working on the two sides of the subject, rigorous theorems on the Schrödinger equation and phenomenology of quarkonium with many collaborators. Actually sometimes these "collaborators" took the initiative of the work themselves and I just helped them. I thought that it was appropriate to give here, together with a short historical account, a "progress report" to indicate where we are now, since the Ettore Majorana Centre for Scientific Culture played an important rôle in the beginnings of this line of research.

2. The Progress Report

As I said before, the first rigorous theorems on the order of energy levels were obtained in 1977 by Harald Grosse and myself[2,3]. One of them concerned the comparison of the energy of the first *radial* excitation of a two-body system with the energy of the first *angular* excitation which coincide for a pure Coulomb potential. However, for charmonium, both in models, and in reality, the radial excitation is *above* the angular excitation. A condition on the potential was obtained under which it was the case but it was complicated and it is obsolete as we shall see in one moment.

In 1984, Baumgartner, Grosse and I found a much more satisfactory answer, which I have already described here in Erice in my lectures on "Toponium Physics"[4] (poor toponium! Maybe the top quark is too heavy and too rapidly decaying to make the notion of a top-antitop bound state meaningful!). I remind you the Schrödinger equation for a central potential and an angular momentum ℓ (here we take $\hbar = 2\mu = 1, \mu$ being the reduced mass)

$$\left[-\frac{d^2}{dr^2} + \frac{\ell(\ell+1)}{r^2} + V(r) - E(n,\ell)\right]U_{n,\ell}(r) = 0 \tag{1}$$

where $U_{n,\ell}$ is the reduced wave function and E the energy is characterized by the number of nodes of the radial wave function, n, and ℓ the orbital angular momentum. A state $(n,0)$ will be a purely radial excitation, while a state $(0,\ell)$ will be a purely angular excitation. Between the energy levels there is a rather obvious ordering:

- the energy increases with n for fixed ℓ. This is a consequence of the Sturm-Liouville theory of second-order differential equations;
- the energy increases with ℓ for fixed n. This is because in (1) ℓ can be treated, following T. Regge, as a continuous variable and the Feynman-Hellmann theorem gives:

$$\frac{dE(n,\ell)}{d\ell} = \langle\frac{2\ell+1}{r^2}\rangle > 0 \tag{2}$$

The question raised by Baqui Bég was more subtle: why is $E(n=1,\ell=0)$ larger than $E(n=0,\ell=1)$ for charmonium. For a pure Coulomb potential we know that the energy levels depend only on the "Principal" quantum number $N = n+\ell+1$, and so

$$E(n+1,\ell) = E(n,\ell+1) \tag{3}$$

What we succeeded to prove[5] is that if the Laplacian of the potential is positive (outside the origin) then we get Theorem I

$$E(n+1,\ell) > E(n,\ell+1) \quad (\Delta V > 0) \tag{4}$$

and the reverse if the Laplacian is negative:

$$E(n+1,\ell) < E(n,\ell+1) \quad (\Delta V < 0) \tag{5}$$

Of course the Coulomb potential, $-1/r$, is such that

$$\Delta V(r) = 4\pi\ \delta^3(\vec{r}) \tag{6}$$

and the Laplacian therefore zero outside the origin.

This answer was essentially the best possible one and allowed to understand a certain number of facts in particle and atomic physics.

In the case of quarkonium, the central potential (we average over spin) is governed by the property of asymptotic freedom. If V is the potential, its Laplacian is $1/r^2 \; d/dr \; (r^2(dV/dr))$. Now dV/dr is nothing but the force (with a minus sign) which in the pure Coulomb case reduces to $-\alpha/r^2$. In the SU_3 colour gauge theory it is $\alpha(r)/r^2$ where $\alpha(r)$ is the running coupling constant (this is one possible definition. Another one is the coefficient of $1/r$ in the potential), which should increase with r because of asymptotic freedom, i.e.,

$$\frac{d}{dr} \alpha(r) = \frac{d}{dr} r^2 \frac{dV}{dr} > 0 \tag{7}$$

which means that the Laplacian of V is positive. On Fig. 1, we see the energy levels of the $c\bar{c}$ system, and if we average over spins of the P states, we see that indeed the χ states ($n = 0, \ell = 1$) are below the ψ' state ($n = 1, \ell = 0$). Similarly, on Fig. 2 we see that the χ_b states ($n = 0, \ell = 1$) lie between the $\Upsilon(n = 0, \ell = 0)$ and the $\Upsilon'(n = 1, \ell = 0)$ and that the χ_b' states ($n = 1, \ell = 1$) lies between the $\Upsilon'(n = 1, \ell = 0)$ and the $\Upsilon''(n = 2, \ell = 0)$.

Another beautiful application is that of muonic atoms. In muonic atoms, especially those with large Z, the nucleus cannot any more be regarded as pointlike compared to the size of the Bohr orbit, and the Coulomb degeneracy is seriously broken. To the extent that the nucleus can be regarded as a simple assembly of protons and neutrons (which is true only if pion clouds or the underlying quark structure is disregarded), it has an extended positive charge distribution, and therefore by Gauss's law, produces a potential with a positive Laplacian. If we take for instance μ^- Ba^{138} [6], we have, using the standard spectroscopic notation:

$$\left.\begin{array}{ll} 2s_{1/2} - 2p_{1/2} & = 405.41 \; keV \\ 3p_{3/2} - 3d_{3/2} & = \;\; 8.69 \; keV \end{array}\right\} \tag{8}$$

Of course, pure QED effects like the Lamb shift go in the same direction but have a smaller order of magnitude. As we shall see later, the sign of the energy differences is also the right one if one replaces the Schrödinger equation by the Dirac equation.

Another situation where the Laplacian has a given sign is that of alkaline atoms and more generally of isoelectronic sequences of alkaline atoms, for instance, lithium and its isoelectronic sequence, Be^+, B^{++}, C^{+++}, etc., or sodium and its isoelectronic sequence Mg^+, Al^{++}, Si^{+++}, etc. In these systems the outer electron is, in the Hartree approximation at least, submitted to the potential of the pointlike nucleus and of the electron cloud of the inner shells, which, again by Gauss's law, has a negative Laplacian. Figure 3, taken from the book of Chpolski "Atomic Physics"[7] shows how the Coulomb degeneracy is broken for lithium; the $2p$ level is above the $2s$ level, and the energy increases from $3s$ to $3p$ and to $3d$. More examples can be found in the Tables of Bashkin[8].

In parallel to all this, it was also interesting to compare the energy level to another well-known solvable case, the anharmonic oscillator potential. For $V = r^2$, the energy levels depend on $N_h = 2n + \ell + 1$. Hence $E(n+1, \ell) = E(n, \ell + 2)$.

There is a change of variables which allows to go from the Coulomb potential to the harmonic oscillator potential which is just $z = r^2$. By using this change of variable we get Theorem II[9]

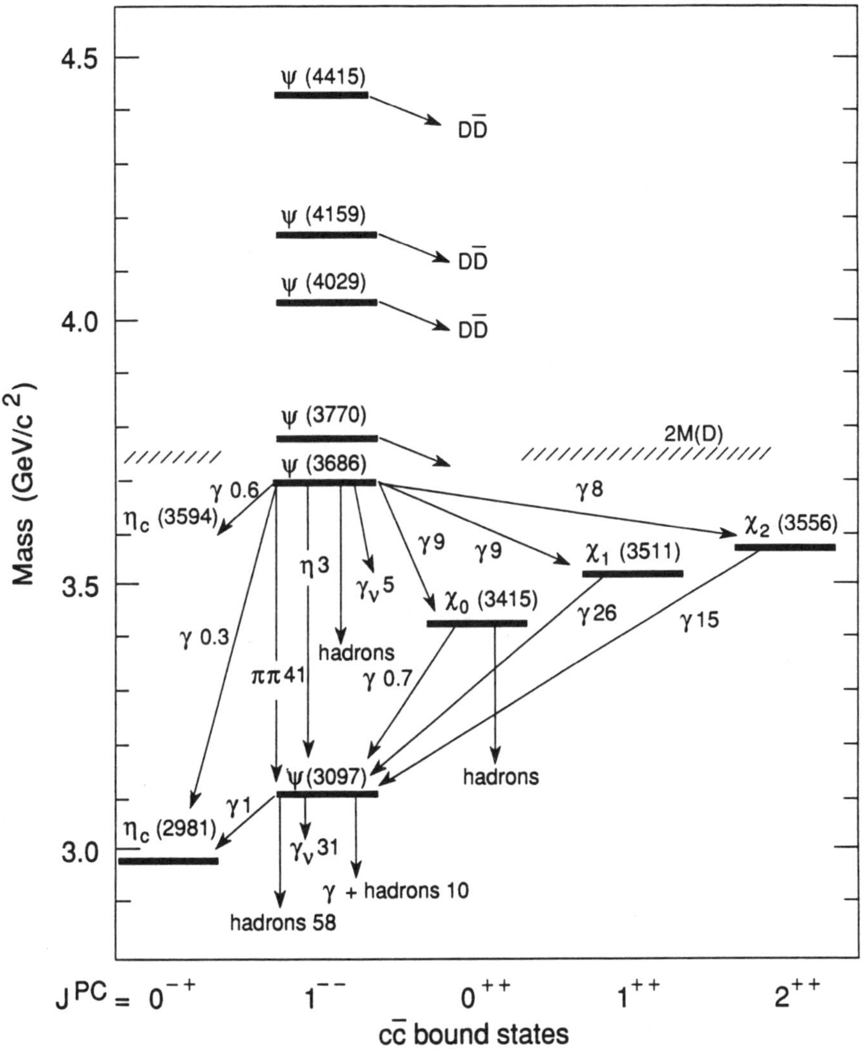
4.5
4.0
3.5
3.0
Mass (GeV/c 2)
ψ (4415)
DD̄
ψ (4159)
DD̄
ψ (4029)
DD̄
ψ (3770)
2M(D)
ψ (3686)
γ 0.6
η c (3594)
γ 8
χ 2 (3556)
γ 9
γ 9
χ 1 (3511)
η 3
γ ν 5
χ 0 (3415)
γ 26
γ 15
γ 0.3
hadrons
π π 41
γ 0.7
hadrons
ψ (3097)
γ 1
η c (2981)
γ ν 31
γ + hadrons 10
hadrons 58
J PC = 0 −+
1 −−
0 ++
1 ++
2 ++
cc̄ bound states

Figure 1

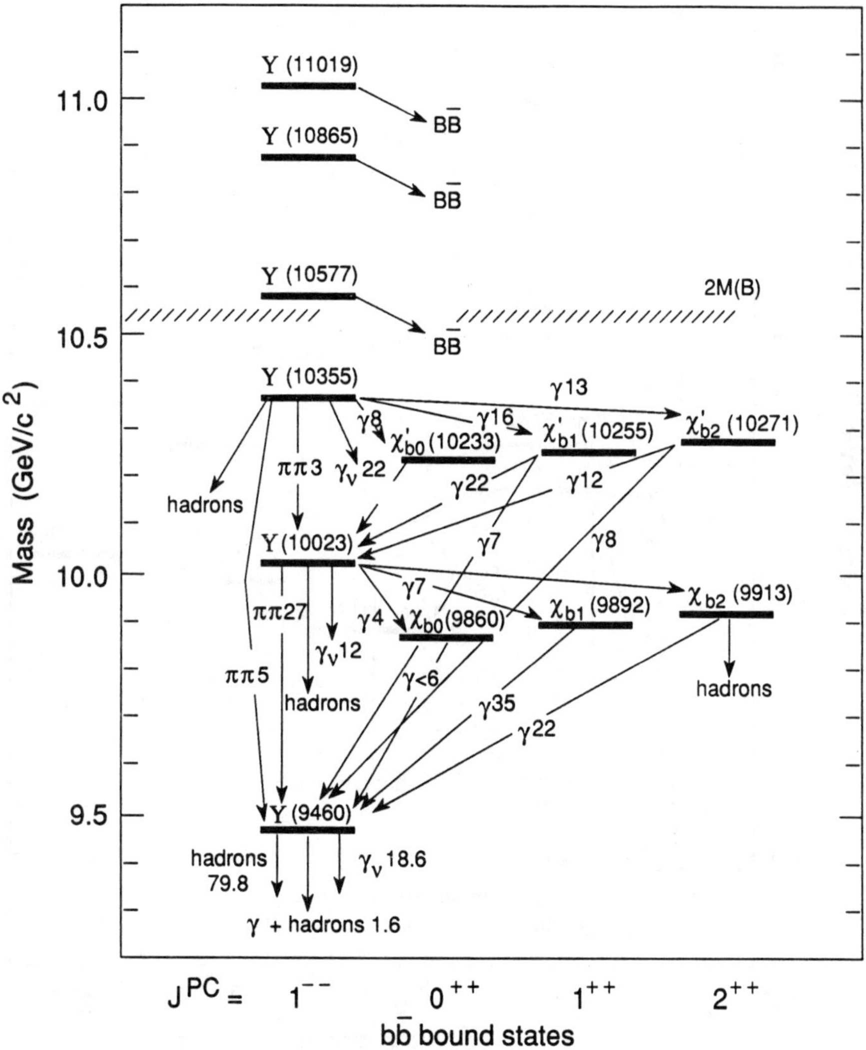
Mass (GeV/c 2)
11.0
10.5
10.0
9.5
Υ (11019)
Υ (10865)
Υ (10577)
Υ (10355)
Υ (10023)
Υ (9460)
BB̄
BB̄
BB̄
2M(B)
χ'b0 (10233)
χ'b1 (10255)
χ'b2 (10271)
χb0 (9860)
χb1 (9892)
χb2 (9913)
γ13
γ8
γ16
ππ3
γν 22
hadrons
γ22
γ12
γ7
γ8
ππ27
γ7
γ4
γν12
ππ5
hadrons
γ<6
γ35
γ22
hadrons
hadrons 79.8
γν 18.6
γ + hadrons 1.6
J PC = 1−− 0++ 1++ 2++
bb̄ bound states

Figure 2

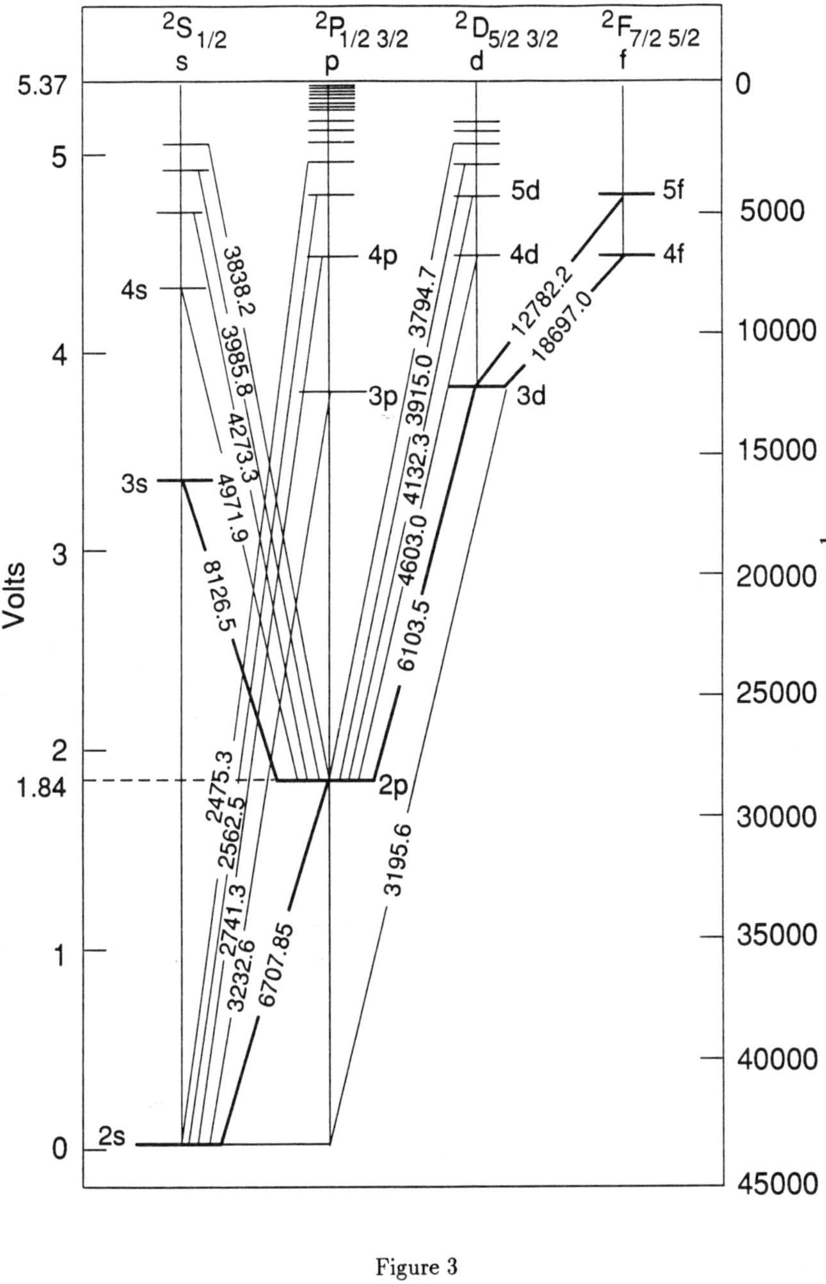

$^2S_{1/2}$
s
$^2P_{1/2\ 3/2}$
p
$^2D_{5/2\ 3/2}$
d
$^2F_{7/2\ 5/2}$
f
5.37
5
4
3
2
1.84
1
0
Volts
0
5000
10000
15000
20000
25000
30000
35000
40000
45000
ν,cm^{-1}
2s
3s
4s
2p
3p
4p
3d
4d
5d
4f
5f
3838.2
3985.8
4273.3
4971.9
8126.5
3794.7
3915.0
4132.3
4603.0
6103.5
12782.2
18697.0
2475.3
2562.5
2741.3
3232.6
6707.85
3195.6

Figure 3

If

$$\frac{d}{dr}\frac{1}{r}\frac{dV}{dr} \gtrless 0$$

i.e., if V is a convex (concave) function of r^2,

$$E(n+1,\ell) \gtrless E(n,\ell+2)$$

A simple application of this is the location of the $\ell = 2$ state of charmonium: it has been established from lattice QCD that the quark-antiquark potential is concave[10] and monotonous increasing, and hence $d/dr \ \ 1/r \ \ dV/dr < 0$. Hence $E(1,0 <$ $E(0,2)$ and hence the ψ'' state, supposed to be an $\ell = 2$ state is above the ψ' state ($n = 1, \ell = 0$) which is the case (3.77 GeV > 3.68 GeV).

In 1985 a big step forward was made by Alan Common who stressed that a lemma used to establish the previous theorems[5,9] was in itself interesting and could lead to non-trivial inequalities in the moments of the ground state wave function for arbitrary ℓ, i.e., the states ($n = 0, \ell$ arbitrary). This lemma[11] says that the logarithm of the ground state reduced wave function divided by $r^{\ell+1}$ is concave (convex) if the Laplacian of the potential is positive (negative), i.e.,

$$\frac{d^2}{dr^2}(\log \frac{U_{0,\ell}(r)}{r^{\ell+1}}) \gtrless 0 \quad if \quad \Delta V \gtrless 0 \tag{9}$$

Similarly, by the same change of variables:

$$\frac{d}{dr}\frac{1}{r}\frac{d}{dr}(\log \frac{U_{0,\ell}(r)}{r^{\ell+1}}) \gtrless 0 \quad if \quad \frac{d}{dr}\frac{1}{r}\frac{dV}{dr} \gtrless 0 \tag{10}$$

Common has realized that these results enabled us to find very interesting properties of purely angular excitations and in 1986, Common and I obtained the following Theorem III [12]
If V is convex (concave) in r^2, $E(0,\ell)$ is convex (concave) in r^2, $E(0,\ell)$ is convex (concave) in ℓ. Hence

$$\frac{E(\ell+1) - E(\ell)}{E(\ell) - E(\ell-1)} \gtrless 1 \quad if \quad \frac{d}{dr}\frac{1}{r}\frac{dV}{dr} \gtrless \tag{11}$$

Here we use the abbreviation $E(\ell) = E(0,\ell)$.

Of course this theorem incorporates the well-known fact that for a pure harmonic oscillator E is linear in ℓ. An obvious application to charmonium due to the concavity of the potential, is

$$M_{\chi_c}(\ell = 1, n = 0) > \frac{1}{2}\Big[M_{\psi''}(\ell = 2, n = 0) + M_{J/\psi}(\ell = 0, n = 0)\Big]$$

which is true since

$$3.51 \ GeV \ > \frac{1}{2}\Big[3.77 + 3.10\Big] \ GeV \ .$$

Remark: Inequality (23), further in the text, allows to improve this slightly, and to replace the right-hand side by 3.47.

It took the arrival of Joachim Stubbe in 1990 with his incredible mathematical agility to derive the analogue result for potentials with given sign of the Laplacian[13].

Theorem IV: If $E(\ell)$ is the energy of purely angular excitations, i.e., $E(\ell) = E(0,\ell)$ and $E_c(\ell) = -1/(\ell+1)^2$ is the same quantity for a pure Coulomb potential

$$\frac{E(\ell+1)-E(\ell)}{E(\ell)-E(\ell-1)} \gtrless \frac{E_c(\ell+1)-E_c(\ell)}{E_c(\ell)-E_c(\ell-1)} = \frac{(2\ell+3)(\ell)^2}{(2\ell+1)(\ell+2)^2} \tag{12}$$

if $\Delta V \gtrless 0 \; \forall r > 0$.

This very interesting theorem can be tested on the same objects, i.e., quark-antiquark systems, muonic atoms (both have $\Delta V > 0$) and alcaline atoms and their isoelectronic sequences (for which $\Delta V < 0$ in the Hartree approximation). Figure 4 shows the ratios $3d-2p/2p-1s$, $4f-3d/3d-2p$ and $5g-4f/4f-3d$ (in spectroscopic notation; we have used a clever average over spins to eliminate most of the "Dirac" effects), for muonic atoms with increasing Z. As one can see, there are lots of experimental data because these energy differences are precisely those measured in favoured electric dipole transitions of the muon. As Z increases, we see that the deviations from Coulomb are more and more marked and, naturally, in the right direction.

Table 1 shows the same ratios as for the muonic atoms for the lithium sequence and the sodium sequence, using the Bashkin tables[8]. There are also inequalities on the ionisation energies which can be obtained and used as a test of the claimed accuracy of the experimental data. Some of these are presented in Table 2, where I designates the experimental ionisation energy and $I(i,j)$ is an upper bound constructed from energy levels of angular momenta i and j.

lithium sequence				sodium sequence			
		$\frac{4f-3d}{3d-2p}$	$\frac{5g-4f}{4f-3d}$			$\frac{5g-4f}{4f-3d}$	$\frac{6h-5g}{5g-4f}$
Coulomb		0.35	0.4638	Coulomb		0.4638	0.5432
Li	I	0.326					
Be	II	0.324	0.462				
B	III	0.326	0.462				
C	IV	0.329	0.462				
N	V	0.331	0.462				
O	VI	0.331					
F	VII	0.334	0.462				
Ne	VIII	0.336					
Na	IX			Na	I	0.456	0.5426
Mg	X	0.338		Mg	II	0.444	0.5419
Al	XI	0.339		Al	IV	0.432	0.5413
Si	XII	0.339		Si	IV	0.423	0.5409
P	XIII	0.339		P	V	0.418	0.5406
S	XIV	0.340		S	VI	0.416	

Table 1: The lithium sequence and the sodium sequence

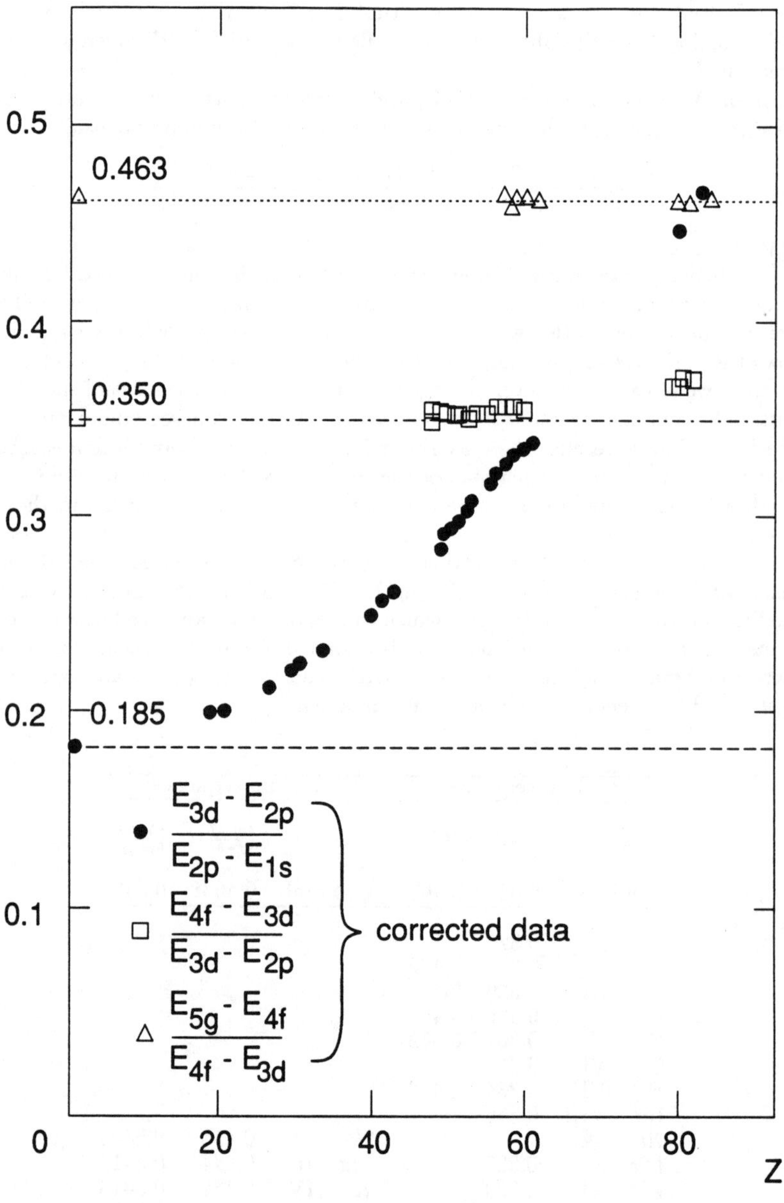

0.5
0.463
0.4
0.350
0.3
0.2
0.185
0.1
0
20
40
60
80
Z
$\frac{E_{3d} - E_{2p}}{E_{2p} - E_{1s}}$
$\frac{E_{4f} - E_{3d}}{E_{3d} - E_{2p}}$
$\frac{E_{5g} - E_{4f}}{E_{4f} - E_{3d}}$
corrected data

Figure 4

		lithium sequence		
		I	I(2,1)	I(3,2) (10^6 cm^{-1})
Li	I	0.04349	0.04439	0.04351
Be	II	0.14688	0.15095	0.14695
B	III	0.30593	0.31421	0.30608
P	XIII	4.935	4.992	4.932
S	XIV	5.704	5.773	5.704
		sodium sequence		
		I	I(3,2)	I(4,3) (10^6 cm^{-1})
Na	I	0.04145	0.04155	0.04146
Mg	II	0.12127	0.12248	0.12214
Al	III	0.22945	0.23403	0.22960
S	VI	0.71019	0.73336	0.71113
Cl	VII	0.92105	0.96191	0.92256
Cr	XIV	3.0985	3.2434	3.1051

Table 2: Upper bounds on ionisation energies

One can even go further and incorporate the spin of the muon or of the outer electron and get interesting inequalities on the fine structure splittings of the purely angular excitations treated in a semi-relativistic way. In particular one has the following theorems, with

$$\delta(\ell) = E(n=0, j=\ell+1/2) - E(n=0, j=\ell-1/2) = \frac{(2\ell+1)}{4m^2} \langle\frac{dV/dr}{r}\rangle \tag{13}$$

A)

$$\frac{\delta(L)}{\delta(\ell)} \gtrless \frac{(\ell+1)^4\ell}{(L+1)^4 L} \quad \text{if} \quad L \geq \ell \quad \text{and} \quad \delta(\ell) > 0 \quad \text{if} \quad \Delta V \gtrless 0 \tag{14}$$

B)

$$\delta(\ell) \gtrless \Delta(\ell) = \frac{2}{m\ell} \frac{(\ell+2)^2}{(2\ell+3)^2} \Big[E(\ell+1) - E(\ell)\Big]^2 \quad \text{if} \quad \Delta V \gtrless 0 \tag{15}$$

Table 3 presents an application to muonic atoms, which shows that for Z less than 40 the actual values and the bounds differ by less than 20%. In the case of alcaline atoms the inequalities are marginally satisfied for the lithium sequence but

Element	Z	$\delta(1)$ KeV	$\Delta(1)$ KeV	$\delta(2)$ KeV	$\frac{\delta(2)}{\delta(1)} > 0.0988$
^{56}Fe	26	4.20 ± 0.09	4.40	0.47 ± 0.07	0.112 ± 0.02
^{75}As	33	11.10 ± 0.8	11.45	2.00 ± 0.8	0.18 ± 0.08
^{92}Mo	42	25.50 ± 0.8	29.60	2.70 ± 0.4	0.106 ± 0.02
^{138}Ba	56	65.9 ± 0.6	94.80	9.81 ± 0.30	0.149 ± 0.015

Table 3: Fine splittings of muonic atoms (δ), their upper bounds (Δ) and their ratios.

fail for the sodium sequence. This is an indication of the failure of the Hartree approximation for these systems and should have been expected, since everybody knows (or should know as my friend J.L. Basdevant taught me!) that the famous sodium doublet producing the ugly yellow light in the freeway tunnels is inverted, i.e., the state with higher J is lower. This cannot be cured by relativistic "Dirac" effects which preserve the level order, but definitely means that the one electron approximation is not sufficient.

In parallel to the study of the spacing of angular excitations, one can also look at the spacing of radial excitations for fixed ℓ, in particular for $\ell = 0$. Then $E(n) = E(n, \ell = 0)$. Only partial, but interesting results have been obtained[14]:

- for perturbations around the harmonic oscillator (for which the levels $E(n, \ell = 0)$ are equally spaced) the spacing increases (or decreases) with n, i.e., $E(n+1) - E(n) \gtrless E(n) - E(n-1)$ if

$$X = \frac{d}{dr} r^5 \frac{d}{dr} \frac{1}{r} \frac{dV}{dr} \gtrless 0 \quad \forall r > 0 \tag{16}$$

  This strange looking condition is such that X vanishes if

$$V = r^2 + \lambda r^2 + \mu + \frac{\nu}{r^2} \ .$$

  However, we have examples of violations of this property outside the perturbation regime, for instance if V is not monotonous;

- we believe that if V is concave, i.e., $V'' < 0$, the level spacing decreases, i.e.,

$$E(n+1) - E(n) < E(n) - E(n-1)$$

  This is strictly true for $V'' = 0$, and established in the framework of the WKB approximation for $V'' < 0$. This property shows that for the $b\bar{b}$ system, the potential models must breakdown for high excitation. Indeed, as can be seen on Fig. 2, $E_{\gamma'} - E_{\gamma} > E_{\gamma''} - E_{\gamma'} > E_{\gamma'''} - E_{\gamma''}$ but $E_{\gamma^{IV}} - E_{\gamma'''} > E_{\gamma'''} - E_{\gamma''}$, a fact which cannot accommodated by any concave potential, and yet we believe that the quark-antiquark potential is concave. Only a coupling to open channels, like $B\bar{B}, B^*\bar{B}, B^*\bar{B}^*$ can explain this.

To finish this progress report, I should say that in 1991 and 1992 we have also investigated timidly relativistic effects, i.e., looked at particles satisfying the Klein-Gordon and Dirac equations.

In the Klein-Gordon equation which is, formally, the simplest because it looks like the Schrödinger equation once the energy is fixed:

$$\left(-\frac{d^2}{dr^2} + \frac{\ell(\ell+1)}{r^2} + m^2 - V^2 + 2EV - E^2\right)U = 0 \ , \tag{17}$$

the degeneracy of the levels is already removed for a pure Coulomb potential. The solution is known and one has

$$E(n+1, \ell) < E(n, \ell+1) \ ,$$

i.e., the levels are ordered like those of a Schrödinger equation with a negative Laplacian. The simplest theorem we could prove is the following[15]:

In the Klein-Gordon equation, if V is attractive and $\Delta V \leq 0$, one has

$$E(n+1,\ell) < E(n,\ell+1)$$

There is a converse theorem for $\Delta V > 0$ but it is more sophisticated: one has to replace ℓ by another "effective" angular momentum and I cannot enter into these details.

Concerning the Dirac equation, Harald Grosse had already obtained a remarkable result in 1986 [16]. It said that, at least for perturbations of the Coulomb potentials, i.e., for

$$V = -\frac{c}{r} + v, \quad v \text{ small} ,$$

the levels with the same J and different ℓ's are such that

$$E(N,J,\ell = J-1/2) \gtrless E(N,J,\ell = J+1/2) ,$$

N being the principal quantum number, if

$$\Delta V \gtrless 0 \quad \forall r > 0 .$$

This means that if one accepts the validity of this result even for larger v, the order of levels found in inequality (8) holds not only in the Schrödinger case but also in the Dirac case. It means also that the Lamb shift effect is equivalent to replacing the source of the Coulomb potential by an extended structure. It remains a challenge to find a non-perturbative version of this theorem.

On the other hand, we have been able to compare levels of the Dirac equation for the same N, the same <u>orbital</u> angular momentum and different total angular momenta[17] and found:

$$E(N, J = \ell + 1/2, \ell) > E(N, J = \ell - 1/2, \ell) \text{ if } \frac{dV}{dr} > 0 . \tag{18}$$

In the semi-relativistic approximation [see Eq. (13)] it is obvious, but it is strictly true for the Dirac equation.

When one combines the two results, one sees that a Coulomb multiplet is completely ordered if $\Delta V < 0$ and $dV/dr > 0$ (the second condition might be superfluous as suggested by the semi-relativistic approximation).

Then, in the multiplet, energies increase for fixed J and increasing L, fixed L and increasing J. Hence we have for instance, for $N = 3$

$$3S_{1/2} < 3P_{1/2} < 3P_{3/2} < 3D_{3/2} < 3D_{5/2} .$$

Future progress on the one particle Dirac equation seems possible. However, level ordering of genuine relativistic two-body systems seems for the time being hopeless.

3. Some Successes of the Quark Model of Hadrons

Here I restrict myself to potential models, in which mesons are made of a quark and an antiquark and baryons are made of three quarks, interacting by flavour independent potentials. They satisfy the Schrödinger equation or possibly a semi-relativistic version in which $m + p^2/2m$ is replaced by $\sqrt{p^2+m^2}$. We do not try to justify theoretically this approach but rather try to exploit it to its extreme limits and see how well it works. There is - or there was - a controversy on the validity of these potential models and sometimes I felt shaken by the attacks of my friends the worshippers of the QCD sum rules, but as time passes and as potential models continue to accumulate successes as experimental results come up, I feel more and more confident that they contain a lot of truth, though - I must confess - I do not really understand why.

Let us start with mesons made of heavy quarks: $c\bar{c}, b\bar{b}, s\bar{s}, c\bar{s}$, and $b\bar{s}$ ($b\bar{s}$ has been added after the lectures because of its recent observation at LEP). There are several potentials, with various motivations, fitting the $c\bar{c}$ and $b\bar{b}$ system. I have taken two extremes: the Buchmüller-Grunberg-Tye potential, QCD inspired[18] and my own phenomenological model[19], $V = A+Br^{0.1}$. Both reproduce the data extremely well as shown on Fig. 5, where only spin averaged states have been indicated for the P states. Following a suggestion of Gell-Mann, I have included the strange quark as "heavy" and this is why I get a fit including the $s\bar{s}, c\bar{s}, b\bar{s}$ and $b\bar{c}$ systems. The parameters are

$$\left.\begin{array}{l} V = -8.064 + 6.870r^{0.1} \\ in\ units\ of\ \ (GeV)^x\ . \\ m_c = 1.8 \quad m_b = 5.174 \quad m_s = 0.518 \end{array}\right\} \tag{19}$$

In addition, an effective pointlike spin-spin interaction is adjusted to fit the $J/\psi = \eta_c$ separation. The m_s mass is adjusted to reproduce the ϕ mass and the $\phi\prime$ mass is predicted to be 1.634 GeV (exp. 1.650), the D_s mass 1.99 GeV (exp. 1.97), the D_s^* mass 2.11 GeV (exp. 2.11), the D_s^{**} mass 2.537 (exp. (Argus) 2.536). The B_s mass is predicted to be[20] between 5.354 and 5.374 GeV (exp. Aleph collaboration[21] 5.3746 ± 0.0075 ± 0.005 GeV). The B_s^* mass is predicted to be between 5.408 and 5.410 GeV and expecting observation. Similarly my predictions for B_c and B_c^* are respectively 6.25 and 6.32 GeV[20].

Application of the potential model to ordinary hadrons has also met lots of success even though it is still more difficult to understand why[22]. would like to give two illustrations.

First of all, at a time where deep inelastic scattering experiments might indicate that the spin of the proton is not carried by quarks, I would like to come back on the question of magnetic moments. Here, we deal with static, effective quarks, dressed by lots of things, which might have little to do with the quarks seen in hard processes. There are many of explanations for the -3/2 ratio of the magnetic moments of the proton and the nucleon[23], but I would like to stress that the overall fit of the magnetic moments of baryons, with and without strangeness, necessitates a ratio of $(m_u \simeq m_d)/m_s$ effective masses of about 0.6 as shown by Table 4 when we compare with a fit with $m_u = m_d = m_s$.

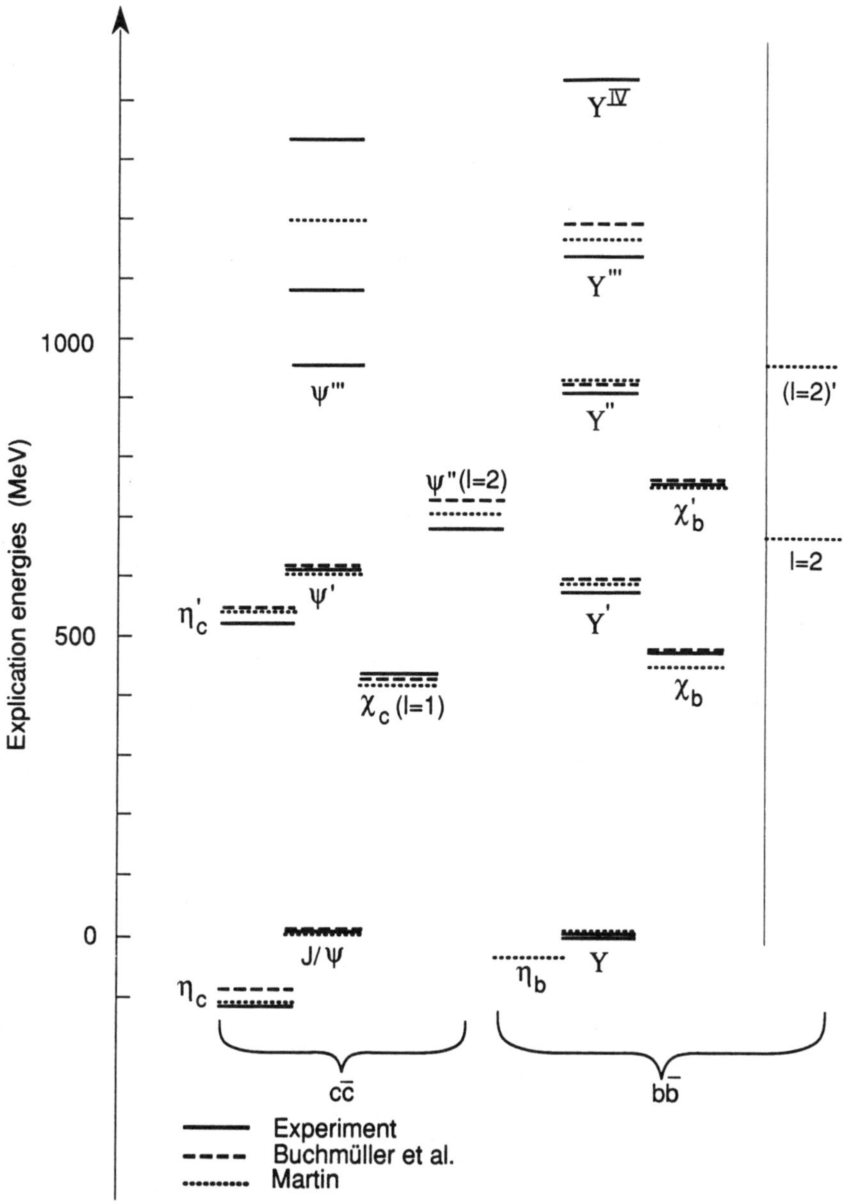
Explication energies (MeV)
1000
500
0
ψ'''
ψ'
J/ψ
ψ"(l=2)
η_c'
η_c
χ_c (l=1)
Y^IV
Y'''
Y"
Y'
Y
η_b
χ_b'
χ_b
(l=2)'
l=2
$c\bar{c}$
$b\bar{b}$
Experiment
Buchmüller et al.
Martin

Figure 5

	Experiment	μ/μ_P Quark Model $m_u/m_s = 1$	Quark Model $m_u/m_s = 0.625$
N	- 0.68	- 0.67	-0.67
Λ	-0.22 ± 0.02	- 0.333	0.207
Σ^+	0.85 ± 0.01	1	0.958
Σ^-	-0.50 ± 0.09	-0.333	-0.37
Ξ^-	-0.24 ± 0.01	-0.333	-0.17
Ξ^0	-0.45 ± 0.05	-0.666	-0.50
Ω^0	-0.72	-1 ±0.07	-0.625

Table 4: Magnetic moments of baryons

In the Table we notice the new result of K.B. Luk and collaborators[24] on the magnetic moment of the Ω^- particle, completely compatible with $\mu_\Omega = 3\mu_{\Lambda_0}$. Though imperfect (and not perfectible very easily!) I find the fit very impressive, and it conforts me in the idea that the description of static baryons in terms of three effective quarks is very good. Another indication in that direction is a fit of baryon masses with a soft, flavour-independent central potential, by J.M. Richard and P. Taxil. To measure the success of the model one must compare it with the predictions using only spin-spin hyperfine splitting by Federman, Rubinstein and Talmi[25], De Rújula, Georgi, Glashow[26] and Sakharov and Zeldovitch[27]. For instance, the quantity $M_{\Xi^*} - M_\Xi / M_{\Sigma^*} - M_\Sigma$ is exactly unity in the latter approach but differs from it with a central potential. A systematic comparison is given in Tables V and VI.

Now I would like to turn to the $c\bar{c}$ singlet P state, which, very likely, has been observed at Fermilab by an American-Italian collaboration, using a gas-jet proton target in the antiproton accumulator[28], and which seems to coincide with the early signal observed at the ISR.

About six months before the observation of this state Joachim Stubbe and I tried to make a prediction which, in a certain way, would be model independent. But let us start by more naïve considerations. The P states are split by spin-spin, spin orbit, and tensor forces. If these spin-dependent forces are treated in perturbation, the weighted average of the triplet P state ($S = 1$)

$$\bar{M} = \frac{1}{Q}\left[5(2^{++}) + 3(1^{++}) + 1(0^{++})\right] \tag{20}$$

does not contain the spin orbit or tensor forces anymore. So the difference between $\bar{M}$ and the mass of the singlet P state $1^{+-}(J^{PC})$ comes only from the spin-spin forces. If, in analogy with what Fermi does in quantum electrodynamics, we take the

	Theory	Experiment
N	input	.939
Δ	input	1.232
$\Lambda^{\bullet}$	1.111	1.115
Σ	1.176	1.193
Ξ	1.304	1.318
Σ^*	1.392	1.383
Ξ^*	1.538	1.533
Ω^-	input	1.672
Λ_c	input	2.282
Σ_c	2.443	2.450
Σ_c^*	2.542	
$\Xi_c = A$	2.457	2.460
S	2.558	
S^*	2.663	
$\Omega_c = T$	2.664	2.740 (3 events!)
T^*	2.775	

Table 5: Baryon masses

	DeRujula Georgi Glashow Sakharov Zeldovitch	Richard Taxil	Experiment
$\frac{M_{\Xi^*}-M_{\Xi}}{M_{\Sigma^*}-M_{\Sigma}}$	1	1.08	1.12
$\frac{2M_{\Sigma^*}+M_{\Sigma}-3M_{\Lambda}}{2(M_{\Delta}-M_{N})}$	1	1.07	1.05
$\frac{3M_{\Lambda}+M_{\Sigma}}{2M_{N}+2M_{\Xi}}$ GMO octet	1	1.005	1.005
$\frac{M_{\Sigma^*}-M_{\Delta}}{M_{\Xi^*}-M_{\Sigma^*}}$ GMO	1	1.10	1.007
$\frac{M_{\Xi^*}-M_{\Sigma^*}}{M_{\Omega^-}-M_{\Xi^*}}$ decuplet	1	1.09	1.08

Table 6: Mass difference ratios

spin-spin force to be proportional to the Laplacian of the one-gluon exchange potential, which is crudely -const/r we get a delta function, the expectation value of which is zero in a P state and so, the super-naïve prediction is

$$M(^1P_1) = \bar{M} = 3525 \; MeV \; .$$

What Joachim Stubbe and I have tried to do is to avoid making detailed assumptions but only assume that the spin splittings are given by the Fermi-Breit Hamiltonian and that the central potential contains only vector-like and scalar-like components[29]. This is true for many models, but not always, and in particular some perturbative QCD calculations[30]. However, if one accepts these assumptions one gets

$$M(^1P_1) - \tilde{M} = -\frac{1}{2m^2}\langle\frac{V'}{r^0}\rangle \; , \tag{21}$$

where V is the sum of the scalar and vector potentials, with

$$\tilde{M} = \frac{1}{12}\Big[-5M(^3P_2) + 27M(^3P_1) - 10M(^3P_0)\Big] \tag{22}$$

and $\langle V'/r\rangle$ can be bounded using the same strategy as for muonic atoms: for $V = v_s + v_v$ concave,

$$\frac{5}{8}m(E_D - E_P)^2 < \langle\frac{V'}{r}\rangle < \frac{m}{2}\Big[E_P - E_S\Big]^2 \; , \tag{23}$$

using this time spin-averaged energies.

With this strategy, using experimental numbers and allowing for some relativistic corrections, we get

$$3536 \pm 12 < M(^1P_1) < 3559 \pm 12 \; MeV \tag{24}$$

for the $c\bar{c}$ system and similarly, for the $b\bar{b}$ system

$$9900.3 \pm 2.8 < M(^1P_1) < 9908.9 \pm 2.8 \; MeV \tag{25}$$

the experimental result for the $c\bar{c}$ $^1P_1$ state (which, incidentally, is compatible with the early indications from the ISR[31]) is 3526.1 MeV, compatible with our bounds and also, ironically, with the supernaïve prediction!

Let us notice that in 1986, the CLEO collaboration had proposed a candidate for the $^1P_1 b\bar{b}$ state[32], which since had disappeared. If we had had the bound at the time (i.e., if Joachim Stubbe had been working with me!) we would have been able to cast some suspicion on this experimental result! The $b\bar{b}$ singlet P state remains to be seen but it is an uneasy entreprise. Voloshin[33] is very sceptical on the possibility of looking at the process $\Upsilon'' \rightarrow ^1 P_1(b\bar{b}) + 2\pi$'s and has been shown to be right in the case of the $^1P_1(c\bar{c})$ which does not decay into $2\pi + J/\psi$ [28].

Next we turn to the Λ_b and more generally to predictions on the masses of strange, charmed, and beautiful baryons. As we shall see, upper bounds on the masses can be obtained by using a concavity argument. This has been already presented here in Erice years ago[34]. Also lower bounds can be obtained from a comparison of mesons and baryon masses[35] which has been recently refined[36].

The Λ_b must of course exist, but it is a rather elusive particle. It was observed years ago by the Zichichi group but not seen by others. For a discussion of the situation at the time, see for instance the proceedings of the Lisbon Conference[37]. Now the UA1 Collaboration claims to have seen the decay[38] $\Lambda_b \to J\psi + \Lambda$ with a mass $5640 \pm 50 \pm 30$ GeV/c^2. However, LEP experiments, and in particular Aleph[39], have seen Λ_b's from their semi-leptonic decay and a correlation between negative high-energy leptons and ordinary lambda's, but they have not seen the above-mentioned two-body decay which makes a measurement of the mass possible.

The upper bounds on beautiful baryons are a consequence of the concavity of the binding energy of a baryon containing a heavy quark with respect to the inverse of the mass of the heavy quark. This is a consequence of a very general theorem. Let H be a Hamiltonian of the form

$$H = A + \lambda B$$

then the ground state energy is a concave function of λ, i.e., $d^2E/d\lambda^2 < 0$. In the particular case

$$H = \frac{p_1^2}{2M} + \frac{p_2^2}{2m} + \frac{p_3^2}{2m} + W \ ,$$

where W is a flavour-independent interaction, for instance a sum of flavour-independent two-body potentials. m is the constituent mass of the light quarks and M the mass of the heavy quark. The binding energy is concave in $1/M$. If we have the masses of the baryon containing strange and charmed quarks and if we have - for instance from the fit (19) - a set of constituent quark masses, we can, by extrapolation, get an upper bound on the beautiful hadrons (Fig. 6).

In this way, we get[35], using $m_s = 0.518, m_c = 1.8, m_b - m_c = 3.405$

$$\begin{cases} B < 5.292 \quad B^* < 5.370 \text{ GeV} \\ B_s^* < 5.463 \\ \Lambda_b < 5.659 \quad \Sigma_b < 5.856 \end{cases}$$

from

$$\begin{cases} K = 0.496 \quad K^* = 0.892 \\ D = 1.869 \quad D^* = 2.009 \\ \qquad\qquad\quad \phi = 1.020 \\ \qquad\qquad\quad D_s^* = 2.110 \\ \Lambda = 1.116 \quad \Sigma = 1.193 \\ \Lambda_c = 2.282 \quad \Sigma_c = 2.450 \end{cases}$$

Notice that the experimental B mass is 5.279 and that the upper bound on the B_s^* mass is only 50 MeV above the result of an explicit model.

Lower bounds on meson masses for mixed mesons have been obtained long ago[40], also by using the concavity theorem, but exploiting the fact that only the reduced mass enters in the Hamiltonian. Then it is very easy to get the following inequality

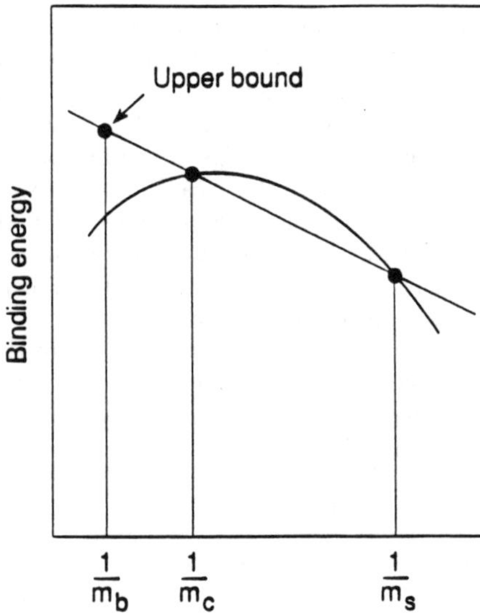

Figure 6

$$M(Q_1\bar{Q}_2) > \frac{1}{@}\Big[M(Q_1\bar{Q}_1) + M(Q_2\bar{Q}_2)\Big] \tag{26}$$

In this way one gets for spin 1 states

$$M_{B_s^*} \quad > \tfrac{1}{2}(M_\Upsilon + M_\phi) \quad = 5.24 \text{ GeV}$$

$$M_{D_s^*} \quad > \tfrac{1}{2}(M_{J/\psi} + M_\phi) \quad = 2.06 \text{ GeV}$$

$$M_{(b\bar{c})^*} \quad > \tfrac{1}{2}(M_{J/\psi} + M_\Upsilon) \quad = 6.28 \text{ GeV}$$

These bounds are not very far from the explicit model that we have seen and, also, not very far from experiment for the D_s^* state which has a mass 2.11 GeV. Lower bounds on baryon masses can be obtained by a completely different strategy which in itself is completely rigorous but which uses, as a starting point, a belief which is not totally established. The point is that one assumes that the quark-quark potential, inside a baryon, is half the corresponding quark-antiquark potential inside a meson. This is true for the one-gluon exchange term. This is also true for a situation where, inside a baryon, two quarks are very close to one another. They form a colour antitriplet (since the baryon is an overall colour singlet) and exert the same attraction as an antiquark on the third quark. But we must divide this by two since we assume two-body forces, to get the attraction exerted by each of the two quarks on the third quark. Then we can split the baryon Hamiltonian into three pieces:

$$
\begin{aligned}
H &= \frac{p_1^2}{2M_1} + \frac{p_2^2}{2M_2} + \frac{p_3^2}{2M_3} + V_{Q_1Q_2} + V_{Q_1Q_3} + V_{Q_2Q_3} \\
&= \frac{1}{2}\left[\frac{p_1^2}{2M_1} + \frac{p_2^2}{2M_2} + V_{Q_1\bar{Q}_2}\right] \\
&+ \frac{1}{2}\left[\frac{p_1^2}{2M_1} + \frac{p_3^2}{2M_3} + V_{Q_1\bar{Q}_3}\right] \\
&+ \frac{1}{2}\left[\frac{p_2^2}{2M_2} + \frac{p_3^2}{2M_3} + V_{Q_2\bar{Q}_3}\right] ,
\end{aligned}
$$

since $V_{Q_1Q_2} = 1/2 V_{Q_1\bar{Q}_2}$, etc. Each of the three pieces is an operator which has, as a lower bound, the energy of the corresponding meson.

Adding the masses to get the total energies, we get

$$
M(Q_1Q_2Q_3) > \frac{1}{2}\left[M_{Q_1\bar{Q}_2} + M_{Q_1\bar{Q}_3} + M_{Q_2\bar{Q}_3}\right] \tag{27}
$$

For comparing spin 3/2 baryons with spin one mesons, where all spins are parallel, we can apply this rule blindly and get

$$
\left\{
\begin{array}{lllr}
\Omega^- & > & \frac{3}{2}\phi & 1.672 > 1.530 \ , \\
\Delta & > & \frac{3}{2}\rho & 1.232 > 1.155 \ , \\
\Sigma^* & > & K^* + \frac{1}{2}\rho & 1.385 > 1.275 \ , \\
\Xi^* & > & \rho + \frac{1}{2}\phi & 1.532 > 1.280 \ , \\
\Sigma_b^* & > & B^* + \frac{1}{2}\rho = & 5.710 \text{ GeV} \ .
\end{array}
\right. \tag{28}
$$

For other spin states there are technical complications that I shall not describe here and one gets

$$
\left\{
\begin{array}{lllr}
N & > & \frac{3}{4}\rho + \frac{3}{4}\pi & 0.938 > 0.681 \ , \\
\Sigma & > & \frac{1}{2}\rho + \frac{3}{4}K + \frac{1}{4}K^* & 1.190 > 0.979 \ , \\
\Xi & > & \frac{1}{2}\phi + \frac{3}{4}K + \frac{1}{4}K^* & 1.319 > 1.104 \ , \\
\Sigma_c & > & \frac{1}{2}\rho + \frac{3}{4}D + \frac{3}{4}D^* & 2.450 > 2.288 \ , \\
\Lambda & > & \frac{1}{2}\pi + \frac{3}{4}K^* + \frac{1}{4}K & 1.116 > 0.863 \ , \\
\Lambda_c & > & \frac{1}{2}\pi + \frac{3}{4}D^* + \frac{1}{4}D & 2.282 > 2.042 \ , \\
\Lambda_b & > & \frac{1}{2}\pi + \frac{3}{4}B^* + \frac{1}{4}B & 5.640 > 5.379 \ .
\end{array}
\right. \tag{29}
$$

We see that the difference between the actual mass, whenever known, and the lower bound oscillates between 100 and 250 MeV. This is because, inside the baryon, each pair of quarks is not at rest. This has recently been corrected for by

J.L. Basdevant, J.M. Richard and myself[36]. Then one gets, for instance,

$$M(sss) > \frac{3}{2}\left[\begin{array}{l} M(s\bar{s}, L=0) \\ +\frac{3}{16}(M(s\bar{s}, L=1) - M(s\bar{s}, L=0)) \end{array}\right] \tag{30}$$

which gives a lower bound

$$M(sss) > 1.659\ MeV$$

while the actual mass is 1.672 MeV.

We have now the technique to generalize this to baryons made of quarks with unequal masses[41], but what is missing is experimental data on mesons. It is interesting to return to beautiful baryons and to compare explicit models and bounds given in Table 7.

Finally, let us say a word on mass differences between different spin states. Our predictions based either on extrapolations or generalizations of the De Rùjula-Georgi-Glashow model are

$$\Sigma_b - \Lambda_b \simeq 190\ MeV$$

$$\Sigma_b^* - \Sigma_b \simeq 20\ MeV\ ,$$

so that both Σ_b and Σ_b^* will be broad states, with the one-pion decay into Λ_b allowed.

REFERENCES

1. A. Martin, in "The Why's of Subnuclear Physics", International School of Subnuclear Physics, Erice 1977, A. Zichichi Editor, Plenum Press, New York and London (1979), p. 395.
2. A. Martin, *Phys.Lett.* **67B** (1977) 330.
3. H. Grosse, *Phys.Lett.* **68B** (1977) 343.
4. A. Martin, in "Quarks, Leptons and Their Constituents", International School of Subnuclear Physics, Erice 1984, A. Zichichi Editor, Plenum Press, New York (1988), p. 447.
5. B. Baumgartner, H. Grosse and A. Martin, *Phys.Lett.* **B146** (1984) 363.
6. R. Engfer, H. Schneuwly, J.L. Vuilleumier, H.K. Walter and A. Zehnder, *At.Data Nucl.Data Tables* **14** (1974) 509.
7. E. Chpolsky, Physique Atomique, Tome II (French translation), Mir, Moscow 1978, p. 217.
8. S. Bashkin and J.O. Stoner, Jr., Tables of Atomic Energy Levels and Grotrian Diagrams, Vols. I, II (North-Holland, Amsterdam, 1978).
9. B. Baumgartner, H. Grosse and A. Martin, *Nucl.Phys.* **B254** (1985) 528.
10. E. Seiler, *Phys.Rev.* **D18** (1978) 133;
    C. Boys and E. Seiler, *Comm.Math.Phys.* **91** (1983) 39;
    C. Bachas, *Phys.Rev.* **D33** (1986) 2723.

5. B. Baumgartner, H. Grosse and A. Martin, *Phys.Lett.* **B146** (1984) 363.
6. R. Engfer, H. Schneuwly, J.L. Vuilleumier, H.K. Walter and A. Zehnder, *At.Data Nucl.Data Tables* **14** (1974) 509.
7. E. Chpolsky, Physique Atomique, Tome II (French translation), Mir, Moscow 1978, p. 217.
8. S. Bashkin and J.O. Stoner, Jr., Tables of Atomic Energy Levels and Grotrian Diagrams, Vols. I, II (North-Holland, Amsterdam, 1978).
9. B. Baumgartner, H. Grosse and A. Martin, *Nucl.Phys.* **B254** (1985) 528.
10. E. Seiler, *Phys.Rev.* **D18** (1978) 133;
    C. Boys and E. Seiler, *Comm.Math.Phys.* **91** (1983) 39;
    C. Bachas, *Phys.Rev.* **D33** (1986) 2723.
11. Besides the original proof in Ref. 5, proofs with increasing simplicity have been obtained:
    M.S. Ashbaugh and R. Benguria, *Phys.Lett.* **A131** (1988) 273;
    A. Martin, *Phys.Lett.* **A147** (1990) 1.
12. A.K. Common and A. Martin, *Europhysics Lett.* **4** (1987) 1349.
13. A. Martin and J. Stubbe, *Europhysics Lett.* **14** (1991) 287 and *Nucl.Phys.* **B367** (1991) 158.
14. A. Martin, J.M. Richard and P. Taxil, *Nucl.Phys.* **B329** (1990) 327.
15. H. Grosse, A. Martin and J. Stubbe, *Phys.Lett.* **225B** (1991) 563.
16. H. Grosse, *Phys.Lett.* **146B** (1987) 413.
17. H. Grosse, A. Martin and J. Stubbe, *Phys.Lett.* **284B** (1992) 347.
18. W. Buchmüller, G. Grunberg and S.H.-H. Tye, *Phys.Rev.Lett.* **45** (1980) 103;
    W. Buchmüller and S.H.-H. Tye, *Phys.Rev.* **D24** (1981) 132.
19. A. Martin, *Phys.Lett.* **100B** (1981) 511.
20. A. Martin, in "Heavy Flavours and High-Energy Collisions in the 1-100 TeV Range", A. Ali and L. Cifarelli Editors, Plenum Press, New York and London (1989), p. 141.
21. ALEPH Collaboration, CERN Preprint PPE 93-97 (1993), submitted to *Phys.Lett.B.*
22. D.P. Stanley and D. Robson, *Phys.Rev.* **D21** (1980) 3180 and *Phys.Rev.Lett.* **45** (1980) 235;
    N. Isgur and P. Kociuk, *Phys.Rev.* **D21** (1980) 1868.
23. See, for instance:
    J. Ellis and M. Karliner, *Phys.Lett.* **B213** (1988) 73;
    and
    J. Ellis, Y. Frishman, A. Hanany and M. Karliner, *Nucl.Phys.* **B382** (1992) 189.
24. H.T. Diehl et al., *Phys.Rev.Lett.* **67** (1991) 804.
25. P. Federman, H.R. Rubinstein and I. Talmi, *Phys.Lett.* **22** (1966) 208.

26. A. De Rújula, H. Georgi and S.L. Glashow, *Phys.Rev.* **D12** (1980) 147.
27. A.D. Sakharov and Y.B. Zeldovitch, *Acta Physica Hungarica* **22** (1967) 153; A.D. Sakharov, *Zh.Ehksp.Teor.Fiz.* **78** (1980) 2112.
28. E760 Collaboration, T.A. Armstrong et al., *Phys.Lett.* **69** (1992) 2337.
29. J. Stubbe and A. Martin, *Phys.Lett.* **B271** (1991) 208.
30. J. Pantaleone and S.H.-H. Tye, *Phys.Rev.* **D37** (1988) 3337.
31. R704 Collaboration, C. Baglin et al., *Phys.Lett.* **171B** (1986) 135.
32. CLEO Collaboration, T. Bowcock et al., *Phys.Rev.Lett.* **58** (1987) 307.
33. M.B. Voloshin, *Sov.J.Nucl.Phys.* **43** (1986) 1011.
34. A. Martin, in "The Search for Charm, Beauty and Truth at High Energies", G. Bellini and S.C.C. Ting Editors, Plenum Press, New York and London (1984), p. 501.
35. A. Martin and J.-M. Richard, *Phys.Lett.* **185B** (1987) 428.
36. J.-L. Basdevant, A. Martin and J.-M. Richard, *Nucl.Phys.* **B343** (1990) 69.
37. Proceedings of the International Conference on High Energy Physics, Lisbon 1981, J. Dias De Deus and J. Soffer Editors, European Physical Society (1982), pp. 879, 885 and 1167.
38. UA1 Collaboration, C. Albajar et al., *Phys.Lett.* **273B** (1991) 540.
39. ALEPH Collaboration, D. Buskulic et al., *Phys.Lett.* **249B** (1992) 145.
40. R. Bertlmann and A. Martin, *Nucl.Phys.* **B168** (1980) 111; S. Nussinov, *Z.Phys.* **C3** (1979) 165.
41. J.-L. Basdevant, A. Martin, J.-M. Richard and T.T. Wu, *Nucl.Phys.* **B393** (1993) 111.
42. D.P. Stanley and D. Robson, *Phys.Rev.* **D21** (1980) 3180; *Phys.Rev.Lett.* **45** (1980) 235.
43. K. Maltman and N. Isgur, *Phys.Rev.* **D22** (1980) 1701.
44. J.-L. Basdevant and S. Boukraa, *Z.Phys.* **C30** (1986) 103.
45. J.-M. Richard and P. Taxil, *Phys.Lett.* **B128** (1983) 453.
46. D. Izatt et al., *Nucl.Phys.* **B199** (1982) 269.

CHAIRMAN: A. Martin

Scientific Secretaries: P.G. Baillie, F. Sannino, R. Sehgal

DISCUSSION

– *Baillie:*

1) Could you explain how you know that the number of narrow $Q\overline{Q}$ resonances is proportional to $M_Q^{1/2}$?

2) In the Υ system you said that above the $B\overline{B}$ threshold you can no longer use a simple potential. Are there ways to adapt potential models to account for the nearby $B\overline{B}$ states?

– *Martin:*

1) Using the WKB approximation:

$$n - \frac{1}{4} = \frac{1}{\pi} \int dr \sqrt{2M_Q(E - V(r))}$$

where n is the number of bound states with binding energy less than E. The potential V(r) is flavour-independent and therefore the same for all $Q\overline{Q}$ systems. The narrow bound states are those which cannot decay into pairs of mesons. For example, in the charmonium case the narrow states cannot decay into $D\overline{D}$, while for Υ they cannot decay into $B\overline{B}$. The mass of a $Q\bar{q}$ meson where Q is heavy and q light, is:

$$M_{Q\bar{q}} = M_Q + c \ ,$$

where c is universal because the $Q\bar{q}$ potential is flavour-independent. So for the number of narrow bound states we must put $E = 2M_{Q\bar{q}} - 2M_Q = 2c$ in the WKB formula, and so:

$$n = M_Q^{1/2} \times (\text{universal quantities})$$

2) Yes, there are ways. For example, the Cornell group has a model in which you take into account open channels. There are various recipes which are relatively successful.

– *Cho:*

1) Can errors and uncertainties be sensibly assessed in potential models? Can predictions from such models be systematically improved?

2) You mentioned that at one point you had to include spin effects. This approach seems somewhat "ad hoc", there being no systematic way to improve your predictions, would you say this is a fair criticism?

3) For a state like Λ_b, with one heavy and two light quarks, the light quarks are relativistic. Why should a potential description, which is non-relativistic, be valid?

– *Martin:*

1) The accuracy of a potential model can only be assessed a posteriori. For instance in the case of charmonium and upsilon systems, only some of the states were known. The χ_b and the Υ^{III} were predicted successfully at the right place by Buchmüller - the predictive power was very good, but there was no a priori recipe.

2) My potential cannot be improved systematically. The important - short-distance - spin effects have been calculated in QCD, with relative success. The spacing of the χ_b and χ_c triplet states have been successfully reproduced while there remain some problems with the χ_b' triplet. Naturally, my phenomenological model cannot predict the spacing inside a multiplet.

3) I agree with you that this is very difficult to justify. Many of the considerations, such as lower bounds remain valid if you replace $p^2/2m$ by $\sqrt{p^2+m^2}-m$. There are on the market a number of "semi-relativistic" calculations, where you hope you can take into account the fact that some quarks are light. The hope of several people, in particular Nussinov, is that results derived in the framework of a potential model are more general than their derivation.

– *Cho:*

Do you know of any way of evaluating the number of bound states a given 3-dimensional potential admits, without explicitly calculating the eigenvalue spectrum? Is there some kind of integral relation that will do this?

– *Martin:*

Yes. There is an inequality which has been derived by amongst others, Elliot Lieb. The number of bound states N satisfies:

$$N \leq c \int d^3x |V^-(x)|^{3/2}$$

where V^- is the attractive part of the potential, and c an undetermined constant. This constant has been estimated in a semi-classical approach to be $1/6\pi^2$, but the rigorous bound is given by a somewhat larger value.

– *Hanany:*

How is your previous inequality for the number of bound states in a 3-dimensional potential related to Levinson's theorem?

– *Martin:*

The inequality I wrote down is much more general; it involves the total number of bound states, for an arbitrary potential, not just spherically symmetric. Levinson's theorem states that the number of bound states of given angular momentum is equal to the variation of the phase shift, from ϕ to ∞ times π.

– *do Couto e Silva:*

1) Is your model equally good for baryons and mesons?
2) What are your predictions for the B_s mass?

– *Martin:*

1) The model works well for heavy mesons, and for reasons we do not fully understand, for heavy baryons, even when they contain light quarks. For light mesons it does not work - the pion plays a dual role as Goldstone boson and as $q\bar{q}$ bound state, for example.

2) For a start:

$$M_{B_s} \geq \frac{1}{2}(M_\Upsilon + M_\phi) \approx 5.24 \text{ GeV}$$

due to the convexity property I discussed in my lecture. I have completed a more precise calculation which gives:

$$M_{B_S} = 5354 \text{ MeV} \ , \ M^*_{B_S} = 5408 \text{ MeV} \ .$$

You will notice that both B_S and B^*_S can be produced by Υ^{IV} decay.

– *Strogova:*

I did not quite understand whether you took into account non-perturbative corrections in your theory?

– *Martin:*

The potential is basically non-perturbative. It originates from lattice QCD, where you have non-perturbative effects.

– *Malik:*

Spacing of the $\ell = 0$ levels changes for V when $\frac{d}{dr} r^5 \frac{d}{dr} \frac{1}{r} \frac{dV}{dr} \neq 0$. Is there any physical reason for that?

– *Martin:*

I do not know of any physical reason. There is one potential that gives equal spacing of levels - the harmonic oscillator. The equal spacing persists if one changes the oscillator strength, or if you shift the energy. Adding c/r^2 is just like changing the angular momentum, so you still get equal spacing:

$$\frac{d}{dr}\, r^5 \frac{d}{dr}\, \frac{1}{r}\, \frac{dV}{dr} \text{ vanishes for } V = \text{ const},\ V = Ar^2 \text{ or } V = c/r^2 \ ,$$

which is why this quantity appears in the inequality. By the way, there is also a criterion, which we have not yet proved, which says that the spacing decreases if the potential is concave (this is believed to be true in lattice QCD, from the work of E. Seiler and others).

– *Leike:*

For the detection of Λ_b and b-mesons the branching ratios of the decay into J/ψ are important. Are there theoretical estimates for the branching ratios of $(bcx) \to J/\psi$ + anything and $(b\bar{c}) \to J/\psi$ + anything?

– *Martin:*

I don't know. All I know is that the branching ratio I quoted for Λ_b is suspiciously large. I expect that the branching ratio of $(b\bar{c}) \to J/\psi + X$ is quite small because the anticharm quark decays 10 times faster than the b quark.

– *Ren:*

Can you give an estimate for the radius of the proton, based on the range of your potential?

– *Martin:*

I have not done any proton calculations, so I do not know the size. I have been told that you can calculate the binding energy and r.m.s. radius using potential models, and that the value of the radius turns out to be a little bit too small compared to the value from the electromagnetic form factor.

– *Ball:*

1) Did you apply your method to the $b\bar{c}$ bound states and what mass do you get for it?

2) Do you use the same value of the strong coupling for calculating heavy-heavy and heavy-light bound states, respectively?

– *Martin:*

1) I have done the calculation, which gives $M_{B_c}=6250$ MeV, $M^*_{B_c}=6318$ MeV. As was the case for the B_s, you can derive a lower bound as the average of $b\bar{b}$ and $c\bar{c}$ bound state masses, which works even better here, which gives 6.28 GeV for the 1^- state.

2) I have not done that, but Buchmüller et al. have considered the running of $\alpha_s(r)$ in the Coulombic part. This running constant, which behaves as $1/\ell n(r)$, can be used in any bound state calculation. To calculate bound states, there are two philosophies. One is to change α_s when you change your system. The other is to use a potential containing a running α_s, which can be used universally, which is what I prefer.

– *Misiak:*

1) Is the potential obtained by fitting the meson masses used to estimate the magnetic moments?

2) How do the potential parameters for baryons differ from those used for mesons?

3) Does the fact that the potential models work well mean that the sea quarks are unimportant, or that they are important, but their effect can be described in terms of a potential?

– *Martin:*

1) Calculation of the magnetic moments is trivial - you do not need a potential model as long as you neglect orbital angular momentum contributions which arise only through mixing of the L=0 ground state with other states by tensor forces. I presented the magnetic moments not because they were new, but rather because people tend to forget the successes of the naïve quark model. The assumption that the magnetic moments of baryons arise purely from the moments of the constituent quarks gives good results to about 10%. In particular, the recently measured Ω^- magnetic moment fits into this picture.

2) They are a little different.

3) I believe that the quarks that appear in these potential models are not the same as the objects seen in deep-inelastic scattering. These effective quarks contain gluons and sea quarks. This is like in nuclear physics, where one can describe a ^{12}C nucleus as containing three α particles, even though each α particle contains four nucleons. [Recently, there has been an attempt by Frishman, Ellis, Karliner and Hanany to understand quarks as a type of soliton, although this has

been restricted to two dimensions.] I believe there is no contradiction between the two different descriptions of quarks.

– *Zichichi:*

The most striking effect is the fact that the constituent quarks agree with our naïve understanding of the spin of the proton, and the current quarks make the spin disappear. Personally I do not believe this, but we will be doing an experiment at HERA to test this.

– *Sopczak:*

We learned from Prof. Stirling's talk that the b quark mass depends strongly on the QCD energy scale μ. How much does this scale uncertainty affect your predictions of the bound state energies?

– *Martin:*

I do not take this into account, but I believe that there is not much effect for bound states; you are effectively working at the same scale. In determinations of α_s from $c\bar{c}$ or $b\bar{b}$ annihilation the scale is set by the total mass of the system and changes a lot. But what counts if you calculate $q\bar{q}$ bound states is the average relative momentum of the two quarks, which changes little from $c\bar{c}$ to $b\bar{b}$. So I do not think the scale uncertainty is important in this case.

– *Hanany:*

Could you discuss what happens if you use the Dirac equation instead of the Schrödinger equation?

– *Martin:*

For years we have mainly used the Schrödinger equation, and only occasionally the Dirac equation. There is a result due to H. Grosse, which is unfortunately only proved for small perturbations around a Coulomb potential. It says that:

$$E\left(N, J = \ell + \frac{1}{2}\right) > E\left(N, J = \ell - \frac{1}{2}\right),$$

if $\Delta V > 0$, where V is a vector potential in the Dirac equation, like the Coulomb. More recently, Harald Grosse, Joachim Stubbe and I have found out that if $\Delta V \leq 0$ and the potential is increasing, then if you take a given Coulomb multiplet it is completely ordered. For example, if N=3:

$$3s_{1/2} < 3p_{1/2} < 3p_{3/2} < 3d_{3/2} < 3d_{5/2}$$

(see Phys. Lett. **B284** (1992) 347). We do not have any theorem for $\Delta V > 0$ as there are conflicting factors. So if you take an alkaline atom and you assume that it is fair to consider it as a one electron system submitted to a system which is negative laplacian, the levels are as shown above.

– *Liu:*

When we deal with atomic systems, there are two fine corrections. One is spin, the other is relativistic correction. Did you ever think about relativistic corrections like the Lamb shift?

– *Martin:*

Radiative corrections have not been included. I do not think they are relevant at the level of accuracy at which we are working.

– *Azimov:*

You apply two-body potential approach to quarkonium states above the threshold of open flavour ($D\overline{D}$ or $B\overline{B}$). How reliable are such applications? What would be the effect of including 4 quark states?

– *Martin:*

Above the meson threshold the potential deteriorates. This was already shown in the morning. For example I showed that the spacing between Υ^{III} and Υ^{IV} is much larger than what it should be if you have a potential description. If you mean by "4 quark states" the $(Q\bar{q})(q\overline{Q})$ states, yes they have been included, with the same success by the Cornell group.

As a response to the plot of $\Lambda_b \to J/\phi\Lambda$ measured by UA1 in which they gave a mass of $5640 \pm 50 \pm 30$ MeV/c^2 Professor Zichichi showed plots depicting the channels in which the first signal for Λ_b was seen in 1981 at the ISR. The new particle was reported in "Nuovo Cimento" with a six standard deviation significance (see E. Fisk discussion fig. 2, 3, 4, 5).

SOLAR NEUTRINOS IN REAL TIME

Sheldon L. Glashow

Department of Physics

Harvard University

Introduction

Is there life after the standard model? Surprises in high–energy physics have become few and far between. The discovery of the third family of quarks and leptons in the late 70s astounded everyone but Kobayashi and Maskawa. Since then, the prediction of a Great Desert has been fulfilled. LEPs null result of 1990, that there are three and only three conventional neutrino states, astounded me and established the desert. I still believe that nature has some tricks we haven't seen. Some of them may have to do with neutrinos. Two potential neutrino surprises have come and gone: the glitch in the spectrum corresponding to a 17 keV neutrino, and the unpublished but bruited–about double vertex events at Fermilab. Tant pis — all we've got left in the neutrino wastebasket is the trouble with solar neutrinos.

1. The Solar Neutrino Issue

The Sun gets most of its energy by burning H to ^{4}He. This process is initiated by the neutrino–producing reaction:

$$p + p \to d + e^+ + \nu, \tag{1}$$

which produces most of the solar neutrinos. The easiest neutrinos to detect, however, are the rare but energetic ones coming from the β process:

$$^8\mathrm{B} \to {}^8\mathrm{Be} + e^+\nu. \tag{2}$$

For over two decades, the Davis experiment has monitored the flux of these neutrinos [1]. Its surprising data torments us as the so–called solar neutrino problem.

The problem is that we don't really know what the problem is. At least three contrary points of view have been expressed repeatedly and emphatically: (a) that the problem is solved by neutrino masses and mixing effects producing oscillations in the Sun (or between the Sun and Earth) explaining the perceived deficit of solar neutrinos; (b) that standard solar model calculations ignore effects that are difficult to compute (*e.g.*, departures from hydrostatic equilibrium due to the ^{3}He instability) or use poorly–known input parameters (*e.g.*, the cross section for $p + {}^7\mathrm{Be} \rightarrow {}^8\mathrm{B} + \gamma$), and therefore yield unreliable predictions of the ^{8}B neutrino flux, or (c) in the opinion of a conservative Scot from Cern, that experiment and theory are actually in tolerable (if not perfect) agreement with one another and consequently, that there is no problem at all. The recently announced (and preliminary!) GALLEX result can be stretched to support any of these points of view. This issue, unlike the late and unlamented 17 KeV neutrino, is still very much with us.

Recent experiments confirm that there is a discrepancy between theory and experiment, and tend to eliminate explanation (c) [2][3][4]. Standard solar models have been used to determine the solar neutrinos fluxes [5][6]. The predicted result ^{8}B neutrino flux is larger the Davis result a factor of two or three. Similar discrepancies pertain to the fluxes reported by Kamioka and the Gallium experiments.

There is a conflict between experiment and theory, but it remains unclear whether the emphasis should be on the word *solar* or *neutrino*. There is something missing from one or another (but probably not both) of our standard models: that of particle physics, or that of the Sun. Which way the ball will bounce is unknown at this time, but we shall soon find out. This brings us to the question we pose: Does the solar neutrino flux vary with time? If it does, what is the relevant time scale. There are many possible answers to this question, ranging from minutes to billons of years.

2. Solar Variation in Big Time

It goes without saying that the solar constant is not constant, and astronomers believe that the Sun was once cooler than it is now. Unfortunately, nobody was doing neutrino experiments back then. Did the Sun heat up monotonically, or did it have its ups and downs. Some stellar physicists question whether the Sun has always been in hydrostatic equilibrium. As the sun evolves, the ^{3}He abundance in its core develops a positive outward gradient. This leads [7] [8] to a hydrostatic instability that is often ignored in standard solar models. Furthermore, Gough [9] points out some evidence of a rotational discontinuity at the solar core. Perhaps quasi–periodic instabilities of the solar core cause Earth's occasional periods of glaciation. The last ice age took place is the Pleistocene era, just a few hundred thousand years ago. The core of the Sun today could be hotter or cooler than its temperature deduced from standard solar models.

Wick Haxton once suggested a means by which we could determine the time–averaged solar neutrino flux over the past million years. His proposal involved the search for neutrino–produced technetium in ancient molybdenum ores. Unfortunately, this experiment may not be feasible. Let us hope that a practicable experiment of this kind may be devised and deployed. The relative constancy of the solar core should be determined by experiment, rather by appeal to the principle of simplicity.

3. Real Time Solar Neutrino Experiments

The Chlorine and Gallium experiments are radiochamical. They measure the average solar neutrino flux over a period of several weeks. Eventually, experiments of this kind could reveal annual flux variations. There must certainly be a 3% January–June variation from perihelion to aphelion. There could also be an 11 year periodicity of the solar neutrino flux associated with the sunspot cycle. Two highly speculative suggestions have been made in this connection: (a) oscillations induced by the solar magnetic field and associated with large (and inexplicable) magnetic interactions of neutrinos; and (b) quasi–periodic instabilities in the solar core which affect both the Sun's external magnetic behaviour and the neutrino flux. However, radiochemical experiments, by their very nature, cannot reveal short–period fluctuations of the solar neutrino flux.

The Kamioka experiment is a real–time experiment measuring the arrival times of ^{8}B neutrinos with great precision. Its statistics are poor. However, three solar neutrino experiments that are in planning or constructions will do much better: Super Kamioka [10], the Sudbury Neutrino Observatory [11], and Icarus [12], each will time the arrivals of thousands of these neutrinos. Furthermore, the Borexino Experiment at the Gran Sasso [13] hopes to detect the essentially monoenergetic neutrinos coming from the solar reaction:

$$e^- + {}^7\mathrm{Be} \to {}^7\mathrm{Li} + \nu. \tag{3}$$

These neutrinos are much more copious than the ^{8}B neutrinos, but their energies are a mere 862 KeV. However, if Borexino succeeds, tens of thousands of these events will be collected per year.

Even more sensitive real–time solar neutrino experiments have been dreamed about. A real–time *pp* neutrino detector that could collect 10^5 events has been considered (Ypsilantis, T. & Seguinot, J., priv. com.). Furthermore, the proposed BOREX experiment (Raghavan, R. *et al.*, Bell Laboratory Report No. ATT-BX-88-01 (1988)) for which Borexino is the prototype may detect a million ^{7}Be neutrinos.

As soon as one or more of the experiments mentioned above are deployed, the search for short–term variations of the solar neutrino flux may begin in earnest. Is there anything there to be found? Alvaro De Rújula and I have been pondering this question.

4. Annual Variations of the Solar Neutrino Flux

Pakvasa and Pantaleone [14] pointed out that Earth's eccentric orbit about the sun offers us a moving laboratory from which we can find or exclude hypothetical neutrino oscillations over a wide range of truly astronomical wavelengths. The proposed Borexino solar neutrino experiment will record the arrival times of thousands of almost monochromatic neutrinos arising from electron capture by ^{7}Be nuclei in the solar core. We develop a simple and efficient algorithm with which to exhibit in advance its reach in neutrino masses and mixing angles and determine these parameters if a signal is present in the data. Oscillations can be observed for neutrino mass differences satisfying $10^{-8}\,\mathrm{eV}^2 > \delta m^2 > 10^{-11}\,\mathrm{eV}^2$, an interesting domain that is otherwise difficult to access.

The distance d from sun's center to the laboratory varies. Neutrino masses and mixing may lead to observable time variations of the Borexino event rate, a point stressed elsewhere [13][15]. When data become available, they can be compared to various hypotheses. In the meantime, it is prudent to know what region of neutrino–oscillation parameter space may be explored. For expository clarity, assume that N events, all due to ^{7}Be neutrinos, are collected in the course of a year with both the detector's acceptance and the sun's neutrino luminosity remaining constant.

Oscillations of astronomical length may pertain to the solar neutrino problem, whose solution (if, indeed, a solution is called for) could involve unknown properties of the sun or of neutrinos. In agreement with Pakvasa and Pantaleone, we find an almost three–decade range of detectable oscillation lengths. It lies above those needed for the MSW effect [16] but includes those relevant to just–so neutrino oscillations [17]. Quite aside from this issue, few can doubt the importance of a search for long–wavelength neutrino oscillations of which we have neither *a priori* knowledge nor ready alternative access.

The largest change in d results from earth's elliptical orbit, whose semimajor axis (the astronomical unit) is $a \simeq 1.5 \times 10^{11}$ m and whose eccentricity is $\epsilon \simeq 0.0167$. The distance to the sun is expressed in terms ψ, the eccentric anomaly:

$$d = a\,(1 - \epsilon\cos\psi). \tag{4}$$

Its annual excursion defines the baseline variation of the experiment,

$$L_1 \equiv 2\epsilon a \simeq 5.0 \times 10^9 \text{ m}.$$

The angle ψ is related to time through Kepler's equation:

$$\frac{2\pi t}{\tau} = \psi - \epsilon\sin\psi,$$

with τ a sidereal year and t measured from perihelion. We incur a negligible error in our study of neutrino oscillations by ignoring the effect of eccentricity on the relation between t and ψ. Accordingly, we put

$$\psi = 2\pi t/\tau \quad \text{and define} \quad z \equiv \cos\psi.$$

Were detection independent of neutrino flavor (or were there no neutrino oscillations), the event rate would be affected only by the inverse–square law:

$$\left(\frac{dN}{d\psi}\right)_0 = \frac{\hat{N}}{2\pi}\,\frac{(1-\epsilon^2)^{3/2}}{(1-\epsilon z)^2}. \tag{5}$$

$\hat{N}$ is the number of neutrinos detected yearly under this circumstance. In the presence of oscillations, $\hat{N}$ is to be compared with predictions of solar models. Since it is not known with certainty, we phrase our tests for neutrino oscillations in terms of N, the number of events descried in a year.

Vacuum neutrino oscillations lead to additional time dependence and to an overall reduction of the event rate. Let δm^2 be the relevant neutrino mass splitting and θ the mixing angle. The oscillation length appropriate to ^{7}Be solar neutrinos with $E = 0.862$ MeV is:

$$L = \frac{4\pi E}{\delta m^2} \simeq \frac{2.14\,\mathrm{eV}^2}{\delta m^2}\ \text{meters.}$$

Borexino detects neutrino–electron elastic scattering, to which all neutrino flavors contribute. The cross section is smaller for ν_μ or ν_τ than it is for ν_e. We define the parameter r according to:

$$1 - r \equiv \frac{\sigma(\nu_{\mu,\tau} + e \to \nu_{\mu,\tau} + e)}{\sigma(\nu_e + e \to \nu_e + e)} = \frac{1 - 4\sin^2\theta_w + 8\sin^4\theta_w}{1 + 4\sin^2\theta_w + 8\sin^4\theta_w},$$

where θ_w is the weak mixing angle. We find $r \simeq 0.79$.

If all the neutrinos were created with the same energy at sun's center, the Borexino event rate (including contributions from all neutrino flavors) would be:

$$\left(\frac{dN}{d\psi}\right)_1 = \left(1 - \frac{r}{2}\sin^2 2\theta\right)\left[1 + \lambda\cos\frac{2\pi d}{L}\right]\left(\frac{dN}{d\psi}\right)_0, \tag{6}$$

where

$$\lambda = \frac{r\sin^2 2\theta}{2 - r\sin^2 2\theta}$$

measures the strength of the observed oscillations. But, the neutrinos are neither exactly monochromatic nor produced at a point.

Neutrino pathlengths vary because their production is distributed in the sun's core, thereby suppressing oscillations. For ^{7}Be neutrinos, the source density is $\propto \exp -(z/s)^2$, where z is projected along the earth–sun direction and $s \simeq R_\odot/17 \simeq 4.1 \times 10^7$ m [18]. Since $R_\oplus \ll s \ll L_1$, we neglect the small daily and monthly perturbations of d in estimating the suppression. Even more important are the thermal motions that broaden the neutrino line — a Doppler shift due to nuclear motion and a contribution of the electron's kinetic energy. The mean solar temperature in the region that ^{7}Be neutrinos are produced is $kT \simeq 1.15$ KeV and the nuclear mass M is $\sim$6.5 GeV/c^2.

The size and temperature of the source reduce oscillations by a multiplicative factor $S(L)$ given roughly by:

$$S(L) = \frac{e^{-(L_0/L)^2}}{\sqrt{1+(L_0'/L)^2}}, \tag{7}$$

where:

$$L_0 = \pi\sqrt{s^2 + 2a^2\,kT/M} \simeq 3.1\times 10^8 \text{ m}, \quad \text{and} \quad L_0' = 2\pi\, a\, kT/E \simeq 1.25\times 10^9 \text{ m}.$$

Thus, there is a lower limit to L (or upper limit to δm^2) for which oscillations can be seen [15]. At $L = L_0$ (or $\delta m^2 \simeq 10^{-8}$ eV^2), they are suppressed by $S(L) = 0.09$ and are will not be detectable by Borexino even if $\sin^2 2\theta = 1$.

An upper limit to detectable values of L (or lower limit to δm^2) is not sharply defined. Less than one oscillation is completed if $L > L_1$. Such oscillations are detectable sensitivity that falls erratically with L [14].

The observed event rate, (6) with the suppression factor taken into account, may be written:

$$\frac{dN}{d\psi} = \left(1 - \frac{r}{2}\sin^2 2\theta\right)\left[1 + \hat\lambda\cos(p - p\epsilon z)\right]\left(\frac{dN}{d\psi}\right)_0, \tag{8}$$

where

$$\hat\lambda = S(L)\,\lambda \quad \text{and} \quad p = 2\pi a/L.$$

Integration of (8) over the year yields the relation between N and $\hat N$:

$$N = \zeta\hat N, \quad \text{where} \quad \zeta = \left(1 - \frac{r}{2}\sin^2 2\theta\right)(1 + \hat\lambda\, J_0(\epsilon p)\cos p). \tag{9}$$

J_0 is a Bessel function of the first kind.

Were Borexino data in hand, we might fit them to (8) with $\hat\lambda$ and p as free parameters. Absent the data, we show how the sensitivity of the proposed experiment may be determined in advance without performing a plethora of Monte–Carlo simulations. dN/dz (which is periodic in the interval $1 \geq z \geq -1$) is the product of a sinusoidal oscillation in z with the slowly–varying function $(dN/d\psi)_0$ and the Jacobian $d\psi/dz$. We search for neutrino oscillations by performing a weighted

Fourier transform of a hypothetical data set of N events. To this end, z_j is computed for each event and the following Fourier sums are evaluated:

$$Q_m(N) = \frac{1}{N}\sum_{j=1}^{N} G(z_j)\,\cos\pi m z_j \quad \text{for} \quad m = 0,\, 1,\, \ldots$$

$$R_m(N) = \frac{1}{N}\sum_{j=1}^{N} G(z_j)\,\sin\pi m z_j \quad \text{for} \quad m = 1,\, 2,\, \ldots$$

The positive weight function G cancels the Jacobian and undoes the small direct effect of earth's annual motion:

$$G(z) \equiv \frac{\pi\,(1-\epsilon z)^2\,\sqrt{1-z^2}}{2\,(1-\epsilon^2)^{3/2}}. \tag{10}$$

The sums' expected values with no oscillations ($\theta = 0$ and $N = \hat{N}$) are given by their continuum limits:

$$\langle Q_m\rangle = \frac{1}{N}\int_0^{2\pi} G(z)\,\cos\pi m z\,\left(\frac{dN}{d\psi}\right)_0 d\psi,$$

and similarly for $\langle R_m\rangle$. With the event rate given by (5), we find $\langle Q_0\rangle = 1$, and $\langle Q_m\rangle = \langle R_m\rangle = 0$ for $m > 0$.

Our choice for G ensures that any non–zero sum with $m > 0$ signals neutrino oscillations, as does a departure of Q_0 from unity. To assess the the statistical significance of a potential signal, we need the rms fluctuations of the sums from their expected values:

$$\Delta Q_m = \sqrt{(\langle Q_m^2\rangle - \langle Q_m\rangle^2)/N},$$

where

$$\langle Q_m^2\rangle = \frac{1}{N}\int_0^{2\pi} [G(z)\,\cos\pi m z]^2\,\left(\frac{dN}{d\psi}\right)_0 d\psi,$$

and similarly for R_m. We obtain:

$$\Delta Q_0 = \sqrt{(\pi^2-8)/8N},$$

$$\Delta Q_m = \frac{\pi}{4\sqrt{N}}\,\sqrt{1 + J_1(2m\pi)/m\pi},$$

$$\Delta R_m = \frac{\pi}{4\sqrt{N}}\,\sqrt{1 - J_1(2m\pi)/m\pi}.$$

Expected values of Q_m and R_m in the presence of neutrino oscillations are indicated by square brackets:

$$[Q_m] = \frac{1}{N} \int_0^{2\pi} G(z) \cos \pi m z \, \frac{dN}{d\psi} \, d\psi.$$

Making use of (5), (8) and (10), we obtain:

$$[Q_m] = \frac{\hat{N}}{2N} \left(1 - \frac{r}{2} \sin^2 2\theta\right) \int_{-1}^{+1} \cos m\pi z \, \left(1 + \hat{\lambda} \cos(p - \epsilon p z)\right) dz,$$

and the same for $[R_m]$ with $\cos m\pi z$ replaced by $\sin m\pi z$. We use (9) to express the results in terms of $\hat{\lambda}$ and p:

$$\begin{aligned}
[Q_0] &= \left(1 + \hat{\lambda} \cos p \, \frac{\sin \epsilon p}{\epsilon p}\right) (1 + \hat{\lambda} \cos p \, J_0(\epsilon p))^{-1}, \\
[Q_m] &= \hat{\lambda} (-1)^m \cos p \, \frac{\epsilon p \sin \epsilon p}{(\epsilon p)^2 - (m\pi)^2} \, (1 + \hat{\lambda} \, J_0(\epsilon p) \cos p)^{-1}, \qquad (11) \\
[R_m] &= \hat{\lambda} (-1)^m \sin p \, \frac{m\pi \sin \epsilon p}{(\epsilon p)^2 - (m\pi)^2} \, (1 + \hat{\lambda} \, J_0(\epsilon p) \cos p)^{-1}.
\end{aligned}$$

We have shown that there is a lower limit $\sim 10^8$ m to detectable oscillation lengths. Thus, only a finite number of sums need be examined. Retrospectively, we find that only those with $m \leq 15$ (corresponding to $L \leq L_1/15$) are required.

Suppose the data were in hand, the sums Q_m and R_m evaluated. The relevant signal–to–noise ratios,

$$\frac{Q_0 - \langle Q_0 \rangle}{\Delta Q_0}, \quad \frac{Q_m}{\Delta Q_m}, \quad \frac{R_m}{\Delta R_m}, \qquad \text{for} \quad 1 \leq m \leq 15, \qquad (12)$$

are defined as the 31 random variables S_j where m is the integral part of $(j+1)/2$ and $0 \leq j \leq 30$. Detectable neutrino oscillations cause one or more of the S_j to depart significantly from zero. The value(s) of j for which this happens offers a rough but immediate determination of the oscillation length, $L \simeq 2L_1/j$.

The S_j correspond to different Fourier components of the data. In the absence of neutrino oscillations, each of them has a normal probability distribution about zero with unit variance. Let $P(s)$ be the probability that $|S_j| > s$ for any fixed j and $Q(s)$ be the probability that $|S_j| > s$ for at least one j. We find $Q(s) = 1 - (1 - P(s))^{31}$ and fix s to be that value for which $Q(s) \leq 0.0027$, corresponding to a three standard deviation effect. It is straightforward to show that $s \simeq 3.9$.

The appearance of at least one S_j exceeding 3.9 is taken as a positive signal for vacuum neutrino oscillations.

To proceed further, we first compute the expected values of S_j in the presence of oscillations as a function of δm^2 and $\sin^2\theta$. This is done by replacing Q and R in (12) by their square–bracketed values as given by (11). The results are denoted by $[S_j]$. Oscillations are detectable if and only if δm^2 and $\sin^2\theta$ are such that $[S_j] \geq 3.9$ for at least one value of j. If $N = 18,000$ ^{7}Be neutrinos are collected, a large domain of possible neutrino squared mass differences, extending between 10^{-11} eV^2 and 10^{-8} eV^2 can be explored.

5. G Modes in the Sun

It has been pointed out by Gough and others that contemporary fluid motions and thermal fluctuations in the sun's core may modify nuclear reaction rates and restore agreement [19] [20]. To test this notion, we propose a search for short–term variations of the solar neutrino flux. Models of the sun fit its radius $R_\odot$ and luminosity $L_\odot$ to an assumed initial ^{4}He abundance and a convective mixing length [21]–[22]. While challenged by solar–neutrino observations, they are supported by solar-surface measurements [23] of the frequencies of thousands of p–modes (pressure modes). These are inverted to yield the sound velocity at depth [24] [25]. Whilst the result agrees with solar models to better than 1%, helioseismology provides scant information about the solar core, where p modes are damped [26].

Solar g modes (for which gravity is the restoring force) may be present in the Sun. Their periods ($2\pi/\omega$) are of order one hour [26] [27]. Since energy–tranport times are much larger, g modes correspond to quasi–adiabatic temperature fluctuations about a radial mean:

$$T(r,\, t,\, \theta,\, \phi) = \overline{T}(r)\left[1 + A\, g(r)\, Y_{lm}(\theta,\, \phi)\, \sqrt{2}\, \cos(\omega t)\right], \tag{13}$$

where A is the amplitude of an oscillation whose angular dependence is that of a spherical harmonic with $\int |Y|^2 d\Omega = 1$ and whose radial eigenfunction $g(r)$ has a maximum of one.

Any g mode present in the sun affects its neutrino–producing processes. We label the pp neutrinos produced via (1) by $a = 1$, the ^{7}Be neutrinos produced by

(3) by $a = 7$, and the ^{8}B neutrinos produced by (2) by $a = 7$. Their angularly–averaged rates $\hat{\epsilon}_a(r,t)$ are:

$$\hat{\epsilon}_a(r,t) = \epsilon_a(r)\,\langle\,(T/\overline{T})^{N_a}\,\rangle_\Omega, \tag{14}$$

where the $\epsilon_a(r)$ depend on the local density, nuclear abundances and $\overline{T}(r)$. The exponents in (14), $N_1 = 4$, $N_7 \simeq -0.5$ and $N_8 \simeq 13$, reflect the T dependences of the reaction rates at fixed abundances [20]. Expanding (14) in powers of A^2, we exhibit the time dependence of the rates:

$$\hat{\epsilon}_a(r,t) = \epsilon_a(r)\,\left(1 + \frac{1}{2}\,A^2\,N_a(N_a-1)\,g^2(r)\,[1+\cos(2\omega t)] + O(A^4)\right). \tag{15}$$

The constant in square brackets affects the time–averaged neutrino fluxes; the cosine generates oscillations with twice the g mode frequency.

We integrate (15) over r using an $n = 1$ mode with $g(r) = x\exp(1-x)$, $x = r/(0.15\,R_\odot)$ [20]. In solar models [5]–[22], $\epsilon_a(r)$ are roughly of the form $\epsilon_a(r) = y^2\,\exp(-y^2)$ with $y = r/(f_a\,R_\odot)$ and $f_a \simeq 25$, 17, 10 for $a = 1$, 7, 8. For the oscillations of the various components of the solar neutrino flux, we predict:

$$F_a \simeq \overline{F}_a\,[1 + \lambda_a\,\cos(2\omega t)], \qquad \lambda_{1,7,8} \simeq (4.8,\,0.21,\,28.5)\,A^2, \tag{16}$$

where $\overline{F}_a$ are time–averaged fluxes. Notice that $\lambda_8 > \lambda_1 >> \lambda_7$.

According to (14), the reaction rates $\hat{\epsilon}_a$ exceed those in a steady sun with temperature profile $\overline{T}(r)$. To keep $L_\odot$ fixed, the solar model must be modified to lower $\overline{T}$. Gough [20] estimates how the time–averaged neutrino fluxes depart, in the presence of an $n = 1$ g-mode, from those of the standard model:

$$\begin{aligned} \overline{F}_1 &\simeq F_1^{SSM}, \\ \overline{F}_7 &\simeq F_7^{SSM}\,[(1 - 33\,A^2 + 267\,A^4], \\ \overline{F}_8 &\simeq F_8^{SSM}\,[1 - 57\,A^2 + 1067\,A^4]. \end{aligned} \tag{17}$$

The effects of the g mode on the time–averaged flux (17) and its fluctuations (16) are greatest for ^{8}B neutrinos. Gough shows that if $A = 0.1$, the ^{8}B flux is reduced by 0.54 and that of ^{7}Be by 0.70, thus removing the discrepancy between experiment and theory. We choose this value of A to set the scale for anticipated neutrino flux oscillations. However, before proceeding, it is necessary to say that

$A = 0.1$ represents a very large amplitude oscillation. If the linear theory of solar oscillations is correct, g mode amplitudes of this magnitude would have been detected by helioseismic observations. The remainder of this discussion presumes — with no justification — that the linear model fails and that the excited g modes are somehow confined to the solar core.

Future experiments will measure arrival times t_i of thousands of neutrinos. Assume that a g mode of frequency ω modulates the neutrino fluxes, as in (16). The precise frequency of the g mode is unknown, but its effect can be found by Fourier transforming the data over a frequency range $f_{\rm min} < f < f_{\rm max}$. Suppose that n neutrinos are detected in a run of duration τ. Let:

$$P(f) \equiv \left| \sum_{j=1}^{n} e^{ift_j} \right| . \tag{18}$$

The signature of a g mode is a peak in $P(f)$ at $f = 2\omega$, emerging even if 2ω exceeds τ/n, the mean counting rate. The peak's expected magnitude is $P_s = \lambda n/2$. Its half-width at half maximum, $\Delta\omega = \sqrt{6}/\tau$, sets the required Fourier resolution. Away from the peak, $P(f)$ fluctuates about $\sqrt{n}$, exceeding P_s with probability $\exp(-P_s^2/2n)$.

The minimum significant signal (with confidence level C.L.) corresponds to a g mode of amplitude:

$$\lambda_{\rm min} = \left[\frac{8}{n} \ln \frac{(f_{\rm max} - f_{\rm min})\,\tau}{1 - \rm C.L.} \right]^{1/2} , \tag{19}$$

With $\tau \sim 1$ year and $f_{\rm max} - f_{\rm min} \sim$ inverse minutes, the logarithm's argument iss large and its precise value immaterial. We see from (16) and (19) that for $A \simeq 0.1$, an experiment sensitive to the *pp* flux must observe $\sim 6 \times 10^4$ events to find a 99%-confidence effect. However, an experiment gathering a mere 3000 (30,000) ^{8}B neutrinos can find an $A = 0.08$ (0.05) signal with 99% confidence, decisively testing whether the suppression of its flux is due to a single confined g mode. If neutrino experiments were to detect the sun's heartbeat, it is the sun that oscillates, not the neutrino.

References

[1] Davis Jr., R. *et al.* in *Proceedings of the 21st International Cosmic Ray Conference* (ed Protheroe, R.J.) 155–158 (University of Adelaide Press, Adelaide 1990).

[2] Hirata, K.S. *et al.,* Phys. Rev. **D44,** 2241–2260 (1991).

[3] GALLEX Collaboration (Anselman, P. *et al.,*) Phys. Lett. **B285,** 376–389 (1992).

[4] Abazov, A.I. *et al.,* Phys. Rev. Lett. **67,** 3332–3335 (1991).

[5] Bahcall, J.N. & Ulrich, R.K., Rev. Mod. Phys. **60,** 297-372 (1988).

[6] Turk–Chièze, S. in *Inside the Sun* (eds Berthomieu, G. & Cribier, M.) 125–132 (Kluwer Acad. Pub., Dordrecht 1990).

[7] Dilke, F.W.W. & Gough, D.O., Nature **240,** 262–264 cont'd 293–294 (1972).

[8] Merryfield, W.J., Toomre J. & Gough, D.O., Astroph. J. **367,** 658–665 (1991).

[9] D.O. Gough, Loeb Lectures at Harvard (1993).

[10] Totsuka, Y., in *Proc. of the International Symposium on Underground Physics Experiments* (ed Nakamura, K.) (Institute for Cosmic Ray Research, University of Tokyo 1990).pp 129–164.

[11] Beier, E. in *Proc. of the International Symposium on Underground Physics Experiments* (ed Nakamura, K.) 165–170 (Institute for Cosmic Ray Research, University of Tokyo 1990).

[12] Baldo–Ceolin, M. in *Massive Neutrinos in Particle Physics and Astrophysics* (ed Fackler, O.W. & Tran Thanh Van, J.) 159–164 (Editions Frontières, Gif sur Yvette 1986).

[13] Borexino at Gran Sasso, a proposal ed. by G. Bellini, R. Raghavan *et al.* (1991).

[14] Sanjib Pakvasa and James Pantaleone, Phys. Rev. Letters **65,** 2479 (1990).

[15] L. Krauss & F. Wilczek, Phys. Rev. Letters, **55** (1985) 122);
See also: A. Loeb, Phys. Rev. **D39** (1989) 1009.

[16] S.P. Mikheyev & A.Y. Smirnov, Nuovo Cim. **19** (1986) 17;
L. Wolfenstein, Phys. Rev. **D17** (1987) 2369.

[17] *E.g.,* S. L. Glashow & L. Krauss, Phys. Lett. 190B (1987) 199.

[18] Bahcall, J.N. & Ulrich, R.K., Rev. Mod. Phys. **60** (1988) 297-372.

[19] Roxburgh, I.W. in *The Internal Solar Angular Velocity* (ed Durney, B.R. & Sofia, S.) 1–5 (Reidel, Dordrecht 1987).

[20] Gough, D.O., Ann. N.Y. Acad. Sci. (in press).

[21] Schwarzchild, M., Howard, R. & Härm, R., Astroph. J. **125,** 233–241 (1957).

[22] Christensen–Dalsgaard, J., Gough, D.O. & Thompson, M.J., Astroph. J. **378,** 413–437 (1991).

[23] Duvall Jr., T. L. in *Inside the Sun, op. cit.* 253-264.

[24] Gough, D.O., Solar Physics **100,** 65–99 (1985).

[25] Gough, D.O. & Thompson, M.J. in *Solar Interior and Atmosphere,* (ed Cox, A.N., Livingstone, W.C. & Mathews, M.) 401–478 (Space Science Series, University of Arizona Press, Tuscon 1991).

[26] Christensen–Dalsgaard, J. & Berthomieu, G. in *Solar Interior and Atmosphere, ibid.* 519-561.

[27] Hill, H. *et al.* in *Solar Interior and Atmosphere, op. cit.* 562–617.

CLOSING CEREMONY

The closing ceremony took place on Sunday 21st July 1992.

The Prizes and Scholarships were awarded as specified below.

PRIZES AND SCHOLARSHIPS

• Prize for **Best Student** awarded to:

Peter CHO, Harvard University, USA.

• Prize for **Best Scientific Secretary** awarded ex-equo to:

Frank BLOCK, CERN/LAA-FBLJA, Geneva, Switzerland.

Gary WORD, Rutgers University, USA.

• Twelve **Scholarships** were open for competition among the participants. They were awarded as follows:

• **J.S. BELL** Scholarship to:
Herbert REINER, Oxford University, UK.

• **Patrick M.S. BLACKETT** Scholarship to:
Rongzhi LIU, CALTEC, USA.

• **James CHADWICK** Scholarship to:
Ofer RIND, MIT, Boston, USA.

• **Amos DE-SHALIT** Scholarship to:
Amihay HANANY, Weizman Institute of Science, Israel.

• **Paul A.M. DIRAC** Scholarship to:
Peter CHO, Harvard University, USA.

• **Isidor I. RABI** Scholarship to:
André SOPZAK, MIT, Boston, USA.

• **Robert HOFSTADTER** Scholarship to:
Eduardo de COUTO e SILVA, Indiana University, Bloomington, USA.

• **Andreij D. SAKHAROV** Scholarship to:
Arnd LEIKE, Hamburg, Germany.

• **Jun John SAKURAI** Scholarship to:
Grant BAILLIE, UCLA., Los Angeles, USA.

• **Gunnar KÄLLEN** Scholarship to:
Karim BENAKLI, Ecole Politechnique, Paris, France.

• **André LAGARRIGUE** Scholarship to:
Rukhsana MALIK, World Laboratory-FBLJA, Geneva, Switzerland.

• **Giulio RACAH** Scholarship to:
Susan SHOTKIN, MIT, Boston, USA.

• Two **EPS** Scholarships were awarded to:
Marian FECKO, Comenius University, Czechoslovachia.
Mikolaj MISIAK, Warsaw University, Poland.

• Two **WFS** Scholarships were awarded to:
Horia Iosif PETRACHE, Brasow, Rumania.
Lybov VASSILEVSKAYA, ITEP, Moscow, Russia.

The following participants gave their collaboration in the Scientific Secretarial work:

P. Grant BAILLIE	*Uzi PAZ*
Karim BENAKLI	*Horia Iosif PETRACHE*
Frank BLOCK	*Zhong Liang REN*
Peter CHO	*Ofer RIND*
Eduardo de COUTO e SILVA	*Francesco SANNINO*
Maria Teresa DOVA	*Rajesh Kumar SEHGAL*
Eda ESKUT	*Susan SHOTKIN*
Amihay HANANY	*Jayant SHUKLA*
Vadim KHODYREV	*André SOPCZAK*
Arnd LEIKE	*Zhijian TAO*
Rongzhi LIU	*Maneesh WADHWA*
Rukhsana MALIK	*Gary WORD*
Nikolaj MISIAK	*Nadeem YUNUS*
Nasr Eddine MOULAI	*Yuri ZAMORA GARCIA*

PARTICIPANTS

Oscar ADRIANI Italy	I N F N Largo Enrico Fermi, 2 50125 FIRENZE, Italy
Guido ALTARELLI Italy	C E R N TH Division 1211 GENEVE 23, Switzerland
Alexei ANISOVICH Russia	Nuclear Physics Institute Gatchina ST. PETERSBURG, Russia
P. Grant BAILLIE South Africa	Department of Physics U C L A 405 Hilgard Avenue LOS ANGELES, CA 90024, USA
Patricia BALL Germany	Institut Theoretische Physik Universitat Philosophenweg 16 6900 HEIDELBERG, Germany
Karim BENAKLI Algeria	Centre de Physique Théorique Ecole Polytechnique 91128 PALAISEAU Cedex, France
Maurizio BIASINI Italy	Dipartimento di Fisica Università Via Pascoli 06100 PERUGIA, Italy
Cesare BINI Italy	Dipartimento di Fisica Università La Sapienza Piazzale Aldo Moro, 2 00185 ROMA, Italy

Emir BLANCO
Argentina

Ettore Majorana Centre
for Scientific Culture - EMCSC
Via Guarnotta, 26
91016 ERICE, Italy

Frank BLOCK
Germany

C E R N
PPE Division
1211 GENEVE 23, Switzerland

Philippe BRAX
France

CEA-DSM
Service de Physique Théorique
CEA - Saclay
91191 GIF-sur-YVETTE, France

Richard BUTTERY
UK

Department of Physics
Schuster Laboratory
Brunswick Street
MANCHESTER, UK

Peter CHO
USA

Lyman Laboratory
Harvard University
CAMBRIDGE, MA 02138, USA

Alessandro CULATTI
Italy

Dipartimento di Fisica
Università
Via Marzolo, 8
35100 PADOVA, Italy

Richard H. DALITZ
UK

Department of Physics
University of Oxford
OXFORD, UK

Eduardo do COUTO e SILVA
Brasil

C E R N
PPE Division
1211 GENEVE 23, Switzerland

Maria Teresa DOVA
Argentina

C E R N
PPE Division
1211 GENEVE 23, Switzerland

Herbert DREINER Germany	Department of Theoretical Physics Oxford University 1 Keble Road OXFORD, OX1 3NP, UK
John ELLIS UK	C E R N TH Division 1211 GENEVE 23, Switzerland
Eda ESKUT Turkey	Ettore Majorana Centre for Scientific Culture - EMCSC Via Guarnotta, 26 91016 ERICE, Italy
Marian FECKO Czechoslovachia	Dept. of Theoretical Physics Comenius University Mlynska Dolina F2 84215 BRATISLAVA, Czechoslovachia
Sergio FERRARA Italy	C E R N TH Division 1211 GENEVE 23, Switzerland
Fausto FIORANI Italy	Dipartimento di Fisica Università Viale delle Scienze 43100 PARMA, Italy
Eugene FISK USA	F N A L BATAVIA, IL, USA
Marco GENOVESE Italy	I N F N Via P. Giuria, 1 10125 TORINO, Italy
Nicola GIGLIETTO Italy	Dipartimento di Fisica Università Via Amendola, 173 70126 BARI, Italy

Sheldon L. GLASHOW USA	Department of Physics Harvard University CAMBRIDGE, MA, USA
Peter Wilhem GOSDZINSKY Germany	Departament de Fisica Universitat Autònoma de Barcelona 08913 BELLATERRA, Spain
Michel GOURDIN France	Dept. de Physique Université P. et M. Curie 4 Place Jussieu 75252 PARIS, France
Dirk GRAUDENZ Germany	Institut für Theoretische Physik RWTH 5100 AACHEN, Germany
Chuang GU China	C E R N PPE Division 1211 GENEVE 23, Switzerland
Thomas HAMBYE Belgium	Inst. de Physique Théorique Université Catholique Chemin du Cyclotron, 2 LOUVAIN LA NEUVE, Belgium
Amihay HANANY Israel	Department of Physics The Weizmann Inst. of Sciences 76100 REHOVOT, Israel
Mans HENNINGSON Sweden	Institute of Theoretical Physics University 41296 GÖTEBORG, Sweden
George KATSILIERIS Greece	National Centre for Scientific Research "Demokritos" P.O. Box 60228 15310 AGHIA PARASKEVI, Greece

Vadim KHODYREV Russia	Inst. High Energy Physics PROTVINO Moscow Region 142284, Russia
Bjarte KILENG Norway	Department of Physics University of Bergen N-5007 BERGEN, Norway
Arnd LEIKE Germany	IFH Zeuthen D E S Y 1615 ZEUTHEN, Germany
Rongzhi LIU China	Laboratori Nazionali del Gran Sasso 67010 ASSERGI, AQ, Italy
Wenjie LIU China	Ettore Majorana Centre for Scientific Culture - EMCSC Via Guarnotta, 26 91016 ERICE, Italy
Maurizio LO VETERE Italy	I N F N Via Dodecaneso, 33 16146 GENOVA, Italy
Gerard John LOWE UK	Dept. of Physics and Astronomy University of Glasgow GLASGOW, G12 8QQ, UK
Johannes MAINUSCH Germany	D E S Y Notkestrasse 85 2000 HAMBURG 52, Germany
Rukhsana MALIK Pakistan	C E R N PPE Division 1211 GENEVE 23, Switzerland
André MARTIN France	C E R N TH Division 1211 GENEVE 23, Switzerland

Mikolaj MISIAK Poland	Warsaw University Hoza 68 WARSAW, Poland
Douglas R.O. MORRISON UK	C E R N PPE Division 1211 GENEVE 23, Switzerland
Nasr Eddine MOULAI Algeria	C E R N PPE Division 1211 GENEVE 23, Switzerland
Carlo MUSSO Italy	Dipartimento di Fisica Università Via Dodecaneso, 33 16132 GENOVA, Italy
Dimitri NANOPOULOS Greece	Center for Theoretical Physics Physics Department Texas A&M University COLLEGE STATION, TX, USA and C E R N TH Division 1211 GENEVE 23, Switzerland
Pran NATH USA	Department of Physics Northeastern University BOSTON, MA, USA
Jay OREAR USA	Department of Physics Cornell University ITHACA, NY, USA
Maura PAVAN Italy	Dipartimento di Fisica Università Via Celoria, 16 20122 MILANO, Italy

Uzi PAZ
Israel
Physics Department
Ben-Gurion University
P.O. Box 653
BEER-SHEVA 84105, Israel

Roberto PECCEI
Italy
Department of Physics
University of California
LOS ANGELES, CA, USA

Horia Iosif PETRACHE
Romania
Crinului 62
BRASOW 2200, Romania

Thomas PORTMANN
Germany
Gesamthochschule
Universitat Siegen
Postfach 101240
5900 SIEGEN, Germany

Fabio RACITI
Italy
Dipartimento di Fisica
Università
Corso Italia 57
95129 CATANIA, Italy

Gerhard RAVEN
The Netherlands
C E R N
PPE Division
1211 GENEVE 23, Switzerland

Zheng Liang REN
China
C E R N
PPE Division
1211 GENEVE 23, Switzerland

Ofer RIND
USA
C E R N
PPE Division
1211 GENEVE 23, Switzerland

Roman ROGALYOV
Russia
Inst. of High Energy Physics
P.O. Box 35
SERPUKHOV, Moscow Region, Russia

Paolo RONCHESE Italy	Dipartimento di Fisica Università Via Marzolo, 8 35131 PADOVA, Italy
Graham G. ROSS UK	Department of Physics University of Oxford OXFORD, UK
Norma SANCHEZ Argentina	Observatoire de Paris Section d'Astrophysique, Demirm 5 Place Jules Janssen 92125 MEUDON, Cedex, France
Francesco SANNINO Italy	Dipartimento di Scienze Fisiche Università Pad. 19 - Mostra d'Oltremare 80125 NAPOLI, Italy
Rui SANTOS Portuga	Departamento de Fisica Faculdade de Ciencias Universidade LISBOA, Portugal
Jochen SCHWENK Germany	Max-Planck-Institut für Physik P.O. Box 40 12 12 8000 MUNCHEN 40, Germany
Rajesh Kumar SEHGAL India	C E R N PPE Division 1211 GENEVE 23, Switzerland
Yuly SHABELSKI Russia	World Laboratory Eloisatron Project Ettore Majorana Centre - EMCSC Via Guarnotta, 26 91016 ERICE, Italy

Susan SHOTKIN USA	C E R N PPE Division 1211 GENEVE 23, Switzerland
Jayant SHUKLA India	C E R N PPE Division 1211 GENEVE 23, Switzerland
Serguei SIVOKLOKOV Russia	Institute of Nuclear Physics Moscow State University 119899 MOSCOW, Russia
Andre SOPCZAK Germany	C E R N PPE Division 1211 GENEVE 23, Switzerland
W. James STIRLING UK	Department of Physics University of Durham DURHAM, UK
Olga STROGOVA Russia	Institute of Nuclear Physics Moscow State University 119899 MOSCOW, Russia
David SUMMERS UK	Department of Physics University of Durham DURHAM, DH1 3LE, UK
Zhijian TAO China	I C T P P.O. Box 586 34100 TRIESTE, Italy
Lybov VASSILEVSKAYA Russia	I T E P B. Cheremushkinskaya 25 MOSCOW, 117259, USSR
José VELHINHO Portugal	Departamento de Fisica Faculdade de Ciencias Universidade LISBOA Portugal

Iiro VILJA Finland	NORDITA Blegdamsvej 17 2100 COPENHAGEN, Denmark
Bertrand VUARIDEL Switzerland	Dep. de Physique Nucléaire et Corpusculaire Université 24, quai Ernest-Ansermet 1211 GENEVE, Switzerland
Maneesh WADHWA India	C E R N PPE Division 1211 GENEVE 23, Switzerland
Gary B. WORD USA	Dept. of Physics and Astronomy Serin Physics Laboratory Rutgers University PISCATAWAY, NJ 08855, USA
Bjorn WIIK Norway	D E S Y Notkestrasse 85 HAMBURG, Germany
Nadeem YUNUS India	C E R N PPE Division 1211 GENEVE 23, Switzerland
Yuri ZAMORA GARCIA Perù	C E R N PPE Division 1211 GENEVE 23, Switzerland
Fabio ZWIRNER Italy	C E R N TH Division 1211 Geneva 23, Switzerland